SOLUTIONS TO BLACK EXERCISES

ROXY WILSON

THE UNIVERSITY OF ILLINOIS, URBANA-CHAMPAIGN

TENTH EDITION

CHEMISTRY

THE CENTRAL SCIENCE

BROWN | LEMAY | BURSTEN

Upper Saddle River, New Jersey 07458

Project Manager: Kristen Kaiser
Executive Editor: Nicole Folchetti
Executive Managing Editor: Kathleen Schiaparelli
Assistant Managing Editor: Becca Richter
Production Editor: Diane Hernandez
Supplement Cover Designer: Elizabeth Wright
Manufacturing Buyer: Ilene Kahn

Pearson Prentice Hall
Pearson Education, Inc.
Upper Saddle River, NJ 07458

Printed in the United States of America

10 9 8 7 6 5 4 3

ISBN 0-13-146485-X

Pearson Education Ltd., *London*
Pearson Education Australia Pty. Ltd., *Sydney*
Pearson Education Singapore, Pte. Ltd.
Pearson Education North Asia Ltd., *Hong Kong*
Pearson Education Canada, Inc., *Toronto*
Pearson Educación de Mexico, S.A. de C.V.
Pearson Education—Japan, *Tokyo*
Pearson Education Malaysia, Pte. Ltd.

Contents

Introduction

Chemistry: The Central Science, 10th edition, contains nearly 2600 end-of-chapter exercises. Considerable attention has been given to these exercises because one of the best ways for students to master chemistry is by solving problems. Grouping the exercises according to subject matter is intended to aid the student in selecting and recognizing particular types of problems. Within each subject matter group, similar problems are arranged in pairs. This provides the student with an opportunity to reinforce a particular kind of problem. There are also a substantial number of general exercises in each chapter to supplement those grouped by topic. Visualizing Concepts, a new feature of the 10th edition, is a set of general exercises that requires students to analyze visual data in order to formulate conclusions about chemical concepts. Integrative exercises, which require students to integrate concepts from several chapters, are a continuing feature of the 10th edition. Answers to the odd numbered topical exercises plus selected general and integrative exercises, about 1200 in all, are provided in the text. These appendix answers help to make the text a useful self-contained vehicle for learning.

This manual, **Solutions to Black Exercises** in **Chemistry: The Central Science, 10th edition,** was written to enhance the end-of-chapter exercises by providing documented solutions for those problems _not_ answered in the appendix of the text. The manual assists the instructor by saving time spent generating solutions for assigned problem sets and aids the student by offering a convenient independent source to check their understanding of the material. Most solutions have been worked in the same detail as the in-chapter sample exercises to help guide students in their studies.

When using this manual, keep in mind that the numerical result of any calculation is influenced by the precision of the numbers used in the calculation. In this manual, for example, atomic masses and physical constants are typically expressed to four significant figures, or at least as precisely as the data given in the problem. If students use slightly different values to solve problems, their answers will differ slightly from those listed in the appendix of the text or this manual. This is a normal and a common occurrence when comparing results from different calculations or experiments.

Rounding methods are another source of differences between calculated values. In this manual, when a solution is given in steps, intermediate results will be rounded to the correct number of significant figures; however, unrounded numbers will be used in subsequent calculations. By following this scheme, calculators need not be cleared to re-enter rounded intermediate results in the middle of a calculation sequence. The final answer will appear with the correct number of significant figures. This may result in a small discrepancy in the last significant digit between student-calculated answers and those given in this manual. Variations due to rounding can occur in any analysis of numerical data.

The first step in checking your solution and resolving differences between your answer and the listed value is to look for similarities and differences in problem-solving methods. Ultimately, resolving the small numerical differences described above is less important than understanding the general method for solving a problem. The goal of this manual is to provide a reference for sound and consistent problem-solving methods in addition to accurate answers to text exercises.

Extraordinary efforts have been made to keep this manual as error-free as possible. All exercises were worked and proof-read by at least three chemists to ensure clarity in methods and accuracy in mathematics. The work and advice of Dr. Angela Manders Cannon and Mr. David Shinn have been invaluable to this project. However, in a written work as technically challenging as this manual, typos and errors inevitably creep in. Please help us find and eliminate them. We hope that both instructors and students will find this manual accurate, helpful and instructive.

Roxy B. Wilson
University of Illinois
School of Chemical Sciences
505 S. Mathews Ave., Box 49-1
Urbana, IL 61801
rbwilson@uiuc.edu

1 Introduction: Matter and Measurement

Visualizing Concepts

1.2 After a *physical change*, the identities of the substances involved are the same as their identity before the change. That is, molecules retain their original composition. During a *chemical change*, at least one new substance is produced; rearrangement of atoms into new molecules occurs.

The diagram represents a **chemical change**, because the molecules after the change are different than the molecules before the change.

1.3 (a) time (b) density (c) length (d) area (e) temperature
(f) volume (g) temperature

1.4 Measurements (darts) that are close to each other are *precise*. Measurements that are close to the "true value" (the bull's eye) are *accurate*.

(a) Figure ii represents data that are both accurate and precise. The darts are close to the bull's eye and each other.

(b) Figure i represents data that are precise but inaccurate. The darts are near each other but their center point (average value) is far from the bull's eye.

(c) Figure iii represents data that are imprecise but their average value is accurate. The darts are far from each other, but their average value, or geometric center point, is close to the bull's eye.

1.6 The determined age of the artifact, 1,900 years, has two significant figures. There is uncertainty in the hundreds place, indicating that the minimum uncertainty in age is 100 years. The 20-year period since the age was determined is not significant relative to the determined age.

1.7 In order to cancel units, the conversion factor must have the unit being canceled opposite the starting position. For example, if the unit cm starts in the numerator, then the conversion factor must have cm in its denominator. However, if the unit cm starts in the denominator, the conversion factor must have cm in the numerator. Ideally, this will lead to the desired units in the appropriate location, numerator or denominator. However, the inverse of the answer can be taken when necessary.

Classification and Properties of Matter

1.10 (a) homogeneous mixture

(b) heterogeneous mixture (particles in liquid)

(c) pure substance

(d) heterogeneous mixture

1.12 (a) C (b) Na (c) F (d) Fe (e) P (f) Ar (g) Ni (h) Ag

1.14 (a) cobalt (b) iodine (c) krypton (d) mercury (e) arsenic

(f) titanium (g) potassium (h) germanium

1.16 Before modern instrumentation, the classification of a pure substance as an element was determined by whether it could be broken down into component elements. Scientists subjected the substance to all known chemical means of decomposition, and if the results were negative, the substance was an element. Classification by negative results was somewhat ambiguous, since an effective decomposition technique might exist but not yet have been discovered.

1.18 *Physical properties*: silver-grey (color); melting point = 420°C; hardness = 2.5 Mohs; density = 7.13 g/cm^3 at 25°C. *Chemical properties*: metal; reacts with sulfuric acid to produce hydrogen gas; reacts slowly with oxygen at elevated temperatures to produce ZnO.

1.20 (a) chemical

(b) physical

(c) physical (The production of H_2O is a chemical change, but its **condensation** is a physical change.)

(d) physical (The production of soot is a chemical change, but its **deposition** is a physical change.)

1.22 Take advantage of differences in physical properties to separate the components of a mixture. First heat the liquid to 100°C to evaporate the water. This is conveniently done in a distillation apparatus (Figure 1.13) so that the water can be collected. After the water is completely evaporated and if there is a residue, measure the physical properties of the residue such as color, density, and melting point. Compare the observed properties of the residue to those of table salt, NaCl. If the properties match, the colorless liquid contained table salt. If the properties don't match, the liquid contained a different dissolved solid. If there is no residue, no dissolved solid is present.

Units and Measurement

1.24 (a) $6.35 \times 10^{-2}\ \text{L} \times \dfrac{1\ \text{mL}}{1 \times 10^{-3}\ \text{L}} = 63.5\ \text{mL}$

(b) $6.5 \times 10^{-6}\ \text{s} \times \dfrac{1\ \mu\text{s}}{1 \times 10^{-6}\ \text{s}} = 6.5\ \mu\text{s}$

(c) $9.5 \times 10^{-4}\ \text{m} \times \dfrac{1\ \text{mm}}{1 \times 10^{-3}\ \text{m}} = 0.95\ \text{mm}$

(d) $4.23 \times 10^{-9}\ \text{m}^3 \times \dfrac{1^3\ \text{mm}^3}{(1 \times 10^{-3})^3\ \text{m}^3} = 4.23\ \text{mm}^3$

$$4.23\ \text{mm}^3 \times \frac{(10^{-1})^3\ \text{cm}^3}{1^3\ \text{mm}^3} \times \frac{1\ \text{mL}}{1\ \text{cm}^3} \times \frac{1 \times 10^{-3}\ \text{L}}{1\ \text{mL}} \times \frac{1\ \mu\text{L}}{1 \times 10^{-6}\ \text{L}} = 4.23\ \mu\text{L}$$

(e) $12.5 \times 10^{-8}\ \text{kg} \times \dfrac{1 \times 10^3\ \text{g}}{1\ \text{kg}} \times \dfrac{1\ \text{mg}}{1 \times 10^{-3}\ \text{g}} = 0.125\ \text{mg}\ (125\ \mu\text{g})$

(f) $3.5 \times 10^{-10}\ \text{g} \times \dfrac{1\ \text{ng}}{1 \times 10^{-9}\ \text{g}} = 0.35\ \text{ng}$

(g) $6.54 \times 10^9\ \text{fs} \times \dfrac{1 \times 10^{-15}\ \text{s}}{1\ \text{fs}} \times \dfrac{1\ \mu\text{s}}{1 \times 10^{-6}\ \text{s}} = 6.54\ \mu\text{s}$

1.26 (a) $9.5 \times 10^{-2}\ \text{kg} \times \dfrac{1 \times 10^3\ \text{g}}{1\ \text{kg}} = 95\ \text{g}$

(b) $0.0023\ \mu\text{m} \times \dfrac{1 \times 10^{-6}\ \text{m}}{1\ \mu\text{m}} \times \dfrac{1\ \text{nm}}{1 \times 10^{-9}\ \text{m}} = 2.3\ \text{nm}$

(c) $7.25 \times 10^{-4}\ \text{s} \times \dfrac{1\ \text{ms}}{1 \times 10^{-3}\ \text{s}} = 0.725\ \text{ms}$

1.28 (a) volume = length^3 (cm^3); density = mass/volume (g/cm^3)

volume = $(1.500)^3\ \text{cm}^3 = 3.375\ \text{cm}^3$

density = $\dfrac{76.31\ \text{g}}{3.375\ \text{cm}^3} = 22.61\ \text{g/cm}^3$ osmium

(b) $65.8\ \text{mL} \times \dfrac{1\ \text{cm}^3}{1\ \text{mL}} \times \dfrac{4.51\ \text{g}}{1\ \text{cm}^3} = 296.758 = 297\ \text{g}$ titanium

(c) $0.1500\ \text{L} \times \dfrac{1\ \text{mL}}{1 \times 10^{-3}\ \text{L}} \times \dfrac{0.8787\ \text{g}}{1\ \text{mL}} = 131.8\ \text{g}$ benzene

1.30 (a) $\dfrac{21.95\ \text{g}}{25.0\ \text{mL}} = 0.878\ \text{g/mL}$

The tabulated value has four significant figures, while the experimental value has three. The tabulated value rounded to three figures is 0.879. The values agree within one in the last significant figure of the experimental value; the two results agree. The liquid could be benzene.

(b) $15.0\ \text{g} \times \dfrac{1\ \text{mL}}{0.7781\ \text{g}} = 19.3\ \text{mL}$ cyclohexane

(c) $r = d/2 = 5.0\ \text{cm}/2 = 2.5\ \text{cm}$

$V = 4/3\ \pi r^3 = 4/3 \times \pi \times (2.5)^3\ \text{cm}^3 = 65\ \text{cm}^3$

$65.4498\ \text{cm}^3 \times \dfrac{11.34\ \text{g}}{\text{cm}^3} = 7.4 \times 10^2\ \text{g}$

(The answer has two significant figures because the diameter had only two figures.)

Note: This is the first exercise where "intermediate rounding" occurs. In this manual, when a solution is given in steps, the intermediate result will be rounded to the correct number of significant figures. However, the **unrounded** number will be used in subsequent calculations. The final answer will appear with the correct number of significant figures. That is, calculators need not be cleared and new numbers entered in the middle of a calculation sequence. This may result in a small discrepancy in the last significant digit between student-calculated answers and those given in the manual. These variations occur in any analysis of numerical data.

For example, in this exercise the volume of the sphere, 65.4498 cm^3, is rounded to 65 cm^3, but 65.4498 is retained in the subsequent calculation of mass, 7.4×10^2 g. In this case, 65 $cm^3 \times 11.34$ g/cm^3 also yields 7.4×10^2 g. In other exercises, the correctly rounded results of the two methods may not be identical.

1.32 Calculate the volume of the rod:

$$2.17 \text{ kg} \times \frac{1000 \text{ g}}{1 \text{ kg}} \times \frac{1 \text{ cm}^3}{2.33 \text{ g}} = 931.3 = 931 \text{ cm}^3$$

$$V = \pi r^2 h; d = 2r, r = d/2; \quad V = \pi\left(\frac{d}{2}\right)^2 h; \quad d^2 = \frac{4V}{\pi h}; d = \left(\frac{4V}{\pi h}\right)^{1/2}$$

$$d = \left(\frac{4(931.3) \text{ cm}^3}{\pi (16.8) \text{ cm}}\right)^{1/2} = 8.401 = 8.40 \text{ cm}$$

1.34 (a) °C = 5/9 (87°F – 32°) = 31°C

(b) K = 25°C + 273 = 298 K; °F = 9/5 (25°C) + 32 = 77°F

(c) °C = 5/9 (175°F – 32°) = 79.444 = 79.4°C

K = °C + 273.15 = 79.444°C + 273.15 = 352.6 K

(d) °F = 9/5 (755°C) + 32 = 1391°F; K = 755°C + 273.15 = 1028 K

(It could be argued that the result of 9/5 (755) has 3 sig figs, so the final Fahrenheit temperature should have 3 sig figs, 1390°F.)

(e) melting point = –248.6°C + 273.15 = 24.6 K
boiling point = –246.1°C + 273.15 = 27.1 K

Uncertainty In Measurement

1.36 Exact: (b), (e) (The number of students is exact on any given day.)

1.38 (a) 5 (b) 3 (c) 4 (d) 4 (e) 6

1.40 (a) 7.93×10^3 mi (b) 4.001×10^4 km

1.42 (a) [320.55 – 6104.5/2.3] = -2.3×10^3 (The intermediate result has two significant figures, so only the thousand and hundred places in the answer are significant.)

(b) [$285.3 \times 10^5 - 0.01200 \times 10^5$] $\times$ 2.8954 = 8.260×10^7 (Since subtraction depends on decimal places, both numbers must have the same exponent to determine decimal places/sig figs. The intermediate result has 1 decimal place and 4 sig figs, so the answer has 4 sig figs.)

(c) (0.0045 × 20,000.0) + (2813 × 12) = 3.4×10^4
2 sig figs / 0 dec pl   2 sig figs / first 2 digits

(d) 863 × [1255 − (3.45 × 108)] = 7.62×10^5
3 sig figs / 0 dec pl

3 sig figs × 0 dec pl/3 sig figs = 3 sig figs

Dimensional Analysis

1.44 (a) $\frac{1.6093 \text{ km}}{1 \text{ mi}}$; when converting miles to kilometers, miles goes in the denominator so that it cancels the original unit, leaving km in the numerator.

(b) $\frac{1 \text{ L}}{1.0567 \text{ qt}}$

1.46 (a) $\frac{2.998 \times 10^8 \text{ m}}{\text{s}} \times \frac{1 \text{ km}}{1000 \text{ m}} \times \frac{60 \text{ s}}{1 \text{ min}} \times \frac{60 \text{ min}}{1 \text{ hr}} = 1.079 \times 10^9 \text{ km/hr}$

(b) $1454 \text{ ft} \times \frac{1 \text{ yd}}{3 \text{ ft}} \times \frac{1 \text{ m}}{1.0936 \text{ yd}} = 443.18 = 443.2 \text{ m}$

(c) $3{,}666{,}500 \text{ m}^3 \times \frac{1^3 \text{ dm}^3}{(1 \times 10^{-1})^3 \text{ m}^3} \times \frac{1 \text{ L}}{1 \text{ dm}^3} = 3.6665 \times 10^9 \text{ L}$

(d) $\frac{232 \text{ mg cholesterol}}{100 \text{ mL blood}} \times \frac{1 \text{ mL}}{1 \times 10^{-3} \text{ L}} \times 5.2 \text{ L} \times \frac{1 \times 10^{-3} \text{ g}}{1 \text{ mg}} = 12 \text{ g cholesterol}$

1.48 (a) $0.105 \text{ in} \times \frac{2.54 \text{ cm}}{\text{in}} \times \frac{1 \times 10^{-2} \text{ m}}{\text{cm}} \times \frac{1 \text{ mm}}{1 \times 10^{-3} \text{ m}} = 2.667 = 2.67 \text{ mm}$

(b) $0.870 \text{ qt} \times \frac{1 \text{ L}}{1.057 \text{ qt}} \times \frac{1 \text{ mL}}{1 \times 10^{-3} \text{ L}} = 823.08 = 823 \text{ mL}$

(c) $\frac{8.75 \, \mu \text{m}}{\text{s}} \times \frac{1 \times 10^{-6} \text{ m}}{1 \, \mu\text{m}} \times \frac{1 \text{ km}}{1 \times 10^3 \text{ m}} \times \frac{60 \text{ s}}{1 \text{ min}} \times \frac{60 \text{ min}}{1 \text{ hr}} = 3.15 \times 10^{-5} \text{ km/hr}$

(d) $4.733 \text{ yd}^3 \times \frac{1 \text{ m}^3}{(1.0936)^3 \text{ yd}^3} = 3.61877 = 3.619 \text{ m}^3$

(e) $\frac{\$3.99}{\text{lb}} \times \frac{2.205 \text{ lb}}{1 \text{ kg}} = 8.798 = \$8.80/\text{kg}$

(f) $\frac{8.75 \text{ lb}}{\text{ft}^3} \times \frac{453.59 \text{ g}}{1 \text{ lb}} \times \frac{1 \text{ ft}^3}{12^3 \text{ in}^3} \times \frac{1 \text{ in}^3}{2.54^3 \text{ cm}^3} \times \frac{1 \text{ cm}^3}{1 \text{ mL}} = 0.140 \text{ g/mL}$

1.50 (a) $1486 \text{ mi} \times \frac{1 \text{ km}}{0.62137 \text{ mi}} \times \frac{\text{charge}}{225 \text{ km}} = 10.6 \text{ charges}$

Since charges are integral events, 11 charges are required.

(b) $\frac{14\text{ m}}{\text{s}} \times \frac{1\text{ km}}{1\times10^3\text{ m}} \times \frac{1\text{ mi}}{1.6093\text{ km}} \times \frac{60\text{ s}}{1\text{ min}} \times \frac{60\text{ min}}{1\text{ hr}} = 31\text{ mi/hr}$

(c) $450\text{ in}^3 \times \frac{(2.54)^3\text{ cm}^3}{1\text{ in}^3} \times \frac{1\text{ mL}}{1\text{ cm}^3} \times \frac{1\times10^{-3}\text{L}}{1\text{ mL}} = 7.37\text{ L}$

(d) $2.4\times10^5\text{ barrels} \times \frac{42\text{ gal}}{1\text{ barrel}} \times \frac{4\text{ qt}}{1\text{ gal}} \times \frac{1\text{ L}}{1.057\text{ qt}} = 3.8\times10^7\text{ L}$

1.52 $9.0\text{ ft} \times 14.5\text{ ft} \times 18.8\text{ ft} = 2453.4 = 2.5 \times 10^3\text{ ft}^3$

$$2453.4\text{ ft}^3 \times \frac{(1\text{ yd})^3}{(3\text{ ft})^3} \times \frac{(1\text{ m})^3}{(1.094\text{ yd})^3} \times \frac{48\ \mu\text{g CO}}{1\text{ m}^3} \times \frac{1 \times 10^{-6}\text{ g}}{1\ \mu\text{g}} = 3.3 \times 10^{-3}\text{ g CO}$$

1.54 Select a common unit for comparison, in this case the kg.

$1\text{ kg} > 2\text{ lb}$, $1\text{ L} \approx 1\text{ qt}$

$5\text{ lb potatoes} < 2.5\text{ kg}$

$5\text{ kg sugar} = 5\text{ kg}$

$1\text{ gal} = 4\text{ qt} \approx 4\text{ L}$. $1\text{ mL } H_2O = 1\text{ g } H_2O$. $1\text{ L} = 1000\text{ g}$, $4\text{ L} = 4000\text{ g} = 4\text{ kg}$

The order of mass from lightest to heaviest is 5 lb potatoes < 1 gal water < 5 kg sugar.

1.56 A wire is a very long, thin cylinder of volume, $V = \pi r^2 h$, where h is the length of the wire and πr^2 is the cross-sectional area of the wire.

Strategy: 1) Calculate total volume of copper in cm^3 from mass and density

2) $h\text{ (length in cm)} = \frac{V}{\pi r^2}$

3) Change cm → ft

$$150\text{ lb Cu} \times \frac{453.6\text{ g}}{1\text{ lb Cu}} \times \frac{1\text{ cm}^3}{8.94\text{ g}} = 7610.7 = 7.61 \times 10^3\text{ cm}^3$$

$$r = d/2 = 8.25\text{ mm} \times \frac{1\text{ cm}}{10\text{ mm}} \times \frac{1}{2} = 0.4125 = 0.413\text{ cm}$$

$$h = \frac{V}{\pi r^2} = \frac{7610.7\text{ cm}^3}{\pi(0.4125)^2\text{ cm}^2} = 1.4237 \times 10^4 = 1.42 \times 10^4\text{ cm}$$

$$1.4237 \times 10^4\text{ cm} \times \frac{1\text{ in}}{2.54\text{ cm}} \times \frac{1\text{ ft}}{12\text{ in}} = 467\text{ ft}$$

(too difficult to estimate)

Additional Exercises

1.58 (a) A gold coin is probably a *solid solution*. Pure gold (element 79) is too soft and too valuable to be used for coinage, so other metals are added. However, the simple term "gold coin" does not give a specific indication of the other metals in the mixture.

A cup of coffee is a *solution* if there are no suspended solids (coffee grounds). It is a heterogeneous mixture if there are grounds. If cream or sugar is added, the homogeneity of the mixture depends on how thoroughly the components are mixed.

A wood plank is a *heterogeneous mixture* of various cellulose components. The different domains in the mixture are visible as wood grain or knots.

(b) The ambiguity in each of these examples is that the name of the substance does not provide a complete description of the material. We must rely on mental images, and these vary from person to person.

1.59 (a) A *hypothesis* is a possible explanation for certain phenomena based on preliminary experimental data. A *theory* may be more general, and has a significant body of experimental evidence to support it; a theory has withstood the test of experimentation.

(b) A scientific *law* is a summary or statement of natural behavior; it tells how matter behaves. A *theory* is an explanation of natural behavior; it attempts to explain why matter behaves the way it does.

1.61 (a) I. $(22.52 + 22.48 + 22.54)/3 = 22.51$

II. $(22.64 + 22.58 + 22.62)/3 = 22.61$

Based on the average, set I is more accurate. That is, it is closer to the true value of 22.52%.

(b) Average deviation = Σ | value – average | /3

I. $|22.52 - 22.51| + |22.48 - 22.51| + |22.54 - 22.51|/3 = 0.02$

II. $|22.64 - 22.61| + |22.58 - 22.61| + |22.62 - 22.61|/3 = 0.02$

The two sets display the same precision, even though set I is more accurate.

1.62 (a) Inappropriate. The circulation of a widely read publication would vary over a year's time, and could simply not be counted to the nearest single subscriber. Probably about four significant figures would be appropriate.

(b) Inappropriate. In a county with 5 million people, the population surely fluctuates with moves, births, and deaths during a month. The population cannot be known precisely to the nearest person over this time period. There would be uncertainty in at least hundreds, probably thousands place in the population.

(c) Inappropriate. Rainfall can be measured to within 0.02 in., but it is probably not possible to record an entire year's rainfall to the nearest 0.01 in. Further, the variation from year to year is sufficiently large that it does not make much sense to report the annual average to this number of significant figures. Probably two significant figures would be appropriate.

(d) Appropriate. The percentage has three significant figures. In a population as large as the United States, the number of people named Brown can surely be counted by census data or otherwise to a precision of three significant figures.

1.64 (a) $\frac{m}{s^2}$ (b) $\frac{kg \bullet m}{s^2}$ (c) $\frac{kg \bullet m}{s^2} \times m = \frac{kg \bullet m^2}{s^2}$

(d) $\frac{kg \bullet m}{s^2} \times \frac{1}{m^2} = \frac{kg}{m \bullet s^2}$ (e) $\frac{kg \bullet m^2}{s^2} \times \frac{1}{s} = \frac{kg \bullet m^2}{s^3}$

1.65 The most dense liquid, Hg, will sink; the least dense, cyclohexane, will float; H_2O will be in the middle.

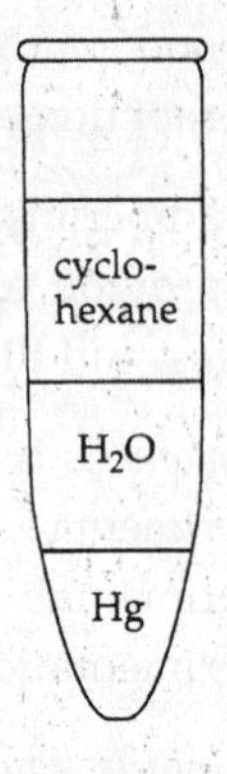

1.67 Density is the ratio of mass and volume. For substances with different densities, the greater the density the smaller the volume of substance that will contain a certain mass. Since volume is directly related to diameter ($V = 4/3\ \pi\ r^3 = 1/6\ \pi\ d^3$), the more dense the substance, the smaller the diameter of a ball that contains a certain mass. The order of the sphere diameters is the reverse order of densities: Pb < Ag < Al. Mathematically, assume 10.0 g of material.

$$\text{Pb: } 10.0\,g \times \frac{1\,cm^3}{11.3\,g} = 0.88496 = 0.885\,cm^3;\ d = (6\,V/\pi)^{1/3} = 1.19\,cm$$

$$\text{Ag: } 10.0\,g \times \frac{1\,cm^3}{10.5\,g} = 0.95238 = 0.952\,cm^3;\ d = 1.22\,cm$$

$$\text{Al: } 10.0\,g \times \frac{1\,cm^3}{2.70\,g} = 3.7037 = 3.70\,cm^3;\ d = 1.92\,cm$$

Note that Pb and Ag, with similar densities, have similar diameters; Al, with a much smaller density, has a much larger diameter.

1.68 (a) $23.2 \times 10^9\ lb \times \frac{453.6\,g}{1\,lb} = 1.05235 \times 10^{13} = 1.05 \times 10^{13}\,g$ NaOH

(b) $1.05235 \times 10^{13}\ g \times \frac{1\,cm^3}{2.130\,g} \times \frac{1\,m^3}{(100)^3\,cm^3} \times \frac{1\,km^3}{(1000)^3\,m^3} = 4.94 \times 10^{-3}\ km^3$

1.70 $0.500\,L \text{ battery acid} \times \frac{1000\,mL}{L} \times \frac{1.28\,g}{mL} = 640\,g \text{ battery acid}$

$640\,g \text{ battery acid} \times \frac{38.1\,g \text{ sulfuric acid}}{100\,g \text{ battery acid}} = 243.84 = 244\,g \text{ sulfuric acid}$

1.71 mass of toluene = 58.58 g – 32.65 g = 25.93 g

volume of toluene = $25.93\,g \times \frac{1\,mL}{0.864\,g} = 30.0116 = 30.0\,mL$

volume of solid = 50.00 mL – 30.0116 mL = 19.9884 = 20.0 mL

$$\text{density of solid} = \frac{32.65\text{ g}}{19.9884\text{ mL}} = 1.63\text{ g/mL}$$

1.72 There are 209.1 degrees between the freezing and boiling points on the Celsius (C) scale and 100 degrees on the glycol (G) scale. Also, –11.5°C = 0°G. By analogy with °F and °C,

$$°G = \frac{100}{209.1}(°C + 11.5) \text{ or } °C = \frac{209.1}{100}(°G) - 11.5$$

These equations correctly relate the freezing point (and boiling point) of ethylene glycol on the two scales.

$$\text{f.p. of } H_2O: \; °G = \frac{100}{209.1}(0°C + 11.5) = 5.50°G$$

1.74 (a) $2.4 \times 10^5 \text{ mi} \times \frac{1.609\text{ km}}{1\text{ mi}} \times \frac{1000\text{ m}}{1\text{ km}} = 3.9 \times 10^8\text{ m}$

(b) $2.4 \times 10^5 \text{ mi} \times \frac{1.609\text{ km}}{1\text{ mi}} \times \frac{1\text{ hr}}{2.4 \times 10^3\text{ km}} \times \frac{60\text{ min}}{1\text{ hr}} \times \frac{60\text{ s}}{1\text{ min}} = 5.8 \times 10^5\text{ s}$

1.75 (a) $575\text{ ft} \times \frac{12\text{ in}}{1\text{ ft}} \times \frac{2.54\text{ cm}}{1\text{ in}} \times \frac{10\text{ mm}}{1\text{ cm}} \times \frac{1\text{ quarter}}{1.55\text{ mm}} = 1.1307 \times 10^5 = 1.13 \times 10^5\text{ quarters}$

(b) $1.1307 \times 10^5\text{ quarters} \times \frac{5.67\text{ g}}{1\text{ quarter}} = 6.41 \times 10^5\text{ g (641 kg)}$

(c) $1.1307 \times 10^5\text{ quarters} \times \frac{1\text{ dollar}}{4\text{ quarters}} = \$28{,}268 = \$2.83 \times 10^4$

(d) $\$7.2 \times 10^{12} \times \frac{1\text{ stack}}{\$28{,}268} = 2.5 \times 10^8$ stacks (approximately 250 million stacks)

1.77 $8.0\text{ oz} \times \frac{1\text{ lb}}{16\text{ oz}} \times \frac{453.6\text{ g}}{\text{lb}} \times \frac{1\text{ cm}^3}{2.70\text{ g}} = 84.00 = 84\text{ cm}^3$

$$\frac{84\text{ cm}^3}{50\text{ ft}^2} \times \frac{1^2\text{ ft}^2}{12^2\text{ in}^2} \times \frac{1^2\text{ in}^2}{2.54^2\text{ cm}^2} \times \frac{10\text{ mm}}{1\text{ cm}} = 0.018\text{ mm}$$

1.78 $11.86\text{ g ethanol} \times \frac{1\text{ cm}^3}{0.789\text{ g ethanol}} = 15.0317 = 15.03\text{ cm}^3$, volume of cylinder

$$V = \pi r^2 h;\; r = (V/\pi h)^{1/2} = \left[\frac{15.0317\text{ cm}^3}{\pi \times 15.0\text{ cm}}\right]^{1/2} = 0.5648 = 0.565\text{ cm}$$

$d = 2r = 1.13$ cm

1.79 (a) Let x = mass of Au in jewelry

9.85 - x = mass of Ag in jewelry

The total volume of jewelry = volume of Au + volume of Ag

$$0.675\text{ cm}^3 = x\text{ g} \times \frac{1\text{ cm}^3}{19.3\text{ g}} + (9.85 - x)\text{g} \times \frac{1\text{ cm}^3}{10.5\text{ g}}$$

$$0.675 = \frac{x}{19.3} + \frac{9.85 - x}{10.5}$$ (To solve, multiply both sides by (19.3)(10.5))

$$0.675\,(19.3)(10.5) = 10.5\,x + (9.85 - x)(19.3)$$

$$136.79 = 10.5\,x + 190.105 - 19.3\,x$$

$$-53.315 = -8.8\,x$$

$$x = 6.06 \text{ g Au};\ 9.85 \text{ g total} - 6.06 \text{ g Au} = 3.79 \text{ g Ag}$$

$$\text{mass \% Au} = \frac{6.06 \text{ g Au}}{9.85 \text{ g jewelry}} \times 100 = 61.5\% \text{ Au}$$

(b) $24 \text{ carats} \times 0.615 = 15$ carat gold

1.80 A solution can be separated into components by physical means, so separation would be attempted. If the liquid is a solution, the solute could be a solid or a liquid; these two kinds of solutions would be separated differently. Therefore, divide the liquid into several samples and do different tests on each. Try evaporating the solvent from one sample. If a solid remains, the liquid is a solution and the solute is a solid. If the result is negative, try distilling a sample to see if two or more liquids with different boiling points are present. If this result is negative, the liquid is probably a pure substance, but negative results are never entirely conclusive. We might not have tried the appropriate separation technique.

1.82 The densities are:

carbon tetrachloride (methane, tetrachloro) - 1.5940 g/cm^3

hexane - 0.6603 g/cm^3

benzene - 0.87654 g/cm^3

methylene iodide (methane, diiodo) - 3.3254 g/cm^3

Only methylene iodide will separate the two granular solids. The undesirable solid (2.04 g/cm^3) is less dense than methylene iodide and will float; the desired material is more dense than methylene iodide and will sink. The other three liquids are less dense than both solids and will not produce separation.

1.83 Study (a) is likely to be both precise and accurate, because the errors are carefully controlled. The secondary weight standard will be resistant to chemical and physical changes, the balance is carefully calibrated, and weighings are likely to be made by the same person. The relatively large number of measurements is likely to minimize the effect of random errors on the average value. The accuracy and precision of study (b) depend on the veracity of the participants' responses, which cannot be carefully controlled. It also depends on the definition of "comparable lifestyle." The percentages are not precise, because the broad definition of lifestyle leads to a range of results (scatter). The relatively large number of participants improves the precision and accuracy. In general, controlling errors and maximizing the number of data points in a study improves precision and accuracy.

2 Atoms, Molecules, and Ions

Visualizing Concepts

2.2 In general, metals occupy the left side of the chart, and nonmetals the right side.

metals: red and green *nonmetals*: blue and yellow

alkaline earth metal: red *noble gas*: yellow

2.4 In a solid, particles are close together and their relative positions are fixed. In a liquid, particles are close but moving relative to each other. In a gas, particles are far apart and moving. All ionic compounds are solids because of the strong forces among charged particles. Molecular compounds can exist in any state: solid, liquid, or gas.

Since the molecules in *ii* are far apart, *ii* must be a molecular compound. The particles in *i* are near each other and exist in a regular, ordered arrangement, so *i* is likely to be an ionic compound.

2.6 Cations (red spheres) have positive charges; anions (blue spheres) have negative charges. There are twice as many anions as cations, so the formula has the general form CA_2. Only $Ca(NO_3)_2$, calcium nitrate, is consistent with the diagram.

Atomic Theory and the Discovery of Atomic Structure

2.8 (a) 6.500 g compound – 0.384 g hydrogen = 6.116 g sulfur

(b) *Conservation of mass*

(c) According to postulate 3 of the atomic theory, atoms are neither created nor destroyed during a chemical reaction. If 0.384 g of H are recovered from a compound that contains only H and S, the remaining mass must be sulfur.

2.10 (a) 1: $\frac{3.56\ \text{g fluorine}}{4.75\ \text{g iodine}} = 0.749\ \text{g fluorine/1 g iodine}$

2: $\frac{3.43\ \text{g fluorine}}{7.64\ \text{g iodine}} = 0.449\ \text{g fluorine/1 g iodine}$

3: $\frac{9.86\ \text{g fluorine}}{9.41\ \text{g iodine}} = 1.05\ \text{g fluorine/1 g iodine}$

(b) To look for integer relationships among these values, divide each one by the smallest.

If the quotients aren't all integers, multiply by a common factor to obtain all integers.

1: $0.749/0.449 = 1.67; 1.67 \times 3 = 5$

2: $0.449/0.449 = 1.00; 1.00 \times 3 = 3$

3: $1.05/0.449 = 2.34; 2.34 \times 3 = 7$

The ratio of g fluorine to g iodine in the three compounds is 5:3:7. These are in the ratio of small whole numbers and, therefore, obey the *law of multiple proportions.* This integer ratio indicates that the combining fluorine "units" (atoms) are indivisible entities.

2.12 Since the unknown particle is deflected in the opposite direction from that of a negatively charged beta (β) particle, it is attracted to the (–) plate and repelled by the (+) plate. The unknown particle is positively charged. The magnitude of the deflection is less than that of the β particle, or electron, so the unknown particle has greater mass than the electron. The unknown is a positively charged particle of greater mass than the electron.

2.14 (a) The droplets carry different total charges because there may be 1, 2, 3, or more electrons on the droplet.

(b) The electronic charge is likely to be the lowest common factor in all the observed charges.

(c) Assuming this is so, we calculate the apparent electronic charge from each drop as follows:

A: $1.60 \times 10^{-19} / 1 = 1.60 \times 10^{-19}$ C

B: $3.15 \times 10^{-19} / 2 = 1.58 \times 10^{-19}$ C

C: $4.81 \times 10^{-19} / 3 = 1.60 \times 10^{-19}$ C

D: $6.31 \times 10^{-19} / 4 = 1.58 \times 10^{-19}$ C

The reported value is the average of these four values. Since each calculated charge has three significant figures, the average will also have three significant figures.

$(1.60 \times 10^{-19}\text{ C} + 1.58 \times 10^{-19}\text{ C} + 1.60 \times 10^{-19}\text{ C} + 1.58 \times 10^{-19}\text{ C}) / 4 = 1.59 \times 10^{-19}\text{ C}$

Modern View of Atomic Structure; Atomic Weights

2.16 (a) $r = d/2; r = \frac{2.8 \times 10^{-8}\text{ cm}}{2} \times \frac{1\text{ Å}}{1 \times 10^{-8}\text{ cm}} = 1.4\text{ Å}$

$$r = \frac{2.8 \times 10^{-8}\text{ cm}}{2} \times \frac{1\text{ m}}{100\text{ cm}} = 1.4 \times 10^{-10}\text{ m}$$

(b) Aligned Sn atoms have **diameters** touching. $d = 2.8 \times 10^{-8}\text{ cm} = 2.8 \times 10^{-10}\text{ m}$

$$6.0\ \mu\text{m} \times \frac{1 \times 10^{-6}\text{ m}}{1\ \mu\text{m}} \times \frac{1\text{ Sn atom}}{2.8 \times 10^{-10}\text{ m}} = 2.1 \times 10^{4}\text{ Sn atoms}$$

(c) $V = 4/3\ \pi\ r^3$; $r = 1.4 \times 10^{-10}$ m

$V = (4/3)[(\pi(1.4 \times 10^{-10})^3]\ m^3 = 1.149 \times 10^{-29} = 1.1 \times 10^{-29}\ m^3$

2.18 (a) The nucleus has most of the mass **but occupies very little** of the volume of an atom.

(b) True

(c) The number of electrons in an atom is equal to the number of **protons** in the atom.

(d) True

2.20 (a) ${}^{31}_{16}X$ and ${}^{32}_{16}X$ are isotopes of the same element, because they have identical atomic numbers.

(b) These are isotopes of the element sulfur, S, atomic number = 16.

2.22 (a) ${}^{32}P$ has 15 p, 17 n (b) ${}^{51}Cr$ has 24 p, 27 n

(c) ${}^{60}Co$ has 27 p, 33 n (d) ${}^{99}Tc$ has 43 p, 56 n

(e) ${}^{131}I$ has 53 p, 78 n (f) ${}^{201}Tl$ has 81 p, 120 n

2.24

Symbol	${}^{121}Sb$	${}^{103}Rh$	${}^{88}Sr$	${}^{127}Te$	${}^{239}Pu$
Protons	51	45	38	52	94
Neutrons	70	58	50	75	145
Electrons	51	45	38	52	94
Mass No.	121	103	88	127	239

2.26 Since the two nuclides are atoms of the same element, by definition they have the same number of protons, 54. They differ in mass number (and mass) because they have different numbers of neutrons. ${}^{129}Xe$ has 75 neutrons and ${}^{130}Xe$ has 76 neutrons.

2.28 (a) 12 amu

(b) The atomic weight of carbon reported on the front-inside cover of the text is the abundance-weighted average of the atomic masses of the two naturally occurring isotopes of carbon, ${}^{12}C$, and ${}^{13}C$. The mass of a ${}^{12}C$ atom is exactly 12 amu, but the atomic weight of 12.011 takes into account the presence of some ${}^{13}C$ atoms in every natural sample of the element.

2.30 Atomic weight (average atomic mass) = Σ fractional abundance $\times$ mass of isotope

Atomic weight = 0.014(203.97302) + 0.241(205.97444) + 0.221(206.97587) + 0.524(207.97663) = 207.22 = 207 amu

(The result has 0 decimal places and 3 sig figs because the fourth term in the sum has 3 sig figs and 0 decimal places.)

2.32 (a) The purpose of the magnet in the mass spectrometer is to change the path of the moving ions. The magnitude of the deflection is inversely related to mass, which is the basis of the discrimination by mass.

(b) The atomic weight of Cl, 35.5, is an average atomic mass. It is the average of the masses of two naturally occurring isotopes, weighted by their abundances.

(c) The single peak at mass 31 in the mass spectrum of phosphorus indicates that the sample contains a single isotope of P, and the mass of this isotope is 31 amu.

2.34 (a) Three peaks: $^1H - {}^1H$, $^1H - {}^2H$, $^2H - {}^2H$

(b) $^1H - {}^1H = 2(1.00783) = 2.01566$ amu

$^1H - {}^2H = 1.00783 + 2.01410 = 3.02193$ amu

$^2H - {}^2H = 2(2.01410) = 4.02820$ amu

The mass ratios are 1 : 1.49923 : 1.99845 or 1 : 1.5 : 2.

(c) $^1H - {}^1H$ is largest, because there is the greatest chance that two atoms of the more abundant isotope will combine.

$^2H - {}^2H$ is the smallest, because there is the least chance that two atoms of the less abundant isotope will combine.

The Periodic Table; Molecules and Ions

2.36 (a) sodium (metal) (b) titanium (metal) (c) gallium (metal)

(d) uranium (metal) (e) palladium (metal) (f) selenium (nonmetal)

(g) krypton (nonmetal)

2.38 C, carbon, nonmetal; Si, silicon, metalloid; Ge, germanium, metalloid; Sn, tin, metal; Pb, lead, metal

2.40 Compounds with the same empirical but different molecular formulas differ by the integer number of empirical formula units in the respective molecules. Thus, they can have very different molecular structure, size, and mass, resulting in very different physical properties.

2.42 A molecular formula contains all atoms in a molecule. An empirical formula shows the simplest ratio of atoms in a molecule or elements in a compound.

(a) molecular formula: C_6H_6; empirical formula: CH

(b) molecular formula: $SiCl_4$; empirical formula: $SiCl_4$ (1:4 is the simplest ratio)

(c) molecular: B_2H_6; empirical: BH_3

(d) molecular: $C_6H_{12}O_6$; empirical: CH_2O

2.44 (a) 4 (b) 6 (c) 9

2.46

(a) C_2H_5Br
$$\begin{array}{ccccccc} & & H & & H & & \\ & & | & & | & & \\ H & - & C & - & C & - & Br \\ & & | & & | & & \\ & & H & & H & & \end{array}$$

(b) C_2H_7N
$$\begin{array}{ccccccccc} & & H & & & & H & & \\ & & | & & & & | & & \\ H & - & C & - & N & - & C & - & H \\ & & | & & | & & | & & \\ & & H & & H & & H & & \end{array}$$

(c) CH_2Cl_2
$$\begin{array}{ccccc} & & H & & \\ & & | & & \\ H & - & C & - & Cl \\ & & | & & \\ & & Cl & & \end{array}$$

(d) NH_2OH
$$\begin{array}{ccccc} H & - & N & - & O \\ & & | & & | \\ & & H & & H \end{array}$$

2.48

Symbol	$^{75}As^{3-}$	$^{59}Ni^{2+}$	$^{127}I^{-}$	$^{197}Au^{3+}$
Protons	33	28	53	79
Neutrons	42	31	74	118
Electrons	36	26	54	76
Net Charge	3−	2+	1−	3+

2.50 (a) Sr^{2+} (b) Sc^{2+} or Sc^{3+} (c) P^{3-} (d) I^- (e) Se^{2-}

2.52 (a) AgI (b) Ag_2S (c) AgF

2.54 (a) Cu_2S (b) Fe_2O_3 (c) Hg_2CO_3 (d) $Ca_3(AsO_4)_2$ (e) $(NH_4)_2CO_3$

2.56 Molecular (all elements are nonmetals):

(a) PF_5 (c) SCl (h) N_2O_4

Ionic (formed from ions, usually contains a metal cation):

(b) NaI (d) $Ca(NO_3)_2$ (e) $FeCl_3$ (f) LaP (g) $CoCO_3$

Naming Inorganic Compounds; Organic Molecules

2.58 (a) selenate (b) selenide (c) hydrogen selenide (biselenide)
(d) hydrogen selenite (biselenite)

2.60 (a) lithium oxide (b) sodium hypochlorite
(c) strontium cyanide (d) chromium(III) hydroxide (chromic hydroxide)
(e) iron(III) carbonate (ferric carbonate) (f) cobalt(II) nitrate (cobaltous nitrate)
(g) ammonium sulfite (h) sodium dihydrogen phosphate
(i) potassium permanganate (j) silver dichromate

2.62 (a) Na_3PO_4 (b) $Co(NO_3)_2$ (c) $Ba(BrO_3)_2$ (d) $Cu(ClO_4)_2$

(e) $Mg(HCO_3)_2$ (f) $Cr(C_2H_3O_2)_3$ (g) $K_2Cr_2O_7$

2.64 (a) HBr (b) H_2S (c) HNO_2

(d) carbonic acid (e) chloric acid (f) acetic acid

2.66 (a) dinitrogen monoxide (b) nitrogen monoxide (c) nitrogen dioxide

(d) dinitrogen pentoxide (e) dinitrogen tetroxide

2.68 (a) $NaHCO_3$ (b) $Ca(ClO)_2$ (c) HCN

(d) $Mg(OH)_2$ (e) SnF (f) CdS, H_2SO_4, H_2S

2.70 *-ane*

(b) Hexane has 6 carbons in its chain.

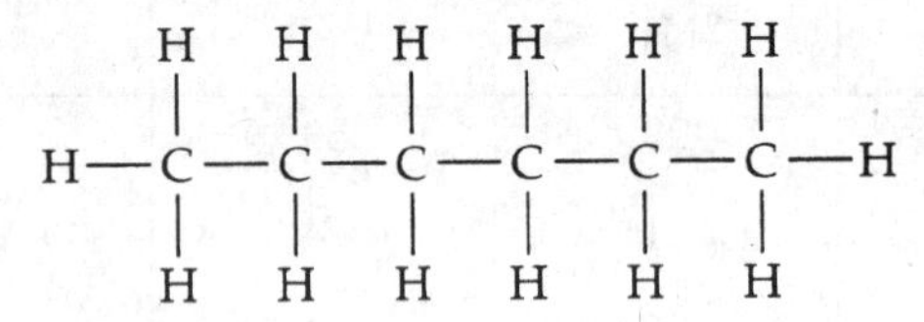

molecular: C_6H_{14}

empirical: C_3H_7

2.72 (a) They both have two carbon atoms in their molecular backbone, or chain.

(b) In 1-propanol one of the H atoms on an outer (terminal) C atom has been replaced by an –OH group.

Additional Exercises

2.73 (a) Based on data accumulated in the late eighteenth century on how substances react with one another, *Dalton* postulated the atomic theory. Dalton's theory is based on the indivisible atom as the smallest unit of an element that can combine with other elements.

(b) By determining the effects of electric and magnetic fields on cathode rays, *Thomson* measured the mass-to-charge ratio of the electron. He also proposed the "plum pudding" model of the atom in which most of the space in an atom is occupied by a diffuse positive charge in which the tiny negatively charged electrons are imbedded.

(c) By observing the rate of fall of oil drops in and out of an electric field, *Millikan* measured the charge of an electron.

(d) After observing the scattering of alpha particles at large angles when the particles struck gold foil, *Rutherford* postulated the nuclear atom. In Rutherford's atom, most of the mass of the atom is concentrated in a small dense region called the nucleus and the tiny negatively charged electrons are moving through empty space around the nucleus.

2.75 (a) Most of the volume of an atom is empty space in which electrons move. Most alpha particles passed through this space. The path of the massive alpha particle would not be significantly altered by interaction with a "puny" electron.

(b) Most of the mass of an atom is contained in a very small, dense area called the nucleus. The few alpha particles that hit the massive, positively charged gold nuclei were strongly repelled and essentially deflected back in the direction they came from.

(c) The Be nuclei have a much smaller volume and positive charge than the Au nuclei; the charge repulsion between the alpha particles and the Be nuclei will be less, and there will be fewer direct hits because the Be nuclei have an even smaller volume than the Au nuclei. Fewer alpha particles will be scattered in general and fewer will be strongly back scattered.

2.76 (a) Droplet D would fall most slowly. It carries the most negative charge, so it would be most strongly attracted to the upper (+) plate and most strongly repelled by the lower (-) plate. These electrostatic forces would provide the greatest opposition to gravity.

(b) Calculate the lowest common factor.

A: $3.84 \times 10^{-8} / 2.88 \times 10^{-8} = 1.33; 1.33 \times 3 = 4$

B: $4.80 \times 10^{-8} / 2.88 \times 10^{-8} = 1.67; 1.67 \times 3 = 5$

C: $2.88 \times 10^{-8} / 2.88 \times 10^{-8} = 1.00; 1.00 \times 3 = 3$

D: $8.64 \times 10^{-8} / 2.88 \times 10^{-8} = 3.00; 3.00 \times 3 = 9$

The total charge on the drops is in the ratio of 4:5:3:9. Divide the total charge on each drop by the appropriate integer and average the four values to get the charge of an electron in warmombs.

A: $3.84 \times 10^{-8} / 4 = 9.60 \times 10^{-9}$ wa

B: $4.80 \times 10^{-8} / 5 = 9.60 \times 10^{-9}$ wa

C: $2.88 \times 10^{-8} / 3 = 9.60 \times 10^{-9}$ wa

D: $8.64 \times 10^{-8} / 9 = 9.60 \times 10^{-9}$ wa

The charge on an electron is 9.60×10^{-9} wa

(c) The number of electrons on each drop are the integers calculated in part (b). A has 4 e^-, B has 5 e^-, C has 3 e^- and D has 9 e^-.

(d) $$\frac{9.60 \times 10^{-9} \text{ wa}}{1 e^-} \times \frac{1 e^-}{1.60 \times 10^{-16} \text{ C}} = 6.00 \times 10^7 \text{ wa/C}$$

2.78 (a) 2 protons and 2 neutrons

(b) the nuclear strong force

(c) The charge of an α particle is twice the magnitude of the charge of an electron, with the opposite sign. That is, $2 (+1.6022 \times 10^{-19})$ C = $+3.2044 \times 10^{-19}$ C.

(d) $\dfrac{3.2044 \times 10^{-19}\ \text{C}}{4.8224 \times 10^{4}\ \text{g/C}} = 6.6448 \times 10^{-24}\ \text{g}$

$6.6448 \times 10^{-24}\ \text{g} \times \dfrac{1\ \text{amu}}{1.66054 \times 10^{-24}\ \text{g}} = 4.0016\ \text{amu}$

(e) The sum of the particle masses in an α particle is 2(1.0073) amu and 2(1.0087) amu = 4.0320 amu. The actual particle mass, 4.0016 amu, is less than the sum of the masses of the components. The difference is the nuclear binding energy, the energy released when protons and neutrons combine to form a nucleus. Mass and energy are interchangeable according to the Einstein relationship $E = mc^2$.

2.80 (a) In arrangement A the number of atoms in 1 cm^2 is just the square of the number that fit linearly in 1 cm.

$1.0\ \text{cm} \times \dfrac{1\ \text{atom}}{4.95\ \text{Å}} \times \dfrac{1 \times 10^{10}\ \text{Å}}{1\ \text{m}} \times \dfrac{1\ \text{m}}{100\ \text{cm}} = 2.02 \times 10^{7} = 2.0 \times 10^{7}\ \text{atoms/cm}$

$1.0\ \text{cm}^2 = (2.02 \times 10^7)^2 = 4.081 \times 10^{14} = 4.1 \times 10^{14}\ \text{atoms/cm}^2$

(b) In arrangement B, the atoms in the horizontal rows are touching along their diameters, as in arrangement A. The number of Rb atoms in a 1.0 cm row is then 2.0×10^7 Rb atoms. Relative to arrangement A, the vertical rows are offset by 1/2 of an atom. Atoms in a "column" are no longer touching along their vertical diameter. We must calculate the vertical distance occupied by a row of atoms, which is now less than the diameter of one Rb atom.

Consider the triangle shown below. This is an isosceles triangle (equal side lengths, equal interior angles) with a side-length of 2d and an angle of 60°. Drop a bisector to the uppermost angle so that it bisects the opposite side.

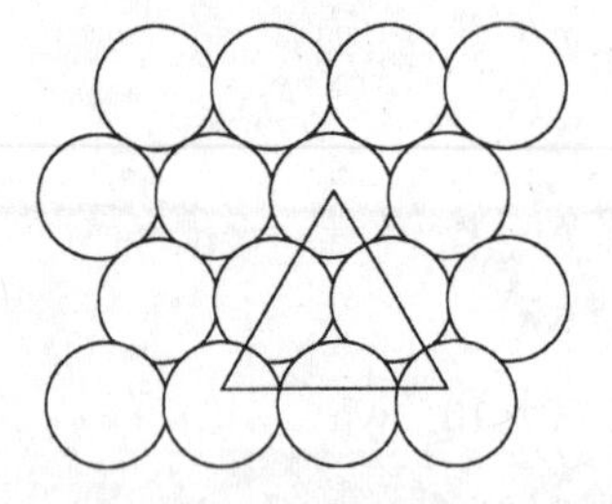

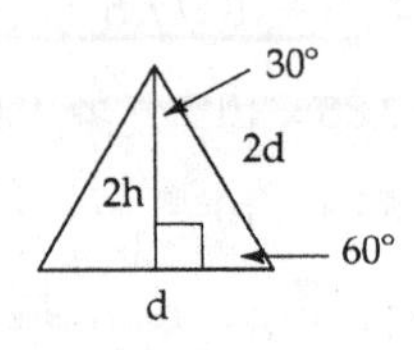

The result is a right triangle with two known side lengths. The length of the unknown side (the angle bisector) is 2h, two times the vertical distance occupied by a row of atoms. Solve for h, the "height" of one row of atoms.

$(2h)^2 + d^2 = (2d)^2;\ 4h^2 = 4d^2 - d^2 = 3d^2;\ h^2 = 3d^2/4$

$h = (3d^2/4)^{1/2} = (3(4.95\ \text{Å})^2/4)^{1/2} = 4.2868 = 4.29\ \text{Å}$

The number of rows of atoms in 1 cm is then

$1.0\ \text{cm} \times \dfrac{1\ \text{row}}{4.2868\ \text{Å}} \times \dfrac{1 \times 10^{10}\ \text{Å}}{1\ \text{m}} \times \dfrac{1\ \text{m}}{100\ \text{cm}} = 2.333 \times 10^{7} = 2.3 \times 10^{7}$

The number of atoms in a 1.0 cm^2 square area is then

$$\frac{2.020 \times 10^7 \text{ atoms}}{1 \text{ row}} \times 2.333 \times 10^7 \text{ rows} = 4.713 \times 10^{14} = 4.7 \times 10^{14}$$

Note that we have ignored the loss of "1/2" atom at the end of each horizontal row. Out of 2.0×10^7 atoms per row, one atom is not significant.

(c) The ratio of atoms in arrangement B to arrangement A is then 4.713×10^{14} atoms/4.081×10^{14} = 1.555 = 1.2:1. Clearly, arrangement B results in less empty space per unit area or volume. If extended to three dimensions, arrangement B would lead to a greater density for Rb metal.

2.81 (a) diameter of nucleus = 1×10^{-4} Å; diameter of atom = 1 Å

$V = 4/3\,\pi\, r^3$; $r = d/2$; $r_n = 0.5 \times 10^{-4}$ Å; $r_a = 0.5$ Å

volume of nucleus = $4/3\,\pi\,(0.5 \times 10^{-4})^3$ Å^3

volume of atom = $4/3\,\pi\,(0.5)^3$ Å^3

$$\text{volume fraction of nucleus} = \frac{\text{volume of nucleus}}{\text{volume of atom}} = \frac{4/3\,\pi\,(0.5 \times 10^{-4})^3\ \text{Å}^3}{4/3\,\pi\,(0.5)^3\ \text{Å}^3} = 1 \times 10^{-12}$$

diameter of atom = 5 Å, r_a = 2.5 Å

$$\text{volume fraction of nucleus} = \frac{4/3\,\pi\,(0.5 \times 10^{-4})^3\ \text{Å}^3}{4/3\,\pi\,(2.5)^3\ \text{Å}^3} = 8 \times 10^{-15}$$

Depending on the radius of the atom, the volume fraction of the nucleus is between 1×10^{-12} and 8×10^{-15}, that is, between 1 part in 10^{12} and 8 parts in 10^{15}.

(b) mass of proton = 1.0073 amu

1.0073 amu × 1.66054×10^{-24} g/amu = 1.6727×10^{-24} g

$$\text{diameter} = 1.0 \times 10^{-15} \text{ m, radius} = 0.50 \times 10^{-15} \text{ m} \times \frac{100 \text{ cm}}{1 \text{ m}} = 5.0 \times 10^{-14} \text{ cm}$$

Assuming a proton is a sphere, $V = 4/3\,\pi\, r^3$.

$$\text{density} = \frac{\text{g}}{\text{cm}^3} = \frac{1.6727 \times 10^{-24} \text{ g}}{4/3\,\pi\,(5.0 \times 10^{-14})^3 \text{ cm}^3} = 3.2 \times 10^{15} \text{ g/cm}^3$$

2.83 $F = k Q_1 Q_2 / d^2$; $k = 9.0 \times 10^9$ N m^2/C^2; $d = 0.53 \times 10^{-10}$ m;
Q (electron) = -1.6×10^{-19} C; Q (proton) = $-$Q (electron) = 1.6×10^{-19} C

$$F = \frac{\dfrac{9.0 \times 10^9 \text{ N m}^2}{\text{C}^2} \times -1.6 \times 10^{-19} \text{ C} \times 1.6 \times 10^{-19} \text{ C}}{(0.53 \times 10^{-10})^2 \text{ m}^2} = 8.202 \times 10^{-8} = 8.2 \times 10^{-8} \text{ N}$$

2.84 (a) The 68.926 amu isotope has a mass number of 69, with 31 protons, 38 neutrons and the symbol $^{69}_{31}$Ga. The 70.925 amu isotope has a mass number of 71, 31 protons, 40 neutrons, and symbol $^{71}_{31}$Ga. (All Ga atoms have 31 protons.)

(b) The average mass of a Ga atom is 69.72 amu. Let x = abundance of the lighter isotope, 1–x = abundance of the heavier isotope. Then x(68.926) + (1–x)(70.925) = 69.72; x = 0.6028 = 0.603, ^{69}Ga = 60.3%, ^{71}Ga = 39.7%.

2.85 (a) There are 24 known isotopes of Ni, from ^{51}Ni to ^{74}Ni.

(b) The five most abundant isotopes are

^{58}Ni, 57.935346 amu, 68.077%

^{60}Ni, 59.930788 amu, 26.223%

^{62}Ni, 61.928346 amu, 3.634%

^{61}Ni, 60.931058 amu, 1.140%

^{64}Ni, 63.927968 amu, 0.926%

Data from *Handbook of Chemistry and Physics*, 74th Ed. [Data may differ slightly in other editions.]

2.87 (a) Five significant figures. $^1H^+$ is a bare proton with mass 1.0073 amu. ^1H is a hydrogen atom, with 1 proton and 1 electron. The mass of the electron is 5.486 × 10^{-4} or 0.0005486 amu. Thus the mass of the electron is significant in the fourth decimal place or fifth significant figure in the mass of ^1H.

(b) Mass of ^1H = 1.0073 amu (proton)

0.0005486 amu (electron)

1.0078 amu (We have not rounded up to 1.0079 since 49 < 50 in the final sum.)

$$\text{Mass \% of electron} = \frac{\text{mass of e}^-}{\text{mass of }^1\text{H}} \times 100 = \frac{5.486 \times 10^{-4}\text{ amu}}{1.0078\text{ amu}} \times 100 = 0.05444\%$$

2.88 copper: Cu, 1B (coinage metals, transition metal)

tin: Sn, 4A

zinc: Zn, 2B

phosphorus: P, 5A

lead: Pb, 4A

2.90 (a) $^{266}_{106}Sg$ has 106 protons, 160 neutrons and 106 electrons

(b) Sg is in Group 6B (or 6) and immediately below tungsten, W. We expect the chemical properties of Sg to most closely resemble those of W.

2.91 (a) chlorine gas, Cl_2: ii (b) propane, C_3H_8: v (c) nitrate ion, NO_3^- : i

(d) sulfur trioxide, SO_3: iii (e) methylchloride, CH_3Cl: iv

2.93 (a) IO_3^- (b) IO_4^- (c) IO (d) HIO (e) HIO_4 or (H_5IO_6)

2.94 (a) perbromate ion (b) selenite ion

(c) AsO_4^{3-} (d) $HTeO_4^-$

2.96 (a) potassium nitrate (b) sodium carbonate (c) calcium oxide
(d) hydrochloric acid (e) magnesium sulfate (f) magnesium hydroxide

2.98

	Formula	Name	Density, g/mL	Melting Point, °C	Boiling Point, °C
(a)	PF_3	phosphorus trifluoride	3.907	-151.5	-101.5
(b)	$SiCl_4$	silicon tetrachloride	1.483	-70	57.57
(c)	C_2H_6O (C_2H_5OH)	ethanol	0.7893	-117.3	78.5

2.99 (a) In an alkane, all C atoms have 4 single bonds, so each C in the partial structure needs 2 more bonds. All alkanes are hydrocarbons, so 2 H atoms will bind to each C atom in the ring.

```
         H    H
          \  /
    H      C      H
     \   /   \   /
H  —  C         C  —  H
      |         |
H  —  C         C  —  H
     /   \   /   \
    H      C      H
          / \
         H   H
```

(b) The molecular formula of cyclohexane is C_6H_{12}; the molecular formula of *n*-hexane is C_6H_{14} (see Solution 2.64(d)). Cyclohexane can be thought of as *n*-hexane in which the two outer (terminal) C atoms are joined to each other. In order to form this C–C bond, each outer C atom must lose 1 H atom. The number of C atoms is unchanged, and each C atom still has 4 single bonds. The resulting molecular formula is $C_6H_{14-2} = C_6H_{12}$.

(c) On the structure in part (a), replace 1 H atom with an OH group.

```
         H    H
          \  /
    H      C      OH
     \   /   \   /
H  —  C         C  —  H
      |         |
H  —  C         C  —  H
     /   \   /   \
    H      C      H
          / \
         H   H
```

2.100 Elements are arranged in the periodic table by increasing atomic number and so that elements with similar chemical and physical properties form a vertical column or group. By its position in the periodic chart, we know whether an element is a metal, nonmetal, or metalloid, and the common charge of its ion. Members of a group have the same common ionic charge and combine in similar ways with other elements.

3 Stoichiometry: Calculations with Chemical Formulas and Equations

Visualizing Concepts

3.2

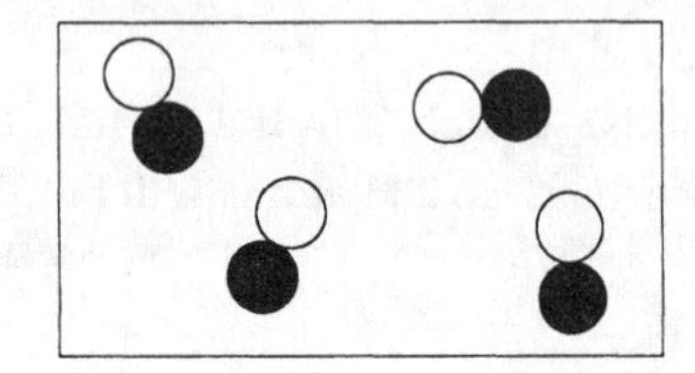

Write the balanced equation for the reaction.

$$2H_2 + CO \rightarrow CH_3OH$$

The combining ratio of H_2: CO is 2:1. If we have 8 H_2 molecules, 4 CO molecules are required for complete reaction. Alternatively, you could examine the atom ratios in the formula of CH_3OH, but the balanced equation is most direct.

3.3 (a) There are twice as many O atoms as N atoms, so the empirical formula of the original compound is NO_2.

(b) No, because we have no way of knowing whether the empirical and molecular formulas are the same. NO_2 represents the simplest ratio of atoms in a molecule but not the only possible molecular formula.

3.5 (a) *Analyze.* Given the molecular model, write the molecular formula.

Plan. Use the colors of the atoms (spheres) in the model to determine the number of atoms of each element.

Solve. Observe 2 gray C atoms, 5 white H atoms, 1 blue N atom, 2 red O atoms. $C_2H_5NO_2$

(b) *Plan.* Follow the method in Sample Exercise 3.9. Calculate formula weight in amu and molar mass in grams.

2 C atoms = 2(12.0 amu) = 24.0 amu

5 H atoms = 5(1.0 amu) = 5.0 amu

1 N atoms = 1(14.0 amu) = 14.0 amu

2 O atoms = 2(16.0 amu) = 32.0 amu

75.0 amu

Formula weight = 75.0 amu, molar mass = 75.0 g/mol

(c) *Plan.* Use the definition of mass % and the results from parts (a) and (b) above to find mass % N in glycine.

Solve. $\text{mass \% N} = \frac{\text{g N}}{\text{g } C_2H_5NO_2} \times 100$

Assume 1 mol $C_2H_5NO_2$. From the molecular formula of glycine [part (a)], there is 1 mol N/mol glycine.

$$\text{mass \% N} = \frac{1 \times (\text{molar mass N})}{\text{molar mass glycine}} \times 100 = \frac{14.0\text{ g}}{75.0\text{ g}} \times 100 = 18.7\%$$

3.6 *Analyze.* Given: 4.0 mol CH_4. Find: mol CO and mol H_2

Plan. Examine the boxes to determine the CH_4:CO mol ratio and CH_4:H_2O mole ratio.

Solve. There are 2 CH_4 molecules in the reactant box and 2 CO molecules in the product box. The mole ratio is 2:2 or 1:1. Therefore, 4.0 mol CH_4 can produce 4.0 mol CO. There are 2 CH_4 molecules in the reactant box and 6 H_2 molecules in the product box. The mole ratio is 2:6 or 1:3. So, 4.0 mol CH_4 can produce 12:0 mol H_2.

Check. Use proportions. 2 mol CH_4/2 mol CO = 4 mol CH_4/4 mol CO;

2 mol CH_4/6 mol H_2 = 4 mol CH_4/12 mol H_2.

3.8 (a) $2NO + O_2 \rightarrow 2NO_2$, O_2 = OO, NO_2 = O●O

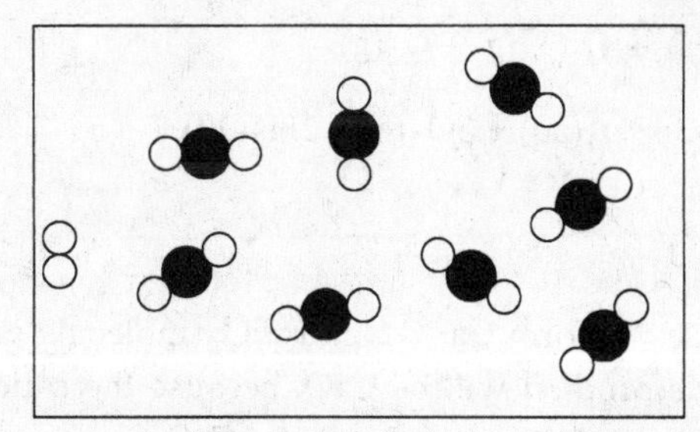

Each NO molecule reacts with 1 O atom (1/2 of an O_2 molecule) to produce 1 NO_2 molecule. Eight NO molecules react with 8 O atoms (4 O_2 molecules) to produce 8 NO_2 molecules. One O_2 molecule doesn't react (is in excess). NO is the limiting reactant.

(b) $\%\text{ yield} = \frac{\text{actual yield}}{\text{theoretical yield}} \times 100$; $\text{actual yield} = \frac{\%\text{ yield}}{100} \times \text{theoretical yield}$

The theoretical yield from part (a) is 8 NO_2 molecules. If the percent yield is 75%, then 0.75(8) = 6 NO_2 would appear in the products box.

Balancing Chemical Equations

3.10 (a) In a CO molecule, there is one O atom bound to C. 2CO indicates that there are **two CO molecules**, each of which contains one C and one O atom. Adding a subscript 2 to CO to form CO_2 means that there are **two O atoms** bound to one C in a CO_2 molecule. The composition of the different molecules, CO_2 and CO, is different and the physical and chemical properties of the two compounds they constitute are very different. The subscript 2 changes molecular composition and thus properties of the compound. The prefix 2 indicates how many molecules (or moles) of the original compound are under consideration.

(b) Yes. There are the same number and kinds of atoms on the reactants side and the products side of the equation.

3.12 (a) $6Li(s) + N_2(g) \rightarrow 2Li_3N(s)$

(b) $La_2O_3(s) + 3H_2O(l) \rightarrow 2La(OH)_3(aq)$

(c) $2NH_4NO_3(s) \rightarrow 2N_2(g) + O_2(g) + 4H_2O(g)$

(d) $Ca_3P_2(s) + 6H_2O(l) \rightarrow 3Ca(OH)_2(aq) + 2PH_3(g)$

(e) $3Ca(OH)_2(aq) + 2H_3PO_4(aq) \rightarrow Ca_3(PO_4)_2(s) + 6H_2O(l)$

(f) $2AgNO_3(aq) + Na_2SO_4(aq) \rightarrow Ag_2SO_4(s) + 2NaNO_3(aq)$

(g) $4CH_3NH_2(g) + 9O_2(g) \rightarrow 4CO_2(g) + 10H_2O(g) + 2N_2(g)$

3.14 (a) $SO_3(g) + H_2O(l) \rightarrow H_2SO_4(aq)$

(b) $B_2S_3(s) + 6H_2O(l) \rightarrow 2H_3BO_3(aq) + 3H_2S(g)$

(c) $4PH_3(g) + 8O_2(g) \rightarrow 6H_2O(g) + P_4O_{10}(s)$

(d) $2Hg(NO_3)_2(s) \xrightarrow{\Delta} 2HgO(s) + 4NO_2(g) + O_2(g)$

(e) $Cu(s) + 2H_2SO_4(aq) \rightarrow CuSO_4(aq) + SO_2(g) + 2H_2O(l)$

Patterns of Chemical Reactivity

3.16 (a) Neutral Ca atom loses $2e^-$ to form Ca^{2+}. Neutral O_2 molecule gains $4e^-$ to form $2O^{2-}$. The formula of the product will be CaO, because the cationic and anionic charges are opposite and equal. $2Ca(s) + O_2(g) \rightarrow 2CaO$

(b) The products are $CO_2(g)$ and $H_2O(l)$. $C_3H_6O(l) + 4O_2(g) \rightarrow 3CO_2(g) + 3H_2O(l)$

3.18 (a) $2Al(s) + 3O_2(g) \rightarrow Al_2O_3(s)$

(b) $Cu(OH)_2(s) \xrightarrow{\Delta} CuO(s) + H_2O(g)$

(c) $C_7H_{16}(l) + 11O_2(g) \rightarrow 7CO_2(g) + 8H_2O(l)$

(d) $2C_5H_{12}O(l) + 15O_2(g) \rightarrow 10CO_2(g) + 12H_2O(l)$

3.20 (a) $2C_3H_6(g) + 9O_2(g) \rightarrow 6CO_2(g) + 6H_2O(l)$ combustion

(b) $NH_4NO_3(s) \rightarrow N_2O(g) + 2H_2O(l)$ decomposition

(c) $C_5H_6O(l) + 6O_2(g) \rightarrow 5CO_2(g) + 3H_2O(l)$ combustion

(d) $N_2(g) + 3H_2(g) \rightarrow 2NH_3(g)$ combination

(e) $K_2O(s) + H_2O(l) \rightarrow 2KOH(aq)$ combination

Formula Weights

3.22 Formula weight in amu to 1 decimal place.

(a) N_2O: FW = 2(14.0) + 1(16.0) = 44.0 amu

(b) $HC_7H_5O_2$: 7(12.0) + 6(1.0) + 2(16.0) = 122.0 amu

(c) $Mg(OH)_2$: 1(24.3) + 2(16.0) + 2(1.0) = 58.3 amu

(d) $(NH_2)_2CO$: 2(14.0) + 4(1.0) + 1(12.0) + 1(16.0) = 60.0 amu

(e) $CH_3CO_2C_5H_{11}$: 7(12.0) + 14(1.0) + 2(16.0) = 130.0 amu

3.24 (a) C_2H_2: FW = 2(12.0) + 2(1.0) = 26.0 amu

$$\% C = \frac{2(12.0)\text{ amu}}{26.0\text{ amu}} \times 100 = 92.3\%$$

(b) $HC_6H_7O_6$: FW = 6(12.0) + 8(1.0) + 6(16.0) = 176.0 amu

$$\% H = \frac{8(1.0)\text{ amu}}{176.0\text{ amu}} \times 100 = 4.5\%$$

(c) $(NH_4)_2SO_4$: FW = 2(14.0) + 8(1.0) + 1(32.1) + 4(16.0) = 132.1 amu

$$\% H = \frac{8(1.0)\text{ amu}}{132.1\text{ amu}} \times 100 = 6.1\%$$

(d) $PtCl_2(NH_3)_2$: FW = 1(195.1) + 2(35.5) + 2(14.0) + 6(1.0) = 300.1 amu

$$\% Pt = \frac{1(195.1)\text{ amu}}{300.1\text{ amu}} \times 100 = 65.01\%$$

(e) $C_{18}H_{24}O_2$: FW = 18(12.0) + 24(1.0) + 2(16.0) = 272.0 amu

$$\% O = \frac{2(16.0)\text{ amu}}{272.0\text{ amu}} \times 100 = 11.8\%$$

(f) $C_{18}H_{27}NO_3$: FW = 18(12.0) + 27(1.0) + 1(14.0) + 3(16.0) = 305.0 amu

$$\% C = \frac{18(12.0)\text{ amu}}{305.0\text{ amu}} \times 100 = 70.8\%$$

3.26 (a) CO_2: FW = 1(12.0) + 2(16.0) = 44.0 amu

$$\% C = \frac{12.0\text{ amu}}{44.0\text{ amu}} \times 100 = 27.3\%$$

(b) CH_3OH: FW = 1(12.0) + 4(1.0) + 1(16.0) = 32.0 amu

$$\% C = \frac{12.0\text{ amu}}{32.0\text{ amu}} \times 100 = 37.5\%$$

(c) C_2H_6: FW = 2(12.0) + 6(1.0) = 30.0 amu

$$\% C = \frac{2(12.0)\text{ amu}}{30.0\text{ amu}} \times 100 = 80.0\%$$

(d) $CS(NH_2)_2$: FW = 1(12.0) + 1(32.1) + 2(14.0) + 4(1.0) = 76.1 amu

$$\% C = \frac{12.0 \text{ amu}}{76.1 \text{ amu}} \times 100 = 15.8\%$$

Avogadro's Number and the Mole

3.28 (a) exactly 12 g (b) 6.0221421×10^{23}, Avogadro's number

3.30 3.0×10^{23} H_2O_2 molecules contains (4 atoms × 0.5 mol) = 2 mol atoms

32 g O_2 contains (2 atoms × 1 mol) = 2 mol atoms

2.0 mol CH_4 contains (5 atoms × 2 mol) = 10 mol atoms

3.32 292 million = $292 \times 10^6 = 2.92 \times 10^8$ people

$$\frac{6.022 \times 10^{23} ¢}{2.92 \times 10^8 \text{ people}} \times \frac{\$1}{100 ¢} = \frac{\$6.022 \times 10^{21}}{2.92 \times 10^8 \text{ people}} = \$2.06 \times 10^{13}/\text{person}$$

$$\$7.0 \text{ trillion} = \$7.0 \times 10^{12} \qquad \frac{\$2.06 \times 10^{13}}{\$7.0 \times 10^{12}} = 2.9$$

Each person would receive an amount that is 2.9 times the dollar amount of the national debt.

3.34 (a) molar mass = 1(137.33) + 2(126.904) = 39.14 g

$$1.906 \times 10^{-2} \text{ mol BaI}_2 \times \frac{391.14 \text{ g}}{1 \text{ mol}} = 7.455 \text{ g BaI}_2$$

(b) molar mass = 1(14.01) + 4(1.008) + 1(35.45) = 53.49 g/mol

$$48.3 \text{ g } NH_4Cl \times \frac{1 \text{ mol}}{53.49 \text{ g}} = 0.903 \text{ mol } NH_4Cl$$

(c)
$$0.05752 \text{ mol } HCHO_2 \times \frac{6.02214 \times 10^{23} \text{ molecules}}{1 \text{ mol}} = 3.464 \times 10^{22} \text{ } HCHO_2 \text{ molecules}$$

(d)
$$4.88 \times 10^{-3} \text{ mol } Al(NO_3)_3 \times \frac{9 \text{ mol O}}{1 \text{ mol } Al(NO_3)_3} \times \frac{6.022 \times 10^{23} \text{ O atoms}}{1 \text{ mol}}$$
$$= 2.64 \times 10^{22} \text{ O atoms}$$

3.36 (a) $Fe_2(SO_4)_3$ molar mass = 2(55.845) + 3(32.07) + 12(16.00) = 399.900 = 399.9 g/mol

$$0.0714 \text{ mol } Fe_2(SO_4)_3 \times \frac{399.9 \text{ g } Fe_2(SO_4)_3}{1 \text{ mol}} = 28.553 = 28.6 \text{ g } Fe_2(SO_4)_3$$

(b) $(NH_4)_2CO_3$ molar mass = 2(14.007) + 8(1.008) + 12.011 + 3(15.9994) = 96.0872

= 96.087 g/mol

$$8.776 \text{ g } (NH_4)_2CO_3 \times \frac{1 \text{ mol}}{96.087 \text{ g } (NH_4)_2CO_3} \times \frac{2 \text{ mol } NH_4^+}{1 \text{ mol } (NH_4)_2CO_3} = 0.1827 \text{ mol } NH_4^+$$

(c) $C_9H_8O_4$ molar mass = 9(12.01) + 8(1.008) + 4(16.00) = 180.154 = 180.2 g/mol

$$6.52 \times 10^{21} \text{ molecules} \times \frac{1 \text{ mol}}{6.022 \times 10^{23} \text{ molecules}} \times \frac{180.2 \text{ g } C_9H_8O_4}{1 \text{ mol aspirin}} = 1.95 \text{ g } C_9H_8O_4$$

(d) $\frac{15.86\text{ g Valium}}{0.05570\text{ mol}} = 284.7\text{ g Valium/mol}$

3.38 (a) $C_{14}H_{18}N_2O_5$ molar mass = 14(12.01) + 18(1.008) + 2(14.01) + 5(16.00)

= 294.30 g/mol

(b) $1.00\text{ mg aspartame} \times \frac{1\times10^{-3}\text{ g}}{1\text{ mg}} \times \frac{1\text{ mol}}{294.3\text{ g}} = 3.398\times10^{-6} = 3.40\times10^{-6}\text{ mol aspartame}$

(c) $3.398\times10^{-6}\text{ mol aspartame} \times \frac{6.022\times10^{23}\text{ molecules}}{1\text{ mol}} = 2.046\times10^{18}$

$= 2.05\times10^{18}$ aspartame molecules

(d) $2.046\times10^{18}\text{ aspartame molecules} \times \frac{18\text{ H atoms}}{1\text{ aspartame molecule}} = 3.68\times10^{19}\text{ H atoms}$

3.40 (a) $7.08\times10^{20}\text{ H atoms} \times \frac{19\text{ C atoms}}{28\text{ H atoms}} = 4.80\times10^{20}\text{ C atoms}$

(b) $7.08\times10^{20}\text{ H atoms} \times \frac{1\ C_{19}H_{28}O_2\text{ molecule}}{28\text{ H atoms}} = 2.529\times10^{19}$

$= 2.53\times10^{19}\ C_{19}H_{28}O_2$ molecules

(c) $2.529\times10^{19}\ C_{19}H_{28}O_2\text{ molecules} \times \frac{1\text{ mol}}{6.022\times10^{23}\text{ molecules}} = 4.199\times10^{-5}$

$= 4.20\times10^{-5}\text{ mol } C_{19}H_{28}O_2$

(d) $C_{19}H_{28}O_2$ molar mass = 19(12.01) + 28(1.008) + 2(16.00) = 288.41 = 288.4 g/mol

$4.199\times10^{-5}\text{ mol } C_{19}H_{28}O_2 \times \frac{288.4\text{ g } C_{19}H_{28}O_2}{1\text{ mol}} = 0.0121\text{ g } C_{19}H_{28}O_2$

3.42 $25\times10^{-6}\text{ g } C_{21}H_{30}O_2 \times \frac{1\text{ mol } C_{21}H_{30}O_2}{314.5\text{ g } C_{21}H_{30}O_2} = 7.95\times10^{-8} = 8.0\times10^{-8}\text{ mol } C_{21}H_{30}O_2$

$7.95\times10^{-8}\text{ mol } C_{21}H_{30}O_2 \times \frac{6.022\times10^{23}\text{ molecules}}{1\text{ mol}} = 4.8\times10^{16}\ C_{21}H_{30}O_2\text{ molecules}$

Empirical Formulas

3.44 (a) Calculate the simplest ratio of moles.

0.104 mol K / 0.052 = 2

0.052 mol C / 0.052 = 1

0.156 mol O / 0.052 = 3

The empirical formula is K_2CO_3.

(b) Calculate moles of each element present, then the simplest ratio of moles.

$$5.28\text{ g Sn} \times \frac{1\text{ mol Sn}}{118.7\text{ g Sn}} = 0.04448\text{ mol Sn};\ 0.04448 / 0.04448 = 1$$

$$3.37\text{ g F} \times \frac{1\text{ mol F}}{19.00\text{ g FSn}} = 0.1774\text{ mol F}; 0.1774 / 0.04448 \approx 4$$

The integer ratio is 1 Sn : 4 F; the empirical formula is SnF_4.

(c) Assume 100 g sample, calculate moles of each element, find the simplest ratio of moles.

$$87.5\%\text{ N} = 87.5\text{ g N} \times \frac{1\text{ mol N}}{14.01\text{ g}} = 6.25\text{ mol N};\ \ 6.25 / 6.25 = 1$$

$$12.5\%\text{ H} = 12.5\text{ g H} \times \frac{1\text{ mol}}{1.008\text{ g}} = 12.4\text{ mol H};\ \ 12.4 / 6.25 \approx 2$$

The empirical formula is NH_2.

3.46 See Solution 3.45 for stepwise problem-solving approach.

(a) $$55.3\text{ g K} \times \frac{1\text{ mol K}}{39.10\text{ g K}} = 1.414\text{ mol K}; 1.414/0.4714 \approx 3$$

$$14.6\text{ g P} \times \frac{1\text{ mol P}}{30.97\text{ g P}} = 0.4714\text{ mol P};\ 0.4714/0.4714 = 1$$

$$30.1\text{ g O} \times \frac{1\text{ mol O}}{16.00\text{ g O}} = 1.881\text{ mol O};\ 1.881/0.4714 \approx 4$$

The empirical formula is K_3PO_4.

(b) $$24.5\text{ g Na} \times \frac{1\text{ mol Na}}{22.99\text{ g Na}} = 1.066\text{ mol Na};\ 1.066/0.5304 \approx 2$$

$$14.9\text{ g Si} \times \frac{1\text{ mol Si}}{28.09\text{ Si}} = 0.5304\text{ mol si};\ 0.5304/0.5304 = 1$$

$$60.6\text{ g F} \times \frac{1\text{ mol F}}{19.00\text{ g F}} = 3.189\text{ mol F}; 3.189/0.5304 \approx 6$$

The empirical formula is Na_2SiF_6.

(c) $$62.1\text{ g C} \times \frac{1\text{ mol C}}{12.01\text{ g C}} = 5.17\text{ mol C};\ \ 5.17 / 0.864 \approx 6$$

$$5.21\text{ g H} \times \frac{1\text{ mol H}}{1.008\text{ g H}} = 5.17\text{ mol O};\ \ 5.17 / 0.864 \approx 6$$

$$12.1\text{ g N} \times \frac{1\text{ mol N}}{14.01\text{ g N}} = 0.864\text{ mol N};\ \ 0.864 / 0.864 = 1$$

$$20.7\text{ g O} \times \frac{1\text{ mol O}}{16.00\text{ g O}} = 1.29\text{ mol O};\ \ 1.29 / 0.864 \approx 1.5$$

Multiplying by two, the empirical formula is $C_{12}H_{12}N_2O_3$.

3.48 (a) $HCHO_2$ FW $12.01 + 1.008 + 2(16.00) = 45.0$ $\frac{MM}{FW} = \frac{90.0}{45.0} = 2$

The molecular formula is $H_2C_2O_4$.

(b) C_2H_4O FW $= 2(12) + 4(1) + 16 = 44.$ $\frac{MM}{FW} = \frac{88}{44} = 2$

The molecular formula is $C_4H_8O_2$.

3.50 Assume 100 g in the following problems.

(a) $75.69 \text{ g C} \times \frac{1 \text{ mol C}}{12.01 \text{ g C}} = 6.30 \text{ mol C};\ 6.30/0.969 = 6.5$

$8.80 \text{ g H} \times \frac{1 \text{ mol H}}{1.008 \text{ g H}} = 8.73 \text{ mol H};\ 8.73/0.969 = 9.0$

$15.51 \text{ g O} \times \frac{1 \text{ mol O}}{16.00 \text{ g O}} = 0.969 \text{ mol O}; 0.969/0.969 = 1$

Multiply by 2 to obtain the integer ratio 13:18:2. The empirical formula is $C_{13}H_{18}O_2$, FW = 206 g. Since the empirical formula weight and the molar mass are equal (206 g), the empirical and molecular formulas are $C_{13}H_{18}O_2$.

(b) $58.55 \text{ g C} \times \frac{1 \text{ mol C}}{12.01 \text{ g C}} = 4.875 \text{ mol C};\ 4.875/1.956 \approx 2.5$

$13.81 \text{ g H} \times \frac{1 \text{ mol H}}{1.008 \text{ g H}} = 13.700 \text{ mol H};\ 13.700/1.956 \approx 7.0$

$27.40 \text{ g N} \times \frac{1 \text{ mol N}}{14.01 \text{ g N}} = 1.956 \text{ mol N};\ 1.956/1.956 = 1.0$

Multiply by 2 to obtain the integer ratio 5:14:2. The empirical formula is $C_5H_{14}N_2$; FW = 102. Since the empirical formula weight and the molar mass are equal
(102 g), the empirical and molecular formulas are $C_5H_{14}N_2$.

(c) $59.0 \text{ g C} \times \frac{1 \text{ mol C}}{12.01 \text{ g C}} = 4.91 \text{ mol C};\ 4.91 / 0.550 \approx 9$

$7.1 \text{ g H} \times \frac{1 \text{ mol H}}{1.008 \text{ g H}} = 7.04 \text{ mol H};\ 7.04 / 0.550 \approx 13$

$26.2 \text{ g O} \times \frac{1 \text{ mol O}}{16.00 \text{ g O}} = 1.64 \text{ mol O};\ 1.64 / 0.550 \approx 3$

$7.7 \text{ g N} \times \frac{1 \text{ mol N}}{14.01 \text{ g N}} = 0.550 \text{ mol N};\ 0.550 / 0.550 = 1$

The empirical formula is $C_9H_{13}O_3N$, FW = 183 amu (or g). Since the molecular weight is approximately 180 amu, the empirical formula and molecular formula are the same, $C_9H_{13}O_3N$.

3.52 (a) *Plan.* Calculate mol C and mol H, then g C and g H; get g O by subtraction.

Solve.

$$6.32\times10^{-3}\ g\ CO_2 \times \frac{1\ mol\ CO_2}{44.01\ g\ CO_2} \times \frac{1\ mol\ C}{1\ mol\ CO_2} = 1.436\times10^{-4} = 1.44\times10^{-4}\ mol\ C$$

$$2.58\times10^{-3}\ g\ H_2O \times \frac{1\ mol\ H_2O}{18.02\ g\ H_2O} \times \frac{2\ mol\ H}{1\ mol\ H_2O} = 2.863\times10^{-4} = 2.86\times10^{-4}\ mol\ H$$

$$1.436\times10^{-4}\ mol\ C \times \frac{12.01\ g\ C}{1\ mol\ C} = 1.725\times10^{-3}\ g\ C = 1.73\ mg\ C$$

$$2.863\times10^{-4}\ mol\ H \times \frac{1.008\ g\ H}{1\ mol\ H} = 2.886\times10^{-4}\ g\ H = 0.289\ mg\ H$$

mass of O = 2.78 mg sample – (1.725 mg C + 0.289 mg H) = 0.77 mg O

$$0.77\times10^{-3}\ g\ O \times \frac{1\ mol\ O}{16.00\ g\ O} = 4.81\times10^{-5}\ mol\ O.$$ Divide moles by 4.81×10^{-5}.

$$C: \frac{1.44\times10^{-4}}{4.81\times10^{-5}} \approx 3;\quad H: \frac{2.86\times10^{-4}}{4.81\times10^{-5}} \approx 6;\quad O: \frac{4.81\times10^{-5}}{4.81\times10^{-5}} = 1$$

The empirical formula is C_3H_6O.

(b) *Plan.* Calculate mol C and mol H, then g C and g H. In this case, get N by subtraction. *Solve.*

$$14.242\times10^{-3}\ g\ CO_2 \times \frac{1\ mol\ CO_2}{44.01\ g\ CO_2} \times \frac{1\ mol\ C}{1\ mol\ CO_2} = 3.2361 \times 10^{-4}\ mol\ C$$

$$4.083\times10^{-3}\ g\ H_2O \times \frac{1\ mol\ H_2O}{18.02\ g\ H_2O} \times \frac{2\ mol\ H}{1\ mol\ H_2O} = 4.5136 \times 10^{-4} = 4.532\times10^{-4}\ mol\ H$$

$$3.2361\times10^{-4}\ g\ mol\ C \times \frac{12.01\ g\ C}{1\ mol\ H} = 3.8866\times10^{-3}\ g\ C = 3.8866\ mg\ C$$

$$4.532\times10^{-4}\ mol\ H \times \frac{1.008\ g\ H}{1\ mol\ H} = 0.45683\times10^{-3}\ g\ H = 0.4568\ mg\ H$$

mass of N = 5.250 mg sample – (3.8866 mg C + 0.4568 mg H) = 0.9066

= 0.907 mg N

$$0.9066\times10^{-3}\ g\ N \times \frac{1\ mol\ N}{14.01\ g\ N} = 6.47\times10^{-5}\ mol\ N.$$ Divide moles by 6.47×10^{-5}.

$$C: \frac{3.24\times10^{-4}}{6.47\times10^{-5}} \approx 5;\quad H: \frac{4.53\times10^{-4}}{6.47\times10^{-5}} \approx 7;\quad N: \frac{6.47\times10^{-5}}{6.47\times10^{-5}} = 1$$

The empirical formula is C_5H_7N, FW = 81. A molar mass of 160 ± 5 indicates a factor of 2 and a molecular formula of $C_{10}H_{14}N_2$.

3.54 The reaction involved is $MgSO_4 \bullet xH_2O(s) \rightarrow MgSO_4(s) + xH_2O(g)$. First, calculate the number of moles of product $MgSO_4$; this is the same as the number of moles of starting hydrate.

$$2.472\ g\ MgSO_4 \times \frac{1\ mol\ MgSO_4}{120.4\ g\ MgSO_4} \times \frac{1\ mol\ MgSO_4 \bullet xH_2O}{1\ mol\ MgSO_4} = 0.02053\ mol\ MgSO_4 \bullet xH_2O$$

Thus, $$\frac{5.061\ g\ MgSO_4 \bullet xH_2O}{0.02053} = 246.5\ g/mol = FW\ of\ MgSO_4 \bullet xH_2O$$

FW of $MgSO_4 \bullet xH_2O$ = FW of $MgSO_4$ + x(FW of H_2O).

246.5 = 120.4 + x(18.02). x = 6.998. The hydrate formula is $MgSO_4 \bullet \underline{7}H_2O$.

Alternatively, we could calculate the number of moles of water represented by weight loss: (5.061 – 2.472) = 2.589 g H_2O lost.

$$2.589\text{ g } H_2O \times \frac{1\text{ mol } H_2O}{18.02\text{ g } H_2O} = 0.1437\text{ mol } H_2O; \quad \frac{\text{mol } H_2O}{\text{mol } MgSO_4} = \frac{0.1437}{0.02053} = 7.000$$

Again the correct formula is $MgSO_4 \bullet \underline{7}H_2O$.

Calculations Based on Chemical Equations

3.56 The **integer coefficients** immediately preceding each molecular formula in a chemical equation give information about relative numbers of moles of reactants and products involved in a reaction.

3.58 $C_6H_{12}O_6(aq) \rightarrow 2C_2H_5OH(aq) + 2CO_2(g)$

(a) $$0.400\text{ mol } C_6H_{12}O_6 \times \frac{2\text{ mol } CO_2}{1\text{ mol } C_6H_{12}O_6} = 0.800\text{ mol } CO_2$$

(b) $$7.50\text{ g } C_2H_5OH \times \frac{1\text{ mol } C_2H_5OH}{46.07\text{ g } C_2H_5OH} \times \frac{1\text{ mol } C_6H_{12}O_6}{2\text{ mol } C_2H_5OH} \times \frac{180.2\text{ g } C_6H_{12}O_6}{1\text{ mol } C_6H_{12}O_6} = 14.7\text{ g } C_6H_{12}O_6$$

(c) $$7.50\text{ g } C_2H_5OH \times \frac{1\text{ mol } C_2H_5OH}{46.07\text{ g } C_2H_5OH} \times \frac{2\text{ mol } CO_2}{2\text{ mol } C_2H_5OH} \times \frac{44.01\text{ g } CO_2}{1\text{ mol } CO_2} = 7.16\text{ g } CO_2$$

3.60 (a) $Fe_2O_3(s) + 3CO(g) \rightarrow 2Fe(s) + 3CO_2(g)$

(b) $$0.150\text{ kg } Fe_2O_3 \times \frac{1000\text{ g}}{1\text{ kg}} \times \frac{1\text{ mol } Fe_2O_3}{159.688\text{ g } Fe_2O_3} = 0.9393 = 0.939\text{ mol } Fe_2O_3$$

$$0.9393\text{ mol } Fe_2O_3 \times \frac{3\text{ mol CO}}{1\text{ mol } Fe_2O_3} \times \frac{28.01\text{ g CO}}{1\text{ mol CO}} = 78.929 = 78.9\text{ g CO}$$

(c) $$0.9393\text{ mol } Fe_2O_3 \times \frac{2\text{ mol Fe}}{1\text{ mol } Fe_2O_3} \times \frac{55.845\text{ g Fe}}{1\text{ mol Fe}} = 104.914 = 105\text{ g Fe}$$

$$0.9393\text{ mol } Fe_2O_3 \times \frac{3\text{ mol } CO_2}{1\text{ mol } Fe_2O_3} \times \frac{44.01\text{ g } CO_2}{1\text{ mol } CO_2} = 124.015 = 124\text{ g } CO_2$$

(d) reactants: 150 kg Fe_2O_3 + 78.9 g CO = 228.9 = 229 g

products: 104.9 g Fe + 124.0 g CO_2 = 228.9 = 229 g

Mass is conserved.

3.62 (a) $CaH_2(s) + 2H_2O(l) \rightarrow Ca(OH)_2(aq) + 2H_2(g)$

(b) $$8.500\text{ g } H_2 \times \frac{1\text{ mol } H_2}{2.016\text{ g } H_2} \times \frac{1\text{ mol } CaH_2}{2\text{ mol } H_2} \times \frac{42.10\text{ g } CaH_2}{1\text{ mol } CaH_2} = 88.75\text{ g } CaH_2$$

3.64 $2C_8H_{18}(l) + 25O_2(g) \rightarrow 16CO_2(g) + 18H_2O(l)$

(a) $1.25 \text{ mol } C_8H_{18} \times \frac{25 \text{ mol } O_2}{2 \text{ mol } C_8H_{18}} = 15.625 = 15.6 \text{ mol } O_2$

(b) $10.0 \text{ g } C_8H_{18} \times \frac{1 \text{ mol } C_8H_{18}}{114.2 \text{ g } C_8H_{18}} \times \frac{25 \text{ mol } O_2}{2 \text{ mol } C_8H_{18}} \times \frac{32.00 \text{ g } O_2}{1 \text{ mol } O_2} = 35.0 \text{ g } O_2$

(c) $1.00 \text{ gal } C_8H_{18} \times \frac{3.7854 \text{ L}}{1 \text{ gal}} \times \frac{1000 \text{ mL}}{1 \text{ L}} \times \frac{0.692 \text{ g}}{1 \text{ mL}} = 2619.5 = 2.62 \times 10^3 \text{ g } C_8H_{18}$

$$2.6195 \times 10^3 \text{ g } C_8H_{18} \times \frac{1 \text{ mol } C_8H_{18}}{114.2 \text{ g } C_8H_{18}} \times \frac{25 \text{ mol } O_2}{2 \text{ mol } C_8H_{18}} \times \frac{32.00 \text{ g } O_2}{1 \text{ mol } O_2} = 9{,}175.1 \text{ g}$$

$$= 9.18 \times 10^3 \text{ g } O_2$$

3.66 (a) *Plan.* Calculate a "mole ratio" between nitroglycerine and total moles of gas produced. (12 + 6 + 1 + 10) = 29 mol gas; 4 mol nitro: 29 total mol gas. *Solve.*

$$2.00 \text{ mL nitro} \times \frac{1.592 \text{ g}}{\text{mL}} \times \frac{1 \text{ mol nitro}}{227.1 \text{ g nitro}} \times \frac{29 \text{ mol gas}}{4 \text{ mol nitro}} = 0.10165 = 0.102 \text{ mol gas}$$

(b) $0.10165 \text{ mol gas} \times \frac{55 \text{ L}}{\text{mol}} = 5.5906 = 5.6 \text{ L}$

(c) $2.00 \text{ mL nitro} \times \frac{1.592 \text{ g}}{\text{mL}} \times \frac{1 \text{ mol nitro}}{227.1 \text{ g nitro}} \times \frac{6 \text{ mol } N_2}{4 \text{ mol nitro}} \times \frac{28.01 \text{ g } N_2}{1 \text{ mol } N_2} = 0.589 \text{ g } N_2$

Limiting Reactants; Theoretical Yields

3.68 (a) *Theoretical yield* is the maximum amount of product possible, as predicted by stoichiometry, assuming that the limiting reactant is converted entirely to product.

Actual yield is the amount of product actually obtained, less than or equal to the theoretical yield. *Percent yield* is the ratio of (actual yield to theoretical yield) × 100.

(b) No reaction is perfect. Not all reactant molecules come together effectively to form products; alternative reaction pathways may produce secondary products and reduce the amount of desired product actually obtained, or it might not be possible to completely isolate the desired product from the reaction mixture. In any case, these factors reduce the actual yield of a reaction.

3.70 (a) $40{,}875 \text{ L beverage} \times \frac{1 \text{ bottle}}{0.355 \text{ L}} = 115{,}140.85 = 1.15 \times 10^5$ portions of beverage

(The uncertainty in 355 mL limits the precision of the number of portions we can reasonably expect to deliver to three significant figures.)

121,515 bottles; 122,500 caps; 1.15×10^5 bottles can be filled and capped.

(b) 122,500 caps – 115,141 portions = 7,359 = 7×10^3 caps remain

121,515 empty bottles – 115,141 portions = 6374 = 6×10^3 bottles remain

(Uncertainty in the number of portions delivered limits the results to 1 sig fig.)

(c) The volume of beverage limits production.

3.72 $0.500 \text{ mol Al(OH)}_3 \times \frac{3 \text{ mol H}_2\text{SO}_4}{2 \text{ mol Al(OH)}_3} = 0.750 \text{ mol H}_2\text{SO}_4$ needed for complete reaction

Only 0.500 mol H_2SO_4 available, so H_2SO_4 limits.

$$0.500 \text{ mol H}_2\text{SO}_4 \times \frac{1 \text{ mol Al}_2(\text{SO}_4)_3}{3 \text{ mol H}_2\text{SO}_4} = 0.1667 = 0.167 \text{ mol Al}_2(\text{SO}_4)_3 \text{ can form}$$

$$0.500 \text{ mol H}_2\text{SO}_4 \times \frac{2 \text{ mol Al(OH)}_3}{3 \text{ mol H}_2\text{SO}_4} = 0.3333 = 0.333 \text{ mol Al(OH)}_3 \text{ react}$$

0.500 mol $Al(OH)_3$ initial – 0.333 mol react = 0.167 mol $Al(OH)_3$ remain

3.74 $4NH_3(g) + 5O_2(g) \rightarrow 4NO(g) + 6H_2O(g)$

(a) Follow the approach in Sample Exercise 3.19.

$$1.50 \text{ g NH}_3 \times \frac{1 \text{ mol NH}_3}{17.03 \text{ g NH}_3} = 0.08808 = 0.0881 \text{ mol NH}_3$$

$$2.75 \text{ g O}_2 \times \frac{1 \text{ mol O}_2}{32.00 \text{ g O}_2} = 0.08594 = 0.0859 \text{ mol O}_2$$

$$0.08594 \text{ mol O}_2 \times \frac{4 \text{ mol NH}_3}{5 \text{ mol O}_2} = 0.06875 = 0.0688 \text{ mol NH}_3 \text{ required}$$

More than 0.0688 mol NH_3 is available, so O_2 is the limiting reactant.

(b) $0.08594 \text{ mol O}_2 \times \frac{4 \text{ mol NO}}{5 \text{ mol O}_2} \times \frac{30.01 \text{ g NO}}{1 \text{ mol NO}} = 2.063 = 2.06 \text{ g NO produced}$

$$0.08594 \text{ mol O}_2 \times \frac{6 \text{ mol H}_2\text{O}}{5 \text{ mol O}_2} \times \frac{18.02 \text{ g H}_2\text{O}}{1 \text{ mol H}_2\text{O}} = 1.8583 = 1.86 \text{ g H}_2\text{O}$$

(c) 0.08808 mol NH_3 – 0.06875 mol NH_3 reacted = 0.01933 = 0.0193 mol NH_3 remain

$$0.01933 \text{ mol NH}_3 \times \frac{17.03 \text{ g NH}_3}{1 \text{ mol NH}_3} = 0.32919 = 0.329 \text{ g NH}_3 \text{ remain}$$

(d) mass products = 2.06 g NO + 1.86 g H_2O + 0.329 g NH_3 remaining = 4.25 g products

mass reactants = 1.50 g NH_3 + 2.75 g O_2 = 4.25 g reactants

(For comparison purposes, the mass of excess reactant can be either added to the products, as above, or subtracted from reactants.)

3.76 *Plan.* Write balanced equation; determine limiting reactant; calculate amounts of excess reactant remaining and products, based on limiting reactant.

Solve. $H_2SO_4(aq) + Pb(C_2H_3O_2)_2(aq) \rightarrow PbSO_4(s) + 2HC_2H_3O_2(aq)$

$$7.50 \text{ g } H_2SO_4 \times \frac{1 \text{ mol } H_2SO_4}{98.09 \text{ g } H_2SO_4} = 0.07646 = 0.0765 \text{ mol } H_2SO_4$$

$$7.50 \text{ g } Pb(C_2H_3O_2)_2 \times \frac{1 \text{ mol } Pb(C_2H_3O_2)_2}{325.3 \text{ g } Pb(C_2H_3O_2)_2} = 0.023056 = 0.0231 \text{ mol } Pb(C_2H_3O_2)_2$$

1 mol H_2SO_4:1 mol $Pb(C_2H_3O_2)_2$, so $Pb(C_2H_3O_2)_2$ is the limiting reactant.

0 mol $Pb(C_2H_3O_2)_2$, (0.07646 – 0.023056) = 0.0534 mol H_2SO_4, 0.0231 mol $PbSO_4$,

(0.023056 × 2) = 0.0461 mol $HC_2H_3O_2$ are present after reaction

0.053405 mol H_2SO_4 × 98.09 g/mol = 5.2385 = 5.24 g H_2SO_4

0.023056 mol $PbSO_4$ × 303.3 g/mol = 6.9928 = 6.99 g $PbSO_4$

0.046111 mol $HC_2H_3O_2$ × 60.05 g/mol = 2.7690 = 2.77 g $HC_2H_3O_2$

Check. The initial mass of reactants was 15.00 g; and the final mass of excess reactant and products is 15.00 g; mass is conserved.

3.78 (a) $C_2H_6 + Cl_2 \rightarrow C_2H_5Cl + HCl$

$$125 \text{ g } C_2H_6 \times \frac{1 \text{ mol } C_2H_6}{30.07 \text{ g } C_2H_6} = 4.157 = 4.16 \text{ mol } C_2H_6$$

$$255 \text{ g } Cl_2 \times \frac{1 \text{ mol } Cl_2}{70.91 \text{ g } Cl_2} = 3.596 = 3.60 \text{ mol } Cl_2$$

Since the reactants combine in a 1:1 mole ratio, Cl_2 is the limiting reactant. The theoretical yield is:

$$3.596 \text{ mol } Cl_2 \times \frac{1 \text{ mol } C_2H_5Cl}{1 \text{ mol } Cl_2} \times \frac{64.51 \text{ g } C_2H_5Cl}{1 \text{ mol } C_2H_5Cl} = 231.98 = 232 \text{ g } C_2H_5Cl$$

(b) $$\% \text{ yield} = \frac{206 \text{ g } C_2H_5Cl \text{ actual}}{232 \text{ g } C_2H_5Cl \text{ theoretical}} \times 100 = 88.8\%$$

3.80 $H_2S(g) + 2NaOH(aq) \rightarrow Na_2S(aq) + 2H_2O(l)$

$$1.50 \text{ g } H_2S \times \frac{1 \text{ mol } H_2S}{34.08 \text{ g } H_2S} = 0.04401 = 0.0440 \text{ mol } H_2S$$

$$2.00 \text{ g NaOH} \times \frac{1 \text{ mol NaOH}}{40.00 \text{ g NaOH}} = 0.0500 \text{ mol NaOH}$$

By inspection, twice as many mol NaOH as H_2S are needed for exact reaction, but mol NaOH given is less than twice mol H_2S, so NaOH limits.

$$0.0500 \text{ mol NaOH} \times \frac{1 \text{ mol } Na_2S}{2 \text{ mol NaOH}} \times \frac{78.05 \text{ g } Na_2S}{1 \text{ mol } Na_2S} = 1.95125 = 1.95 \text{ g } Na_2S \text{ theoretical}$$

$$\frac{92.0\%}{100} \times 1.95125 \text{ g } Na_2S \text{ theoretical} = 1.7951 = 1.80 \text{ g } Na_2S \text{ actual}$$

Additional Exercises

3.82 The formulas of the fertilizers are NH_3, NH_4NO_3, $(NH_4)_2SO_4$ and $(NH_2)_2CO$. Qualitatively, the more heavy, non-nitrogen atoms in a molecule, the smaller the mass % of N. By inspection, the mass of NH_3 is dominated by N, so it will have the greatest % N, $(NH_4)_2SO_4$ will have the least. In order of increasing % N:

$(NH_4)_2SO_4 < NH_4NO_3 < (NH_2)_2CO < NH_3$.

Check by calculation:

$(NH_4)_2SO_4$: FW = 2(14.0) + 8(1.0) + 1(32.1) + 4(16.0) = 132.1 amu

% N = [2(14.0)/132.1] × 100 = 21.2%

NH_4NO_3: FW = 2(14.0) + 4(1.0) + 3(16.0) = 80.0 amu

% N = [2(14.0)/80.0] × 100 = 35.0%

$(NH_2)_2CO$: FW = 2(14.0) + 4(1.0) = 1(12.0) + 1(16.0) = 60.0 amu

% N = [2(14.0)/60.0] × 100 = 46.7% N

NH_3: FW = 1(14.0) + 3(1.0) = 17.0

% N = [14.0/17.0] × 100 = 82.4 % N

3.84 (a) $$\frac{5.342\times10^{-21}\text{ g}}{1\text{ molecule penicillin G}}\times\frac{6.0221\times10^{23}\text{ molecules}}{1\text{ mol}}=3217\text{ g/mol penicillin G}$$

(b) 1.00 g hemoglobin (hem) contains 3.40×10^{-3} g Fe.

$$\frac{1.00\text{ g hem}}{3.40\times10^{-3}\text{ g Fe}}\times\frac{55.85\text{ g Fe}}{1\text{ mol Fe}}\times\frac{4\text{ mol Fe}}{1\text{ mol hem}}=6.57\times10^{4}\text{ g/mol hemoglobin}$$

3.85 (a) $$1.000\times10^{4}\text{ Si atoms}\times\frac{1\text{ mol}}{6.022\times10^{23}\text{ atoms}}\times\frac{28.0855\text{ g Si}}{1\text{ mol Si}}=4.6638\times10^{-19}\text{ g Si}$$

(b) $$4.6638\times10^{-19}\text{ g Si}\times\frac{1\text{ cm}^3\text{ Si}}{2.3\text{ g Si}}=2.03\times10^{-19}=2.0\times10^{-19}\text{ cm}^3$$

(c) $V = l^3$; $l = (V)^{1/3} = (2.03 \times 10^{-19}\text{ cm}^3)^{1/3} = 5.9 \times 10^{-7}$ cm (= 5.9 nm)

3.87 *Plan.* Assume 1.000 g and get mass O by subtraction. *Solve.*

(a) $$0.7787\text{ g C}\times\frac{1\text{ mol C}}{12.01\text{ g C}}=0.06484\text{ mol C}$$

$$0.1176\text{ g H}\times\frac{1\text{ mol H}}{1.008\text{ g H}}=0.1167\text{ mol H}$$

$$0.1037\text{ g O}\times\frac{1\text{ mol C}}{16.00\text{ g O}}=0.006481\text{ mol O}$$

Dividing through by the smallest of these values we obtain $C_{10}H_{18}O$.

(b) The formula weight of $C_{10}H_{18}O$ is 154. Thus, the empirical formula is also the molecular formula.

3.88 Since all the C in the vanillin must be present in the CO_2 produced, get g C from g CO_2.

$$2.43\,g\,CO_2 \times \frac{1\,mol\,CO_2}{44.01\,g\,CO_2} \times \frac{12.01\,g\,C}{1\,mol\,C} = 0.6631 = 0.663\,g\,C$$

Since all the H in vanillin must be present in the H_2O produced, get g H from g H_2O.

$$0.50\,g\,H_2O \times \frac{1\,mol\,H_2O}{18.02\,g\,H_2O} \times \frac{2\,mol\,H}{1\,mol\,H_2O} \times \frac{1.008\,g\,H}{1\,mol\,H} = 0.0559 = 0.056\,g\,H$$

Get g O by subtraction. (Since the analysis was performed by combustion, an unspecified amount of O_2 was a reactant, and thus not all the O in the CO_2 and H_2O produced came from vanillin.) 1.05 g vanillin – 0.663 g C – 0.056 g H = 0.331 g O

$$0.6631\,g\,C \times \frac{1\,mol\,C}{12.01\,g\,C} = 0.0552\,mol\,C;\ 0.0552 / 0.0207 = 2.67$$

$$0.0559\,g\,H \times \frac{1\,mol\,H}{1.008\,g\,H} = 0.0555\,mol\,C;\ 0.0555 / 0.0207 = 2.68$$

$$0.331\,g\,O \times \frac{1\,mol\,O}{16.00\,g\,O} = 0.0207\,mol\,O;\ 0.0207 / 0.0207 = 1.00$$

Multiplying the numbers above by **3** to obtain an integer ratio of moles, the empirical formula of vanillin is $C_8H_8O_3$.

3.90 The mass percentage is determined by the relative number of atoms of the element times the atomic weight, divided by the total formula mass. Thus, the mass percent of bromine in $KBrO_x$ is given by $0.5292 = \frac{79.91}{39.10 + 79.91 + x(16.00)}$. Solving for x, we obtain x = 2.00. Thus, the formula is $KBrO_2$.

3.91 (a) Let AW = the atomic weight of X.

According to the chemical reaction, moles XI_3 reacted = moles XCl_3 produced

$$0.5000\,g\,XI_3 \times 1\,mol\,XI_3 / (AW + 380.71)\,g\,XI_3$$
$$= 0.2360\,g\,XCl_3 \times \frac{1\,mol\,XCl_3}{(AW + 106.36)\,g\,XCl_3}$$

0.5000 (AW + 106.36) = 0.2360 (AW + 380.71)

0.5000 AW + 53.180 = 0.2360 AW + 89.848

0.2640 AW = 36.67; AW = 138.9 g

(b) X is lanthanum, La, atomic number 57.

3.93 $O_3(g) + 2NaI(aq) + H_2O(l) \rightarrow O_2(g) + I_2(s) + 2NaOH(aq)$

(a) $3.8 \times 10^{-5}\,mol\,O_3 \times \frac{2\,mol\,NaI}{1\,mol\,O_3} = 7.6 \times 10^{-5}\,mol\,NaI$

(b) $0.550 \text{ mg } O_3 \times \frac{1\times10^{-3} \text{ g}}{1 \text{ mg}} \times \frac{1 \text{ mol } O_3}{48.00 \text{ g } O_3} \times \frac{2 \text{ mol NaI}}{1 \text{ mol } O_3} \times \frac{149.9 \text{ g NaI}}{1 \text{ mol NaI}}$

$= 3.4352 \times 10^{-3} = 3.44 \times 10^{-3}$ g NaI = 3.44 mg NaI

3.94 $2NaCl(aq) + 2H_2O(l) \rightarrow 2NaOH(aq) + H_2(g) + Cl_2(g)$

Calculate mol Cl_2 and relate to mol H_2, mol NaOH.

$$1.5\times10^6 \text{ kg} \times \frac{1000 \text{ g}}{1 \text{ kg}} \times \frac{1 \text{ mol } Cl_2}{70.91 \text{ g } Cl_2} = 2.115\times10^7 = 2.1\times10^7 \text{ mol } Cl_2$$

$$2.115\times10^7 \text{ mol } Cl_2 \times \frac{1 \text{ mol } H_2}{1 \text{ mol } Cl_2} \times \frac{2.016 \text{ g } H_2}{1 \text{ mol } H_2} = 4.26\times10^7 \text{ g } H_2 = 4.3\times10^4 \text{ kg } H_2$$

$$4.3\times10^7 \text{ g} \times \frac{1 \text{ metric ton}}{1\times10^6 \text{ g (1 Mg)}} = 43 \text{ metric tons } H_2$$

$$2.115\times10^7 \text{ mol } Cl_2 \times \frac{2 \text{ mol NaOH}}{1 \text{ mol } Cl_2} \times \frac{40.0 \text{ g NaOH}}{1 \text{ mol NaOH}} = 1.69\times10^9 = 1.7\times10^9 \text{ g NaOH}$$

1.7×10^9 g NaOH = 1.7×10^6 kg NaOH = 1.7×10^3 metric tons NaOH

3.96 (a) *Plan.* Calculate the total mass of C from g CO and g CO_2. Calculate the mass of H from g H_2O. Calculate mole ratios and the empirical formula. *Solve.*

$$0.467 \text{ g CO} \times \frac{1 \text{ mol CO}}{28.01 \text{ g CO}} \times \frac{1 \text{ mol C}}{1 \text{ mol CO}} \times 12.01 \text{ g C} = 0.200 \text{ g C}$$

$$0.733 \text{ g } CO_2 \times \frac{1 \text{ mol } CO_2}{44.01 \text{ g } CO_2} \times \frac{1 \text{ mol C}}{1 \text{ mol } CO_2} \times 12.01 \text{ g C} = 0.200 \text{ g C}$$

Total mass C is 0.200 g + 0.200 g = 0.400 g C.

$$0.450 \text{ g } H_2O \times \frac{1 \text{ mol } H_2O}{18.02 \text{ g } H_2O} \times \frac{2 \text{ mol H}}{1 \text{ mol } H_2O} \times \frac{1.008 \text{ g H}}{1 \text{ mol H}} = 0.0503 \text{ g H}$$

(Since hydrocarbons contain only the elements C and H, g H can also be obtained by subtraction: 0.450 g sample - 0.400 g C = 0.050 g H.)

$$0.400 \text{ g C} \times \frac{1 \text{ mol C}}{12.01 \text{ g C}} = 0.0333 \text{ mol C}; \ 0.0333 / 0.0333 = 1.0$$

$$0.0503 \text{ g H} \times \frac{1 \text{ mol H}}{1.008 \text{ g H}} = 0.0499 \text{ mol H}; \ 0.0499 / 0.0333 = 1.5$$

Multiplying by a factor of 2, the empirical formula is C_2H_3.

(b) Mass is conserved. Total mass products - mass sample = mass O_2 consumed.

0.467 g CO + 0.733 g CO_2 + 0.450 g H_2O - 0.450 g sample = 1.200 g O_2 consumed

(c) For complete combustion, 0.467 g CO must be converted to CO_2.

$2CO(g) + O_2(g) \rightarrow 2CO_2(g)$

$$0.467 \text{ g CO} \times \frac{1 \text{ mol CO}}{28.01 \text{ g C}} \times \frac{1 \text{ mol } O_2}{2 \text{ mol CO}} \times \frac{32.00 \text{ g } O_2}{1 \text{ mol } O_2} = 0.267 \text{ g } O_2$$

The total mass of O_2 required for complete combustion is

$1.200\ g + 0.267\ g = 1.467\ g\ O_2$.

3.97 $N_2(g) + 3H_2(g) \rightarrow 2NH_3(g)$

Determine the moles of N_2 and H_2 required to form the 2.0 moles of NH_3 present after the reaction has stopped.

$$2.0\ \text{mol NH}_3 \times \frac{3\ \text{mol H}_2}{2\ \text{mol NH}_3} = 3.0\ \text{mol H}_2\ \text{reacted}$$

$$2.0\ \text{mol NH}_3 \times \frac{1\ \text{mol N}_2}{2\ \text{mol NH}_3} = 1\ \text{mol N}_2\ \text{reacted}$$

mol H_2 initial = 2.0 mol H_2 remain + 3.0 mol H_2 reacted = 5.0 mol H_2

mol N_2 initial = 2.0 mol N_2 remain + 1.0 mol N_2 reacted = 3.0 mol N_2

In tabular form:	$N_2(g)$	+	$3H_2(g)$	$\rightarrow$	$2NH_3(g)$
initial	3.0 mol		5.0 mol		0 mol
reaction	-1.0 mol		-3.0 mol		+2.0 mol
final	2.0 mol		2.0 mol		2.0 mol

(Tables like this will be extremely useful for solving chemical equilibrium problems in Chapter 15.)

3.99 (a) $2C_2H_2(g) + 5O_2(g) \rightarrow 4CO_2(g) + 2H_2O(g)$

(b) Following the approach in Sample Exercise 3.19,

$$10.0\ \text{g C}_2\text{H}_2 \times \frac{1\ \text{mol C}_2\text{H}_2}{26.04\ \text{g C}_2\text{H}_2} \times \frac{5\ \text{mol O}_2}{2\ \text{mol C}_2\text{H}_2} \times \frac{32.00\ \text{g O}_2}{1\ \text{mol O}_2} = 30.7\ \text{g O}_2\ \text{required}$$

Only 10.0 g O_2 are available, so O_2 limits.

(c) Since O_2 limits, 0.0 g O_2 remain.

Next, calculate the g C_2H_2 consumed and the amounts of CO_2 and H_2O produced by reaction of 10.0 g O_2.

$$10.0\ \text{g O}_2 \times \frac{1\ \text{mol O}_2}{32.00\ \text{g O}_2} \times \frac{2\ \text{mol C}_2\text{H}_2}{5\ \text{mol O}_2} \times \frac{26.04\ \text{g C}_2\text{H}_2}{1\ \text{mol C}_2\text{H}_2} = 3.26\ \text{g C}_2\text{H}_2\ \text{consumed}$$

10.0 g C_2H_2 initial – 3.26 g consumed = 6.74 = 6.7 g C_2H_2 remain

$$10.0\ \text{g O}_2 \times \frac{1\ \text{mol O}_2}{32.00\ \text{g O}_2} \times \frac{4\ \text{mol CO}_2}{5\ \text{mol O}_2} \times \frac{44.01\ \text{g CO}_2}{1\ \text{mol CO}_2} = 11.0\ \text{g CO}_2\ \text{produced}$$

$$10.0\ \text{g O}_2 \times \frac{1\ \text{mol O}_2}{32.00\ \text{g O}_2} \times \frac{2\ \text{mol H}_2\text{O}}{5\ \text{mol O}_2} \times \frac{18.02\ \text{g H}_2\text{O}}{1\ \text{mol H}_2\text{O}} = 2.25\ \text{g H}_2\text{O produced}$$

3.100 (a) $$1.5 \times 10^5\ \text{g C}_9\text{H}_8\text{O}_4 \times \frac{1\ \text{mol C}_9\text{H}_8\text{O}_4}{180.2\ \text{g C}_9\text{H}_8\text{O}_4} \times \frac{1\ \text{mol C}_7\text{H}_6\text{O}_3}{1\ \text{mol C}_9\text{H}_8\text{O}_4} \times \frac{138.1\ \text{g C}_7\text{H}_6\text{O}_3}{1\ \text{mol C}_7\text{H}_6\text{O}_3}$$

$$= 1.1496 \times 10^5\ \text{g} = 1.1 \times 10^2\ \text{kg C}_7\text{H}_6\text{O}_3$$

(b) If only 80 percent of the acid reacts, then we need 1/0.80 = 1.25 times as much to obtain the same mass of product: $1.25 \times 1.15 \times 10^2$ kg = 1.4×10^2 kg $C_7H_6O_3$

(c) Calculate the number of moles of each reactant:

$$1.85 \times 10^5 \text{ g } C_7H_6O_3 \times \frac{1 \text{ mol } C_7H_6O_3}{138.1 \text{ g } C_7H_6O_3} = 1.340 \times 10^3 = 1.34 \times 10^3 \text{ mol } C_7H_6O_3$$

$$1.25 \times 10^5 \text{ g } C_4H_6O_3 \times \frac{1 \text{ mol } C_4H_6O_3}{102.1 \text{ g } C_4H_6O_3} = 1.224 \times 10^3 = 1.22 \times 10^3 \text{ mol } C_4H_6O_3$$

We see that $C_4H_6O_3$ limits, because equal numbers of moles of the two reactants are consumed in the reaction.

$$1.224 \times 10^3 \text{ mol } C_4H_6O_3 \times \frac{1 \text{ mol } C_9H_8O_4}{1 \text{ mol } C_7H_6O_3} \times \frac{180.2 \text{ g } C_9H_8O_4}{1 \text{ mol } C_9H_8O_4} = 2.206 \times 10^5$$

$$= 2.21 \times 10^5 \text{ g} C_9H_8O_4$$

(d) $$\text{percent yield} = \frac{1.82 \times 10^5 \text{ g}}{2.206 \times 10^5 \text{ g}} \times 100 = 82.5\%$$

Integrative Exercises

3.102 (a) *Plan.* volume of Ag cube $\xrightarrow{\text{density}}$ mass of Ag → mol Ag → Ag atoms

Solve. $$(1.000)^3 \text{ cm}^3 \text{ Ag} \times \frac{10.49 \text{ g Ag}}{1 \text{ cm}^3 \text{ Ag}} \times \frac{1 \text{ mol Ag}}{107.87 \text{ g Ag}} \times \frac{6.022 \times 10^{23} \text{ atoms}}{1 \text{ mol}}$$

$$= 5.8562 \times 10^{22} = 5.856 \times 10^{22} \text{ Ag atoms}$$

(b) 1.000 cm^3 cube volume, 74% is occupied by Ag atoms

0.7400 cm^3 = volume of 5.856×10^{22} Ag atoms

$$\frac{0.7400 \text{ cm}^3}{5.8562 \times 10^{22} \text{ Ag atoms}} = 1.2636 \times 10^{-23} = 1.264 \times 10^{-23} \text{ cm}^3 / \text{Ag atom}$$

Since atomic dimensions are usually given in Å, we will show this conversion.

$$1.264 \times 10^{-23} \text{ cm}^3 \times \frac{(1 \times 10^{-2})^3 \text{ m}^3}{1 \text{ cm}^3} \times \frac{1 \text{ Å}^3}{(1 \times 10^{-10})^3 \text{ m}^3} = 12.64 \text{ Å}^3 / \text{Ag atom}$$

(c) $V = 4/3\, \pi r^3$; $r^3 = 3V/4\pi$; $r = (3V/4\pi)^{1/3}$

$r_A = (3 \times 12.636 \text{ Å}^3 / 4\pi)^{1/3} = 1.4449 = 1.445$ Å

3.104 *Plan.* We can proceed by writing the ratio of masses of Ag to $AgNO_3$, where y is the atomic mass of nitrogen. *Solve.*

$$\frac{Ag}{AgNO_3} = 0.634985 = \frac{107.8682}{107.8682 + 3(15.9994) + y}$$

Solve for y to obtain y = 14.0088. This is to be compared with the currently accepted value of 14.0067.

3.106 *Analyze.* Given: 2.0 in × 3.0 in boards, 5000 boards, 0.65 mm thick Cu; 8.96 g/cm^3 Cu; 85% Cu removed; 97% yield for reaction. Find: mass $Cu(NH_3)_4Cl_2$, mass NH_3.

Plan. vol Cu/board × density → mass Cu/board → 5000 boards × 85% = total Cu removed = actual yield; actual yield/0.97 = theoretical yield Cu.

mass Cu → mol Cu → mol $Cu(NH_3)_4$ Cl or NH_3 → desired masses.

Solve. $2.0\text{ in}\times 3.0\text{ in}\times\frac{(2.54)^2\text{ cm}^2}{\text{in}^2}\times 0.65\text{mm}\times\frac{1\text{ cm}}{10\text{ mm}}=2.516=2.5\text{ cm}^3\text{ Cu/board}$

$$\frac{2.516\text{ cm}^3\text{ Cu}}{\text{board}}\times\frac{8.96\text{ g}}{\text{cm}^3}\times 5000\text{ boards}\times 0.85\text{ removed}=95{,}814\text{ g}=96\text{ kg Cu removed}$$

$$\frac{95{,}814\text{ g Cu actual yield}}{0.97}=98{,}777\text{ g}=99\text{ kg Cu theoretical}$$

$$98{,}777\text{ g Cu}\times\frac{1\text{ mol Cu}}{63.546\text{ g Cu}}\times\frac{1\text{ mol Cu(NH}_3)_4\text{Cl}_2}{1\text{mol Cu}}\times\frac{202.575\text{ g}}{1\text{ mol Cu(NH}_3)_4\text{Cl}_2}=314{,}887\text{ g}$$
$$=3.1\times 10^2\text{ kg Cu(NH}_3)_4\text{Cl}_2$$

$$98{,}777\text{ g Cu}\times\frac{1\text{ mol Cu}}{63.546\text{ g Cu}}\times\frac{4\text{ mol NH}_3}{1\text{ mol Cu}}\times\frac{17.03\text{ g NH}_3}{\text{mol NH}_3}=105{,}891\text{ g}=1.1\times 10^2\text{ kg NH}_3$$

3.107 (a) *Plan.* Calculate the kg of air in the room and then the mass of HCN required to produce a dose of 300 mg HCN/kg air. *Solve.*

12 ft × 15 ft × 8.0 ft = 1440 = 1.4 × 10^3 ft^3 of air in the room

$$1440\text{ ft}^3\text{ air}\times\frac{(12\text{ in})^3}{1\text{ ft}^3}\times\frac{(2.54\text{ cm})^3}{1\text{ in}^3}\times\frac{0.00118\text{ g air}}{1\text{ cm}^3\text{ air}}\times\frac{1\text{ kg}}{1000\text{ g}}=48.12=48\text{ kg air}$$

$$48.12\text{ kg air}\times\frac{300\text{ mg HCN}}{1\text{ kg air}}\times\frac{1\text{ g}}{1000\text{ mg}}=14.43=14\text{ g HCN}$$

(b) $2NaCN(s) + H_2SO_4(aq) \rightarrow Na_2SO_4(aq) + 2HCN(g)$

The question can be restated as: What mass of NaCN is required to produce 14 g of HCN according to the above reaction?

$$14.43\text{ g HCN}\times\frac{1\text{ mol HCN}}{27.03\text{ g HCN}}\times\frac{2\text{ mol NaCN}}{2\text{ mol HCN}}\times\frac{49.01\text{ g NaCN}}{1\text{ mol NaCN}}=26.2=26\text{ g NaCN}$$

(c) $$12\text{ ft}\times 15\text{ ft}\times\frac{1\text{ yd}^2}{9\text{ ft}^2}\times\frac{30\text{ oz}}{1\text{ yd}^2}\times\frac{1\text{ lb}}{16\text{ oz}}\times\frac{454\text{ g}}{1\text{ lb}}=17{,}025$$
$$=1.7\times 10^4\text{ g acrilan in the room}$$

50% of the carpet burns, so the starting amount of CH_2CHCN is 0.50(17,025) = 8,513 = 8.5 × 10^3 g

$$8{,}513\text{ g CH}_2\text{CHCN}\times\frac{50.9\text{ g HCN}}{100\text{ g CH}_2\text{CHCH}}=4333=4.3\times 10^3\text{ g HCN possible}$$

If the actual yield of combustion is 20%, actual g HCN = 4,333(0.20) = 866.6 = 8.7 × 10^2 g HCN produced. From part (a), 14 g of HCN is a lethal dose. The fire produces much more than a lethal dose of HCN.

4 Aqueous Reactions and Solution Stoichiometry

Visualizing Concepts

4.2 Although CH_3OH and HCl are both molecular compounds, HCl is an acid and strong electrolyte. Strong electrolytes exist in solution almost completely as ions, so an aqueous HCl solution conducts electricity. CH_3OH is a nonelectrolyte that exists as neutral molecules in aqueous solution. Since there are no charge carriers, aqueous solutions of nonelectrolytes such as CH_3OH do not conduct electricity.

4.4 The brightness of the bulb in Figure 4.2 is related to the number of ions per unit volume of solution. If 0.1 *M* $HC_2H_3O_2$ has about the same brightness of 0.001 *M* HBr, the two solutions have about the same number of ions. Since 0.1 *M* $HC_2H_3O_2$ has 100 times more solute than 0.001 *M* HBr, HBr must be dissociated to a much greater extent than $HC_2H_3O_2$. HBr is one of the few molecular acids that is a strong electrolyte. $HC_2H_3O_2$ is a weak electrolyte; if it were a nonelectrolyte, the bulb in Figure 4.2 wouldn't glow.

4.6 Certain pairs of ions form precipitates because their attraction is so strong that they cannot be surrounded and separated by solvent molecules. That is, the attraction between solute particles is greater than the stabilization offered by interaction of individual ions with solvent molecules.

4.8 Use the difference in reactivities with SO_4^{2-} to identify $Pb^{2+}(aq)$ and $Mg^{2+}(aq)$. Test a portion of each solution with $H_2SO_4(aq)$. $Pb^{2+}(aq)$ is an exception to the soluble sulfates rule, so $Pb(NO_3)_2(aq)$ will form a precipitate, while $Mg(NO_3)_2(aq)$ will not.

4.10 Diagram I shows spectator ions but no precipitate; this corresponds to reaction (b). Diagram II shows spectator ions and a 1:1 precipitate; this corresponds to reaction (c). Diagram III shows only precipitate; this corresponds to reaction (a). The second product in reaction (a) is $H_2O(l)$, which is also the solvent. Solvent molecules are not shown in any of the diagrams.

Electrolytes

4.12 When CH_3OH dissolves, neutral CH_3OH molecules are dispersed throughout the solution. These electrically neutral particles do not carry charge and the solution is nonconducting. When $HC_2H_3O_2$ dissolves, mostly neutral molecules are dispersed throughout the solution. A few of the dissolved molecules ionize to form $H^+(aq)$ and $C_2H_3O_2^-(aq)$. These few ions carry some charge and the solution is weakly conducting.

4.14 Ions are hydrated when they are surrounded by H_2O molecules in aqueous solution.

4.16 (a) $MgI_2(aq) \rightarrow Mg^{2+}(aq) + 2I^-(aq)$

(b) $Al(NO_3)_3(aq) \rightarrow Al^{3+}(aq) + 3NO_3^-(aq)$

(c) $HClO_4(aq) \rightarrow H^+(aq) + ClO_4^-(aq)$

(d) $KC_2H_3O_2(aq) \rightarrow K^+(aq) + C_2H_3O_2^-(aq)$

4.18 (a) acetone (nonelectrolyte): $CH_3COCH_3(aq)$ molecules only; hypochlorous acid (weak electrolyte): $HClO(aq)$ molecules, $H^+(aq)$, $ClO^-(aq)$; ammonium chloride (strong electrolyte): $NH_4^+(aq)$, $Cl^-(aq)$

(b) NH_4Cl, 0.2 mol solute particles; HClO, between 0.1 and 0.2 mol particles; CH_3OCH_3, 0.1 mol of solute particles

Precipitation Reactions and Net Ionic Equations

4.20 According to Table 4.1:

(a) $\mathbf{Ni(OH)_2}$: insoluble

(b) $\mathbf{PbBr_2}$: insoluble;

(c) $\mathbf{Ba(NO_3)_2}$: soluble

(d) $\mathbf{AlPO_4}$: insoluble

(e) $\mathbf{AgC_2H_3O_2}$: soluble

4.22 In each reaction, the precipitate is in bold type.

(a) $Ni(NO_3)_2(aq) + 2NaOH(aq) \rightarrow \mathbf{Ni(OH)_2(s)} + 2NaNO_3(aq)$

(b) No precipitate, and, therefore, no reaction. There is no chemical change to any of the reactant ions.

(c) $Na_2S(aq) + Cu(C_2H_3O_2)_2(aq) \rightarrow \mathbf{CuS(s)} + 2NaC_2H_3O_2(aq)$

4.24 Spectator ions are those that do not change during reaction.

(a) $2Cr^{3+}(aq) + 3CO_3^{2-}(aq) \rightarrow Cr_2(CO_3)_3(s)$; spectators: NH_4^+, SO_4^{2-}

(b) $Ba^{2+}(aq) + SO_4^{2-}(aq) \rightarrow BaSO_4(s)$; spectators: K^+, NO_3^-

(c) $Fe^{2+}(aq) + 2OH^-(aq) \rightarrow Fe(OH)_2(s)$; spectators: K^+, NO_3^-

4.26 Br^- and NO_3^- can be ruled out because the Ba^{2+} salts are soluble. (Actually all NO_3^- salts are soluble.) CO_3^{2-} forms insoluble salts with the three cations given; it must be the anion in question.

4.28 (a) $Pb(C_2H_3O_2)_2(aq) + Na_2S(aq) \rightarrow PbS(s) + 2NaC_2H_3O_2(aq)$
net ionic: $Pb^{2+}(aq) + S^{2-}(aq) \rightarrow PbS$
$Pb(C_2H_3O_2)_2(aq) + CaCl_2(aq) \rightarrow (PbCl_2)s + Ca(C_2H_3PO)_2(aq)$
net ionic: $Pb^{2+}(aq) + 2Cl^-(aq) \rightarrow (PbCl_2)(s)$
$Na_2S(aq) + CaCl_2(aq) \rightarrow CaS(aq) + 2NaCl(aq)$
net ionic: no reaction

(b) Spectator ions: Na^+, Ca^{2+}, $C_2H_3O_2^-$

Acid-Base Reactions

4.30 $NH_3(aq)$ is a weak base, while KOH and $Ca(OH)_2$ are strong bases. $NH_3(aq)$ is only slightly ionized, so even 0.5 *M* NH_3 is less basic than 0.1 *M* KOH. $Ca(OH)_2$ has twice as many OH^- per moles as KOH, so 0.1 *M* $Ca(OH)_2$ is more basic than 0.1 *M* KOH. The most basic solution is 0.1 *M* $Ca(OH)_2$.

4.32 (a) NH_3 produces OH^- in aqueous solution by reacting with H_2O (hydrolysis): $NH_3(aq) + H_2O(1) f\ NH_4^+(aq) + OH^-(aq)$. The OH^- causes the solution to be basic.

(b) The term "weak" refers to the tendency of HF to dissociate into H^+ and F^- in aqueous solution, not its reactivity toward other compounds.

(c) H_2SO_4 is a **diprotic** acid; it has two ionizable hydrogens. The first hydrogen completely ionizes to form H^+ and HSO_4^-, but HSO_4^- only **partially** ionizes into H^+ and SO_4^{2-} (HSO_4^- is a weak electrolyte). Thus, an aqueous solution of H_2SO_4 contains a mixture of H^+, HSO_4^- and SO_4^{2-}, with the concentration of HSO_4^- greater than the concentration of SO_4^{2-}.

4.34 All soluble ionic hydroxides from Table 4.1 are listed as strong bases in Table 4.2. Insoluble hydroxides like $Cd(OH)_2$ are not listed as strong bases. "Insoluble" means that less than 1% of the base molecules exist as separated ions and are dissolved. Thus, insoluble hydroxide salts produce too few $OH^-(aq)$ to be considered strong bases.

4.36 Since the solution does conduct some electricity, but less than an equimolar NaCl solution (a strong electrolyte), the unknown solute must be a weak electrolyte. The weak electrolytes in the list of choices are NH_3 and H_3PO_3; since the solution is acidic, the unknown must be $\mathbf{H_3PO_3}$.

4.38 (a) $HClO_4$: strong (b) HNO_3: strong (c) NH_4Cl: strong

(d) CH_3OCH_3: non (e) $CoSO_4$: strong (f) $C_{12}H_{22}O_{11}$: non

4.40 (a) $HC_2H_3O_2(aq) + KOH(aq) \rightarrow KC_2H_3O_2(aq) + H_2O(l)$

$HC_2H_3O_2(aq) + OH^-(aq) \rightarrow C_2H_3O_2^-(aq)\ H_2O(l)$

(b) $Cr(OH)_3(s) + 3HNO_3(aq) \rightarrow Cr(NO_3)_3(aq) + 3H_2O(l)$

$Cr(OH)_3(s) + 3H^+(aq) \rightarrow 3H_2O(l) + Cr^{3+}(aq)$

(c) $Ca(OH)_2(aq) + 2HClO(aq) \rightarrow Ca(ClO)_2(aq) + 2H_2O(l)$

$HClO(aq) + OH^-(aq) \rightarrow ClO^-(aq) + H_2O(l)$

4.42 (a) $FeO(s) + 2H^+(aq) \rightarrow H_2O(l) + Fe^{2+}(aq)$

(b) $NiO(s) + 2H^+(aq) \rightarrow H_2O(l) + Ni^{2+}(aq)$

4.44 $K_2O(aq) + H_2O(l) \rightarrow 2KOH(aq)$, molecular; $O^{2-}(aq) + H_2O(l) \rightarrow 2OH^-(aq)$, net ionic; base: ($H^+$ ion acceptor) $O^{2-}(aq)$; acid: (H^+ ion donor) $H_2O(aq)$; spectator: K^+

Oxidation-Reduction Reactions

4.46 Oxidation and reduction can only occur together, not separately. When a metal reacts with oxygen, the metal atoms lose electrons and the oxygen atoms gain electrons. Free electrons do not exist under normal conditions. If electrons are lost by one substance they must be gained by another, and vice versa.

4.48 Elements (metals) from Table 4.5 in region A include Na, Mg, K, and Ca; those from region C are Hg and Au. Let's consider K and Au. Since metals from region A are more readily oxidized than those from region C, K will be oxidized to K^+ and Au^{3+} will be reduced to Au in the redox reaction. (Choose Au^{3+} because it is the Au ion shown in Table 4.5.)

In a balanced redox reaction, the number of electrons lost must equal the number of electrons gained. Since K loses 1 electron in forming K^+, while Au^{3+} gains 3 electrons when forming Au, 3 K atoms must be oxidized for every 1 Au^{3+} that is reduced. This relationship dictates the coefficients in the balanced redox reaction.

$$3K(s) + Au^{3+}(aq) \rightleftharpoons Au(s) + 3K^+(aq)$$

4.50 (a) +4 (b) +2 (c) +3 (d) -2 (e) +3 (f) +6

4.52 (a) acid-base reaction

(b) oxidation-reduction reaction; Fe is reduced, C is oxidized

(c) precipitation reaction

(d) oxidation-reduction reaction; Zn is oxidized, N is reduced

4.54 (a) $2HCl(aq) + Ni(s) \rightarrow NiCl_2(aq) + H_2(g)$; $Ni(s) + 2H^+(aq) \rightarrow Ni^{2+}(aq) + H_2(g)$

(b) $H_2SO_4(aq) + Fe(s) \rightarrow FeSO_4(aq) + H_2(g)$; $Fe(s) + 2H^+(aq) \rightarrow Fe^{2+}(aq) + H_2(g)$

Products with the metal in a higher oxidation state are possible, depending on reaction conditions and acid concentration.

(c) $2HBr(aq) + Mg(s) \rightarrow MgBr_2(aq) + H_2(g)$; $Mg(s) + 2H^+(aq) \rightarrow Mg^{2+}(aq) + H_2(g)$

(d) $2HC_2H_3O_2(aq) + Zn(s) \rightarrow Zn(C_2H_3O_2)_2(aq) + H_2(g)$;

$Zn(s) + 2HC_2H_3O_2(aq) \rightarrow Zn^{2+}(aq) + 2C_2H_3O_2^-(aq) + H_2(g)$

4.56 (a) $Mn(s) + NiCl_2(aq) \rightarrow MnCl_2(aq) + Ni(s)$

(b) $Cu(s) + Cr(C_2H_3O_2)(aq) \rightarrow NR$

(c) $2Cr(s) + 3NiSO_4(aq) \rightarrow Cr_2(SO_4)_3(aq) + 3Ni(s)$

(d) $Pt(s) + HBr(aq) \rightarrow NR$

(e) $H_2(g) + CuCl_2(aq) \rightarrow Cu(s) + 2HCl(aq)$

4.58 (a) $Br_2 + 2NaI \rightarrow 2NaBr + I_2$ indicates that Br_2 is more easily reduced than I_2.

$Cl_2 + 2NaBr \rightarrow 2NaCl + Br_2$ shows that Cl_2 is more easily reduced than Br_2.

The order for ease of reduction is $Cl_2 > Br_2 > I_2$. Conversely, the order for ease of oxidation is $I^- > Br^- > Cl^-$.

(b) Since the halogens are nonmetals, they tend to form anions when they react chemically. Nonmetallic character decreases going down a family and so does the tendency to gain electrons during a chemical reaction. Thus, the ease of reduction of the halogen, X_2, decreases going down the family and the ease of oxidation of the halide, X^-, increases going down the family.

(c) $Cl_2 + 2KI \rightarrow 2KCl + I_2$; $Br_2 + LiCl \rightarrow$ no reaction

Solution Composition; Molarity

4.60 (a) The concentration of the remaining solution is unchanged, assuming the original solution was thoroughly mixed. Molar concentration is a **ratio** of moles solute to liters solution. Although there are fewer moles solute remaining in the flask, there is also less solution volume, so the ratio of moles solute/solution volume remains the same.

(b) The second solution is five times as concentrated as the first. An equal volume of the more concentrated solution will contain five times as much solute (five times the number of moles and also five times the mass) as the 0.50 *M* solution. Thus, the mass of solute in the 2.50 *M* solution is 5 × 4.5 g = 22.5 g.

Mathematically:

$$\frac{\dfrac{2.50 \text{ mol solute}}{1 \text{ L solution}}}{\dfrac{0.50 \text{ mol solute}}{1 \text{ L solution}}} = \frac{\text{x grams solute}}{4.5 \text{ g solute}}$$

$$\frac{2.50 \text{ mol solute}}{0.50 \text{ mol solute}} = \frac{\text{x g solute}}{4.5 \text{ g solute}}; \ 5.0(4.5 \text{ g solute}) = 23 \text{ g solute}$$

The result has 2 sig figs; 22.5 rounds to 23 g solute.

4.62 (a) $M = \dfrac{\text{mol solute}}{\text{L solution}}; \ \dfrac{0.145 \text{ mol } Na_2SO_4}{0.750 \text{ L}} = 0.193\ M\ Na_2SO_4$

(b) $\text{mol} = M \times \text{L}; \ \dfrac{0.0850 \text{ mol } KMnO_4}{1 \text{ L}} \times 0.125 \text{ L} = 1.06 \times 10^{-2} \text{ mol } KMnO_4$

(c) $\text{L} = \dfrac{\text{mol}}{M}; \ \dfrac{0.255 \text{ mol HCl}}{11.6 \text{ mol HCl/L}} = 2.20 \times 10^{-2} \text{ L or } 22.0 \text{ mL}$

4.64 Calculate the mol of Na^+ at the two concentrations; the difference is the mol NaCl required to increase the Na^+ concentration to the desired level.

$$\frac{0.118 \text{ mol}}{\text{L}} \times 4.6 \text{ L} = 0.5428 = 0.54 \text{ mol } Na^+$$

$$\frac{0.138 \text{ mol}}{\text{L}} \times 4.6 \text{ L} = 0.6348 = 0.63 \text{ mol } Na^+$$

(0.6348 – 0.5428) = 0.092 = 0.09 mol NaCl (2 decimal places and 1 sig fig)

$$0.092 \text{ mol NaCl} \times \frac{58.5 \text{ g NaCl}}{\text{mol}} = 5.38 = 5 \text{ g NaCl}$$

4.66 $M = \frac{mol}{L}$; $mol = \frac{g}{MM}$ (MM is the symbol for molar mass in this manual.)

(a) $\frac{0.360 \text{ mol } K_2Cr_2O_7}{1 \text{ L}} \times 50.0 \text{ mL} \times \frac{1 \text{ L}}{1000 \text{ mL}} \times \frac{294.2 \text{ g } K_2Cr_2O_7}{1 \text{ mol } K_2Cr_2O_7} = 5.30 \text{ g } K_2Cr_2O_7$

(b) $4.28 \text{ g } (NH_4)_2SO_4 \times \frac{1 \text{ mol } (NH_4)_2SO_4}{132.2 \text{ g } (NH_4)_2SO_4} \times \frac{1}{300. \text{ mL}} \times \frac{1000 \text{ mL}}{1 \text{ L}} = 0.108 \, M \, (NH_4)_2SO_4$

(c) $2.25 \text{ g } CuSO_4 \times \frac{1 \text{ mol } CuSO_4}{159.6 \text{ g } CuSO_4} \times \frac{1 \text{ L}}{0.240 \text{ mol } CuSO_4} \times \frac{1000 \text{ mL}}{1 \text{ L}} = 58.7 \text{ mL solution}$

4.68 (a) $0.1 \, M \, CaCl_2 = 0.2 \, M \, Cl^-$; $0.15 \, M \, KCl = 0.15 \, M \, Cl^-$

$0.1 \, M \, CaCl_2$ has the higher Cl^- concentration.

(b) $0.1 \, M$ KCl has a higher Cl^- concentration than $0.080 \, M$ LiCl. Total volume does not affect concentration.

(c) $0.050 \, M \, HCl = 0.050 \, M \, Cl^-$; $0.020 \, M \, CdCl_2 = 0.040 \, M \, Cl^-$

$0.050 \, M$ HCl has the higher Cl^- concentration.

4.70 (a) $H^+ : \frac{0.130 \, M \times 16.0 \text{ mL} + 0.600 \, M \times 12.0 \text{ mL}}{28.0 \text{ mL}} = 0.331 \, M \, H^+$

Cl^-: concentration Cl^- = concentration H^+ = $0.331 \, M \, Cl^-$

(b) $Na^+ : \frac{2(0.200 \, M \times 18.0 \text{ mL})}{33.0 \text{ mL}} = 0.218 \, M$; $K^+ : = \frac{0.150 \, M \times 15.0 \text{ mL}}{33.0 \text{ mL}} = 0.0682 \, M$

$SO_4^{2-} : \frac{0.200 \, M \times 18.0 \text{ mL}}{33.0 \text{ mL}} = 0.109 \, M$; $Cl^- : \frac{0.150 \, M \times 15.0 \text{ mL}}{33.0 \text{ mL}} = 0.0682 \, M$

(c) $Na^+ : \frac{2.38 \text{ g NaCl}}{0.0500 \text{ L}} \times \frac{1 \text{ mol}}{58.44 \text{ g}} = 0.815 \, M$; $Ca^{2+} : 0.400 \, M$

Cl^-: $0.815 \, M$ (from NaCl(s)) + $0.800 \, M$ (from $CaCl_2$(aq)) = $1.615 \, M$

4.72 (a) $V_1 = M_2V_2/M_1$; $\frac{0.400 \, M \, HNO_3 \times 0.350 \text{ mL}}{10.0 \, M \, HNO_3} = 0.0140 \text{ L} = 14.0 \text{ mL conc. } HNO_3$

(b) $M_2 = M_1V_1/V_2$; $\frac{10.0 \, M \, HNO_3 \times 25.0 \text{ mL}}{500 \text{ mL}} = 0.500 \, M \, HNO_3$

4.74 (a) The amount of $AgNO_3$ needed is: $0.150 \, M \times 0.2500 \text{ L} = 0.0375 \text{ mol } AgNO_3$

$0.0375 \text{ mol } AgNO_3 \times \frac{169.88 \text{ g } AgNO_3}{1 \text{ mol } AgNO_3} = 6.3705 = 6.37 \text{ g } AgNO_3$

Add this amount of solid to a 250 mL volumetric flask, dissolve in a small amount of water, bring the total volume to exactly 250 mL, and agitate well.

(b) Dilute the $6.0 \, M \, HNO_3$ to prepare 100 mL of $0.50 \, M \, HNO_3$. To determine the volume of $6.0 \, M \, HNO_3$ needed, calculate the moles HNO_3 present in 100 mL of $0.50 \, M \, HNO_3$ and then the volume of $6.0 \, M$ solution that contains this number of moles.

$0.100 \text{ L} \times 0.50\ M = 0.050 \text{ mol } HNO_3$ needed;

$$L = \frac{\text{mol}}{M};\ \text{L } 6.0\ M\ HNO_3 = \frac{0.050 \text{ mol needed}}{6.0\ M} = 0.00833 \text{ L} = 8.3 \text{ mL}$$

Thoroughly clean, rinse, and fill a buret with the 6.0 M HNO_3, taking precautions appropriate for working with a relatively concentrated acid. Dispense 8.3 mL of the 6.0 M acid into a 100 mL volumetric flask, add water to the mark, and mix thoroughly.

4.76 $$50.000 \text{ mL glycerol} \times \frac{1.2656 \text{ g glycerol}}{1 \text{ mL glycerol}} = 63.280 \text{ g glycerol}$$

$$63.280 \text{ g } C_3H_8O_3 \times \frac{1 \text{ mol } C_3H_8O_3}{92.094 \text{ g } C_3H_8O_3} = 0.687124 = 0.68712 \text{ mol } C_3H_8O_3$$

$$M = \frac{0.687124 \text{ mol } C_3H_8O_3}{0.25000 \text{ L solution}} = 2.7485\ M\ C_3H_8O_3$$

Solution Stoichiometry; Titrations

4.78 *Plan.* $M \times L = \text{mol } Cd(NO_3)_2$; balanced equation → mol ratio → mol NaOH → g NaOH

Solve. $$\frac{0.500 \text{ mol } Cd(NO_3)_2}{1 \text{ L}} \times 0.02500 \text{ L} = 0.0125 \text{ mol } Cd(NO_3)_2$$

$$Cd(NO_3)_2(aq) + 2NaOH(aq) \rightarrow Cd(OH)_2(s) + 2NaNO_3(aq)$$

$$0.0125 \text{ mol } Cd(NO_3)_2 \times \frac{2 \text{ mol NaOH}}{1 \text{ mol } Cd(NO_3)_2} \times \frac{40.00 \text{ g NaOH}}{1 \text{ mol NaOH}} = 1.00 \text{ g NaOH}$$

4.80 (a) $$2HCl(aq) + Ba(OH)_2(aq) \rightarrow BaCl_2(aq) + 2H_2O(l)$$

$$\frac{0.101 \text{ mol } Ba(OH)_2}{1 \text{ L } Ba(OH)_2} \times 0.0500 \text{ L } Ba(OH)_2 \times \frac{2 \text{ mol HCl}}{1 \text{ mol } Ba(OH)_2} \times \frac{1 \text{ L HCl}}{0.120 \text{ mol HCl}} = 0.0842 \text{ L or } 84.2 \text{ mL HCl soln}$$

(b) $$H_2SO_4(aq) + 2NaOH(aq) \rightarrow Na_2SO_4(aq) + 2H_2O(l)$$

$$0.200 \text{ g NaOH} \times \frac{1 \text{ mol NaOH}}{40.00 \text{ g NaOH}} \times \frac{1 \text{ mol } H_2SO_4}{2 \text{ mol NaOH}} \times \frac{1 \text{ L } H_2SO_4}{0.125 \text{ mol } H_2SO_4} = 0.0200 \text{ L or } 20.0 \text{ mL } H_2SO_4 \text{ soln}$$

(c) $$BaCl_2(aq) + Na_2SO_4(aq) \rightarrow BaSO_4(s) + 2NaCl(aq)$$

$$752 \text{ mg} = 0.752 \text{ g } Na_2SO_4 \times \frac{1 \text{ mol } Na_2SO_4}{142.1 \text{ g } Na_2SO_4} \times \frac{1 \text{ mol } BaCl_2}{1 \text{ mol } Na_2SO_4} \times \frac{1}{0.0558 \text{ L}} = 0.0948 \text{ M } BaCl_2$$

(d) $$2HCl(aq) + Ca(OH)_2(aq) \rightarrow CaCl_2(aq) + 2H_2O(l)$$

$$0.0427 \text{ L HCl} \times \frac{0.208 \text{ mol HCl}}{1 \text{ L HCl}} \times \frac{1 \text{ mol } Ca(OH)_2}{2 \text{ mol HCl}} \times \frac{74.10 \text{ g } Ca(OH)_2}{1 \text{ mol } Ca(OH)_2} = 0.329 \text{ g } Ca(OH)_2$$

4.82 See Exercise 4.79(a) for a more detailed approach.

$$\frac{0.115 \text{ mol NaOH}}{1 \text{ L}} \times 0.0425 \text{ L} \times \frac{1 \text{ mol } HC_2H_3O_2}{1 \text{ mol NaOH}} \times \frac{60.05 \text{ g } HC_2H_3O_2}{1 \text{ mol } HC_2H_3O_2}$$

$$= 0.29349 = 0.293 \text{ g } HC_2H_3O_2 \text{ in } 3.45 \text{ mL}$$

$$1.00 \text{ qt vinegar} \times \frac{1 \text{ L}}{1.057 \text{ qt}} \times \frac{1000 \text{ mL}}{1 \text{ L}} \times \frac{0.29349 \text{ g } HC_2H_3O_2}{3.45 \text{ mL vinegar}} = 80.5 \text{ g } HC_2H_3O_2/\text{qt}$$

4.84 The balanced equation for the titration is:

$Sr(NO_3)_2(aq) + Na_2CrO_4(aq) \rightarrow SrCrO_4(s) + 2NaNO_3(aq)$

Beginning with a 0.100 L sample, we can do the following conversions:

volume soln $\rightarrow$ g $Sr(NO_3)_2 \rightarrow$ mol $Sr(NO_3)_2 \rightarrow$ mol $Na_2CrO_4 \rightarrow$ vol Na_2CrO_4 soln

$$0.100 \text{ L soln} \times \frac{6.82 \text{ g } Sr(NO_3)_2}{0.500 \text{ L soln}} \times \frac{1 \text{ mol } Sr(NO_3)_2}{211.6 \text{ g } Sr(NO_3)_2} \times \frac{1 \text{ mol } Na_2CrO_4}{1 \text{ mol } Sr(NO_3)_2}$$

$$\times \frac{1 \text{ L soln}}{0.0335 \text{ mol } Na_2CrO_4} = 0.192 \text{ L } Na_2CrO_4 \text{ soln}$$

4.86 (a) $HNO_3(aq) + NaOH(s) \rightarrow NaNO_3(aq) + H_2O(l)$

(b) Determine the limiting reactant, then the identity and concentration of ions remaining in solution. Assume that the $H_2O(l)$ produced by the reaction does **not** increase the total solution volume.

$$12.0 \text{ g NaOH} \times \frac{1 \text{ mol NaOH}}{40.00 \text{ g NaOH}} = 0.300 \text{ mol NaOH}$$

$0.200\ M\ HNO_3 \times 0.0750 \text{ L } HNO_3 = 0.0150 \text{ mol } HNO_3$.

The mol ratio is 1:1, so HNO_3 is the limiting reactant. No excess H^+ remains in solution. The remaining ions are OH^- (excess reactant), Na^+, and NO_3^- (spectators).

OH^-: 0.300 mol OH^- initial – 0.0150 mol OH^- react = 0.285 mol OH^- remain

0.285 mol OH^-/0.0750 L soln = 3.80 M $OH^-(aq)$

Na^+: 0.300 mol Na^+/0.0750 L soln = 4.00 M $Na^+(aq)$

NO_3^-: 0.0150 mol NO_3^-/0.0750 L = 0.200 M $NO_3^-(aq)$

(c) The resulting solution is **basic** because of the large excess of $OH^-(aq)$.

4.88 *Plan.* $CaCO_3(s) + 2HCl(aq) \rightarrow CaCl_2(aq) + H_2O(l) + CO_2(g)$

$HCl(aq) + NaOH(aq) \rightarrow NaCl(aq) + H_2O(l)$

total mol HCl – excess mol HCl = mol HCl reacted; mol $CaCO_3$ = (mol HCl)/2;

g $CaCO_3$ = mol $CaCO_3$ × molar mass; mass % = (g $CaCO_3$/g sample) × 100

Solve.

$$\frac{1.035 \text{ mol HCl}}{1 \text{ L soln}} \times 0.03000 \text{ L} = 0.031050 = 0.03105 \text{ mol HCl total}$$

$$\frac{1.010 \text{ mol NaOH}}{1 \text{ L soln}} \times 0.01156 \text{ L} = 0.011676 = 0.01168 \text{ mol HCl excess}$$

0.031050 total - 0.011676 excess = 0.019374 = 0.01937 mol HCl reacted

$$0.019374 \text{ mol HCl} \times \frac{1 \text{ mol } CaCO_3}{2 \text{ mol HCl}} \times \frac{100.09 \text{ g } CaCO_3}{1 \text{ mol } CaCO_3} = 0.96959 = 0.9696 \text{ g } CaCO_3$$

$$\text{mass \% } CaCO_3 = \frac{\text{g } CaCO_3}{\text{g rock}} \times 100 = \frac{0.96959}{1.248} \times 100 = 77.69\%$$

Additional Exercises

4.89 As soon as the dispersion formed, ion-pairing would begin, and eventually KBr(s) would form. In water, the solute ions stay separated as in solution because they are stabilized by electrostatic attractive forces with the polar water molecules. In a nonpolar solvent like mineral oil, there are no electrostatic attractive forces between the solute ions and the solvent. The oppositely charged ions are attracted to each other, forming KBr solid.

4.91 The two precipitates formed are due to AgCl(s) and $SrSO_4$(s). Since no precipitate forms on addition of hydroxide ion to the remaining solution, the other two possibilities, Ni^{2+} and Mn^{2+}, are absent.

4.93 (a) $Al(OH)_3(s) + 3H^+(aq) \rightarrow Al^{3+}(aq) + 3H_2O(l)$

(b) $Mg(OH)_2(s) + 2H^+(aq) \rightarrow Mg^{2+}(aq) + 2H_2O(l)$

(c) $MgCO_3(s) + 2H^+(aq) \rightarrow Mg^{2+}(aq) + H_2O(l) + CO_2(g)$

(d) $NaAl(CO_3)(OH)_2(s) + 4H^+(aq) \rightarrow Na^+(aq) + Al^{3+}(aq) + 3H_2O(l) + CO_2(g)$

(e) $CaCO_3(s) + 2H^+(aq) \rightarrow Ca^{2+}(aq) + H_2O(l) + CO_2(g)$

[In (c), (d) and (e), one could also write the equation for formation of bicarbonate, e.g., $MgCO_3(s) + H^+(aq) \rightarrow Mg^{2+} + HCO_3^-(aq)$.]

4.94 (a) $2H^+(aq) + SO_3^{2-}(aq) \rightarrow H_2SO_3(aq)$; sulfurous acid

(b) $H_2SO_3(aq) \rightarrow H_2O(l) + SO_2(g)$; sulfur dioxide

(c) The boiling point of $SO_2(g)$ is -10°C. It is a gas at room temperature (23°C) and pressure (1 atm).

(d) (i) $Na_2SO_3(aq) + 2HCl(aq) \rightarrow 2NaCl(aq) + H_2O(l) + SO_2(g)$

$SO_3^{2-}(aq) + 2H^+(aq) \rightarrow H_2O(l) + SO_2(g)$

(ii) $Ag_2SO_3(s) + 2HCl(aq) \rightarrow 2AgCl(s) + H_2O(l) + SO_2(g)$

$Ag_2SO_3(s) + 2H^+(aq) + 2Cl^-(aq) \rightarrow 2AgCl(s) + H_2O(l) + SO_2(g)$

(iii) $KHSO_3(s) + HCl(aq) \rightarrow KCl(aq) + H_2O(l) + SO_2(g)$

$KHSO_3(s) + H^+(aq) \rightarrow K^+(aq) + H_2O(l) + SO_2(g)$

(iv) $ZnSO_3(aq) + 2HCl(aq) \rightarrow ZnCl_2(aq) + H_2O(l) + SO_2(g)$

$SO_3^{2-}(aq) + 2H^+(aq) \rightarrow H_2O(l) + SO_2(g)$

4.96 A metal on Table 4.5 is able to displace the metal cations below it from their compounds. That is, zinc will reduce the cations below it to their metals.

(a) $Zn(s) + Na^+(aq) \rightarrow$ no reaction

(b) $Zn(s) + Pb^{2+}(aq) \rightarrow Zn^{2+}(aq) + Pb(s)$

(c) $Zn(s) + Mg^{2+}(aq) \rightarrow$ no reaction

(d) $Zn(s) + Fe^{2+}(aq) \rightarrow Zn^{2+}(aq) + Fe(s)$

(e) $Zn(s) + Cu^{2+}(aq) \rightarrow Zn^{2+}(aq) + Cu(s)$

(f) $Zn(s) + Al^{3+}(aq) \rightarrow$ no reaction

4.97 (a) A : La_2O_3 Metals often react with the oxygen in air to produce metal oxides.

B : $La(OH)_3$ When metals react with water (HOH) to form H_2, OH^- remains.

C : $LaCl_3$ Most chlorides are soluble.

D : $La_2(SO_4)_3$ Sulfuric acid provides SO_4^{2-} ions.

(b) $4La(s) + 3O_2(g) \rightarrow 2La_2O_3(s)$

$2La(s) + 6HOH(l) \rightarrow 2La(OH)_3(s) + 3H_2(g)$

(There are no spectator ions in either of these reactions.)

molecular: $La_2O_3(s) + 6HCl(aq) \rightarrow 2LaCl_3(aq) + 3H_2O(l)$

net ionic: $La_2O_3(s) + 6H^+(aq) \rightarrow 2La^{3+}(aq) + 3H_2O(l)$

molecular: $La(OH)_3(s) + 3HCl(aq) \rightarrow LaCl_3(aq) + 3H_2O(l)$

net ionic: $La(OH)_3(s) + 3H^+(aq) \rightarrow La^{3+}(aq) + 3H_2O(l)$

molecular: $2LaCl_3(aq) + 3H_2SO_4(aq) \rightarrow La_2(SO_4)_3(s) + 6HCl(aq)$

net ionic: $2La^{3+}(aq) + 3SO_4^{2-}(aq) \rightarrow La_2(SO_4)_3(s)$

(c) La metal is oxidized by water to produce $H_2(g)$, so La is definitely above H on the activity series. In fact, since an acid is not required to oxidize La, it is probably one of the more active metals.

4.99 (a) $$0.0400 \text{ L soln} \times \frac{0.160 \text{ mol NaCl}}{1 \text{ L soln}} = 6.40 \times 10^{-3} \text{ mol NaCl}$$

$$0.0650 \text{ L soln} \times \frac{0.150 \text{ mol NaCl}}{1 \text{ L soln}} = 9.75 \times 10^{-3} \text{ mol NaCl}$$

Total moles NaCl = $1.615 \times 10^{-2} = 1.62 \times 10^{-2}$

Total volume = 0.0400 L + 0.0650 L = 0.1050 L

$$\text{Molarity} = \frac{1.615 \times 10^{-2} \text{ mol}}{0.1050 \text{ L}} = 0.154\ M$$

(b) Both solutions are the same concentration, 0.750 M, so the resulting solution is 0.750 M.

4.101 Na^+ must replace the total positive (+) charge due to Ca^{2+} and Mg^{2+}. Think of this as moles of charge rather than moles of particles.

$$\frac{0.010 \text{ mol } Ca^{2+}}{1 \text{ L water}} \times 1.0 \times 10^3 \text{ L} \times \frac{2 \text{ mol} + \text{change}}{1 \text{ mol } Ca^{2+}} = 20 \text{ mol of} + \text{charge}$$

$$\frac{0.0050 \text{ mol } Mg^{2+}}{1 \text{ L water}} \times 1.0 \times 10^3 \text{ L} \times \frac{2 \text{ mol} + \text{charge}}{1 \text{ mol } Mg^{2+}} = 10 \text{ mol of} + \text{charge}$$

30 moles of + charge must be replaced; 30 mol Na^+ are needed.

4.102 $H_2C_4H_4O_6 + 2OH^-(aq) \rightarrow C_4H_4O_6^{2-}(aq) + 2H_2O(l)$

$$0.02262 \text{ L NaOH soln} \times \frac{0.2000 \text{ mol NaOH}}{1 \text{ L}} \times \frac{1 \text{ mol } H_2C_4H_4O_6}{2 \text{ mol NaOH}} \times \frac{1}{0.04000 \text{ L } H_2C_4H_4O_6}$$

$$= 0.05655 \ M \ H_2C_4H_4O_6 \text{ soln}$$

4.104 mol OH^- from NaOH(aq) + mol OH^- from $Zn(OH)_2(s)$ = mol H^+ from HBr

mol H^+ = M HBr × L HBr = 0.500 M HBr × 0.400 L HBr = 0.200 mol H^+

mol OH^- from NaOH = M NaOH × L NaOH = 0.500 M NaOH × 0.0985 L NaOH

= 0.04925 = 0.0493 mol OH^-

mol OH^- from $Zn(OH)_2(s)$ = 0.200 mol H^+ – 0.04925 mol OH^- from NaOH = 0.15075

= 0.151 mol OH^- from $Zn(OH)_2$

$$0.15075 \text{ mol } OH^- \times \frac{1 \text{ mol } Zn(OH)_2}{2 \text{ mol } OH^-} \times \frac{99.41 \text{ g } Zn(OH)_2}{1 \text{ mol } Zn(OH)_2} = 7.49 \text{ g } Zn(OH)_2$$

Integrative Exercises

4.106 (a) At the equivalence point of a titration, mol NaOH added = mol H^+ present

$$M_{NaOH} \times L_{NaOH} = \frac{\text{g acid}}{\text{MM acid}} \text{ (for an acid with 1 acidic hydrogen)}$$

$$\text{MM acid} = \frac{\text{g acid}}{M_{NaOH} \times L_{NaOH}} = \frac{0.2053 \text{ g}}{0.1008 \ M \times 0.0150 \text{ L}} = 136 \text{ g/mol}$$

(b) Assume 100 g of acid.

$$70.6 \text{ g C} \times \frac{1 \text{ mol C}}{12.01 \text{ g C}} = 5.88 \text{ mol C}; \ 5.88 / 1.47 \approx 4$$

$$5.89 \text{ g H} \times \frac{1 \text{ mol H}}{1.008 \text{ g H}} = 5.84 \text{ mol H}; \ 5.84 / 1.47 \approx 4$$

$$23.5 \text{ g O} \times \frac{1 \text{ mol O}}{16.00 \text{ g O}} = 1.47 \text{ mol O}; \ 1.47 / 1.47 = 1$$

The empirical formula is C_4H_4O.

$\frac{MM}{FW} = \frac{136}{68.1} = 2$; the molecular formula is $2 \times$ the empirical formula.

The molecular formula is $C_8H_8O_2$.

4.108 *Plan.* Write balanced equation.

$$\text{mass } H_2SO_4 \text{ soln} \xrightarrow{\text{mass \%}} \text{mass } H_2SO_4 \rightarrow \text{mol } H_2SO_4 \rightarrow \text{mol } Na_2CO_3 \rightarrow \text{mass } Na_2CO_3$$

Solve. $H_2SO_4(aq) + Na_2CO_3(s) \rightarrow Na_2SO_4(aq) + H_2O(l) + CO_2(g)$

$$5.0 \times 10^3 \text{ kg conc. } H_2SO_4 \times \frac{0.950 \text{ kg } H_2SO_4}{1.00 \text{ kg conc. } H_2SO_4} = 4.75 \times 10^3 = 4.8 \times 10^3 \text{ kg } H_2SO_4$$

$$4.75 \times 10^3 \text{ kg } H_2SO_4 \times \frac{1 \times 10^3 \text{ g}}{1 \text{ kg}} \times \frac{1 \text{ mol } H_2SO_4}{98.08 \text{ g } H_2SO_4} \times \frac{1 \text{ mol } Na_2CO_3}{1 \text{ mol } H_2SO_4}$$

$$\times \frac{105.99 \text{ g } NaHCO_3}{1 \text{ mol } NaHCO_3} \times \frac{1 \text{ kg}}{1 \times 10^3 \text{ g}} = 5.133 \times 10^3 = 5.1 \times 10^3 \text{ kg } Na_2CO_3$$

4.109 (a) $Mg(OH)_2(s) + 2HNO_3(aq) \rightarrow Mg(NO_3)_2(aq) + 2H_2O(l)$

(b) $5.53 \text{ g } Mg(OH)_2 \times \frac{1 \text{ mol } Mg(OH)_2}{58.32 \text{ g } Mg(OH)_2} = 0.09482 = 0.0948 \text{ mol } Mg(OH)_2$

$0.200\ M\ HNO_3 \times 0.0250 \text{ L} = 0.00500 \text{ mol } HNO_3$

The 0.00500 mol HNO_3 would neutralize 0.00250 mol $Mg(OH)_2$ and much more $Mg(OH)_2$ is present, so HNO_3 is the limiting reactant.

(c) Since HNO_3 limits, 0 mol HNO_3 is present after reaction.

0.00250 mol $Mg(NO_3)_2$ is produced.

0.09482 mol $Mg(OH)_2$ initial – 0.00250 mol $Mg(OH)_2$ react

= 0.0923 mol $Mg(OH)_2$ remain

4.111 *Plan.* Cl^- is present in NaCl and $MgCl_2$; using mass %, calculate mass NaCl and $MgCl_2$ in mixture, mol Cl^- in each, then molarity of Cl^- in 0.500 L solution. *Solve.*

$$7.50 \text{ mixture} \times \frac{0.765 \text{ g NaCl}}{1.00 \text{ g mixture}} \times \frac{1 \text{ mol NaCl}}{58.44 \text{ g NaCl}} \times \frac{1 \text{ mol } Cl^-}{1 \text{ mol NaCl}} = 0.09818 = 0.0982 \text{ mol } Cl^-$$

$$7.50 \text{ mixture} \times \frac{0.065 \text{ g } MgCl_2}{1.00 \text{ g mixture}} \times \frac{1 \text{ mol } MgCl_2}{95.21 \text{ g } MgCl_2} \times \frac{2 \text{ mol } Cl^-}{1 \text{ mol MgCl}} = 0.01024 = 0.010 \text{ mol } Cl^-$$

$$\text{mol } Cl^- = 0.09818 + 0.01024 = 0.10842 = 0.108 \text{ mol } Cl^-;\ M = \frac{0.10842 \text{ mol } Cl^-}{0.5000 \text{ L}} = 0.217\ M\ Cl^-$$

4.112 *Plan.* $M = \frac{\text{mol } Br^-}{\text{L seawater}}$; mg $Br^- \rightarrow$ g $Br^- \rightarrow$ mol Br^-;

$$1 \text{ kg seawater} \rightarrow \text{g} \xrightarrow{\text{density}} \text{mL water} \rightarrow \text{L sea water}$$

Solve. $65 \text{ mg Br}^- \times \dfrac{1 \text{ g Br}^-}{1000 \text{ mg Br}^-} \times \dfrac{1 \text{ mol Br}^-}{79.90 \text{ g Br}^-} = 8.135 \times 10^{-4} = 8.1 \times 10^{-4} \text{ mol Br}^-$

$$1 \text{ kg seawater} \times \frac{1000 \text{ g}}{1 \text{ kg}} \times \frac{1 \text{ mL water}}{1.025 \text{ g water}} \times \frac{1 \text{ L}}{1000 \text{ mL}} = 0.9756 \text{ L}$$

$$M \text{ Br}^- = \frac{8.135 \times 10^{-4} \text{ mol Br}^-}{0.9756 \text{ L seawater}} = 8.3 \times 10^{-4} \; M \text{ Br}^-$$

4.114 (a) AsO_4^{3-}; +5

(b) Ag_3PO_4 is silver phosphate; Ag_3AsO_4 is silver arsenate

(c) $0.0250 \text{ L soln} \times \dfrac{0.102 \text{ mol Ag}^+}{1 \text{ L soln}} \times \dfrac{1 \text{ mol Ag}_3\text{AsO}_4}{3 \text{ mol Ag}^+} \times \dfrac{1 \text{ mol As}}{1 \text{ mol Ag}_3\text{AsO}_4} \times \dfrac{74.92 \text{ g As}}{1 \text{ mol As}}$

$= 0.06368 = 0.0637 \text{ g As}$

$$\text{mass percent} = \frac{0.06368 \text{ g As}}{1.22 \text{ g sample}} \times 100 = 5.22\% \text{ As}$$

4.116 (a) mol HCl initial – mol NH_3 from air = mol HCl remaining

= mol NaOH required for titration

mol NaOH = 0.0588 M × 0.0131 L = $7.703 \times 10^{-4} = 7.70 \times 10^{-4}$ mol NaOH

= 7.70×10^{-4} mol HCl remain

mol HCl initial – mol HCl remaining = mol NH_3 from air

(0.0105 M HCl × 0.100 L) – 7.703×10^{-4} mol HCl = mol NH_3

10.5×10^{-4} mol HCl – 7.703×10^{-4} mol HCl = $2.80 \times 10^{-4} = 2.8 \times 10^{-4}$ mol NH_3

$$2.8 \times 10^{-4} \text{ mol NH}_3 \times \frac{17.03 \text{ g NH}_3}{1 \text{ mol NH}_3} = 4.77 \times 10^{-3} = 4.8 \times 10^{-3} \text{ g NH}_3$$

(b) ppm is defined as molecules of $NH_3/1 \times 10^6$ molecules in air.

Calculate molecules NH_3 from mol NH_3.

$$2.80 \times 10^{-4} \text{ mol NH}_3 \times \frac{0.022 \times 20^{23} \text{ molecules}}{1 \text{ mol}} = 1.686 \times 10^{20}$$

$= 1.7 \times 10^{20}$ NH_3 molecules

Calculate total volume of air processed, then g air using density, then molecules air using molar mass.

$$\frac{10.0 \text{ L}}{1 \text{ min}} \times 10.0 \text{ min} \times \frac{1.20 \text{ g air}}{1 \text{ L air}} \times \frac{1 \text{ mol air}}{29.0 \text{ g air}} \times \frac{6.022 \times 10^{23} \text{ molecules}}{1 \text{ mol}}$$

$= 2.492 \times 10^{24} = 2.5 \times 10^{24}$ air molecules

$$\text{ppm NH}_3 = \frac{1.686 \times 10^{20} \text{ NH}_3 \text{ molecules}}{2.492 \times 10^{24} \text{ air molecules}} \times 1 \times 10^6 = 68 \text{ ppm NH}_3$$

(c) 68 ppm > 50 ppm. The manufacturer is **not** in compliance.

5 Thermochemistry

Visualizing Concepts

5.2 (a) The internal energy, E, of the products is greater than that of the reactants, so the diagram represents an increase in the internal energy of the system.

(b) ΔE for this process is positive, +.

(c) If no work is associated with the process, it is endothermic.

5.3 (a) For an endothermic process, the sign of q is positive; the system gains heat. This is true only for system (iii).

(b) In order for ΔE to be less than 0, there is a net transfer of heat or work from the system to the surroundings. The magnitude of the quantity leaving the system is greater than the magnitude of the quantity entering the system. In system (i), the magnitude of the heat leaving the system is less than the magnitude of the work done on the system. In system (iii), the magnitude of the work done by the system is less than the magnitude of the heat entering the system. None of the systems has $\Delta E < 0$.

(c) In order for ΔE to be greater than 0, there is a net transfer of work or heat to the system from the surroundings. In system (i), the magnitude of the work done on the system is greater than the magnitude of the heat leaving the system. In system (ii), work is done on the system with no change in heat. In system (iii) the magnitude of the heat gained by the system is greater than the magnitude of the work done on the surroundings. $\Delta E > 0$ for all three systems.

5.5 $w = -P\Delta V$. Since ΔV for the process is (–), the sign of w is (+).

$\Delta E = q + w$. At constant pressure, $\Delta H = q$. If the reaction is endothermic, the signs of ΔH and q are (+). The sign of w is (+), so the sign of ΔE is (+). The internal energy of the system increases during the change. (This situation is described by the diagram (ii) in Exercise 5.3.)

5.7 (a) $N_2(g) + O_2(g) \rightarrow 2NO(g)$. Since $\Delta V = 0$, $w = 0$.

(b) $\Delta H = 90.37$ kJ for production of 1 mol of NO(g). The definition of a formation reaction is one where elements combine to form one mole of a single product. The enthalpy change for such a reaction is the enthalpy of formation.

5.8 (a) $\Delta H_A = \Delta H_B + \Delta H_C$. The net enthalpy change associated with going from the initial state to the final state does not depend on path. The change can be accomplished via reaction A, or via two successive reactions, B then C, with the same net enthalpy change.

(b) $\Delta H_Z = \Delta H_X + \Delta H_Y$. The diagram indicates that Reaction Z can be written as the sum of reactions X and Y. Hess's Law states that the enthalpy change for the net reaction Z is the sum of the enthalpy changes of the steps X and Y, regardless of whether the reaction actually occurs via this path. $\Delta H_Z = \Delta H_X + \Delta H_Y$ because ΔH is a state function, independent of path.

The Nature of Energy

5.10 (a) The kinetic energy of the ball **decreases** as it moves higher. As the ball moves higher and opposes gravity, kinetic energy is changed into potential energy.

(b) The potential energy of the ball **increases** as it moves higher.

(c) The heavier ball would go **half as high** as the tennis ball. At the apex of the trajectory, all initial kinetic energy has been changed into potential energy. The magnitude of the change in potential energy is $m\ g\ \Delta h$, which is equal to the energy initially imparted to the ball. If the same amount of energy is imparted to a ball with twice the mass, m doubles so Δh is half as large.

5.12 (a) *Plan.* Convert lb → kg, mi/hr → m/s.

Solve. $950 \text{ lb} \times \dfrac{1 \text{ kg}}{2.205 \text{ lb}} = 430.84 = 431 \text{ kg}$

$$\frac{68 \text{ mi}}{1 \text{ hr}} \times \frac{1.6093 \text{ km}}{1 \text{ mi}} \times \frac{1000 \text{ m}}{1 \text{ km}} \times \frac{1 \text{ hr}}{60 \text{ min}} \times \frac{1 \text{ min}}{60 \text{ sec}} = 30.398 = 30 \text{ m/s}$$

$$E_k = 1/2\ mv^2 = 1/2 \times 430.84 \text{ kg} \times (30.398)^2 \text{ m}^2/\text{s}^2 = 2.0 \times 10^5 \text{ J}$$

(b) E_k is proportional to v^2, so if speed decreases by a factor of 2, kinetic energy decreases by a factor of 4.

(c) Brakes stop a moving vehicle, so the kinetic energy of the motorcycle is primarily transferred to friction between brakes and wheels, and somewhat to deformation of the tire and friction between the tire and road.

5.14 (a) *Analyze.* Given: 1 kwh; 1 watt = 1 J/s; 1 watt • s = 1 J.

Find: conversion factor for joules and kwh.

Plan. kwh → wh → ws → J

Solve. $1 \text{ kwh} \times \dfrac{1000 \text{ w}}{1 \text{ kw}} \times \dfrac{60 \text{ min}}{\text{h}} \times \dfrac{60 \text{ s}}{\text{min}} \times \dfrac{1 \text{ J}}{1 \text{ w} \bullet \text{s}} = 3.6 \times 10^6 \text{ J}$

$1 \text{ kwh} = 3.6 \times 10^6 \text{ J}$

(b) *Analyze.* Given: 100 watt bulb. Find: heat in kcal radiated by bulb or person in 24 hr.

Plan. 1 watt = 1 J/s; 1 kcal = 4.184×10^3 J; watt → J/s → J → kcal. *Solve.*

$$100 \text{ watt} = \frac{100 \text{ J}}{1 \text{ s}} \times \frac{60 \text{ sec}}{\text{min}} \times \frac{60 \text{ min}}{\text{hr}} \times 24 \text{ hr} \times \frac{1 \text{ kcal}}{4.184 \times 10^3 \text{ J}} = 2065 = 2.1 \times 10^3 \text{ kcal}$$

24 hr has 2 sig figs, but 100 watt is ambiguous. The answer to 1 sig fig would be 2×10^3 kcal.

Check. $(1 \times 10^2 \times 6 \times 10^1 \times 6/10^3) \approx 6^3 \times 10 \approx 2000$ kcal

5.16 The air gun imparts a certain amount of kinetic energy to the pellet. As the pellet rises against the force of gravity, kinetic energy is changed to potential energy. When all kinetic energy has been transferred to potential energy (or lost as heat through friction) the pellet stops rising and falls to earth. In principle, if enough kinetic energy could be imparted to the pellet, it could escape the force of gravity and move into space. For an air gun and a pellet, this is practically impossible.

5.18 (a) The system is not closed, because it is exchanging mass with the surroundings. That is, solution flows into and out of the flask.

(b) If the system is defined as shown, it can be closed by blocking the flow in and out, but leaving the flask full of solution.

5.20 (a) Heat is the energy transferred from a hotter object to a colder object.

(b) Heat is transferred from one object (system) to another until the two objects (systems) are at the same temperature.

5.22 (a) Electrostatic attraction; no work is done because the particles are held stationary.

(b) Magnetic attraction; work is done because the nail is moved a distance.

The First Law of Thermodynamics

5.24 (a) $\Delta E_{sys} = -\Delta E_{surr}$; $\Delta E_{sys} = q + w$

(b) The quantities q and w are negative when the system loses heat to the surroundings (it cools), or does work on the surroundings.

5.26 In each case, evaluate q and w in the expression $\Delta E = q + w$. For an exothermic process, q is negative; for an endothermic process, q is positive.

(a) q is positive and w is negative. $\Delta E = 900 \text{ J} - 422 \text{ J} = 478 \text{ J}$. The process is endothermic.

(b) q is negative and w is essentially zero. $\Delta E = -3140$ J. The process is exothermic.

(c) q is negative and w is zero. $\Delta E = -8.65$ kJ. The process is exothermic.

5.28 $E_{el} = \dfrac{\kappa Q_1 Q_2}{r^2}$ For two oppositely charged particles, the sign of E_{el} is negative; the closer the particles, the greater the magnitude of E_{el}.

(a) The potential energy becomes less negative as the particles are separated (r increases).

(b) ΔE for the process is positive; the internal energy of the system increases as the oppositely charged particles are separated.

(c) Work is done on the system to separate the particles so w is positive. We have no direct knowledge of the change in q, except that it cannot be large and negative, because overall $\Delta E = q + w$ is positive.

5.30 (a) Independent. Potential energy is a state function.

(b) Dependent. Some of the energy released could be employed in performing work, as is done in the body when sugar is metabolized; heat is not a state function.

(c) Dependent. The work accomplished depends on whether the gasoline is used in an engine, burned in an open flame, or in some other manner. Work is not a state function.

Enthalpy

5.32 (a) When a process occurs under constant external pressure, the enthalpy change (ΔH) equals the amount of heat transferred. $\Delta H = q_P$.

(b) $\Delta H = q_P$. If the system absorbs heat, q and ΔH are positive and the enthalpy of the system increases.

5.34 (a) $ZnCO_3(s) \rightarrow ZnO(s) + CO_2(g)$

$\Delta H = 71.5$ kJ

(b) $ZnO(s) + CO_2(g)$

$\Delta H = +71.5$ kJ

$ZnCO_3(s)$

5.36 *Plan.* Consider the sign of an enthalpy change that would convert one of the substances into the other. *Solve.*

(a) $CO_2(s) \rightarrow CO_2(g)$. This change is sublimation, which is endothermic, $+\Delta H$. $CO_2(g)$ has the higher enthalpy.

(b) $H_2 \rightarrow 2H$. Breaking the H-H bond requires energy, so the process is endothermic, $+\Delta H$. Two moles of H atoms have higher enthalpy.

(c) $H_2O(g) \rightarrow H_2(g) + 1/2\ O_2(g)$. Decomposing H_2O into its elements requires energy and is endothermic, $+\Delta H$. One mole of $H_2(g)$ and 0.5 mol $O_2(g)$ at 25°C have the higher enthalpy.

(d) $N_2(g)$ at 100° $\rightarrow$ $N_2(g)$ at 300°. An increase in the temperature of the sample requires that heat is added to the system, +q and $+\Delta H$. $N_2(g)$ at 300° has the higher enthalpy.

5.38 (a) The reaction is endothermic, so heat is absorbed by the system during the course of reaction.

(b) $45.0\text{ g } CH_3OH \times \frac{1\text{ mol } CH_3OH}{32.04\text{ g } CH_3OH} \times \frac{90.7\text{ kJ}}{1\text{ mol } CH_3OH} = 127$ kJ heat transferred (absorbed)

(c) $18.5\text{ kJ} \times \frac{2\text{ mol } H_2}{90.7\text{ kJ}} \times \frac{2.016\text{ g } H_2}{1\text{ mol } H_2} = 0.822\text{ g } H_2$ produced

(d) The sign of ΔH is reversed for the reverse reaction: $\Delta H = -90.7$ kJ

$27.0\text{ g CO} \times \frac{1\text{ mol CO}}{28.01\text{ g CO}} \times \frac{-90.7\text{ kJ}}{1\text{ mol CO}} = -87.4$ kJ heat transferred (released)

5.40 (a) $0.855 \text{ mol } O_2 \times \frac{-89.4 \text{ kJ}}{3 \text{ mol } O_2} = -25.48 = -25.5 \text{ kJ}$

(b) $10.75 \text{ g KCl} \times \frac{1 \text{ mol KCl}}{74.55 \text{ g KCl}} \times \frac{-89.4 \text{ kJ}}{2 \text{ mol KCl}} = -6.4457 = -6.45 \text{ kJ}$

(c) Since the sign of ΔH is reversed for the reverse reaction, it seems reasonable that other characteristics would be reversed, as well. If the forward reaction proceeds spontaneously, the reverse reaction is probably not spontaneous. Also, we know from experience that KCl(s) does not spontaneously react with atmospheric O_2(g), even at elevated temperature.

5.42 At constant volume (ΔV = 0), ΔE = q_v. According to the definition of enthalpy, H = E + PV, so ΔH = ΔE + Δ(PV). For an ideal gas at constant temperature and volume, ΔPV = VΔP = RTΔn. For this reaction, there are 2 mol of gaseous product and 3 mol of gaseous reactants, so Δn = -1. Thus VΔP or Δ(PV) is negative. Since ΔH = ΔE + Δ(PV), the negative Δ(PV) term means that ΔH will be smaller or more negative than ΔE.

5.44 The gas is the system. If 418 J of heat is added, q = +418 J. Work done by the system decreases the overall energy of the system, so w = -107 J.

ΔE = q + w = 418 J - 107 J = 311 J. ΔH = q = 418 J (at constant pressure).

5.46 (a) $3C_2H_2(g) \rightarrow C_6H_6(l)$ ΔH = -630 kJ

(b) $C_6H_6(l) \rightarrow 3C_2H_2(g)$ ΔH = 630 kJ
ΔH for the formation of 3 mol of acetylene is 630 kJ. ΔH for the formation of 1 mol of C_2H_2 is then 630 kJ/3 = 210 kJ.

(c) The exothermic reverse reaction is more likely to be thermodynamically favored.

(d)

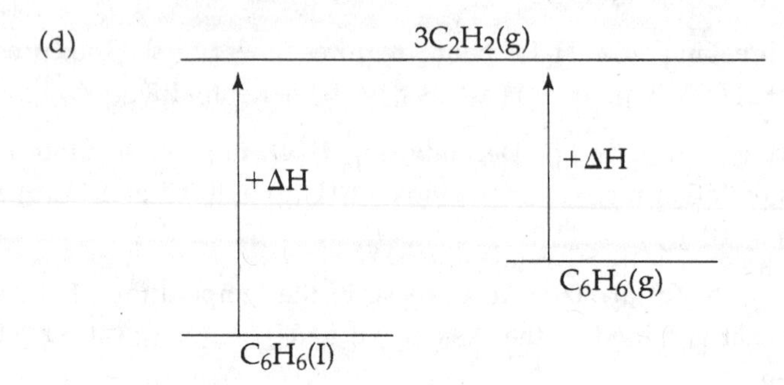

If the reactant is in the higher enthalpy gas phase, the overall ΔH for the reaction has a smaller positive value.

Calorimetry

The specific heat of water to four significant figures, **4.184 J/g • K,** will be used in many of the following exercises; temperature units of K and °C will be used interchangeably.

5.48 *Analyze.* Both objects are heated to 100°C. The two hot objects are placed in the same amount of cold water at the same temperature. Object A raises the water temperature more than object B. *Plan.* Apply the definition of heat capacity to heating the water and heating the objects to determine which object has the greater heat capacity. *Solve.*

(a) Both beakers of water contain the same mass of water, so they both have the same heat capacity. Object A raises the temperature of its water more than object B, so more heat was transferred from object A than from object B. Since both objects were heated to the same temperature initially, object A must have absorbed more heat to reach the 100° temperature. The greater the heat capacity of an object, the greater the heat required to produce a given rise in temperature. Thus, object A has the greater heat capacity.

(b) Since no information about the masses of the objects is given, we cannot compare or determine the specific heats of the objects.

5.50 (a) In Table 5.2, Hg(l) has the smallest specific heat, so it will require the smallest amount of energy to heat 50.0 g of the substance 10 k.

(b) $$50.0\text{ g Hg(l)} \times 10\text{ K} \times \frac{0.14\text{ J}}{\text{g}\bullet\text{K}} = 70\text{ J}$$

5.52 $$62.0\text{ g ethylene glycol} \times \frac{2.42\text{ J}}{\text{g}\bullet\text{K}} \times (40.8^\circ\text{C} - 15.2^\circ\text{C}) = 3.84 \times 10^3\text{ J}$$

5.54 (a) Following the logic in Solution 5.53, the dissolving process is endothermic, ΔH is positive. The total mass of the solution is (60.0 g H_2O + 3.88 g NH_4NO_3) = 63.88 = 63.9 g. The temperature change of the solution is 23.0 - 18.4 = 4.6°C. The heat lost by the solution is

$$63.88\text{ g solution} \times \frac{4.184\text{ J}}{1\text{ g}\bullet^\circ\text{C}} \times 4.6^\circ\text{C} \times \frac{1\text{ kJ}}{1000\text{ J}} = 1.229 = 1.2\text{ kJ}$$

Thus, 1.2 kJ is absorbed when 3.88 g $NH_4NO_3(s)$ dissolves.

$$\frac{+1.229\text{ kJ}}{3.88\text{ NH}_4\text{NO}_3} \times \frac{80.05\text{ g NH}_4\text{NO}_3}{1\text{ mol NH}_4\text{NO}_3} = +25.36 = +25\text{ kJ/mol NH}_4\text{NO}_3$$

(b) This process is endothermic, because the temperature of the surroundings decreases, indicating that heat is absorbed by the system.

5.56 (a) $C_6H_5OH(s) + 7O_2(g) \rightarrow 6CO_2(g) + 3H_2O(l)$

(b) $q_{bomb} = -q_{rxn}$; ΔT = 26.37°C - 21.36°C = 5.01°C

$$q_{bomb} = \frac{11.66\text{ kJ}}{1^\circ\text{C}} \times 5.01^\circ\text{C} = 58.417 = 58.4\text{ kJ}$$

At constant volume, $q_v = \Delta E$. ΔE and ΔH are very similar.

$$\Delta H_{rxn} \approx \Delta E_{rxn} = q_{rxn} = -q_{bomb} = \frac{-58.417\text{ kJ}}{1.800\text{ g C}_6\text{H}_5\text{OH}} = -32.454 = -32.5\text{ kJ/g C}_6\text{H}_5\text{OH}$$

$$\Delta H_{rxn} = \frac{-32.454\text{ kJ}}{1\text{ g C}_6\text{H}_5\text{OH}} \times \frac{94.11\text{ g C}_6\text{H}_5\text{OH}}{1\text{ mol C}_6\text{H}_5\text{OH}} = \frac{-3.054 \times 10^3\text{ kJ}}{\text{mol C}_6\text{H}_5\text{OH}}$$

$$= -3.05 \times 10^3\text{ kJ/mol C}_6\text{H}_5\text{OH}$$

5.58 (a) $$C = 1.640\text{ g HC}_7\text{H}_5\text{O}_2 \times \frac{26.38\text{ kJ}}{1\text{ g HC}_7\text{H}_5\text{O}_2} \times \frac{1}{4.95^\circ\text{C}} = 8.740 = 8.74\text{ kJ/}^\circ\text{C}$$

(b) $\frac{8.740 \text{ kJ}}{^\circ\text{C}} \times 4.68^\circ\text{C} \times \frac{1}{1.320 \text{ g sample}} = 30.99 = 31.0 \text{ kJ/g sample}$

(c) If water is lost from the calorimeter, there is less water to heat, so the same amount of heat (kJ) from a reaction would cause a larger increase in the calorimeter temperature. The calorimeter constant, kJ/°C, would decrease, because °C is in the denominator of the expression.

Hess's Law

5.60 (a) *Analyze/Plan.* Arrange the reactions so that in the overall sum, B, appears in both reactants and products and can be canceled. This is a general technique for using Hess's Law. *Solve.*

$$\begin{array}{ll} A \rightarrow B & \Delta H = +30 \text{ kJ} \\ B \rightarrow C & \Delta H = +60 \text{ kJ} \\ \hline A \rightarrow C & \Delta H = +30 \text{ kJ} \end{array}$$

(b)

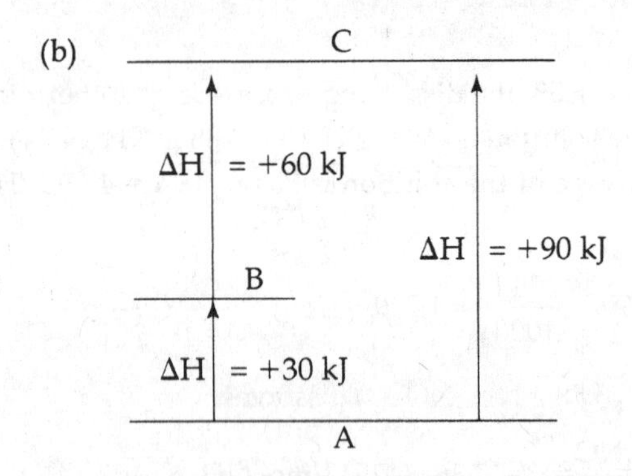

Check. The process of A forming C can be described as A forming B and B forming C.

5.62

$$\begin{array}{lll} 3H_2(g) + 3/2\,O_2(g) & \rightarrow 3H_2O(g) & \Delta H = 3/2(-483.6 \text{ kJ}) \\ O_3(g) & \rightarrow 3/2\,O_2(g) & \Delta H = 1/2(-284.6 \text{ kJ}) \\ \hline 3H_2(g) + O_3(g) & \rightarrow P_4O_{10}(s) & \Delta H = -867.7 \text{ kJ} \end{array}$$

5.64

$$\begin{array}{lll} N_2O(g) & \rightarrow N_2(g) + 1/2\,O_2(g) & \Delta H = 1/2\ (-163.2 \text{ kJ}) \\ NO_2(g) & \rightarrow NO(g) + 1/2\,O_2(g) & \Delta H = 1/2(113.1 \text{ kJ}) \\ N_2(g) + O_2(g) & \rightarrow 2NO(g) & \Delta H = 180.7 \text{ kJ} \\ \hline N_2O(g) + NO_2(g) & \rightarrow 3NO(g) & \Delta H = 155.7 \text{ kJ} \end{array}$$

Enthalpies of Formation

5.66 (a) Tables of ΔH_f° are useful because, according to Hess's law, the standard enthalpy of any reaction can be calculated from the standard enthalpies of formation for the reactants and products.

$$\Delta H^\circ_{rxn} = \Sigma\Delta H_f^\circ \text{ (products)} - \Sigma\Delta H_f^\circ \text{ (reactants)}$$

(b) The standard enthalpy of formation for any element in its standard state is zero. Elements in their standard states are the reference point for the enthalpy of formation scale.

(c) $6C(s) + 6H_2(g) + 3O_2(g) \rightarrow C_6H_{12}O_6(s)$

5.68 (a) $1/2\ H_2(g) + 1/2\ Br_2(l) \rightarrow HBr(g)$ $\Delta H^\circ_f = -36.23\ kJ$

(b) $Ag(s) + 1/2\ N_2(g) + 3/2\ O_2(g) \rightarrow AgNO_3(s)$ $\Delta H^\circ_f = -124.4\ kJ$

(c) $2Hg(l) + Cl_2(g) \rightarrow Hg_2Cl_2(s)$ $\Delta H^\circ_f = -264.9\ kJ$

(d) $2C(s, gr) + 1/2\ O_2(g) + 3H_2(g) \rightarrow C_2H_5OH(l)$ $\Delta H^\circ_f = -277.7\ kJ$

5.70 Use heats of formation to calculate ΔH° for the combustion of butane.

$$C_4H_{10}(l) + 13/2\ O_2(g) \rightarrow 4CO_2(g) + 5H_2O(l)$$

$$\Delta H^\circ_{rxn} = 4\Delta H\ CO_2(g) + 5\Delta H^\circ_f\ H_2O(l) - \Delta H^\circ_f\ C_4H_{10}(l) - 13/2\ \Delta H^\circ_f\ O_2(g)$$

$$\Delta H^\circ_{rxn} = 4(-393.5\ kJ) + 5(-285.83\ kJ) - (-147.6\ kJ) - 13/2(0) = -2855.6 = -2856\ kJ/mol\ C_4H_{10}$$

$$1.0\ g\ C_4H_{10} \times \frac{1\ mol\ C_4H_{10}}{58.123\ g\ C_4H_{10}} \times \frac{-2855.6\ kJ}{1\ mol\ C_4H_{10}} = -49\ kJ$$

5.72 (a) $\Delta H^\circ_{rxn} = 4\Delta H^\circ_f\ H_2O(g) + \Delta H^\circ_f\ N_2(g) - \Delta H^\circ_f\ N_2O_4(g) - 4\Delta H^\circ_f\ H_2(g)$

$= 4(-241.82) + 0 - (9.66) - 4(0) = -976.94\ kJ$

(b) $\Delta H^\circ_{rxn} = \Delta H^\circ_f\ K_2CO_3(s) + \Delta H^\circ_f\ H_2O(g) - 2\Delta H^\circ_f\ KOH(s) - \Delta H^\circ_f\ CO_2(g)$

$= -1150.18\ kJ - 241.82\ kJ - 2(-424.7)\ kJ - (-393.5\ kJ) = -149.1\ Kj$

(c) $\Delta H^\circ_{rxn} = 3/\ 8\Delta H^\circ_f\ S_8(s) + 2\Delta H^\circ_f\ H_2O(g) - \Delta H^\circ_f\ SO_2(g) - 2\Delta H^\circ_f\ H_2S(g)$

$= 3/8(0) + 2(-241.82) - (-269.9) - 2(-20.17) = -173.4\ kJ$

(d) $\Delta H^\circ_{rxn} = 2\Delta H^\circ_f\ FeCl_3(s) + 3\Delta H^\circ_f\ H_2O(g) - \Delta H^\circ_f\ Fe_2O(g) - 6\Delta H^\circ_f\ HCl(g)$

$= 2(-400\ kJ) + 3(-241.82\ kJ) - (-822.16\ kJ) - 6(-92.30\ kJ) = -149.5\ kJ$

5.74 $\Delta H^\circ_{rxn} = \Delta H^\circ_f\ Ca(OH)_2(s) + \Delta H^\circ_f\ C_2H_2(g) - 2\Delta H^\circ_f\ H_2O(l) - \Delta H^\circ_f\ CaC_2(s)$

$-127.2\ kJ = -986.2\ kJ + 226.7\ kJ - 2(-285.83\ kJ) - \Delta H^\circ_f\ CaC_2(s)$

ΔH°_f for $CaC_2(s) = -60.6\ kJ$

5.76 (a)

$2B(s) + 3/2\ O_2(g) \rightarrow B_2O_3(s)$	$\Delta H^\circ = 1/2(-2509.1\ kJ)$
$3H_2(g) + 3/2\ O_2(g) \rightarrow 3H_2O(l)$	$\Delta H^\circ = 3/2(-571.7\ kJ)$
$B_2O_3(s) + 3H_2O(l) \rightarrow B_2H_6(g) + 3O_2(g)$	$\Delta H^\circ = -(-2147.5\ kJ)$
$2B(s) + 3H_2(g) \rightarrow B_2H_6(g)$	$\Delta H^\circ_f = +35.4\ kJ$

(b) If, like B_2H_6, the combustion of B_5H_9 produces B_2O_3 as the boron-containing product, the heat of combustion of B_5H_9 in addition to data given in part (a) would enable calculation of the heat of formation of B_5H_9.

The combustion reaction is: $B_5H_9(l) + 6O_2(g) \rightarrow 5/2\,B_2O_3(s) + 9/2\,H_2O(l)$

$$\begin{array}{ll} 5/4\,[4B(s)+3O_2(g) \rightarrow 2B_2O_3(s)] & \Delta H = 5/4(-2509.1\text{ kJ}) \\ 9/4\,[2H_2(g)+O_2(g) \rightarrow 2H_2O(l)] & \Delta H = 9/4(-571.7\text{ kJ}) \\ 5/2\,B_2O_3(s)+9/2\,H_2O(l) \rightarrow B_5H_9(l)+6O_2(g) & \Delta H = -\text{ (heat of combustion)} \\ \hline 5B(s)+9/2\,H_2(g) \rightarrow B_5H_9(l) & \Delta H_f^\circ \text{ of } B_5H_9(l) \end{array}$$

$[\Delta H_f^\circ\ B_5H_9(l) = -[\text{heat of combustion of } B_5H_9(l)] - 3136.4\text{ kJ} - 1286\text{ kJ}]$

We need to measure the heat of combustion of $B_5H_9(l)$.

5.78 (a) $10C(s) + 4H_2(g) \rightarrow C_{10}H_8(s)$ formation

$C_{10}H_8(s) + 12O_2(g) \rightarrow 10CO_2(g) + 4H_2O(l)$ combustion, $\Delta H^\circ = -5154$ kJ

(b) $\Delta H^\circ_{rxn} = 10\Delta H_f^\circ\ CO_2(g) + 4\Delta H_f^\circ\ H_2O(l) - \Delta H_f^\circ\ C_{10}H_8 - 12\Delta H_f^\circ\ O_2(g)$

$-5154 = 10(-393.5\text{ kJ}) + 4(-285.83\text{ kJ}) - \Delta H_f^\circ\ C_{10}H_8(s) - 12(0)$

$\Delta H_f^\circ\ C_{10}H_8(s) = 10(-393.5\text{ kJ}) + 4(-285.83\text{ kJ}) + 5154\text{ kJ} = 76\text{ kJ}$

Check. The result has 0 decimal places because the heat of combustion has 0 decimal places.

Foods and Fuels

5.80 (a) Fats are appropriate for fuel storage because they are insoluble in water (and body fluids) and have a high fuel value.

(b) For convenience, assume 100 g of chips.

$$12\text{ g protein} \times \frac{17\text{ kJ}}{1\text{ g protein}} \times \frac{1\text{ Cal}}{4.184\text{ kJ}} = 48.76 = 49\text{ Cal}$$

$$14\text{ g fat} \times \frac{38\text{ kJ}}{1\text{ g fat}} \times \frac{1\text{ Cal}}{4.184\text{ kJ}} = 127.15 = 130\text{ Cal}$$

$$74\text{ g carbohydrates} \times \frac{17\text{ kJ}}{1\text{ g carbohydrates}} \times \frac{1\text{ Cal}}{4.184\text{ kJ}} = 300.67 = 301\text{ Cal}$$

total Cal = (48.76 + 127.15 + 300.67) = 476.58 = 480 Cal

$$\%\text{ Cal from fat} = \frac{127.15\text{ Cal fat}}{476.58\text{ total Cal}} \times 100 = 26.68 = 27\%$$

(Since the conversion from kJ to Cal was common to all three components, we would have determined the same percentage by using kJ.)

(c) $25\text{ g fat} \times \frac{38\text{ kJ}}{\text{g fat}} = x\text{ g protein} \times \frac{17\text{ kJ}}{\text{g protein}}$; $x = 56\text{ g protein}$

5.82 Calculate the fuel value in a pound of M&M® candies.

$$96\text{ fat} \times \frac{38\text{ kJ}}{1\text{ g fat}} = 3648\text{ kJ} = 3.6 \times 10^3\text{ kJ}$$

$$320 \text{ g carbohydrate} \times \frac{17 \text{ kJ}}{1 \text{ g carbohydrate}} = 5440 \text{ kJ} = 5.4 \times 10^3 \text{ kJ}$$

$$21 \text{ g protein} \times \frac{17 \text{ kJ}}{1 \text{ g protein}} = 357 \text{ kJ} = 3.6 \times 10^2 \text{ kJ}$$

total fuel value = 3648 kJ + 5440 kJ + 357 kJ = 9445 kJ = 9.4×10^3 kJ/lb

$$\frac{9445 \text{ kJ}}{\text{lb}} \times \frac{1 \text{ lb}}{453.6 \text{ g}} \times \frac{42 \text{ g}}{\text{serving}} = 874.5 \text{ kJ} = 8.7 \times 10^2 \text{ kJ/serving}$$

$$\frac{874.5 \text{ kJ}}{\text{serving}} \times \frac{1 \text{ kcal}}{4.184 \text{ kJ}} \times \frac{1 \text{ Cal}}{1 \text{ kcal}} = 209.0 \text{ Cal} = 2.1 \times 10^2 \text{ Cal/serving}$$

Check. 210 Cal is the approximate food value of a candy bar, so the result is reasonable.

5.84

$$177 \text{ mL} \times \frac{1.0 \text{ g wine}}{1 \text{ mL}} \times \frac{0.106 \text{ g ethanol}}{1 \text{ g wine}} \times \frac{1 \text{ mol ethanol}}{46.1 \text{ g ethanol}} \times \frac{1367 \text{ kJ}}{1 \text{ mol ethanol}} \times \frac{1 \text{ Cal}}{4.184 \text{ kJ}}$$

$$= 133 = 1.3 \times 10^2 \text{ Cal}$$

Check. A "typical" 6 oz. glass of wine has 150-250 Cal, so this is a reasonable result. Note that alcohol is responsible for most of the food value of wine.

5.86

$$\Delta H^\circ_{rxn} = \Delta H^\circ_f\, CO_2(g) + 2\Delta H^\circ_f\, H_2O(g) - \Delta H^\circ_f\, CH_4(g) - \Delta H^\circ_f\, O_2(g)$$

$$= -393.5 \text{ kJ} + 2(-241.82 \text{ kJ}) - (-74.8 \text{ kJ}) - 0 \text{ kJ} = -802.3 \text{ kJ}$$

$$\Delta H^\circ_{rxn} = \Delta H^\circ_f\, CF_4(g) + 4\Delta H^\circ_f\, HF(g) - \Delta H^\circ_f\, CH_4(g) - 4\Delta H^\circ_f\, F_2(g)$$

$$= -679.9 \text{ kJ} + 4(-268.61 \text{ kJ}) - (-74.8 \text{ kJ}) - 0 \text{ kJ} = -1679.5 \text{ kJ}$$

The second reaction is twice as exothermic as the first. The "fuel values" of hydrocarbons in a fluorine atmosphere are approximately twice those in an oxygen atmosphere. Note that the difference in ΔH° values for the two reactions is in the ΔH°_f for the products, since the ΔH°_f for the reactants is identical.

Additional Exercises

5.88 (a) $E_p = mgh = 52.0 \text{ kg} \times 9.81 \text{ m/s}^2 \times 10.8 \text{ m} = 5509.3 \text{ J} = 5.51 \text{ kJ}$

(b) $E_k = 1/2\, mv^2;\ v = (2E_k/m)^{1/2} = \left(\dfrac{2 \times 5509 \text{ kg} \bullet \text{m}^2/\text{s}^2}{52.0 \text{ kg}}\right)^{1/2} = 14.6 \text{ m/s}$

(c) Yes, the diver does work on entering (pushing back) the water in the pool.

5.89 In the process described, one mole of solid CO_2 is converted to one mole of gaseous CO_2. The volume of the gas is much greater than the volume of the solid. Thus the system (that is, the mole of CO_2) must work against atmospheric pressure when it expands. To accomplish this work while maintaining a constant temperature requires the absorption of additional heat beyond that required to increase the internal energy of the CO_2. The remaining energy is turned into work.

5.91 Freezing is an exothermic process (the opposite of melting, which is clearly endothermic). When the system, the soft drink, freezes, it releases energy to the surroundings, the can. Some of this energy does the work of splitting the can.

5.92 (a) $q = 0$, $w > 0$ (work done to system), $\Delta E > 0$

(b) Since the system (the gas) is losing heat, the sign of q is negative. The changes in state described in cases (a) and (b) are identical and ΔE is the same in both cases. The distribution of energy transferred as either work or heat is different in the two scenarios. In case (b), more work is required to compress the gas because some heat is lost to the surroundings. (The moral of this story is that the more energy lost by the system as heat, the greater the work on the system required to accomplish the desired change.)

5.94 If a function sometimes depends on path, then it is simply not a state function. Enthalpy is a state function, so ΔH for the two pathways leading to the same change of state pictured in Figure 5.10 must be the same. However, q is not the same for the both. Our conclusion must be that $\Delta H \neq q$ for these pathways. The condition for $\Delta H = q_p$ (other than constant pressure) is that the only possible work on or by the system is pressure-volume work. Clearly, the work being done in this scenario is not pressure-volume work, so $\Delta H \neq q$, even though the two changes occur at constant pressure.

5.96 (a) $q_{Cu} = \frac{0.385\ J}{g \bullet K} \times 121.0\ g\ Cu \times (30.1^{\circ}C - 100.4^{\circ}C) = -3274.9 = -3.27 \times 10^3\ J$

The negative sign indicates the 3.27×10^3 J are lost by the Cu block.

(b) $q_{H_2O} = \frac{4.184\ J}{g \bullet K} \times 150.0\ g\ H_2O \times (30.1^{\circ}C - 25.1^{\circ}C) = 3138 = 3.1 \times 10^3\ J$

The positive sign indicates that 3.14×10^3 J are gained by the H_2O.

(c) The difference in the heat lost by the Cu and the heat gained by the water is $3.275 \times 10^3\ J - 3.138 \times 10^3\ J = 0.137 \times 10^3\ J = 1 \times 10^2$ J. The temperature change of the calorimeter is 5.0°C. The heat capacity of the calorimeter in J/K is

$$0.137 \times 10^3\ J \times \frac{1}{5.0^{\circ}C} = 27.4 = 3 \times 10\ J/K.$$

Since q_{H_2O} is known to one decimal. place, the difference has one decimal place and the result has 1 sig fig.

If the rounded results from (a) and (b) are used,

$$C_{calorimeter} = \frac{0.2 \times 10^3\ J}{5.0\ ^{\circ}C} = 4 \times 10\ J/K.$$

(d) $q_{H_2O} = 3.275 \times 10^3\ J = \frac{4.184\ J}{g \bullet K} \times 150.0\ g \times (\Delta T)$

$\Delta T = 5.22^{\circ}C$; $T_f = 25.1^{\circ}C + 5.22^{\circ}C = 30.3^{\circ}C$

5.97 (a) From the mass of benzoic acid that produces a certain temperature change, we can calculate the heat capacity of the calorimeter.

$$\frac{0.235\ g\ \text{benzoic acid}}{1.642^{\circ}C\ \text{change observed}} \times \frac{26.38\ kJ}{1\ g\ \text{benzoic acid}} = 3.7755 = 3.78\ kJ/^{\circ}C$$

Now we can use this experimentally determined heat capacity with the data for caffeine.

$$\frac{1.525^{\circ}\text{C rise}}{0.265\text{ g caffeine}} \times \frac{3.7755\text{ kJ}}{1^{\circ}\text{C}} \times \frac{194.2\text{ g caffeine}}{1\text{ mol caffeine}} = 4.22 \times 10^3\text{ kJ/mol caffeine}$$

(b) The overall uncertainty is approximately equal to the sum of the uncertainties due to each effect. The uncertainty in the mass measurement is 0.001/0.235 or 0.001/0.265, about 1 part in 235 or 1 part in 265. The uncertainty in the temperature measurements is 0.002/1.642 or 0.002/1.525, about 1 part in 820 or 1 part in 760. Thus the uncertainty in heat of combustion from each measurement is

$$\frac{4220}{235} = 18\text{ kJ};\quad \frac{4220}{265} = 16\text{ kJ};\quad \frac{4220}{820} = 5\text{ kJ};\quad \frac{4220}{760} = 6\text{ kJ}$$

The sum of these uncertainties is 45 kJ. In fact, the overall uncertainty is less than this because independent errors in measurement do tend to partially cancel.

5.99 (a) For comparison, balance the equations so that 1 mole of CH_4 is burned in each.

$$CH_4(g) + O_2(g) \rightarrow C(s) + 2H_2O(l) \qquad \Delta H^\circ = -496.9\text{ kJ}$$

$$CH_4(g) + 3/2\,O_2(g) \rightarrow CO(g) + 2H_2O(l) \qquad \Delta H^\circ = -607.4\text{ kJ}$$

$$CH_4(g) + 2O_2(g) \rightarrow CO_2(g) + 2H_2O(l) \qquad \Delta H^\circ = -890.4\text{ kJ}$$

(b) $\Delta H^\circ_{rxn} = \Delta H^\circ_f\,C(s) + 2\Delta H^\circ_f\,H_2O(l) - \Delta H^\circ_f\,CH_4(g) - \Delta H^\circ_f\,O_2(g)$

$= 0 + 2(-285.83\text{ kJ}) - (-74.8) - 0 = -496.9\text{ kJ}$

$\Delta H^\circ_{rxn} = \Delta H^\circ_f\,CO(g) + 2\Delta H^\circ_f\,H_2O(l) - \Delta H^\circ_f\,CH_4(g) - 3/2\,\Delta H^\circ_f\,O_2(g)$

$= (-110.5\text{ kJ}) + 2(-285.83\text{ kJ}) - (-74.8\text{ kJ}) - 3/2(0) = -607.4\text{ kJ}$

$\Delta H^\circ_{rxn} = \Delta H^\circ_f\,CO_2(g) + 2\Delta H^\circ_f\,H_2O(l) - \Delta H^\circ_f\,CH_4(g) - 2\Delta H^\circ_f\,O_2(g)$

$= -393.5\text{ kJ} + 2(-285.83\text{ kJ}) - (-74.8\text{ kJ}) - 2(0) = -890.4\text{ kJ}$

(c) Assuming that $O_2(g)$ is present in excess, the reaction that produces $CO_2(g)$ represents the most negative ΔH per mole of CH_4 burned. More of the potential energy of the reactants is released as heat during the reaction to give products of lower potential energy. The reaction that produces $CO_2(g)$ is the most "downhill" in enthalpy.

5.100 For nitroethane:

$$\frac{1368\text{ kJ}}{1\text{ mol }C_2H_5NO_2} \times \frac{1\text{ mol }C_2H_5NO_2}{75.072\text{ g }C_2H_5NO_2} \times \frac{1.052\text{ g }C_2H_5NO_2}{1\text{ cm}^3} = 19.17\text{ kJ/cm}^3$$

For ethanol:

$$\frac{1367\text{ kJ}}{1\text{ mol }C_2H_5OH} \times \frac{1\text{ mol }C_2H_5OH}{46.069\text{ g }C_2H_5OH} \times \frac{0.789\text{ g }C_2H_5OH}{1\text{ cm}^3} = 23.4\text{ kJ/cm}^3$$

For methylhydrazine:

$$\frac{1305\text{ kJ}}{1\text{ mol }CH_6N_2} \times \frac{1\text{ mol }CH_6N_2}{46.072\text{ g }CH_6N_2} \times \frac{0.874\text{ g }CH_6N_2}{1\text{ cm}^3} = 24.8\text{ kJ/cm}^3$$

Thus, **methylhydrazine** would provide the most energy per unit volume, with ethanol a close second.

5.102 The reaction for which we want ΔH is:

$$4NH_3(l) + 3O_2(g) \rightarrow 2N_2(g) + 6H_2O(g)$$

Before we can calculate ΔH for this reaction, we must calculate ΔH_f for $NH_3(l)$.

We know that ΔH_f for $NH_3(g)$ is -46.2 kJ, and that for $NH_3(l) \rightarrow NH_3(g)$, ΔH = 23.2 kJ

Thus, $\Delta H_{vap} = \Delta H_f\ NH_3(g) - \Delta H_f\ NH_3(l)$.

$$23.2\text{ kJ} = -46.2\text{ kJ} - \Delta H_f\ NH_3(l);\ \Delta H_f\ NH_3(l) = -69.4\text{ kJ/mol}$$

Then for the overall reaction, the enthalpy change is:

$$\Delta H_{rxn} = 6\Delta H_f\ H_2O(g) + 2\Delta H_f\ N_2(g) - 4\Delta H_f\ NH_3(l) - 3\Delta H_f\ O_2$$
$$= 6(-241.82\text{ kJ}) + 2(0) - 4(-69.4\text{ kJ}) - 3(0) = -1173.3\text{ kJ}$$

$$\frac{-1173.3\text{ kJ}}{4\text{ mol }NH_3} \times \frac{1\text{ mol }NH_3}{17.0\text{ g }NH_3} = \frac{0.81\text{ g }NH_3}{1\text{ cm}^3} \times \frac{1000\text{ cm}^3}{1\text{ L}} = \frac{1.4\times10^4\text{ kJ}}{\text{L }NH_3}$$

(This result has two significant figures because the density is expressed to two figures.)

$$2CH_3OH(l) + 3O_2(g) \rightarrow 2CO_2(g) + 4H_2O(g)$$

$$\Delta H = 2(-393.5\text{ kJ}) + 4(-241.82\text{ kJ}) - 2(-239\text{ kJ}) - 3(0) = -1276\text{ kJ}$$

$$\frac{-1276\text{ kJ}}{2\text{ mol }CH_3OH} \times \frac{1\text{ mol }CH_3OH}{32.04\text{ g }CH_3OH} \times \frac{0.792\text{ g }CH_3OH}{1\text{ cm}^3} \times \frac{1000\text{ cm}^3}{1\text{ L}} = \frac{1.58\times10^4\text{ kJ}}{\text{L }CH_3OH}$$

In terms of heat obtained per unit volume of fuel, methanol is a slightly better fuel than liquid ammonia.

5.103 **1,3-butadiene,** C_4H_6, / = 54.092 g/mol

(a) $C_4H_6(g) + 11/2\ O_2(g) \rightarrow 4CO_2(g) + 3H_2O(l)$

$$\Delta H^\circ_{rxn} = 4\Delta H^\circ_f\ CO_2(g) + 3\Delta H^\circ_f\ H_2O(l) - \Delta H^\circ_f\ C_4H_6(g) - 11/2\ \Delta H^\circ_f\ O_2(g)$$
$$= 4(-393.5\text{ kJ}) + 3(-285.83\text{ kJ}) - 111.9\text{ kJ} + 11/2\ (0) = -2543.4\text{ kJ/mol }C_4H_6$$

(b) $\dfrac{-2543.4\text{ kJ}}{1\text{ mol }C_4H_6} \times \dfrac{1\text{ mol }C_4H_6}{54.092\text{ g}} = 47.020 \rightarrow 47\text{ kJ/g}$

(c) $\%H = \dfrac{6(1.008)}{54.092} \times 100 = 11.18\%\ H$

1-butene, C_4H_8, / = 56.108 g/mol

(a) $C_4H_8(g) + 6O_2(g) \rightarrow 4CO_2(g) + 4H_2O(l)$

$$\Delta H^\circ_{rxn} = 4\Delta H^\circ_f\ CO_2(g) + 4\Delta H^\circ_f\ H_2O(l) - \Delta H^\circ_f\ C_4H_8(g) - 6\Delta H^\circ_f\ O_2(g)$$
$$= 4(-393.5\text{ kJ}) + 4(-285.83\text{ kJ}) - 1.2\text{ kJ} - 6(0) = -2718.5\text{ kJ/mol }C_4H_8$$

(b) $\dfrac{-2718.5\text{ kJ}}{1\text{ mol }C_4H_8} \times \dfrac{1\text{ mol }C_4H_8}{56.108\text{ g }C_4H_8} = 48.451 \rightarrow 48\text{ kJ/g}$

(c) $\%H = \dfrac{8(1.008)}{56.108} \times 100 = 14.37\%\ H$

n-butane, $C_4H_{10}(g)$, / = 58.124 g/mol

(a) $C_4H_{10}(g) + 13/2\ O_2(g) \rightarrow 4CO_2(g) + 5H_2O(l)$

$$\Delta H^\circ_{rxn} = 4\Delta H^\circ_f\ CO_2(g) + 5\Delta H^\circ_f\ H_2O(l) - \Delta H^\circ_f\ C_4H_{10}(g) - 13/2\ \Delta H^\circ_f\ O_2(g)$$

$$= 4(-393.5\text{ kJ}) + 5(-285.83\text{ kJ}) - (-124.7\text{ kJ}) - 13/2(0) = -2878.5\text{ kJ/mol } C_4H_{10}$$

(b) $$\frac{-2878.5\text{ kJ}}{1\text{ mol } C_4H_{10}} \times \frac{1\text{ mol } C_4H_{10}}{58.124\text{ g } C_4H_{10}} = 49.523 \rightarrow 50\text{ kJ/g}$$

(c) $$\%H = \frac{10(1.008)}{58.124} \times 100 = 17.34\%\ H$$

(d) It is certainly true that as the mass % H increases, the fuel value (kJ/g) of the hydrocarbon increases, given the same number of C atoms. A graph of the data in parts (b) and (c) (see below) suggests that mass % H and fuel value are directly proportional when the number of C atoms is constant.

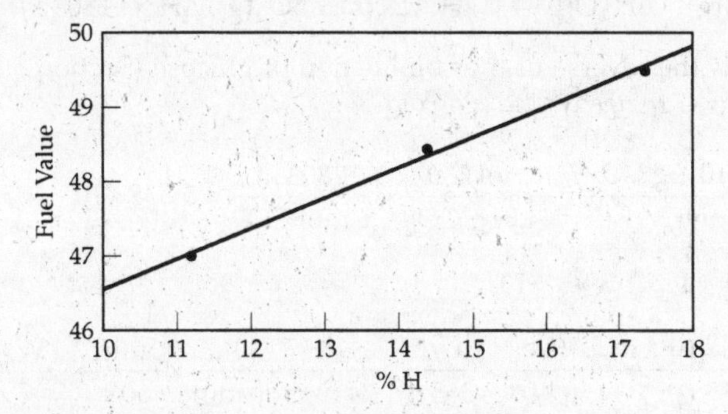

5.104 (a) $C_6H_{12}O_6(s) + 6O_2(g) \rightarrow 6CO_2(g) + 6H_2O(l)$

$$\Delta H^\circ_{rxn} = 6\Delta H^\circ_f\ CO_2(g) + 6\Delta H^\circ_f\ H_2O(l) - \Delta H^\circ_f\ C_6H_{12}O_6(s) - 6\Delta H^\circ_f\ O_2(g)$$

$$= 6(-393.5\text{ kJ}) + 6(-285.83\text{ kJ}) - (-1273\text{ kJ}) - 6(0)$$

$$= -2803\text{ kJ/mol } C_6H_{12}O_6$$

$C_{12}H_{22}O_{11}(s) + 12O_2(g) \rightarrow 12CO_2(g) + 11H_2O(l)$

$$\Delta H^\circ_{rxn} = 12\Delta H^\circ_f\ CO_2(g) + 11\Delta H^\circ_f\ H_2O(l) - \Delta H^\circ_f\ C_{12}H_{22}O_{11}(s) - 12\Delta H^\circ_f\ O_2(g)$$

$$= 12(-393.5\text{ kJ}) + 11(-285.83\text{ kJ}) - (-2221\text{ kJ}) - 12(0)$$

$$= -5645\text{ kJ/mol } C_{12}H_{22}O_{11}$$

(b) $$\frac{-2803\text{ kJ}}{1\text{ mol } C_6H_{12}O_6} \times \frac{1\text{ mol } C_6H_{12}O_6}{180.2\text{ g } C_6H_{12}O_6} = -\frac{15.55\text{ kJ}}{1\text{ g } C_6H_{12}O_6} \rightarrow 16\text{ kJ/g } C_6H_{12}O_6 \text{ (fuel value)}$$

$$\frac{-5645\text{ kJ}}{1\text{ mol } C_{12}H_{22}O_{11}} \times \frac{1\text{ mol } C_{12}H_{22}O_{11}}{342.3\text{ g } C_{12}H_{22}O_{11}} = -\frac{16.49\text{ kJ}}{1\text{ g } C_{12}H_{22}O_{11}} \rightarrow 16\text{ kJ/g } C_{12}H_{22}O_{11}$$

(fuel value)

(c) The average fuel value of carbohydrates (Section 5.8) is 17 kJ/g. These two carbohydrates have fuel values (16 kJ/g), slightly lower but in line with this average. (More complex carbohydrates supply more energy and raise the average value.)

5.106 *Plan.* Use dimensional analysis to calculate the amount of solar energy supplied per m^2 in 1 hr. Use stoichiometry to calculate the amount of plant energy used to produce sucrose per m^2 in 1 hr. Calculate the ratio of energy for sucrose to total solar energy, per m^2 per hr.

Solve. 1 W = 1 J/s, 1 kW = 1 kJ/s

$$\frac{1.0\text{ kW}}{\text{m}^2} = \frac{1.0\text{ kJ/s}}{\text{m}^2} = \frac{1.0\text{ kJ}}{\text{m}^2\bullet\text{s}} \times \frac{60\text{ s}}{1\text{ min}} \times \frac{60\text{ min}}{1\text{ hr}} = \frac{3.6\times10^3\text{ kJ}}{\text{m}^2\bullet\text{hr}}$$

$$\frac{5645\text{ kJ}}{\text{mol sucrose}} \times \frac{1\text{ mol sucrose}}{342.3\text{ g sucrose}} \times \frac{0.20\text{ g sucrose}}{\text{m}^2\bullet\text{hr}} = 3.298 = 3.3\text{ kJ/m}^2\bullet\text{hr}$$

for sucrose production

$$\frac{3.298\text{ kJ for sucrose}}{3.6\times10^3\text{ kJ total solar}} \times 100 = 0.092\%\text{ sulight used to produce sucrose}$$

5.107 (a) $6CO_2(g) + 6H_2O(l) \rightarrow C_6H_{12}O_6(s) + 6O_2(g)$, $\Delta H° = 2803$ kJ

This is the reverse of the combustion of glucose (Section 5.8 and Solution 5.104), so $\Delta H° = -(-2803)$ kJ = +2803 kJ.

$$\frac{5.5\times10^{16}\text{ g CO}_2}{\text{yr}} \times \frac{1\text{ mol CO}_2}{44.01\text{ g CO}_2} \times \frac{2803\text{ kJ}}{6\text{ mol CO}_2} = 5.838\times10^{17} = 5.8\times10^{17}\text{ kJ}$$

(b) 1 W = 1 J/s; 1 W • s = 1 J

$$\frac{5.838\times10^{17}\text{ kJ}}{\text{yr}} \times \frac{1000\text{ J}}{\text{kJ}} \times \frac{1\text{ yr}}{365\text{ d}} \times \frac{1\text{ d}}{24\text{ hr}} \times \frac{1\text{ hr}}{60\text{ min}} \times \frac{1\text{min}}{60\text{ s}} \times \frac{1\text{ W}\bullet\text{s}}{\text{J}}$$

$$\times \frac{1\text{ MW}}{1\times10^5\text{ W}} = 1.851\times10^7\text{ MW} = 1.9\times10^7\text{ MW}$$

$$1.9\times10^7\text{ MW} \times \frac{1\text{ plant}}{10^3\text{ MW}} = 1.9\times10^4 = 19{,}000\text{ nuclear power plants}$$

Integrative Exercises

5.109 The situation in Figure 4.3 is a more complex version of that pictured in Exercise 5.28 and discussed in Solution 5.28. An ionic solid is an orderly arrangement of closely spaced ions, oppositely charged particles. The potential energy of any pair of these ions is described as $E_{el} = \kappa\, Q_1Q_2/r^2$. Separating these oppositely charged particles leads to an increase in the energy of the system, + ΔE. Work is done to the NaCl by the water molecules. Since both NaCl and water are part of the system, the net amount of work is zero. Since $\Delta E = q + w$, $\Delta E = q$ and both are positive. The dissolving process typically takes place at constant atmospheric pressure, so $\Delta H = q$ and ΔH is also positive.

To verify this conclusion, carry out the dissolution of NaCl in a constant pressure calorimeter. Begin with 1 L of H_2O, record the temperature, add 0.1 mol NaCl, and dissolve completely; record the final temperature. If ΔH is positive, the temperature will increase.

5.110 (a),(b) $Ag^+(aq) + Li(s) \rightarrow Ag(s) + Li^+(aq)$

$\Delta H^\circ = \Delta H_f^\circ\ Li^+(aq) - \Delta H_f^\circ\ Ag^+(aq)$

$= -278.5\ kJ - 105.90\ kJ = -384.4\ kJ$

$Fe(s) + 2Na^+(aq) \rightarrow Fe^{2+}(aq) + 2Na(s)$

$\Delta H^\circ = \Delta H_f^\circ\ Fe^{2+}(aq) - 2\Delta H_f^\circ\ Na^+(aq)$

$= -87.86\ kJ - 2(-240.1\ kJ) = +392.3\ kJ$

$2K(s) + 2H_2O(l) \rightarrow 2KOH(aq) + H_2(g)$

$\Delta H^\circ = 2\Delta H_f^\circ\ KOH(aq) - 2\Delta H_f^\circ\ H_2O(l)$

$= 2(-482.4\ kJ) - 2(-285.83\ kJ) = -393.1\ kJ$

(c) Exothermic reactions are more likely to be favorable, so the first and third reactions should be favorable and the second reaction should be unfavorable.

(d) In the activity series of metals, Table 4.5, any metal can be oxidized by the cation of a metal below it on the table.

Ag^+ is below Li, so the first reaction will occur.

Na^+ is above Fe, so the second reaction will not occur.

H^+ (formally in H_2O) is below K, so the third reaction will occur.

These predictions agree with those in part (c).

5.112 (a) mol Cu = $M \times L = 1.00\ M \times 0.0500\ L = 0.0500$ mol

g = mol × / = $0.0500 \times 63.546 = 3.1773 = 3.18$ g Cu

(b) The precipitate is copper(II) hydroxide, $Cu(OH)_2$.

(c) $CuSO_4(aq) + 2KOH(aq) \rightarrow Cu(OH)_2(s) + K_2SO_4(aq)$, complete

$Cu^{2+}(aq) + 2OH^-(aq) \rightarrow Cu(OH)_2(s)$, net ionic

(d) The temperature of the calorimeter rises, so the reaction is exothermic and the sign of q is negative.

$$q = -6.2^\circ C \times 100\ g \times \frac{4.184\ J}{1\ g{\bullet}^\circ C} = -2.6 \times 10^3\ J = -2.6\ kJ$$

The reaction as carried out involves only 0.050 mol of $CuSO_4$ and the stoichiometrically equivalent amount of KOH. On a molar basis,

$$\Delta H = \frac{-2.6\ kJ}{0.050\ mol} = -52\ kJ \text{ for the reaction as written in part (c)}$$

5.113 (a) $AgNO_3(aq) + NaCl(aq) \rightarrow NaNO_3(aq) + AgCl(s)$

net ionic equation: $Ag^+(aq) + Cl^-(aq) \rightarrow AgCl(s)$

$\Delta H^\circ = \Delta H_f^\circ\ AgCl(s) - \Delta H_f^\circ\ Ag^+(aq) - \Delta H_f^\circ Cl^-(aq)$

$\Delta H^\circ = -127.0\ kJ - (105.90\ kJ) - (-167.2\ kJ) = -65.7\ kJ$

(b) $\Delta H°$ for the complete molecular equation will be the same as $\Delta H°$ for the net ionic equation. $Na^+(aq)$ and $NO_3^-(aq)$ are spectator ions; they appear on both sides of the chemical equation. Since the overall enthalpy change is the enthalpy of the products minus the enthalpy of the reactants, the contributions of the spectator ions cancel.

(c) $\Delta H° = \Delta H_f^\circ\ NaNO_3(aq) + \Delta H_f^\circ\ AgCl(s) - \Delta H_f^\circ\ AgNO_3(aq) - \Delta H_f^\circ\ NaCl(aq)$

$\Delta H_f^\circ\ AgNO_3(aq) = \Delta H_f^\circ\ NaNO_3(aq) + \Delta H_f^\circ\ AgCl(s) - \Delta H_f^\circ\ NaCl(aq) - \Delta H°$

$\Delta H_f^\circ\ AgNO_3(aq) = -446.2\ kJ + (-127.0\ kJ) - (-407.1\ kJ) - (-65.7\ kJ)$

$\Delta H_f^\circ\ AgNO_3(aq) = -100.4\ kJ/mol$

5.114 (a) $21.83\ g\ CO_2 \times \frac{1\ mol\ CO_2}{44.01\ g\ CO_2} \times \frac{1\ mol\ C}{1\ mol\ CO_2} \times \frac{12.01\ g\ C}{1\ mol\ C} = 5.9572 = 5.957\ g\ C$

$4.47\ g\ H_2O \times \frac{1\ mol\ H_2O}{18.02\ g\ H_2O} \times \frac{2\ mol\ H}{1\ mol\ H_2O} \times \frac{1.008\ g\ H}{mol\ H} = 0.5001 = 0.500\ g\ H$

The sample mass is (5.9572 + 0.5001) = 6.457 g

(b) $5.957\ g\ C \times \frac{1\ mol\ C}{12.01\ g\ C} = 0.4960\ mol\ C;\ 0.4960/0.496 = 1$

$0.500\ g\ H \times \frac{1\ mol\ H}{1.008\ g\ H} = 0.496\ mol\ H;\ 0.496/0.496 = 1$

The empirical formula of the hydrocarbon is CH.

(c) Calculate the ΔH_f° for 6.457 g of the sample.

$6.457\ g\ sample + O_2(g) \rightarrow 21.83\ g\ CO_2(g) + 4.47\ g\ H_2O(g),\ \Delta H° = -311\ kJ$

$\Delta H_{comb}^\circ = \Delta H_f^\circ\ CO_2(g) + \Delta H_f^\circ\ H_2O(g) - \Delta H_f^\circ\ sample - \Delta H_f^\circ\ O_2(g)$

$\Delta H_f^\circ\ sample = \Delta H_f^\circ\ CO_2(g) + \Delta H_f^\circ\ H_2O(g) - \Delta H_{comb}^\circ$

$\Delta H_f^\circ\ CO_2(g) = 21.83\ g\ CO_2 \times \frac{1\ mol\ CO_2}{44.01\ g\ CO_2} \times \frac{-393.5\ kJ}{mol\ CO_2} = -195.185 = -195.2\ kJ$

$\Delta H_f^\circ\ H_2O(g) = 4.47\ g\ H_2O \times \frac{1\ mol\ H_2O}{18.02\ g\ H_2O} \times \frac{-241.82\ kJ}{mol\ H_2O} = 59.985 = -60.0\ kJ$

$\Delta H_f^\circ\ sample = -195.185\ kJ - 59.985\ kJ - (-311\ kJ) = 55.83 = 56\ kJ$

$\frac{55.83\ kJ}{6.457\ g\ sample} \times \frac{13.02\ g}{CH\ unit} = 112.6 = 1.1 \times 10^2\ kJ/CH\ unit$

(d) The hydrocarbons in Appendix C with empirical formula CH are C_2H_2 and C_6H_6.

substance	ΔH_f°/mol	ΔH_f°/CH unit
$C_2H_2(g)$	226.7 kJ	113.4 kJ
$C_6H_6(g)$	82.9 kJ	13.8 kJ
$C_6H_6(l)$	49.0 kJ	8.17 kJ
sample		1.1×10^2 kJ

The calculated value of ΔH_f°/ CH unit for the sample is a good match with acetylene, $C_2H_2(g)$.

5.115 (a) $CH_4(g) \rightarrow C(g) + 4H(g)$ (i) reaction given

$CH_4(g) \rightarrow C(s) + 2H_2(g)$ (ii) reverse of formation

The differences are: the state of C in the products; the chemical form, atoms, or diatomic molecules, of H in the products.

(b) i. $\Delta H^\circ = \Delta H_f^\circ\, C(g) + 4\Delta H_f^\circ\, H(g) - \Delta H_f^\circ\, CH_4(g)$

$= 718.4\ \text{kJ} + 4(217.94)\ \text{kJ} - (-74.8)\ \text{kJ} = 1665.0\ \text{kJ}$

ii. $\Delta H^\circ = \Delta H_f^\circ\, CH_4 = -(-74.8)\ \text{kJ} = 74.8\ \text{kJ}$

The rather large difference in ΔH° values is due to the enthalpy difference between isolated gaseous C atoms; the orderly, bonded array of C atoms in graphite, C(s); and the enthalpy difference between isolated H atoms and H_2 molecules. In other words, it is due to the difference in the enthalpy stored in chemical bonds in C(s) and $H_2(g)$ versus the corresponding isolated atoms.

(c) $CH_4(g) + 4F_2(g) \rightarrow CF_4(g) + 4HF(g)$ $\Delta H^\circ = -1679.5$ kJ

The ΔH° value for this reaction was calculated in Solution 5.86.

$$3.45\ \text{g}\ CH_4 \times \frac{1\ \text{mol}\ CH_4}{16.04\ \text{g}\ CH_4} \times 0.21509 = 0.215\ \text{mol}\ CH_4$$

$$1.22\ \text{g}\ F_2 \times \frac{1\ \text{mol}\ F_2}{38.00\ \text{g}\ F_2} = 0.03211 = 0.0321\ \text{mol}\ F_2$$

There are fewer mol F_2 than CH_4, but 4 mol F_2 are required for every 1 mol of CH_4 reacted, so clearly F_2 is the limiting reactant.

$$0.03211\ \text{mol}\ F_2 \times \frac{-1679.5\ \text{kJ}}{4\ \text{mol}\ F_2} = -13.48 = -13.5\ \text{kJ heat evolved}$$

6 Electronic Structure of Atoms

Electronic Structure of Atoms

6.1 (a) Speed is distance traveled per unit time. Measure the distance between the point where the stone is dropped and a second reference point, possibly the shore line. Using a stop watch, measure the elapsed time between when the stone is dropped and when the first wave reaches the second reference point. Be sure to drop the stone exactly at the first reference point. Find the ratio of distance to time.

(b) Measure the distance between two wave crests (or troughs or any analogous points on two adjacent waves). Better yet, measure the distance between two crests (or analogous points) that are several waves apart and divide by the number of waves that separate them.

(c) One way to make this distance measurement is to use a standard. Take a photo of the event and include an object of known dimensions in the picture. Measure the distances described above on the picture. Obtain a conversion factor to actual distances by measuring the object on the picture and comparing the photo dimensions to the true dimensions. Since speed is distance/time, and wavelength is distance, we can calculate frequency by dividing speed by wavelength, $\nu = c/\lambda$.

(d) We can measure frequency of the wave by dropping a cork in the water and counting the number of times per second it moves through a complete cycle of motion.

6.3 (a) The glowing stove burner is an example of black body radiation, the observational basis for Planck's quantum theory. The wavelengths emitted are related to temperature, with cooler temperatures emitting longer wavelengths and hotter temperatures emitting shorter wavelengths. At the hottest setting, the burner emits orange visible light. At the cooler low setting, the burner emits longer wavelengths out of the visible region, and the burner appears black.

(b) If the burner had a super high setting, the emitted wavelengths would be shorter than those of orange light and the glow color would be more blue. (See Figure 6.4 for color variation with wavelength.)

6.4 (a) Increase. The rainbow has shorter wavelength blue light on the inside and longer wavelength red light on the outside. (See Figure 6.4.)

(b) Decrease. Wavelength and frequency are inversely related. Wavelength increases so frequency decreases going from the inside to the outside of the rainbow.

(c) The light from the hydrogen discharge take is not a continuous spectrum, so not all visible wavelengths will be in our "hydrogen discharge rainbow." Starting with the shortest wavelengths, it will be violet followed by blue-violet and blue-green on the inside. Then there will be a gap, and finally a red band. (See the H spectrum in Figure 6.12.)

6.6 (a) $\psi^2(x)$ will be positive or zero at all values of x, and have two maxima with larger magnitudes than the maximum in $\psi(x)$.

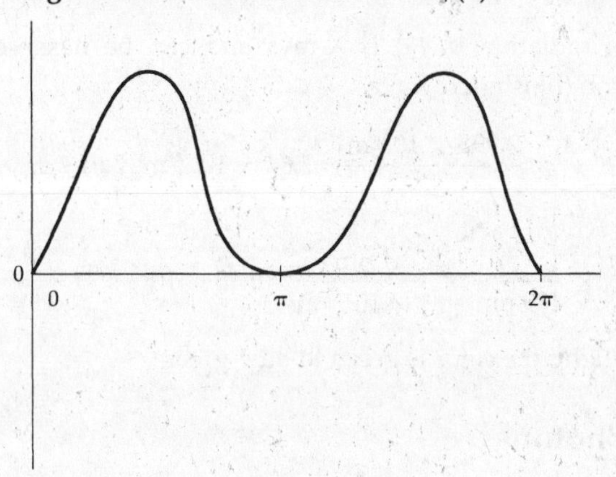

(b) The greatest probability of finding the electron is at the two maxima in $\psi^2(x)$ at $x = \pi/2$ and $3\pi/2$.

(c) There is zero probability of finding the electron at $x = \pi$. This value is called a node.

6.7 (a) 1

(b) p (dumbbell shape, node at the nucleus)

(c) The lobes in the contour representation would extend farther along the y axis. A larger principle quantum number (4p vs. 3p) implies a greater average distance from the nucleus for electrons occupying the orbital.

Radiant Energy

6.10 (a) Wavelength (λ) and frequency (ν) are inversely proportional; the proportionality constant is the speed of light (c). $\nu = c/\lambda$.

(b) Light in the 210–230 nm range is in the ultraviolet region of the spectrum. These wavelengths are slightly shorter than the 400 nm short-wavelength boundary of the visible region.

6.12 (a) False. Electromagnetic radiation passes through water. The fact that you can see objects through a glass of water should make this clear.

(b) True.

(c) False. Infrared light has lower frequencies than visible light.

(d) False. A foghorn blast is a form of sound waves, which are not accompanied by oscillating electric and magnetic fields.

6.14 Wavelength of (a) gamma rays < (d) yellow (visible) light < (e) red (visible) light < (b) 93.1 MHz FM (radio) waves < (c) 680 kHz or 0.680 MHz AM (radio) waves

6.16 (a) $\nu = c/\lambda; \; \frac{2.998 \times 10^8 \text{ m}}{\text{s}} \times \frac{1}{10.0 \text{ Å}} \times \frac{1 \text{ Å}}{1 \times 10^{-10} \text{ m}} = 3.00 \times 10^{17} \text{ s}^{-1}$

(b) $\lambda = c/\nu; \; \frac{2.998 \times 10^8 \text{ m}}{\text{s}} \times \frac{1 \text{ s}}{7.6 \times 10^{10}} = 3.94 \times 10^{-3} \text{ m}$

(c) The 1×10^{-9} m radiation in (a) is X-rays and can be observed by an X-ray detector. Radiation (b) is microwave.

(d) $25.5 \text{ fs} \times \frac{1 \times 10^{-15} \text{ s}}{1 \text{ fs}} \times \frac{2.998 \times 10^8 \text{ m}}{\text{s}} = 7.64 \times 10^{-6} \text{ m } (7.64 \text{ μm})$

6.18 $\nu = c/\lambda; \; \frac{2.998 \times 10^8 \text{ m}}{1 \text{ s}} \times \frac{1}{489 \text{ nm}} \times \frac{1 \text{ nm}}{1 \times 10^{-9} \text{ m}} = 6.13 \times 10^{14} \text{ s}^{-1}$

The laser emits visible light; the color is green to blue-green.

Quantized Energy and Photons

6.20 Planck's original hypothesis was that energy could only be gained or lost in discreet amounts (quanta) with a certain minimum size. The size of the minimum energy change is related to the frequency of the radiation absorbed or emitted, $\Delta E = h\nu$, and energy changes occur only in multiples of $h\nu$.

Einstein postulated that light itself is quantized, that the minimum energy of a photon (a quantum of light) is directly proportional to its frequency, $E = h\nu$. If a photon that strikes a metal surface has less than the threshold energy, no electron is emitted from the surface. If the photon has energy equal to or greater than the threshold energy, an electron is emitted and any excess energy becomes the kinetic energy of the electron.

6.22 (a) $E = hc/\lambda = 6.626 \times 10^{-34} \text{ J} \bullet \text{s} \times \frac{2.998 \times 10^8 \text{ m}}{1 \text{ s}} \times \frac{1}{10.8 \text{ mm}} \times \frac{1 \text{ mm}}{1 \times 10^{-3} \text{ m}}$

$= 1.84 \times 10^{-23} \text{ J}$

(b) $E = h\nu = 6.626 \times 10^{-34} \text{ J} \bullet \text{s} \times \frac{101.1 \times 10^6}{1 \text{ s}} = 6.699 \times 10^{-26} \text{ J}$

(c) The relationship $\nu = E/h$ requires energy in J/photon. Change kJ/mol to J/photon and divide by h.

$$\frac{24.7 \text{ kJ}}{\text{mol}} \times \frac{1 \times 10^3 \text{ J}}{\text{kJ}} \times \frac{1 \text{ mol}}{6.022 \times 10^{23} \text{ photons}} \times \frac{1}{6.626 \times 10^{-34} \text{ J} \bullet \text{s}} = 6.19 \times 10^{13} \text{ s}^{-1}$$

This radiation is in the infrared region.

6.24 $E = h\nu$

$\text{AM}: 6.626 \times 10^{-34} \text{ J} \bullet \text{s} \times \frac{1010 \times 10^3}{1 \text{ s}} = 6.69 \times 10^{-28} \text{ J}$

$$FM: 6.626 \times 10^{-34} \text{ J}\bullet\text{s} \times \frac{98.3 \times 10^6}{1 \text{ s}} = 6.51 \times 10^{-26} \text{ J}$$

The FM photon has about 100 times more energy than the AM photon.

6.26

$$\frac{941 \times 10^3 \text{ J}}{\text{mol } O_2} \times \frac{1 \text{ mol}}{6.022 \times 10^{23} \text{ photons}} = 1.563 \times 10^{-18} = 1.56 \times 10^{-18} \text{ J/photon}$$

$$\lambda = hc/E = \frac{6.626 \times 10^{-34} \text{ J}\bullet\text{s}}{1.563 \times 10^{-18} \text{ J}} \times \frac{2.998 \times 10^8 \text{ m}}{1 \text{ s}} = 1.27 \times 10^{-7} \text{ m} = 127 \text{ nm}$$

According to Figure 6.4, this is ultraviolet radiation.

6.28 (a) The radiation is microwave.

(b)

$$E_{photon} = hc/\lambda = \frac{6.626 \times 10^{-34} \text{ J}\bullet\text{s}}{3.55 \times 10^{-3} \text{ m}} \times \frac{2.998 \times 10^8 \text{ m}}{1 \text{ s}} = 5.5957 \times 10^{-23}$$

$$= 5.60 \times 10^{-23} \text{ J/photon}$$

$$\frac{5.5957 \times 10^{-23} \text{ J}}{1 \text{ photon}} \times \frac{3.2 \times 10^8 \text{ photons}}{1 \text{ s}} \times \frac{60 \text{ s}}{1 \text{ min}} \times \frac{60 \text{ min}}{1 \text{ hr}} = 6.4463 \times 10^{-11}$$

$$= 6.4 \times 10^{-11} \text{ J/hr}$$

6.30 (a)

$$\nu = E/h = \frac{4.41 \times 10^{-19} \text{ J}}{6.626 \times 10^{-34} \text{ J}\bullet\text{s}} = 6.6556 \times 10^{14} = 6.66 \times 10^{14} \text{ s}^{-1}$$

(b)

$$\lambda = hc/E = \frac{6.626 \times 10^{-34} \text{ J}\bullet\text{s}}{4.41 \times 10^{-19} \text{ J}} \times \frac{2.998 \times 10^8 \text{ m}}{\text{s}} = 4.50 \times 10^{-7} \text{ m} = 450 \text{ nm}$$

(c)

$$E_{439} = hc/\lambda = \frac{6.626 \times 10^{-34} \text{ J}\bullet\text{s}}{439 \times 10^{-9} \text{ m}} \times \frac{2.998 \times 10^8 \text{ m}}{\text{s}} = 4.525 \times 10^{-19} = 4.53 \times 10^{-19} \text{ J}$$

$$E_K = E_{439} - E_{min} = 4.525 \times 10^{-19} \text{ J} - 4.41 \times 10^{-19} \text{ J} = 0.115 \times 10^{-19} = 1.1 \times 10^{-20} \text{ J}$$

(d) One electron is emitted per photon. Calculate the number of 439 nm photons in 1.00 μJ. The excess energy in each photon will become the kinetic energy of the electron; it cannot be "pooled" to emit additional electrons.

$$1.00 \text{ μJ} \times \frac{1 \times 10^{-6} \text{ J}}{\text{μJ}} \times \frac{1 \text{ photon}}{4.525 \times 10^{-19} \text{ J}} \times \frac{1 e^-}{1 \text{ photon}} = 2.21 \times 10^{12} \text{ electrons}$$

Bohr's Model; Matter Waves

6.32 (a) According to Bohr theory, when hydrogen emits radiant energy, electrons are moving from a higher allowed energy state to a lower one. Since only certain energy states are allowed, only certain energy changes can occur. These allowed energy changes correspond ($\lambda = hc/\Delta E$) to the wavelengths of the lines in the emission spectrum of hydrogen.

(b) When a hydrogen atom changes from the ground state to an excited state, the single electron moves further away from the nucleus, so the atom "expands".

6.34 (a) Absorbed. (b) Emitted. (c) Absorbed.

6.36 (a) $\Delta E = -2.18 \times 10^{-18}\ J\left[\frac{1}{n_f^2} - \frac{1}{n_i^2}\right] = -2.18 \times 10^{-18}\ J\,(1/1 - 1/16) = -2.044 \times 10^{-18}$

$= -2.04 \times 10^{-18}\ J$

$$\nu = E/h = \frac{2.044 \times 10^{-18}\ J}{6.626 \times 10^{-34}\ J \bullet s} = 3.084 \times 10^{15} = 3.08 \times 10^{15}\ s^{-1}$$

$$\lambda = c/\nu = \frac{2.998 \times 10^{8}\ m}{1\ s} \times \frac{1\ s}{3.084 \times 10^{15}} = 9.72 \times 10^{-8}\ m$$

Since the sign of ΔE is negative, radiation is emitted.

(b) $\Delta E = -2.18 \times 10^{-18}\ J(1/4 - 1/25) = -4.578 \times 10^{-19} = -4.58 \times 10^{-19}\ J$

$$\nu = \frac{4.578 \times 10^{-19}\ J}{6.626 \times 10^{-34}\ J \bullet s} = 6.909 \times 10^{14} = 6.91 \times 10^{14}\ s^{-1};\ \lambda = \frac{2.998 \times 10^{8}\ m/s}{6.909 \times 10^{14}/s}$$

$\lambda = 4.34 \times 10^{-7}$ m. Visible radiation is emitted.

(c) $\Delta E = -2.18 \times 10^{-18}\ J\ (1/36 - 1/9) = 1.817 \times 10^{-19} = 1.82 \times 10^{-19}\ J$

$$\nu = \frac{1.817 \times 10^{-19}\ J}{6.626 \times 10^{-34}\ J \bullet s} = 2.742 \times 10^{14} = 2.74 \times 10^{14}\ s^{-1};\ \lambda = \frac{2.998 \times 10^{8}\ m/s}{2.742 \times 10^{14}/s}$$

$\lambda = 1.09 \times 10^{-6}$ m. Radiation is absorbed.

6.38 (a) Transitions with $n_f = 1$ have larger ΔE values and shorter wavelengths than those with $n_f = 2$. These transitions will lie in the ultraviolet region.

(b) $n_i = 2, n_f = 1;\ \lambda = hc/E = \dfrac{6.626 \times 10^{-34}\ J \bullet s \times 2.998 \times 10^{8}\ m/s}{-2.18 \times 10^{-18}\ J\,(1/1 - 1/4)} = 1.21 \times 10^{-7}\ m$

$n_i = 3, n_f = 1;\ \lambda = hc/E = \dfrac{6.626 \times 10^{-34}\ J \bullet s \times 2.998 \times 10^{8}\ m/s}{-2.18 \times 10^{-18}\ J\,(1/1 - 1/9)} = 1.03 \times 10^{-7}\ m$

$n_i = 4, n_f = 1;\ \lambda = hc/E = \dfrac{6.626 \times 10^{-34}\ J \bullet s \times 2.998 \times 10^{8}\ m/s}{-2.18 \times 10^{-18}\ J\,(1/1 - 1/16)} = 0.972 \times 10^{-7}\ m$

6.40 (a) $2626\ nm \times \dfrac{1 \times 10^{-9}\ m}{1\ nm} = 2.626 \times 10^{-6} m;$ this line is in the infrared.

(b) Absorption lines with $n_i = 1$ are in the ultraviolet and with $n_i = 2$ are in the visible. Thus, $n_i \geq 3$, but we do not know the exact value of n_i. Calculate the longest wavelength with $n_i = 3$ ($n_f = 4$). If this is less than 2626 nm, $n_i > 3$.

$$\lambda = hc/E = \frac{6.626 \times 10^{-34}\ J \bullet s \times 2.998 \times 10^{8}\ m/s}{-2.18 \times 10^{-18}\ J\,(1/16 - 1/9)} = 1.875 \times 10^{-6}\ m$$

This wavelength is shorter than 2.626×10^{-6} m, so $n_i > 3$; try $n_i = 4$ and solve for n_f as in Solution 6.39.

$$n_f = \left(\frac{1}{n_i^2} - \frac{hc}{\lambda(2.18 \times 10^{-18}\ J)}\right)^{-1/2} = \left(1/16 - \frac{6.626 \times 10^{-34}\ J \cdot s \times 2.998 \times 10^8\ m/s}{2.626 \times 10^{-6}\ m \times 2.18 \times 10^{-18}\ J}\right)^{-1/2} = 6$$

$n_f = 6, n_i = 4$

6.42 $\lambda = h/mv$; change mass to kg and velocity to m/s

$$\text{mass of muon} = 206.8 \times 9.1094 \times 10^{-28}\ g \times \frac{1\ kg}{1000\ g} = 1.8838 \times 10^{-28} = 1.88 \times 10^{-28}\ kg$$

$$\lambda = \frac{6.626 \times 10^{-34}\ kg \bullet m^2 \bullet s}{1\ s^2} \times \frac{1}{1.8838 \times 10^{-28}\ kg} \times \frac{1\ s}{8.85 \times 10^3\ m/s} = 3.97 \times 10^{-10}\ m$$

$$= 3.97\ \text{Å}$$

6.44 $m_e = 9.1094 \times 10^{-31}$ kg (back inside cover of text)

$$\lambda = \frac{6.626 \times 10^{-34}\ kg \bullet m^2 \bullet s}{1\ s^2} \times \frac{1}{9.1094 \times 10^{-31}\ kg} \times \frac{1\ s}{9.38 \times 10^6\ m} = 7.75 \times 10^{-11}\ m$$

$$7.75 \times 10^{-11}\ m \times \frac{1\ \text{Å}}{1 \times 10^{-10}\ m} = 0.775\ \text{Å}$$

Since atomic radii and interatomic distances are on the order of 1–5 Å (Section 2.3), the wavelength of this electron is comparable to the size of atoms.

6.46 $\Delta x \geq = h/4\pi m\Delta v$; use masses in kg, Δv in m/s.

(a) $$\frac{6.626 \times 10^{-34}\ J \bullet s}{4\pi(9.109 \times 10^{-31}\ kg)(0.01 \times 10^5\ m/s)} = 6 \times 10^{-8}\ m$$

(b) $$\frac{6.626 \times 10^{-34}\ J \bullet s}{4\pi(1.675 \times 10^{-27}\ kg)(0.01 \times 10^5\ m/s)} = 3 \times 10^{-11}\ m$$

(c) For particles moving with the same uncertainty in velocity, the more massive neutron has a much smaller uncertainty in position than the lighter electron. In our model of the atom, we know where the massive particles in the nucleus are located, but we cannot know the location of the electrons with any certainty, if we know their speed.

Quantum Mechanics and Atomic Orbitals

6.48 (a) The Bohr model states with 100% certainty that the electron in hydrogen can be found 0.53 Å from the nucleus. The quantum mechanical model, taking the wave nature of the electron and the uncertainty principle into account, is a statistical model that states the probability of finding the electron in certain regions around the nucleus. While 0.53 Å might be the radius with highest probability, that probability would always be less than 100%.

(b) The equations of classical physics predict the instantaneous position, direction of motion, and speed of a macroscopic particle; they do not take quantum theory or the wave nature of matter into account. For macroscopic particles, these are not significant, but for microscopic particles like electrons, they are crucial. Schrödinger's equation takes these important theories into account to produce a statistical model of electron location given a specific energy.

(c) The square of the wave function has the physical significance of an amplitude, or probability. The quantity ψ^2 at a given point in space is the probability of locating the electron within a small volume element around that point at any given instant. The total probability, that is, the sum of ψ^2 over all the space around the nucleus, must equal 1.

6.50 (a) For $n = 3$, there are 3 l values (2, 1, 0) and 9 m_l values ($l = 2$; $m_l = -2, -1, 0, 1, 2$; $l = 1$, $m_l = -1, 0, 1$; $l = 0$, $m_l = 0$).

(b) For $n = 5$, there are 5 l values (4, 3, 2, 1, 0) and 25 m_l values ($l = 4$, $m_l = -4$ to $+4$; $l = 3$, $m_l = -3$ to $+3$; $l = 2$, $m_l = -2$ to $+2$; $l = 1$, $m_l = -1$ to $+1$; $l = 0$, $= 0$).

In general, for each principal quantum number n there are n l-values and n^2 m_l-values. For each shell, there are n kinds of orbitals and n^2 total orbitals.

6.52 (a) 2, 1, 1; 2, 1, 0; 2, 1 -1 (b) 5, 2, 2; 5, 2, 1; 5, 2, 0; 5, 2, -1; 5, 2, -2

6.54 (a) Permissible, 2p. (b) Forbidden, for $l = 0$, m_l can only equal 0.

(c) Permissible, 4d. (d) Forbidden, for n = 3, the largest l value is 2.

6.56

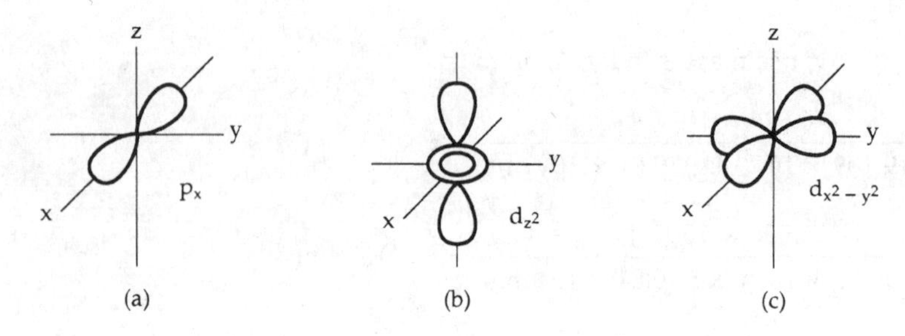

6.58 (a) In an s orbital, there are (n - 1) nodes.

(b) The $2p_x$ orbital has one node (the yz plane passing through the nucleus of the atom). The 3s orbital has two nodes.

(c) Probability density, $\psi^2(r)$, is the probability of finding an electron at a single point, r. The radial probability function, P(r), is the probability of finding an electron at any point that is distance r from the nucleus. Figure 6.18 contains plots of P(r) vs. r for 1s, 2s, and 3s orbitals. The most obvious features of these plots are the radii of maximum probability for the three orbitals, and the number and location of nodes for the three orbitals.

By comparing plots for the three orbitals, we see that as n increases, the number of nodes increases and the radius of maximum probability (orbital size) increases.

(d) 2s = 2p < 3s < 4d < 5s. In the hydrogen atom, orbitals with the same n value are degenerate and energy increases with increasing n value.

Many-Electron Atoms and Electron Configurations

6.60 (a) The electron with the greater average distance from the nucleus feels a smaller attraction for the nucleus and is higher in energy. Thus the 3p is higher in energy than 3s.

(b) Because it has a larger *n* value, a 3s electron has a greater average distance from the chlorine nucleus than a 2p electron. The 3s electron experiences a smaller attraction for the nucleus and requires less energy to remove from the chlorine atom.

6.62 (a) The Pauli exclusion principle states that no two electrons can have the same four quantum numbers.

(b) An alternate statement of the Pauli exclusion principle is that a single orbital can hold a maximum of two electrons. Thus, the Pauli principle limits the maximum number of electrons in a main shell and its subshells, which determines when a new row of the periodic table begins.

6.64 (a) 4 (b) 14 (c) 2 (d) 2

6.66 (a) "Valence electrons" are those beyond the previous noble-gas or core electron configuration.

(b) "Unpaired electrons" are electrons that occupy orbitals singly. That is, when there is only one electron in an orbital, this electron is "unpaired."

(c) A P atom has five valence electrons: $3s^23p^3$. Three of them (those in the degenerate 3p orbitals) are unpaired.

6.68 (a) Ga: $[Ar]4s^23d^{10}4p^1$, 1 unpaired electron

(b) Ca: $[Ar]4s^2$, 0 unpaired electrons

(c) V: $[Ar]4s^23d^3$, 3 unpaired electrons

(d) I: $[Kr]5s^24d^{10}5p^5$, 1 unpaired electron

(e) Y: $[Kr]5s^24d^1$, 1 unpaired electron

(f) Pt: $[Xe]6s^14f^{14}5d^9$, 2 unpaired electrons

(g) Lu: $[Xe]6s^24d^{14}5d^1$, 1 unpaired electron

6.70 (a) $[Rn]7s^25f^{14}6d^{10}7p^6$

(b) None. Element 118 would belong to Group 8A, the Noble Gases. These elements have completely filled sub-shells and orbitals.

(c) Group 8A, the Noble Gases

6.72 (a) 7A (halogens) (b) 4B (c) 3A

(d) the f-block elements Sm and Pu

6.74 Count the total number of electrons to assign the element.

(a) N: $[He]2s^22p^3$ (b) Se: $[Ar]4s^23d^{10}4p^4$ (c) Rh: $[Kr]5s^24d^7$

Additional Exercises

6.76 (a) Elements that emit in the visible: Ba (dark blue), Ca (dark blue), K (dark blue), Na (yellow/orange). (The other wavelengths are in the ultraviolet.)

(b) Au: shortest wavelength, highest energy

Na: longest wavelength, lowest energy

(c) $\lambda = c/\nu = \dfrac{2.998 \times 10^8 \text{ m/s}}{6.59 \times 10^{14}/\text{s}} \times \dfrac{1 \text{ nm}}{1 \times 10^{-9} \text{ m}} = 455 \text{ nm}$, Ba

6.78 (a) $\nu = c/\lambda = \dfrac{2.998 \times 10^8 \text{ m/s}}{320 \text{ nm}} \times \dfrac{1 \text{ nm}}{1 \times 10^{-9} \text{ m}} = 9.37 \times 10^{14} \text{ s}^{-1}$

(b) $E = hc/\lambda = \dfrac{6.626 \times 10^{-34} \text{ J} \bullet \text{s} \times 2.998 \times 10^8 \text{ m/s}}{3.20 \times 10^{-7} \text{ m}} \times \dfrac{1 \text{ kJ}}{1000 \text{ J}} \times \dfrac{6.022 \times 10^{23} \text{ photons}}{\text{mole}}$

$= 374 \text{ kJ/mol}$

(c) UV-B photons have shorter wavelength and higher energy.

(d) Yes. The higher energy UV-B photons would be more likely to cause sunburn.

6.79 $E = hc/\lambda \rightarrow$ J/photon; total energy = power × time; photons = total energy / J / photon

$E = \dfrac{6.626 \times 10^{-34} \text{ J} \bullet \text{s} \times 2.998 \times 10^8 \text{ m/s}}{780 \times 10^{-9} \text{ m}} = 2.5468 \times 10^{-19} = 2.55 \times 10^{-19} \text{ J/photon}$

$0.10 \text{ mW} = \dfrac{0.10 \times 10^{-3} \text{ J}}{1 \text{ s}} \times 69 \text{ min} \times \dfrac{60 \text{ s}}{1 \text{ min}} = 0.4140 = 0.41 \text{ J}$

$0.4140 \text{ J} \times \dfrac{1 \text{ photon}}{2.5468 \times 10^{-19} \text{ J}} = 1.626 \times 10^{18} = 1.6 \times 10^{18} \text{ photons}$

6.81 $\dfrac{2.6 \times 10^{-12} \text{ C}}{1 \text{ s}} \times \dfrac{1 \text{ e}^-}{1.602 \times 10^{-19} \text{ C}} \times \dfrac{1 \text{ photon}}{1 \text{ e}^-} = 1.623 \times 10^7 = 1.6 \times 10^7 \text{ photons/s}$

$\dfrac{E}{\text{photon}} = hc/\lambda = \dfrac{6.626 \times 10^{-34} \text{ J} \bullet \text{s}}{630 \text{ nm}} \times \dfrac{2.998 \times 10^8 \text{ m}}{1 \text{ s}} \times \dfrac{1 \text{ nm}}{1 \times 10^{-9} \text{ m}} \times \dfrac{1.623 \times 10^7 \text{ photon}}{\text{s}}$

$= 5.1 \times 10^{-12} \text{ J/s}$

6.82 (a) $\dfrac{2.00 \times 10^5 \text{ J}}{\text{mol}} \times \dfrac{1 \text{ mol photons}}{6.022 \times 10^{23} \text{ photons}} = 3.321 \times 10^{-19} = 3.32 \times 10^{-19} \text{ J/photon}$

(b) $\lambda = \dfrac{hc}{E} = \dfrac{6.626 \times 10^{-34} \text{ J} \bullet \text{s} \times 2.998 \times 10^8 \text{ m/s}}{3.321 \times 10^{-19} \text{ J}} = 5.98 \times 10^{-7} \text{ m}$

(c) 5.98×10^{-7} m = 598 nm is well within the visible portion of the electromagnetic spectrum and corresponds to yellow or yellow-orange light. Red light, with wavelengths near or greater than 700 nm, does not have sufficient energy to initiate electron transfer and darken the film.

6.83 (a) $\nu = c/\lambda;\ \frac{2.998 \times 10^8 \text{ m}}{\text{s}} \times \frac{1}{680 \text{ nm}} \times \frac{1 \text{ nm}}{1 \times 10^{-9} \text{ m}} = 4.4088 \times 10^{14} = 4.41 \times 10^{14} \text{ s}^{-1}$

(b) Calculate J/photon using E = hc/λ; change to kJ/mol.

$$E_{photon} = \frac{6.626 \times 10^{-34} \text{ J} \cdot \text{s}}{680 \times 10^{-9} \text{ m}} \times \frac{2.998 \times 10^8 \text{ m}}{\text{s}} = 2.9213 \times 10^{-19} = 2.92 \times 10^{-19} \text{ J/photon}$$

$$\frac{2.9213 \times 10^{-19} \text{ J}}{\text{photon}} \times \frac{6.022 \times 10^{23} \text{ photons}}{\text{mol}} \times \frac{1 \text{ kJ}}{1000 \text{ J}} = 175.92 = 176 \text{ kJ/mol}$$

(c) Nothing. The incoming (incident) radiation does not transfer sufficient energy to an electron to overcome the attractive forces holding the electron in the metal.

(d) For frequencies greater than ν_o, any "extra" energy not needed to remove the electron from the metal becomes the kinetic energy of the ejected electron. The kinetic energy of the electron is directly proportional to this extra energy.

(e) Let E_{total} be the total energy of an incident photon, E_{min} be the minimum energy required to eject an electron, and E_k be the "extra" energy that becomes the kinetic energy of the ejected electron.

$E_{total} = E_{min} + E_k$, $E_k = E_{total} - E_{min} = h\nu - h\nu_o$, $E_k = h(\nu - \nu_o)$. The slope of the line is the value of h, Planck's constant.

6.85 (a) Gaseous atoms of various elements in the sun's atmosphere typically have ground state electron configurations. When these atoms are exposed to radiation from the sun, the electrons change from the ground state to one of several allowed excited states. Atoms absorb the wavelengths of light which correspond to these allowed energy changes. All other wavelengths of solar radiation pass through the atmosphere unchanged. Thus, the dark lines are the wavelengths that correspond to allowed energy changes in atoms of the solar atmosphere. The continuous background is all other wavelengths of solar radiation.

(b) The scientist should record the absorption spectrum of pure neon or other elements of interest. The black lines should appear at the same wavelengths regardless of the source of neon.

6.86 (a) He^+ is hydrogen-like because it is a one-electron particle. An He atom has two electrons. The Bohr model is based on the interaction of a single electron with the nucleus, but does not accurately account for additional interactions when two or more electrons are present.

(b) Divide each energy by the smallest value to find the integer relationship.

H: $2.18 \times 10^{-18} / 2.18 \times 10^{-18} = 1;\ Z = 1$

He^+: $8.72 \times 10^{-18} / 2.18 \times 10^{-18} = 4;\ Z = 2$

Li^{2+}: $1.96 \times 10^{-17} / 2.18 \times 10^{-18} = 9;\ Z = 3$

The ground-state energies are in the ratio of 1:4:9, which is also the ratio Z^2, the square of the nuclear charge for each particle.

The ground state energy for hydrogen-like particles is:

$E = R_H Z^2$. (By definition, n = 1 for the ground state of a one-electron particle.)

(c) Z = 6 for C. $E = -2.18 \times 10^{-18}\ J\ (6)^2 = -7.85 \times 10^{-17}\ J$

6.88 *Plan.* Change keV to J/electron. Calculate v from kinetic energy. $\lambda = h/mv$. *Solve.*

$$18.6\ keV \times \frac{1000\ eV}{keV} \times \frac{96.485\ kJ}{1\ eV \bullet mol} \times \frac{1000\ J}{1\ kJ} \times \frac{1\ mol}{6.022 \times 10^{23}\ electrons}$$

$$= 2.980 \times 10^{-15} = 2.98 \times 10^{-15}\ J/electron$$

$$E_k = mv^2/2;\ v^2 = 2E_k/m;\ v = \sqrt{2E_k/m}$$

$$v = \left(\frac{2 \times 2.980 \times 10^{-15}\ kg \bullet m^2/s^2}{9.1094 \times 10^{-31}\ kg}\right)^{1/2} = 8.089 \times 10^7 = 8.09 \times 10^7\ m/s$$

$$\lambda = h/mv = \frac{6.626 \times 10^{-34}\ J \bullet s}{9.1094 \times 10^{-31}\ kg \times 8.089 \times 10^7\ m/s} \times \frac{1\ kg \bullet m^2/s^2}{1\ J} = 8.99 \times 10^{-12}\ m = 8.99\ pm$$

6.89 An *orbit* is an exactly circular path with specified radius for an electron in an allowed energy state. Each allowed orbit is denoted by a principle quantum number n. An *orbital* is a region of space where there is a high probability of finding an electron in an allowed energy state. Orbitals are three-dimensional volumes with characteristic size, shape, and orientation and are denoted by the three quantum numbers n (size), l (shape), and m_l (orientation). An orbital is a statistical prediction while an orbit is a "known" location. The notion of a known orbit violates the *uncertainty principle* (see Solution 6.47).

6.91 (a) Probability density, $[\psi(r)]^2$, is the probability of finding an electron at a single point at distance r from the nucleus. The radial probability function, $4\pi r^2$, is the probability of finding an electron at any point on the sphere defined by radius r. $P(r) = 4\pi r^2[\psi(r)]^2$.

(b) The term $4\pi r^2$ explains the differences in plots of the two functions. Plots of the probability density, $[\psi(r)^2]$ for s orbitals shown in Figure 6.21 each have their maximum value at r = 0, with (n – 1) smaller maxima at greater values of r. The plots of radial probability, P(r), for the same s orbitals shown in Figure 6.18 have values of zero at r = 0 and the size of the maxima increases. P(r) is the product of $[\psi(r)]^2$ and $4\pi r^2$. At r = 0, the value of $[\psi(r)]^2$ is finite and large, but the value of $4\pi r^2$ is zero, so the value of P(r) is zero. As r increases, the values of $[\psi(r)]^2$ vary as shown in Figure 6.21, but the values of $4\pi r^2$ increase continuously, leading to the increasing size of P(r) maxima as r increases.

(c)

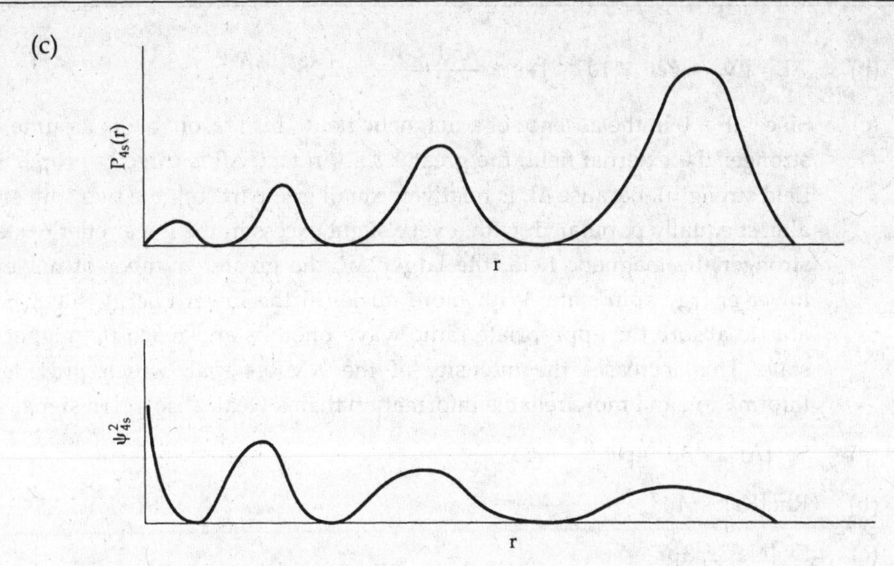

6.92 What the noble gas elements have in common are completed ns and np subshells. Since the Pauli principle limits the number of electrons per orbital to two, this leads to the first three magic numbers, $2(1s^2)$, $10(1s^22s^22p^6)$, and $18(1s^22s^22p^63s^23p^6)$. In the fourth row, (n – 1) d orbitals begin to fill as their energy falls below that of the np orbitals. This leads to the next two magic numbers, $36(1s^22s^22p^63s^23p^64s^23d^{10}4p^6)$ and $54(1s^22s^22p^63s^23p^64s^23d^{10}4p^65s^24d^{10}5p^6)$. In the sixth row, the energy of the 4f orbitals falls below that of the (n – 1)d and np subshells, and it fills. This explains the final magic number, $86(1s^22s^22p^63s^23p^64s^23d^{10}4p^65s^24d^{10}5p^66s^24f^{14}5d^{10}6p^6)$.

6.93 (a) The p_z orbital has a nodal plane where $z = 0$. This is the xy plane.

(b) The d_{xy} orbital has four lobes and two nodal planes, the two planes where $x = 0$ and $y = 0$. These are the yz and xz planes.

(c) The $d_{x^2-y^2}$ has four lobes and two nodal planes, the planes where $x^2 - y^2 = 0$. These are the planes that bisect the x and y axes and contain the z axis.

6.94 (a) In the absence of a magnetic field, electrons with opposite m_s values have the same energy. Because electrons with opposite spins will have oppositely oriented magnetic fields, only the interaction of the magnetic fields of the electrons with an external magnetic field will cause the energies of the electrons to be different and observable.

(b) According to Figure 6.27, the particle with its magnetic field parallel to the external field will have the lower energy. The left electron has its magnetic field oriented parallel to the described magnetic orientation, so it will be lower in energy.

(c) Microwave photons used to excite unpaired electrons in the ESR experiment have higher energy than radio wave photons used to excite nuclei in NMR.

6.95 (a) This is the frequency of microwaves that excite the nuclei from one spin state to the other.

(b) $\Delta E = h\nu = 6.626 \times 10^{-34}\ J{\bullet}s \times \frac{450 \times 10^6}{s} = 2.98 \times 10^{-25}\ J$

(c) Since $\Delta E = 0$ in the absence of a magnetic field, it is reasonable to assume that the stronger the external field, the greater ΔE. (In fact, ΔE is directly proportional to field strength). Because ΔE is relatively small [see part (b)], the two spin states are almost equally populated, with a very slight excess in the lower energy state. The stronger the magnetic field, the larger ΔE, the greater number of nuclei in the lower energy spin state. With more nuclei in the lower energy state, more are able to absorb the appropriate radio wave photons and reach the higher energy state. This increases the intensity of the NMR signal, which provides more information and more reliable information than a weak absorption signal.

6.97 (a) Se: $[Ar]4s^2 3d^{10} 4p^4$

(b) Rh: $[Kr]5s^2 4d^7$

(c) Si: $[Ne]3s^2 3p^2$

(d) Hg: $[Xe]6s^2 4f^{14} 5d^{10}$

(e) Hf: $[Xe]6s^2 4f^{14} 5d^2$

6.98 The core would be the electron configuration of element 118. If no new subshell begins to fill, the condensed electron configuration of element 126 would be similar to those of elements vertically above it on the periodic chart, Pu and Sm. The condensed configuration would be $[118]8s^2 6f^6$. On the other hand, the 5g subshell could begin to fill after 8s, resulting in the condensed configuration $[118]8s^2 5g^6$. Exceptions are also possible (likely).

Integrative Exercises

6.100 $\Delta H^\circ_{rxn} = \Delta H^\circ_f\ O_2(g) + \Delta H^\circ_f\ O(g) - \Delta H^\circ_f\ O_3(g)$

$\Delta H^\circ_{rxn} = 0 + 247.5\ kJ - 142.3\ kJ = +105.2\ kJ$

$$\frac{105.2\ kJ}{mol\ O_3} \times \frac{1\ mol\ O_3}{6.022 \times 10^{23}\ molecules} \times \frac{1000\ J}{1\ kJ} = \frac{1.747 \times 10^{-19}\ J}{O_3\ molecule}$$

$$\Delta E = hc/\lambda;\ \lambda = \frac{hc}{\Delta E} = \frac{6.626 \times 10^{-34}\ J{\bullet}s \times 2.998 \times 10^8\ m/s}{1.747 \times 10^{-19}\ J} = 1.137 \times 10^{-6}\ m$$

Radiation with this wavelength is in the infrared portion of the spectrum. (Clearly, processes other than simple photodissociation cause O_3 to absorb ultraviolet radiation.)

6.101 (a) The electron configuration of Zr is $[Kr]5s^2 4d^2$ and that of Hf is $[Xe]6s^2 4f^{14} 5d^2$. Although Hf has electrons in f orbitals as the rare earth elements do, the 4f subshell in Hf is filled, and the 5d electrons primarily determine the chemical properties of the element. Thus, Hf should be chemically similar to Zr rather than the rare earth elements.

(b) $ZrCl_4(s) + 4Na(l) \rightarrow Zr(s) + 4NaCl(s)$

This is an oxidation-reduction reaction; Na is oxidized and Zr is reduced.

(c) $2ZrO_2(s) + 4Cl_2(g) + 3C(s) \rightarrow 2ZrCl_4(s) + CO_2(g) + 2CO(g)$

$$55.4\,g\,ZrO_2 \times \frac{1\,mol\,ZrO_2}{123.2\,g\,ZrO_2} \times \frac{2\,mol\,ZrCl_4}{2\,mol\,ZrO_2} \times \frac{233.0\,g\,ZrCl_4}{1\,mol\,ZrCl_4} = 105\,g\,ZrCl_4$$

(d) In ionic compounds of the type MCl_4 and MO_2, the metal ions have a 4+ charge, indicating that the neutral atoms have lost four electrons. Zr, $[Kr]5s^24d^2$, loses the four electrons beyond its Kr core configuration. Hf, $[Xe]6s^24f^{14}5d^2$, similarly loses its four 6s and 5d electrons, but not electrons from the "complete" 4f subshell.

6.102 (a) Each oxide ion, O^{2-}, carries a 2- charge. Each metal oxide is a neutral compound, so the metal ion or ions must adopt a total positive charge equal to the total negative charge of the oxide ions in the compound. The table below lists the electron configuration of the neutral metal atom, the positive charge of each metal ion in the oxide, and the corresponding electron configuration of the metal ion.

i.	K: $[Ar]\,4s^1$	1+	[Ar]
ii.	Ca: $[Ar]\,4s^2$	2+	[Ar]
iii.	Sc: $[Ar]\,4s^23d^1$	3+	[Ar]
iv.	Ti: $[Ar]\,4s^23d^2$	4+	[Ar]
v.	V: $[Ar]\,4s^23d^3$	5+	[Ar]
vi.	Cr: $[Ar]\,4s^13d^5$	6+	[Ar]

Each metal atom loses all (valence) electrons beyond the Ar core configuration. In K_2O, Sc_2O_3 and V_2O_5, where the metal ions have odd charges, two metal ions are required to produce a neutral oxide.

(b)
i. potassium oxide
ii. calcium oxide
iii. scandium(III) oxide
iv. titanium (IV) oxide
v. vanadium (V) oxide
vi. chromium (VI) oxide

(Roman numerals are required to specify the charges on the transition metal ions, because more than one stable ion may exist.)

(c) Recall that $\Delta H_f^\circ = 0$ for elements in their standard states. In these reactions, M(s) and $H_2(g)$ are elements in their standard states.

i. $K_2O(s) + H_2(g) \rightarrow 2K(s) + H_2O(g)$

$$\Delta H^\circ = \Delta H_f^\circ\, H_2O(g) + 2\Delta H_f^\circ\, K(s) - \Delta H\, K_2O(s) - \Delta H_f^\circ\, H_2(g)$$

$$\Delta H^\circ = -241.82\,kJ + 2(0) - (-363.2\,kJ) - 0 = 121.4\,kJ$$

ii. $CaO(s) + H_2(g) \rightarrow Ca(s) + H_2O(g)$

$$\Delta H° = \Delta H^\circ_f\ H_2O(g) + \Delta H^\circ_f\ Ca(s) - \Delta H^\circ_f\ CaO(s) - \Delta H^\circ_f\ H_2(g)$$

$$\Delta H° = -241.82\ kJ + 0 - (-635.1\ kJ) - 0 = 393.3\ kJ$$

iii. $TiO_2(s) + 2H_2(g) \rightarrow Ti(s) + 2H_2O(g)$

$$\Delta H° = 2\Delta H^\circ_f\ H_2O(g) + \Delta H^\circ_f\ Ti(s) - \Delta H^\circ_f\ TiO_2(s) - 2\Delta H^\circ_f\ H_2(g)$$

$$= 2(-241.82) + 0 - (-938.7) - 2(0) = 455.1\ kJ$$

iv. $V_2O_5(s) + 5H_2(g) \rightarrow 2V(s) + 5H_2O(g)$

$$\Delta H° = 5\Delta H^\circ_f\ H_2O(g) + 2\Delta H^\circ_f\ V(s) - \Delta H^\circ_f\ V_2O_5(s) - 5\Delta H^\circ_f\ H_2(g)$$

$$= 5(-241.82) + 2(0) - (-1550.6) - 5(0) = 341.5\ kJ$$

(d) ΔH°_f becomes more negative moving from left to right across this row of the periodic chart. Since Sc lies between Ca and Ti, the median of the two ΔH°_f values is approximately –785 kJ/mol. However, the trend is clearly not linear. Dividing the ΔH°_f values by the positive charge on the pertinent metal ion produces the values –363, –318, –235, and –310. The value between Ca^{2+} (–318) and Ti^{4+} (–235) is Sc^{3+} (–277). Multiplying (–277) by 3, a value of approximately –830 kJ results. A reasonable range of values for ΔH°_f of $Sc_2O_3(s)$ is then –785 to –830 kJ/mol.

6.104 (a) ^{238}U: 92 p, 146 n, 92 e; ^{235}U: 92 p, 143 n, 92 e

In keeping with the definition isotopes, only the number of neutrons is different in the two nuclides. Since the two isotopes have the same number of electrons, they will have the same electron configuration.

(b) U: $[Rn]7s^25f^4$

(c) From Figure 6.30, the actual electron configuration is $[Rn]7s^25f^36d^1$. The energies of the 6d and 5f orbitals are very close, and electron configurations of many actinides include 6d electrons.

(d) $^{238}_{92}U \rightarrow ^{234}_{90}Th + ^{4}_{2}He$ $\quad$ ^{234}Th has 90 p, 143 n, 90 e. $\quad$ ^{238}U has lost 2 p, 2 n, 2 e.

These are organized into $^{4}_{2}He$ shown in the nuclear reaction above.

(e) From Figure 6.30, the electron configuration of Th is $[Rn]7s^26d^2$. This is not really surprising because there are so many rare earth electron configurations that are exceptions to the expected orbital filling order. However, Th is the only rare earth that has two d valence electrons. Furthermore, the configuration of Th is different than that of Ce, the element above it on the periodic chart, so the electron configuration is at least interesting.

7 Periodic Properties of the Elements

Visualizing Concepts

7.1 (a) The light bulb itself represents the nucleus of the atom. The brighter the bulb, the more nuclear charge the electron "sees." A frosted glass lampshade between the bulb and our eyes reduces the brightness of the bulb. The shade is analogous to core electrons in the atom shielding outer electrons (our eyes) from the full nuclear charge (the bare light bulb).

(b) Increasing the wattage of the light bulb mimics moving right along a row of the periodic table. The brighter bulb inside the same shade is analogous to having more protons in the nucleus while the core electron configuration doesn't change.

(c) Moving down a family, both the nuclear charge and the core electron configuration changes. To simulate the addition of core electrons farther from the nucleus, we would add larger frosted glass shades as well as increase the wattage of the bulb to show the increase in Z. The effect of the shade should dominate the increase in wattage, so that the brightness of the light decreases moving down a column.

7.3 (a) The bonding atomic radius of A, r_A, is $d_1/2$. The distance d_2 is the sum of the bonding atomic radii of A and X, $r_A + r_X$. Since we know that $r_A = d_1/2$, $d_2 = r_X + d_1/2$, $r_X = d_2 - d_1/2$.

(b) The length of the X-X bond is $2r_X$.

$2r_X = 2\,(d_2 - d_1/2) = 2d_2 - d_1$.

7.4

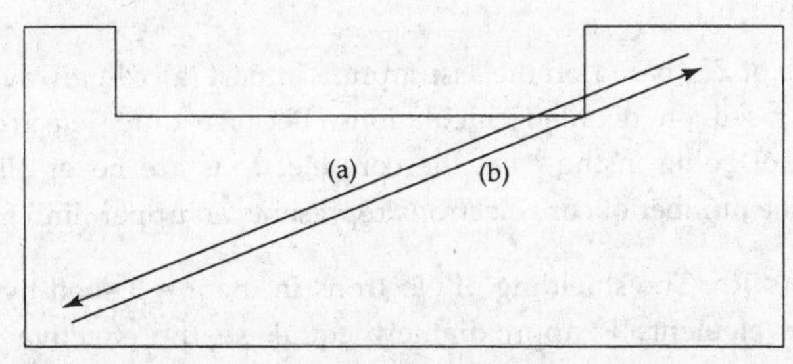

Lines (a) and (b) coincide, but their directions are opposite. Line (a) goes from upper right to lower left, and line (b) from lower left to upper right.

(c) From the diagram, we observe that the trends in bonding atomic radius (size) and ionization energy are opposite each other. As bonding atomic radius increases, ionization energy decreases, and vice versa.

7.6 (a) $X + 2F_2 \rightarrow XF_4$

(b) If X is a nonmetal, XF_4 is a molecular compound. If X is a metal, XF_4 is ionic. For an ionic compound with this formula, X would have a charge of 4+, and a much smaller bonding atomic radius than F^-. X in the diagram has about the same bonding radius as F, so it is likely to be a nonmetal.

Periodic Table; Effective Nuclear Charge

7.8 (a) The verification of the existence of many new elements by accurately measuring their atomic weights spurred interest in a classification scheme. Mendeleev (and Meyer) noted that certain chemical and physical properties recur periodically when the elements are arranged by increasing atomic weight. The accurate atomic weights provided a common property on which to base a classification scheme of the elements.

(b) Moseley realized that the characteristic X-ray frequencies emitted by each element were related to a unique integer that he assigned to each element. We now know this integer as the atomic number, the number of protons in the nucleus of an atom. In general, atomic weight increases as atomic number increases, but there are a few exceptions. If elements are arranged by increasing atomic number, a few seeming contradictions in the Mendeleev table (the positions of Ar and K or Te and I) are eliminated.

7.10 (a) Electrostatic attraction for the nucleus lowers the energy of an electron, while electron-electron repulsions increase this energy. The concept of effective nuclear charge allows us to model this increase in the energy of an electron as a smaller net attraction to a nucleus with a smaller positive charge, Z_{eff}.

(b) In Be (or any element), the 1s electrons are not shielded by any core electrons, so they experience a much greater Z_{eff} than the 2s electrons.

7.12 (a) $Z_{eff} = Z - S$; S: $[Ne]3s^23p^4$; $Z = 16$. If the 10 core electrons provide perfect shielding and the 6 valence electrons provide no shielding, $S = 10$.

$Z_{eff} = 16 - 10 = 6$.

(b) An estimate of Z_{eff} based on the assumption in part (a) will always be lower than the value based on detailed calculations. Because outer electrons have some probability of being in the core, the core electrons are never 100% effective at shielding; the number of core electrons represents an upper limit for S.

7.14 Mg < P < K < Ti < Rh. The shielding of electrons in the $n = 3$ shell by 1s and 2s core electrons in these elements is approximately equal, so the effective nuclear charge increases as Z increases.

Atomic and Ionic Radii

7.16 (a) Since the quantum mechanical description of the atom does not specify the exact location of electrons, there is no specific distance from the nucleus where the last electron can be found. Rather, the electron density decreases gradually as the distance from the nucleus increases. There is no quantum mechanical "edge" of an atom.

(b) When nonbonded atoms touch, it is their electron clouds that interact. These interactions are primarily repulsive because of the negative charges of electrons. Thus, the size of the electron clouds determines the nuclear approach distance of nonbonded atoms.

7.18 The distance between Ge atoms in solid germanium is two times the bonding atomic radius from Figure 7.6. The Ge-Ge distance is 2×1.22 Å = 2.44 Å.

7.20 Bi – I = 2.81 Å = $r_{Bi} + r_I$. From Figure 7.6, r_I = 1.33 Å.

r_{Bi} = [Bi – I] – r_I = 2.81 Å – 1.33 Å = 1.48 Å.

7.22 (a) The vertical difference in radius is due to a change in principal quantum number of the outer electrons. The horizontal difference in radius is due to the change in electrostatic attraction between the outer electron and a nucleus with one more or one fewer proton. Adding or subtracting a proton has a much smaller radius effect than moving from one principal quantum level to the next.

(b) S < Si < Se < Ge. This order is predicted by the trends in increasing atomic radius moving to the left in a row and down a column of the periodic chart, assuming that changes moving down a column are larger [see part (a)]. That is, the order above assumes that the change from S to Se is larger than the change from S to Si. This order is confirmed by the values in Figure 7.5.

7.24 (a) K < Rb < Cs (b) Te < Sn < In (c) Cl < P < Sr

7.26 (a) As Z stays constant and the number of electrons increases, the electron-electron repulsions increase, the electrons spread apart, and the ions become larger.

$I^- > I > I^+$

(b) Going down a column, the increasing average distance of the outer electrons from the nucleus causes the size of particles with like charge to increase.

$Ca^{2+} > Mg^{2+} > Be^{2+}$

(c) Fe: $[Ar]4s^2 3d^6$; Fe^{2+}: $[Ar]3d^6$; Fe^{3+}: $[Ar]3d^5$. The 4s valence electrons in Fe are on average farther from the nucleus than the 3d electrons, so Fe is larger than Fe^{2+}. Since there are five 3d orbitals, in Fe^{2+} at least one orbital must contain a pair of electrons. Removing one electron to form Fe^{3+} significantly reduces repulsion, increasing the nuclear charge experienced by each of the other d electrons and decreasing the size of the ion. Fe > Fe^{2+} > Fe^{3+}

7.28 The order of radii is Ca > Ca^{2+} > Mg^{2+}, so the largest sphere is Ca, the intermediate one is Ca^{2+}, and the smallest is Mg^{2+}.

(b) (i) N^{3-}: Ne (ii) Ba^{2+}: Xe (iii) Se^{2-}: Kr (iv) Bi^{3+}: Hg

7.30 (a) Sr^{2+}, Br^- (b) Y^{3+}, Br^-, Kr (c) P^{3-}, Ti^{4+} (d) Fe^{3+}, Mn^{2+}

7.32 (a) Cl < S < K (b) $K^+ < Cl^- < S^{2-}$

(c) Even though K has the largest Z value, the n-value of the outer electron is larger than the n-value of valence electrons in S and Cl so K atoms are largest. When the 4s electron is removed, K^+ is isoelectronic with Cl^- and S^{2-}. The larger Z value causes the 3p electrons in K^+ to experience the largest effective nuclear charge and K^+ is the smallest ion.

7.34 (a) Cl^- is larger than Cl because the increase in electron repulsions that accompany addition of an electron causes the electron cloud to expand.

(b) S^{2-} is larger than O^{2-}, because for particles with like charges, size increases going down a family.

(c) K^+ is larger than Ca^{2+} because the two ions are isoelectronic and K^+ has the larger Z.

Ionization Energies; Electron Affinities

7.36 (a) $Sn(g) \rightarrow Sn^+(g) + 1e^-$; $Sn^+(g) \rightarrow Sn^{2+}(g) + 1e^-$

(b) $Ti^{3+}(g) \rightarrow Ti^{4+}(g) + 1e^-$

7.38 (a) The effective nuclear charges of Li and Na are similar, but the outer electron in Li has a smaller n-value and is closer to the nucleus than the outer electron in Na. More energy is needed to overcome the greater attraction of the Li electron for the nucleus.

(b) Sc: [Ar] $4s^2 3d^1$; Ti: [Ar] $4s^2 3d^2$. The fourth ionization of titanium involves removing a 4s outer electron, while the fourth ionization of Sc requires removing a 3p electron from the [Ar] core. The effective nuclear charges experienced by the two 4s electrons in Ti are much more similar than the effective nuclear charges of a 4s outer electron and a 3p core electron in Sc. Thus, the difference between the third and fourth ionization energies of Sc is much larger.

(c) The electron configuration of Li^+ is $1s^2$ or [He] and that of Be^+ is [He]$2s^1$. Be^+ has one more valence electron to lose while Li^+ has the stable noble gas configuration of He. It requires much more energy to remove a 1s core electron close to the nucleus of Li^+ than a 2s valence electron farther from the nucleus of Be^+.

7.40 (a) Moving from F to I in group 7A, first ionization energies decrease and atomic radii increase. The greater the atomic radius, the smaller the electrostatic attraction of an outer electron for the nucleus and the smaller the ionization energy of the element.

(b) First ionization energies increase slightly going from K to Kr and atomic sizes decrease. As valence electrons are drawn closer to the nucleus (atom size decreases), it requires more energy to completely remove them from the atom (first ionization energy increases). Each trend has a discontinuity at Ga, owing to the increased shielding of the 4p electrons by the filled 3d subshell.

7.42 (a) Mo. As the effective nuclear charge increases in moving from left to right in the fifth row, the energy required to remove an electron increases.

(b) N. Valence electrons in N are closer to the nucleus (n = 2) and are shielded only by the [He] core, so they experience greater attraction for the nucleus and have a higher ionization energy.

(c) Cl. Effective nuclear charge increases moving both right across a row and up a family. Valence electrons in Cl, which is to the right and above Ga, experience the greater Z_{eff} and have the larger first ionization energy.

(d) Rn. Pb and Rn are in the same row, so Rn with the larger Z experiences a greater effective nuclear charge and a larger ionization energy.

7.44 (a) Mn^{3+}: $[Ar]3d^4$

(b) Se^{2-}: $[Ar]4s^23d^{10}4p^6$ = [Kr], noble-gas configuration

(c) Sc^{3+}: [Ar], noble-gas configuration (d) Ru^{2+}: $[Kr]4d^6$

(e) Tl^+: $[Xe]6s^24f^{14}5d^{10}$ (f) Au^+: $[Xe]4f^{14}5d^{10}$

7.46 (a) Cu^{2+}, 1 unpaired electron (b) Tl^+, 0 unpaired electrons

7.48 Argon is a noble gas, with a very stable core electron configuration. This causes the element to resist chemical change. Positive, endothermic, values for ionization energy and electron affinity mean that energy is required to either remove or add electrons. Valence electrons in Ar experience the largest Z_{eff} of any element in the third row, because the nuclear buildup is not accompanied by an increase in screening. This results in a large, positive ionization energy. When an electron is added to Ar, the n = 3 electrons become core electrons which screen the extra electrons so effectively that Ar^- has a higher energy than an Ar atom and a free electron. This results in a large positive electron affinity.

7.50 Electron affinity $Br(g) + 1e^- \rightarrow Br^-(g)$

$[Ar]4s^23d^{10}4p^5$ $\quad$ $[Ar]4s^23d^{10}4p^6$

When a Br atom gains an electron, the Br^- ion adopts the stable electron configuration of Kr. Since the electron is added to the same 4p subshell as other outer electrons, it experiences essentially the same attraction for the nucleus. Thus, the energy of the Br^- ion is lower than the total energy of a Br atom and an isolated electron, and electron affinity is negative.

Electron affinity: $Kr(g) + 1e^- \rightarrow Kr^-(g)$

$[Ar]4s^23d^{10}4p^6$ $\quad$ $[Ar]4s^23d^{10}4p^65s^1$

Energy is required to add an electron to a Kr atom; Kr^- has a higher energy than the isolated Kr atom and free electron. In Kr^- the added electron would have to occupy the higher energy 5s orbital; a 5s electron is farther from the nucleus and effectively shielded by the spherical Kr core and is not stabilized by the nucleus.

7.52 $Mg^+(g) + 1e^- \rightarrow Mg(g)$
$[Ne]\,3s^1$ $\quad$ $[Ne]\,3s^2$

This process is the reverse of the first ionization of Mg. The magnitude of the energy change for this process is the same as the magnitude of the first ionization energy of Mg, 738 kJ/mol.

Properties of Metals and Nonmetals

7.54 S < Si < Ge < Ca. S is a nonmetal, Si and Ge are metalloids, and Ca is a metal. We expect that electrical conductivity increases as metallic character increases. Since metallic character increases going down a column and to the left in a row, the order of increasing electrical conductivity is as shown above.

7.56 (a) In general, ionization energy decreases as metallic character increases. According to Figure 7.11 the ionization energies of the group 5A elements decrease as atomic weight increases (going down the column). Therefore, metallic character of the group 5A elements increases with increasing atomic weight.

(b) Ag < Sn < Se < C < F. Nonmetallic character increases going up and to the right on the periodic chart. Vertical trends dominate the relationship between Se and C.

7.58 (a) metal oxides: $Li_2O(s) + H_2O(l) \rightarrow 2\,LiOH(aq)$

$BaO(s) + H_2O(l) \rightarrow Ba(OH)_2(aq)$

nonmetal oxides: $P_4O_{10}(s) + 6H_2O(l) \rightarrow 4H_3PO_4(aq)$

$SO_3(g) + H_2O(l) \rightarrow H_2SO_4(aq)$

(b) Metals have lower ionization energies than nonmetals, so they tend to form ionic oxides, while nonmetals form molecular oxides. Ionic compounds, in this case oxides, dissociate into ions when they dissolve in water. The reactive oxide ion ends up as hydroxide, separated from the metal ion. Molecular oxides do not ionize upon dissolution, so the oxygen remains bound to the nonmetal.

7.60 The more nonmetallic the central atom, the more acidic the oxide. In order of increasing acidity: $CaO < Al_2O_3 < SiO_2 < CO_2 < P_2O_5 < SO_3$

7.62 (a) $XCl_4(l) + 2H_2O(l) \rightarrow XO_2(s) + 4HCl(g)$

The second product is HCl(g).

(b) If X were a metal, both the oxide and the chloride would be high melting solids. If X were a nonmetal, XO_2 would be a nonmetallic, molecular oxide and probably gaseous, like CO_2, NO_2, and SO_2. Neither of these statements describes the properties of XO_2 and XCl_4, so X is probably a metalloid.

(c) Use the *Handbook of Chemistry* to find formulas and melting points of oxides, and formulas and boiling points of chlorides of selected metalloids.

metalloid	formula of oxide	m.p. of oxide	formula of chloride	b.p. of chloride
boron	B_2O_3	460°C	BCl_3	12°C
silicon	SiO_2	~1700°C	$SiCl_4$	58°C
germanium	GeO GeO_2	710°C ~1100°C	$GeCl_2$ $GeCl_4$	decomposes 84°C
arsenic	As_2O_3 As_2O_5	315°C 315°C	$AsCl_3$	132°C

Boron, arsenic, and, by analogy, antimony, do not fit the description of X, because the formulas of their oxides and chlorides are wrong. Silicon and germanium, in the same family, have oxides and chlorides with appropriate formulas. Both SiO_2 and GeO_2 melt above 1000°C, but the boiling point of $SiCl_4$ is much closer to that of XCl_4. Element X is silicon.

7.64 (a) $K_2O(s) + H_2O(l) \rightarrow 2KOH(aq)$

(b) $P_2O_3(l) + 3H_2O(l) \rightarrow 2H_3PO_3(aq)$

(c) $Cr_2O_3(s) + 6HCl(aq) \rightarrow 2CrCl_3(aq) + 3H_2O(l)$

(d) $SeO_2(s) + 2KOH(aq) \rightarrow K_2SeO_3(aq) + H_2O(l)$

Group Trends in Metals and Nonmetals

7.66 (a) Rb: $[Kr]5s^1$, r = 2.11 Å   Ag: $[Kr]5s^1 4d^{10}$, r = 1.53 Å

The electron configurations both have a [Kr] core and a single 5s electron; Ag has a completed 4d subshell as well. The radii are very different because the 5s electron in Ag experiences a much greater effective nuclear charge. Ag has a much larger Z (47 vs. 37), and although the 4d electrons in Ag shield the 5s electron somewhat, the increased shielding does not compensate for the large increase in Z.

(b) Ag is much less reactive (less likely to lose an electron) because its 5s electron experiences a much larger effective nuclear charge and is more difficult to remove.

7.68 (a) Cs is much more reactive than Li toward H_2O because its valence electron is less tightly held (greater *n* value), and Cs is more easily oxidized.

(b) The purple flame indicates that the metal is potassium (see Figure 7.23).

(c) $K_2O_2(s) + H_2O(l) \rightarrow H_2O_2(aq) + K_2O(aq)$
potassium peroxide   hydrogen peroxide

7.70 (a) $2K(s) + 2H_2O(l) \rightarrow 2KOH(aq) + H_2(g)$

(b) $Ba(s) + 2H_2O(l) \rightarrow Ba(OH)_2(aq) + H_2(g)$

(c) $6Li(s) + N_2(g) \rightarrow 2Li_3N(s)$

(d) $2Mg(s) + O_2(g) \rightarrow 2MgO(s)$

7.72 (a) The reactions of the alkali metals with hydrogen and with a halogen are redox reactions. In both classes of reaction, the alkali metal loses electrons and is oxidized. Both hydrogen and the halogen gain electrons and are reduced. The product is an ionic solid, where either hydride ion, H^-, or a halide ion, X^-, is the anion and the alkali metal is the cation.

(b) $Ca(s) + F_2(g) \rightarrow CaF_2(s)$ $Ca(s) + H_2(g) \rightarrow CaH_2(s)$

Both products are ionic solids containing Ca^{2+} and the corresponding anion in a 1:2 ratio.

7.74 *Plan.* Predict the physical and chemical properties of At based on the trends in properties in the halogen (7A) family. *Solve.*

(a) F, at the top of the column, is a gas; I, immediately above At, is a solid; the melting points of the halogens increase going down the column. At is likely to be a solid at room temperature.

(b) All halogens form ionic compounds with Na; they have the generic formula NaX. The compound formed by At will have the formula NaAt.

7.76 Xe has a lower ionization energy than Ne. The valence electrons in Xe are much farther from the nucleus than those of Ne (n = 5 vs n = 2) and much less tightly held by the nucleus; they are more "willing" to be shared than those in Ne. Also, Xe has empty 5d orbitals that can help to accommodate the bonding pairs of electrons, while Ne has all its valence orbitals filled.

7.78 (a) $Cl_2(g) + H_2O(l) \rightarrow HCl(aq) + HOCl(aq)$

(b) $Ba(s) + H_2(g) \rightarrow BaH_2(s)$

(c) $2Li(s) + S(s) \rightarrow Li_2S(s)$

(d) $Mg(s) + F_2(g) \rightarrow MgF_2(s)$

Additional Exercises

7.80 (a) *eka – aluminum* is gallium (Ga), Z = 31.

(b) From the *Handbook of Chemistry and Physics*, 74th edition, the physical properties of Ga are:

atomic weight = 69.92 g/mol m.p. = 29.78°C

density = 5.904 g/mL (s) b.p. = 2403°C

= 6.095 g/mL (l) oxide = Ga_2O_3, Ga_2O

The atomic weight, density, and formula of oxide all agree with Mendeleev's predictions. Gallium does have a low melting point and a high boiling point.

7.82 (a) P: (+15 – 10) = +5

(b) The 3s electrons penetrate the [Ne] core electrons (by analogy to Figure 7.4) and experience less shielding than the 3p electrons. That is, S is greater for 3p electrons, owing to the penetration of the 3s electrons, so Z - S (3p) is less than Z - S(3s).

(c) The electron configuration of P is $[Ne]3s^23p^3$. The 3p electrons are the outermost electrons; they experience a smaller Z_{eff} than 3s electrons and thus a smaller attraction for the nucleus, given equal *n*-values. The first electron lost is a 3p electron. Each 3p orbital holds one electron, so there is no preference as to which 3p electron will be lost.

7.83 Close approach by two positively charged nuclei is impossible because of the large electrostatic repulsion between like-charged particles at small distances. The additional space between the nuclei in a molecule like F_2 is occupied by bonding electrons, which are electrostatically stabilized by attraction to both nuclei. The electrons also provide a buffer between the two nuclei.

7.84 (a) Z_{eff} = Z - S. According to our simple model, moving from C to N, causes Z to increase by 1 and S to remain the same, so Z_{eff} is greater by 1 for N than for C.

(b) Our simple model assumed that valence electrons don't screen each other. If there is some mutual screening by valence electrons, S for N would be slightly greater than S for C, because N has one more valence electron than C. The increase in Z_{eff} would be less than the full 1+, as calculated.

(c) In O, $[He]2s^22p^4$, one of the p orbitals is doubly occupied, which significantly increases electron-electron repulsion. This leads to an overall higher energy for the 2p electrons. In our model, this appears as a larger S and a smaller Z_{eff}.

7.86 (a) The estimated distances in the table below are the sum of the radii of the group 5A elements and H from Figure 7.6.

bonded atoms	estimated distance	measured distance
P - H	1.43	1.419
As - H	1.56	1.519
Sb - H	1.75	1.707

In general, the estimated distances are a bit longer than the measured distances. This probably shows a systematic bias in either the estimated radii or in the method of obtaining the measured values.

(b) The principal quantum number of the outer electrons and thus the average distance of these electrons from the nucleus increases from P (n = 3) to As (n = 4) to Sb (n = 5). This causes the systematic increase in M - H distance.

7.87 Ge - H distance = $r_{Ge} + r_H$ = 1.22 + 0.37 = 1.59Å

Ge - Cl distance = $r_{Ge} + r_{Cl}$ = 1.22 + 0.99 = 2.21 Å

7.89 (a) Hg^{2+}

(b) No. According to Figure 2.24, Hg is not essential for life, even in trace amounts.

(c) Zn^{2+}: $[Ar]3d^{10}$, r = 0.74 Å; Cd^{2+}: $[Kr]4d^{10}$, r = 0.95 Å; Hg^{2+}: $[Xe]4f^{14}5d^{10}$, r ≈ 1.05 Å.

If there were no 4f electrons in Hg^{2+}, we would expect an ionic radius of around 1.15 Å, an increase of ~0.20 Å due to the increase in principal quantum number of the valence electrons. However, the increase in Z due to filling of the 4f orbitals is not completely offset by shielding. The 5d valence electrons in Hg^{2+} experience a greater than expected Z_{eff} which largely offsets the increase in principal quantum number; ionic radius of Hg^{2+} is smaller than expected. This phenomenon is known as the *lanthanide contraction* and affects the physical properties of elements in the sixth period and beyond. See also Solution 7.88.

(d) Since the ionic radius of Hg^{2+} is similar to that of Cd^{2+}, Hg^{2+} will be physiologically more similar to Cd^{2+}.

(e) By common knowledge, and verified with WebElements.com™, both Hg and Hg^{2+} are extremely toxic to humans.

7.90 (a) $2Sr(s) + O_2(g) \rightarrow 2SrO(s)$

(b) In the Figure, large spheres represent O^{2-} anions and small ones represent Sr^{2+} cations. According to Figure 7.7, anions are larger than their neutral atoms and cations are smaller. The changes in repulsion and Z_{eff} associated with adding or removing electrons from neutral atoms outweigh the increase in principal quantum number between from O to Sr.

(c) Assume that the corners of the cube are at the centers of the outermost O^{2-} ions, and that the edges pass through the centers of perimeter Sr^{2+} ions. The length of an edge is then $r(O^{2-}) + 2r(Sr^{2+}) + r(O^{2-}) = 2r(O^{2-}) + 2r(Sr^{2+}) = 2(1.40\ Å) + 2(1.13Å) = 5.06\ Å$.

(d) Density is the ratio of mass to volume.

$$d = \frac{\text{mass SrO in cube}}{\text{vol cube}} = \frac{\#\ \text{SrO units} \times \text{mass of SrO}}{\text{vol cube}}$$

Calculate the mass of 1 SrO unit in grams and the volume of the cube in cm^3; solve for number of SrO units.

$$\frac{103.62\ \text{g SrO}}{\text{mol}} \times \frac{1\ \text{mol SrO}}{6.022 \times 10^{23}\ \text{SrO units}} = 1.7207 \times 10^{-22} = 1.721 \times 10^{-22}\ \text{g/SrO unit}$$

$$V = (5.06)^3\ Å^3 \times \frac{(1 \times 10^{-8})^3\ cm^3}{Å^3} = 1.2955 \times 10^{-22} = 1.30 \times 10^{-22}\ cm^3$$

$$d = \frac{\#\ \text{of SrO units} \times 1.7207 \times 10^{-22}\ \text{g/SrO unit}}{1.2955 \times 10^{-22}\ cm^3} = 5.10\ g/cm^3$$

$$\#\ \text{of SrO units} = 5.10\ g/cm^3 \times \frac{1.2955 \times 10^{-22}\ cm^3}{1.7207 \times 10^{-22}\ \text{g/SrO unit}} = 3.84\ \text{units}$$

Since the number of formula units must be an integer, there are four SrO formula units in the cube. Using average values for ionic radii to estimate the edge length probably leads to the discrepancy.

7.92 (a) Ca: [Ar] $4s^2$; Ca^{2+}: [Ar]. Zn: [Ar] $4s^2 3d^{10}$; Zn^{2+}: [Ar] $3d^{10}$. In both cases, forming ions involves losing the only electrons in the n = 4 shell. The average distance from the nucleus of n = 3 electrons is less, so the radii of the ions are smaller.

(b) The atomic radius of Ca is greater than that of Zn because Zn has a significantly greater Z and Z_{eff}, which draws the 4s valence electrons closer to the nucleus.

(c) In both Ca and Zn atoms, the outermost electrons are in the 4s subshell. The much larger Z_{eff} for Zn causes it to have a much smaller radius. In the +2 ions, both have lost their 4s electrons. The outermost electrons in Zn are in the 3d sublevel and are significantly shielded by the [Ar] core electrons. Thus, the radius of Zn^{2+} is closer to the radius of Ca^{2+} than the radii of the neutral ions.

7.93 The statement is somewhat true, but more accurate if changed to read: "A negative value for the electron affinity of an atom occurs when the outermost electrons only incompletely shield the added electron from the nucleus." This new statement totally explains the negative electron affinity of Br and the positive value for Kr. For Br^-, the electron is added to the 4p subshell and is incompletely shielded by the "other" 4s and 4p electrons. For Kr^-, the electron is added to the 5s subshell, which is effectively shielded by the spherical Kr core.

7.95 (a) P: [Ne] $3s^2 3p^3$; S: [Ne] $3s^2 3p^4$. In P, each 3p orbital contains a single electron, while in S one 3p orbital contains a pair of electrons. Removing an electron from S eliminates the need for electron pairing and reduces electrostatic repulsion, so the overall energy required to remove the electron is smaller than in P, even though Z is greater.

(b) C: [He] $2s^2 2p^2$; N: [He] $2s^2 2p^3$; O: [He] $2s^2 2p^4$. An electron added to a N atom must be paired in a relatively small 2p orbital, so the additional electron-electron repulsion more than compensates for the increase in Z and the electron affinity is smaller (less exothermic) than that of C. In an O atom, one 2p orbital already contains a pair of electrons, so the additional repulsion from an extra electron is offset by the increase in Z and the electron affinity is greater (more exothermic). Note from Figure 7.12 that the electron affinity of O is only slightly more exothermic than that of C, although the value of Z has increased by 2.

(c) O^+: [Ne] $2s^2 2p^3$; O^{2+}: [Ne] $2s^2 2p^2$; F^+: [Ne] $2s^2 2p^4$; F^{2+}: [Ne] $2s^2 2p^3$. The decrease in electron-electron repulsion going from F^+ to F^{2+} energetically favors ionization and causes it to be less endothermic than the corresponding process in O, where there is no significant decrease in repulsion.

(d) Mn^{2+}: [Ar]$3d^5$; Mn^{3+}; [Ar] $3d^4$; Cr^{2+}: [Ar] $3d^4$; Cr^{3+}: [Ar] $3d^3$; Fe^{2+}: [Ar] $3d^6$; Fe^{3+}: [Ar] $3d^5$. The third ionization energy of Mn is expected to be larger than that of Cr because of the larger Z value of Mn. The third ionization energy of Fe is less than that of Mn because going from $3d^6$ to $3d^5$ reduces electron repulsions, making the process less endothermic than predicted by nuclear charge arguments.

7.97 (a) For both H and the alkali metals, the added electron will complete an *ns* subshell (1s for H and *ns* for the alkali metals) so shielding and repulsion effects will be similar. For the halogens, the electron is added to an *np* subshell, so the energy change is likely to be quite different.

(b) True. Only He has a smaller estimated "bonding" atomic radius, and no known compounds of He exist. The electron configuration of H is $1s^1$. The single 1s electron experiences no repulsion from other electrons and feels the full unshielded nuclear charge. It is held very close to the nucleus. The outer electrons of all other elements that form compounds are shielded by a spherical inner core of electrons and are less strongly attracted to the nucleus, resulting in larger atomic radii.

(c) Ionization is the process of removing an electron from an atom. For the alkali metals, the *ns* electron being removed is effectively shielded by the core electrons, so ionization energies are low. For the halogens, a significant increase in nuclear charge occurs as the *np* orbitals fill, and this is not offset by an increase in shielding. The relatively large effective nuclear charge experienced by *np* electrons of the halogens is similar to the unshielded nuclear charge experienced by the H 1s electron. Both H and the halogens have large ionization energies.

7.98 Since Xe reacts with F_2, and O_2 has approximately the same ionization energy as Xe, O_2 will probably react with F_2. Possible products would be O_2F_2, analogous to XeF_2, or OF_2.

$$O_2(g) + F_2(g) \rightarrow O_2F_2(g)$$
$$O_2(g) + 2F_2(g) \rightarrow 2OF_2(g)$$

7.100 Moving one place to the right in a horizontal row of the table, for example, from Li to Be, there is an increase in ionization energy. Moving downward in a given family, for example from Be to Mg, there is usually a decrease in ionization energy. Similarly, atomic size decreases in moving one place to the right and increases in moving downward. Thus, two elements such as Li and Mg that are diagonally related tend to have similar ionization energies and atomic sizes. This in turn gives rise to some similarities in chemical behavior. Note, however, that the valences expected for the elements are not the same. That is, lithium still appears as Li^+, magnesium as Mg^{2+}.

7.101 Fr (1A), Ra (2A), Po (6A), At (7A), Rn (8A)

(a) Most metallic character: Fr. (Metallic character decreases from left to right in a row.)

(b) Most nonmetallic character: Rn.

(c) Largest ionization energy: Rn. (Ionization energy increases from left to right in a row.)

(d) Smallest ionization energy: Fr.

(e) Greatest electron affinity: At. (Electron affinity becomes more exothermic from left to right in a row.)

(f) Largest atomic radius: Fr. (Size decreases from left to right in a row.)

(g) Appears least like the element above it: Fr. (According to the trends in melting point, Fr may be a liquid or a gas at room temperature, while Cs is a solid. The others should be in the same state as the element above them.)

(h) Highest melting point: Ra. (The melting points of the group 2A metals are much higher than those of the other groups, even though the values decrease going down the group.)

(i) Reacts most readily with H_2O: Fr. (It has the lowest ionization energy.)

7.102 (a) *Plan.* Use qualitative physical (bulk) properties to narrow the range of choices, then match melting point and density to identify the specific element. *Solve.*

Hardness varies widely in metals and nonmetals, so this information is not too useful. The relatively high density, appearance, and ductility indicate that the element is probably less metallic than copper. Focus on the block of nine main group elements centered around Sn. Pb is not a possibility because it was used as a comparison standard. The melting point of the five elements closest to Pb are:

Tl, 303.5°C; In, 156.1°C; Sn, 232°C; Sb, 630.5°C; Bi, 271.3°C

The best match is In. To confirm this identification, the density of In is 7.3 g/cm^3, also a good match to properties of the unknown element.

(b) In order to write the correct balanced equation, determine the formula of the oxide product from the mass data, assuming the unknown is In.

5.08 g oxide – 4.20 g In = 0.88 g O

4.20 g In/114.82 g/mol = 0.0366 mol In; 0.0366/0.0366 = 1

0.88 g O/16.00 g/mol = 0.0550 mol O; 0.0550/0.0366 = 1.5

Multiplying by 2 produces an integer ratio of 2 In: 3 O and a formula of In_2O_3. The balanced equation is: $4\ In(s) + 3O_2(g) \rightarrow 2\ In_2O_3(s)$

(c) According to Figure 7.2, the element In was discovered between 1843–1886. The investigator who first recorded this data in 1822 could have been the first to discover In.

Integrative Exercises

7.104 (a) $\nu = c/\lambda$; $1 \text{ Hz} = 1 \text{ s}^{-1}$

$$\text{Ne: } \nu = \frac{2.998 \times 10^8 \text{ m/s}}{14.610 \text{ Å}} \times \frac{1 \text{ Å}}{1 \times 10^{-10} \text{ m}} = 2.052 \times 10^{17} \text{ s}^{-1} = 2.052 \times 10^{17} \text{ Hz}$$

$$\text{Ca: } \nu = \frac{2.998 \times 10^8 \text{ m/s}}{3.358 \times 10^{-10} \text{ m}} = 8.928 \times 10^{17} \text{ Hz}$$

$$\text{Zn: } \nu = \frac{2.998 \times 10^8 \text{ m/s}}{1.435 \times 10^{-10} \text{ m}} = 20.89 \times 10^{17} \text{ Hz}$$

$$\text{Zr: } \nu = \frac{2.998 \times 10^8 \text{ m/s}}{0.786 \times 10^{-10} \text{ m}} = 38.14 \times 10^{17} = 38.1 \times 10^{17} \text{ Hz}$$

$$\text{Sn: } \nu = \frac{2.998 \times 10^8 \text{ m/s}}{0.491 \times 10^{-10} \text{ m}} = 61.06 \times 10^{17} = 61.1 \times 10^{17} \text{ Hz}$$

(b)

Element	Z	ν	$\nu^{1/2}$
Ne	10	2.052×10^{17}	4.530×10^8
Ca	20	8.928×10^{17}	9.449×10^8
Zn	30	20.89×10^{17}	14.45×10^8
Zr	40	38.14×10^{17}	19.5×10^8
Sn	50	61.06×10^{17}	24.7×10^8

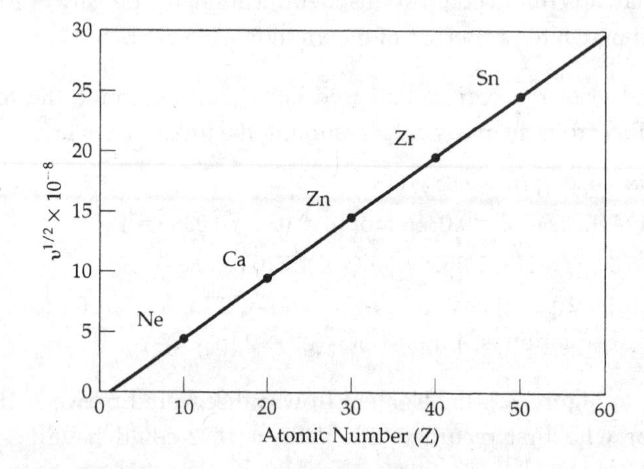

(c) The plot in part (b) indicates that there is a linear relationship between atomic number and the square root of the frequency of the X-rays emitted by an element. Thus, elements with each integer atomic number should exist. This relationship allowed Moseley to predict the existence of elements that filled "holes" or gaps in the periodic table.

(d) For Fe, Z = 26. From the graph, $\nu^{1/2} = 12.5 \times 10^8$, $\nu = 1.56 \times 10^{18}$ Hz.

$$\lambda = c/\nu = \frac{2.998 \times 10^8 \text{ m/s}}{1.56 \times 10^{18} \text{ s}^{-1}} \times \frac{1 \text{ Å}}{1 \times 10^{-10} \text{ m}} = 1.92 \text{ Å}$$

(e) $\lambda = 0.980$ Å $= 0.980 \times 10^{-10}$ m

$$\nu = c/\lambda = \frac{2.998 \times 10^8 \text{ m/s}}{0.980 \times 10^{-10} \text{ m}} = 30.6 \times 10^{17} \text{ Hz};\ \nu^{1/2} = 17.5 \times 10^8$$

From the graph, $\nu^{1/2} = 17.5 \times 10^8$, Z = 36. The element is krypton, Kr.

7.106 (a) $E = hc/\lambda$; 1 nm = 1×10^{-9} m; 58.4 nm = 58.4×10^{-9} m;

1 eV = 96.485 kJ/mol, 1 eV • mol = 96.485 kJ

$$E = \frac{6.626 \times 10^{-34} \text{ J}\cdot\text{s} \times 2.998 \times 10^8 \text{ m/s}}{58.4 \times 10^{-9} \text{ m}} = 3.4015 \times 10^{-18} = 3.40 \times 10^{-18} \text{ J/photon}$$

$$\frac{3.4015 \times 10^{-18} \text{ J}}{\text{photon}} \times \frac{1 \text{ kJ}}{1000 \text{ J}} \times \frac{6.022 \times 10^{23} \text{ photons}}{\text{mol}} \times \frac{1 \text{ eV}\cdot\text{mol}}{96.485 \text{ kJ}} = 21.230 = 21.2 \text{ eV}$$

(b) $Hg(g) \rightarrow Hg^+(g) + 1e^-$

(c) $I_1 = E_{58.4} - E_K = 21.23 \text{ eV} - 10.75 \text{ eV} = 10.48 = 10.5 \text{ eV}$

$$10.48 \text{ eV} \times \frac{96.485 \text{ kJ}}{1 \text{ eV}\cdot\text{mol}} = 1.01 \times 10^3 \text{ kJ/mol}$$

(d) From Figure 7.11, iodine (I) appears to have the ionization energy closest to that of Hg, approximately 1000 kJ/mol.

7.107 (a)

$Na(g) \rightarrow Na^+(g) + 1e^-$ (ionization energy of Na)

$Cl(g) + 1e^- \rightarrow Cl^-(g)$ (electron affinity of Cl)

$Na(g) + Cl(g) \rightarrow Na^+(g) + Cl^-(g)$

(b) $\Delta H = I_1(Na) + E_1(Cl) = +496 \text{ kJ} - 349 \text{ kJ} = +147 \text{ kJ}$, endothermic

(c) The reaction $2Na(s) + Cl_2(g) \rightarrow 2NaCl(s)$ involves many more steps than the reaction in part (a). One important difference is the production of NaCl(s) versus NaCl(g). The condensation $NaCl(g) \rightarrow NaCl(s)$ is very exothermic and is the step that causes the reaction of the elements in their standard states to be exothermic, while the gas phase reaction is endothermic.

7.109 (a) $r_{Bi} = r_{BiBr_3} - r_{Br} = 2.63 \text{ Å} - 1.14 \text{ Å} = 1.49 \text{ Å}$

(b) $Bi_2O_3(s) + 6HBr(aq) \rightarrow 2BiBr_3(aq) + 3H_2O(l)$

(c) Bi_2O_3 is soluble in acid solutions because it act as a base and undergoes acid-base reactions like the one in part (b). It is insoluble in base because it cannot acts as an acid. Thus, Bi_2O_3 is a basic oxide, the oxide of a metal. Based on the properties of its oxide, Bi is characterized as a metal.

(d) Bi: $[Xe]6s^2 4f^{14} 5d^{10} 6p^3$. Bi has five outer electrons in the 6p and 6s subshells. If all five electrons participate in bonding, compounds such as BiF_5 are possible. Also, Bi has a large enough atomic radius (1.49 Å) and low-energy orbitals available to accommodate more than four pairs of bonding electrons.

(e) The high ionization energy and relatively large negative electron affinity of F, coupled with its small atomic radius, make it the most electron withdrawing of the halogens. BiF_5 forms because F has the greatest tendency to attract electrons from Bi. Also, the small atomic radius of F reduces repulsions between neighboring bonded F atoms. The strong electron withdrawing properties of F are also the reason that only F compounds of Xe are known.

7.110 (a) $4KO_2(s) + 2CO_2(g) \rightarrow 2K_2CO_3(s) + 3O_2(g)$

(b) K, +1; O, +1/2 (O_2^- is superoxide ion); C, +4; O, –2 → K, +1; C, +4; O, –2; O, 0

(c) $$18.0 \text{ g } CO_2 \times \frac{1 \text{ mol } CO_2}{44.01 \text{ g } CO_2} \times \frac{4 \text{ mol } KO_2}{2 \text{ mol } CO_2} \times \frac{71.10 \text{ g } KO_2}{1 \text{ mol } KO_2} = 58.2 \text{ g } KO_2$$

$$18.0 \text{ g } CO_2 \times \frac{1 \text{ mol } CO_2}{44.01 \text{ g } CO_2} \times \frac{3 \text{ mol } O_2}{2 \text{ mol } CO_2} \times \frac{32.00 \text{ g } O_2}{1 \text{ mol } O_2} = 19.6 \text{ g } O_2$$

8 Basic Concepts of Chemical Bonding

Visualizing Concepts

8.1 *Analyze/Plan.* Count the number of electrons in the Lewis symbol. This corresponds to the 'A'-group number of the family. *Solve.*

(a) Group 14 or 4A

(b) Group 2 or 2A

(c) Group 15 or 5A

(These are the appropriate groups in the s and p blocks, where Lewis symbols are most useful.)

8.3 *Analyze/Plan.* Count the valence electrons in the orbital diagram, take ion charge into account, and find the element with this orbital electron count on the periodic chart. Write the complete electron configuration for the ion. *Solve.*

(a) This ion has seven 3d electrons. Transition metals, or d-block elements, have valence electrons in d-orbitals. Transition metal ions first lose electrons from the 4s orbital, then from 3d if required by the charge. This 2+ ion has lost two electrons from 4s, none from 3d. The transition metal with seven 3d-electrons is cobalt, Co.

(b) The electron configuration of Co is $[Ar]4s^23d^7$. (The configuration of Co^{2+} is $[Ar]3d^7$).

8.4 *Analyze/Plan.* This question is a "reverse" Lewis structure. Count the valence electrons shown in the Lewis structure. For each atom, assume zero formal charge and determine the number of valence electrons an unbound atom has. Name the element. *Solve.*

A: 1 shared e^- pair = 1 valence electron + 3 unshared pairs = 7 valence electrons, F

E: 2 shared pairs = 2 valence electrons + 2 unshared pairs = 6 valence electrons, O

D: 4 shared pairs = 4 valence electrons, C

Q: 3 shared pairs = 3 valence electrons + 1 unshared pair = 5 valence electrons, N

X: 1 shared pair = 1 valence electron, no unshared pairs, H

Z: same as X, H

Check. Count the valence electrons in the Lewis structure. Does the number correspond to the molecular formula CH_2ONF? 12 e^- pair in the Lewis structure. CH_2ONF = 4 + 2 + 6 + 5 + 7 = 24 e^-, 12 e^- pair. The molecular formula we derived matches the Lewis structure.

8.6 (a) The central atom Xe, is a member of Group 8A, currently known as noble gases. Prior to 1960, this group was known as inert gases, because there were no known compounds of group 8A elements. Noble gas elements have 8 valence electrons, so they satisfy the octet rule without forming chemical bonds. Since forming compounds wasn't necessary for these elements, it was assumed that no compounds of group 8A elements existed.

(b) There are a total of three resonance structures for the given Lewis structure of XeO_3, each with the single bond in a different Xe–O bonding domain.

(c) The given Lewis structure does not satisfy the octet rule, because the central Xe atom has more than 8 (12 actually) electrons.

(d) Below are four possible Lewis structures for XeO_3, with formal charges shown. Several other resonance structures with one or two double bonds can be drawn.

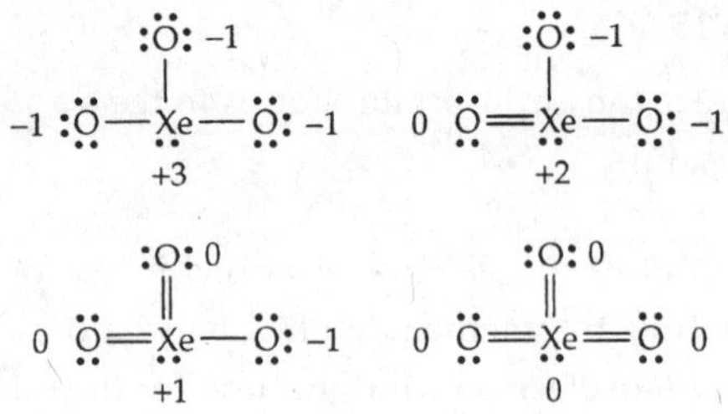

:O: −1 (single bond to Xe); −1 :O—Xe—O: −1; Xe +3

:O: −1 (single bond to Xe); 0 O=Xe—O: −1; Xe +2

:O: 0 (double bond to Xe); 0 O=Xe—O: −1; Xe +1

:O: 0 (double bond to Xe); 0 O=Xe=O 0; Xe 0

The Lewis structure given in this problem is on the lower left. It neither minimizes formal charge nor obeys the octet rule, so it is not the "best" Lewis structure for XeO_3. According to Section 8.7, the best single Lewis structure is usually the one that obeys the octet rule. This structure is shown on the upper left above. (The structure that minimizes formal charge is on the lower right.) The "best" view of bonding in XeO_3 is a composite of all correct Lewis structures, not any single Lewis structure.

Lewis Symbols

8.8 (a) Atoms will gain, lose or share electrons to achieve the nearest noble-gas electron configuration. Except for H and He, this corresponds to eight electrons in the valence shell, thus the term octet rule.

(b) S: $[Ne]3s^2 3p^4$ A sulfur atom has six valence electrons, so it must gain two electrons to achieve an octet.

(c) $1s^2 2s^2 2p^3$ = $[He]2s^2 2p^3$ The atom (N) has five valence electrons and must gain three electrons to achieve an octet.

8.10 Sc: $1s^2 2s^2 2p^6 3s^2 3p^6 4s^2 3d^1$ = $[Ar]4s^2 3d^1$. Scandium has three (3) valence electrons. These valence electrons are available for chemical bonding, while the core electrons do not participate in bonding.

8.12 (a) K· (b) ·Si· (c) $[:Mg:]^{2+}$ or Mg^{2+} (d) $[:P:]^{3-}$ or P^{3-}

Ionic Bonding

8.14 $Mg\cdot + \cdot\ddot{Br}: + \cdot\ddot{Br}: \longrightarrow Mg^{2-} + 2[:\ddot{Br}:]^-$

8.16 (a) BaO (b) RbI (c) Li_2S (d) $MgBr_2$

8.18 (a) Zn^{2+}: $[Ar]3d^{10}$

(b) Te^{2-}: $[Kr]5s^24d^{10}5p^6$ = [Xe], noble-gas configuration

(c) Se^{3+}: $[Ar]4s^23d^{10}4p^1$ (This is a very unlikely ion. A more stable and commonly found ion would be Sc^{3+}.) Sc^{3+}: [Ar], noble-gas configuration

(d) Ru^{2+}: $[Kr]4d^6$

(e) Tl^+: $[Xe]6s^24f^{14}5d^{10}$

(f) Au^+ $[Xe]4f^{14}5d^{10}$

8.20 (a) NaF, 910 kJ/mol; MgO, 3795 kJ/mol

The two factors that affect lattice energies are charge and ionic radii. The Na–F and Mg–O separations are similar (Na^+ is larger than Mg^{2+}, but F^- is smaller than O^{2-}). The charges on Mg^{2+} and O^{2-} are twice those of Na^+ and F^-, so according to Equation 8.4, the lattice energy of MgO is approximately four times that of NaF.

(b) $MgCl_2$, 2326 kJ/mol; $SrCl_2$, 2127 kJ/mol

The two factors that affect lattice energies are charge and ionic radii. The ionic charges are the same in the two compounds. The ionic radius of Mg^{2+} is smaller than that of Sr^{2+} so the Mg–Cl distance is slightly smaller than the Sr–Cl distance. Since lattice energy is inversely proportional to the ion separation, the lattice energy of $MgCl_2$ is slightly larger than that of $SrCl_2$.

8.22 (a) According to Equation 8.4, electrostatic attraction increases with increasing charges of the ions and decreases with increasing radius of the ions. Thus, lattice energy (i) increases as the charges of the ions increase and (ii) decreases as the sizes of the ions increase.

(b) RbBr < NaBr < LiCl < MgO. This order is confirmed by the lattice energies given in Table 8.2. MgO has the highest lattice energy because the ions have 2+ and 2– charges. The other compounds have cations with 1+ charges and anions with 1– charges. They are placed in order of decreasing ionic separation. Rb^+ and Br^- have the largest radii; Na^+ is smaller than Rb^+, Li^+ is smaller than Na^+, and Cl^- is smaller than Br^-.

8.24 (a) In MgO, the magnitude of the charges on both ions is 2; in $MgCl_2$, the magnitudes of the charges are 2 and 1. Also, the Cl^- ion is larger than the O^{2-} ion, so the charge separation is greater in $MgCl_2$. Thus, the lattice energy of MgO is greater, because the product of the ionic charges is greater and the ion separation is smaller.

(b) The ions have 1+ and 1– charges in all three compounds. In NaCl the cationic and anionic radii are smaller than in the other two compounds, so it has the largest lattice energy. In RbBr and CsBr, the anion is the same, but the Cs cation is larger, so CsBr has the smaller lattice energy.

(c) In BaO, the magnitude of the charges of both ions is 2; in KF, the magnitudes are 1. Charge considerations alone predict that BaO will have the higher lattice energy. The distance effect is less clear; O^{2-} and F^- are isoelectronic, so F^-, with the larger Z, has a slightly smaller radius. Ba^{2+} is two rows lower on the periodic chart than K^+, but it has a greater positive charge, so the radii are probably similar. In any case, the ionic separations in the two compounds are not very different, and the charge effect dominates.

8.26 $Ca(s) \rightarrow Ca(g)$; $Br_2(l) \rightarrow 2Br(g)$; $Ca(g) \rightarrow Ca^+(g) + 1e^-$;

$Ca^+(g) \rightarrow Ca^{2+}(g) + 1e^-$; $2Br(g) + 2e^- \rightarrow 2Br^-(g)$, exothermic;

$Ca^{2+}(g) + 2Br^-(g) \rightarrow CaBr_2(s)$, exothermic

8.28 By analogy to Figure 8.4:

$$\Delta H_{latt} = -\Delta H_f^\circ\ CaCl_2 + \Delta H_f^\circ\ Ca(g) + 2\Delta H_f^\circ\ Cl(g) + I_1\ (Ca) + I_2\ (Ca) + 2E\ (Cl)$$
$$= -(-795.8\ kJ) + 179.3\ kJ + 2(121.7\ kJ) + 590\ kJ + 1145\ kJ + 2(-349\ kJ) = +2256\ kJ$$

From Table 8.2, the lattice energy of NaCl, +788 kJ/mol, is considerably less than that of CaF_2. The 2+ charge of Ca^{2+} leads to much greater electrostatic attractions and a higher lattice energy.

Covalent Bonding, Electronegativity, and Bond Polarity

8.30 K and Ar. K is an active metal with one valence electron. It is most likely to achieve an octet by losing this single electron and to participate in ionic bonding. Ar has a stable octet of valence electrons; it is not likely to form chemical bonds of any type.

8.32 :F· + :F· + :F· + ·P: ⟶ PF_3 (Lewis structure: P bonded to three F atoms, each F with three lone pairs, P with one lone pair)

8.34 S=C=S (each S with two lone pairs)

The C–S bonds in CS_2 are double bonds, so the C–S distances will be shorter than a C–S single bond distance.

8.36 (a) The electronegativity of the elements increases going from left to right across a row of the periodic chart.

(b) Electronegativity decreases going down a family of the periodic chart.

(c) Generally, the trends in electronegativity are the same as those in ionization energy and opposite those in electron affinity. That is, the more positive the ionization energy and the more negative the electron affinity (ignoring a few exceptions), the greater the electronegativity of an element.

8.38 Electronegativity increases going up and to the right in the periodic table.

(a) O (b) Al (c) Cl (d) F

8.40 The more different the electronegativity values of the two elements, the more polar the bond.

(a) O–F < C–F < Be–F. This order is clear from the periodic trend.

(b) S–Br < C–P < O–Cl. Refer to the electronegativity values in Figure 8.6 to confirm the order of bond polarity. The 3 pairs of elements all have the same positional relationship on the periodic chart. The more electronegative element is one row above and one column to the left of the less electronegative element. This leads us to conclude that ΔEN is similar for the 3 bonds, which is confirmed by values in Figure 8.6. The most polar bond, O–Cl, involves the most electronegative element, O. Generally, the largest electronegativity differences tend to be between row 2 and row 3 elements. The 2 bonds in this exercise involving elements in row 2 and row 3 do have slightly greater ΔEN than the S–Br bond, between elements in rows 3 and 4.

(c) C–S < N–O < B–F. You might predict that N–O is least polar since the elements are adjacent on the table. However, the big decrease going from the second row to the third means that the electronegativity of S is not only less than that of O, but essentially the same as that of C. C–S is the least polar.

8.42 (a) The more electronegative element, Br, will have a stronger attraction for the shared electrons and adopt a partial negative charge.

(b) Q is the charge at either end of the dipole.

$$Q = \frac{\mu}{r} = \frac{1.21\,\text{D}}{2.49\,\text{Å}} \times \frac{1\,\text{Å}}{1 \times 10^{-10}\,\text{m}} \times \frac{3.34 \times 10^{-30}\,\text{C} \bullet \text{m}}{1\,\text{D}} \times \frac{1\,e}{1.60 \times 10^{-19}\,\text{C}}$$

$$= 0.1014 = 0.101\,e$$

The charges on I and Br are 0.101 *e*.

8.44 Generally, compounds formed by a metal and a nonmetal are described as ionic, while compounds formed from two or more nonmetals are covalent.

(a) MnF_3, ionic (b) CrO_3, ionic

(c) $AsBr_5$, covalent (As is a metalloid, so the bonding model is not obvious. The structure of the compound, a symmetrical trigonal bipyramid, indicates that it is probably covalent.)

(d) sulfur tetrafluoride, covalent

(e) molybdenum(IV) chloride, ionic

(f) scandium(III) chloride, ionic

Lewis Structures; Resonance Structures

8.46 (a) 12 e^-, 6 e^- pairs (b) 14 valence e^-, 7 e^- pairs

H—C=Ö (with H bonded below C) H—Ö—Ö—H

(c) 50 valence e^-, 25 e^- pairs

```
   :F: :F:
    |   |
:F—C—C—F:
    |   |
   :F: :F:
```

(Choose the Lewis structure that obeys the octet rule, Section 8.7)

(d) 26 valence e^-, 13 e^- pairs

```
[ :O—As—O: ]3−
       |
      :O:
```

(e) 26 valence e^- 13 e^- pairs

```
H—O—S—O—H
      |
     :O:
```

(Choose the Lewis structure that obeys they octet rule, Section 8.7.)

(f) 10 e^-, 5 e^- pairs

```
H—C≡C—H
```

8.48 (a) 26 e^-, 13 e^- pairs

```
:F—P—F:
    |
   :F:
```

The octet rule is satisfied for all atoms in the structure.

(b) F is more electronegative than P. Assuming F atoms hold all shared electrons, the oxidation number of each F is –1. The oxidation number of P is +3.

(c) Assuming perfect sharing, the formal charges on all F and P atoms are 0.

(d) The oxidation number on P is +3; the formal charge is 0. These represent extremes in the possible electron distribution, not the best picture. By virtue of their greater electronegativity, the F atoms carry a partial negative charge, and the P atom a partial positive charge.

8.50 Formal charges are given near the atoms, oxidation numbers are listed below the structures.

(a) 18 e^-, 9 e^- pairs

```
:O—S=O
 -1  +1  0
```

ox. #: S, +4; O, –2

(b) 24 e^-, 12 e^- pairs

```
          +2
-1 :O—S=O 0
        |
       :O:
       -1
```

ox. #: S, +6; O, –2

(c) 26 e^-, 13 e^- pairs

```
        +1
[ -1:O—S—O:-1 ]2−
        |
       :O:
       -1
```

ox. #: S, +4; O, –2

(d) 32 e^-, 16 e^- pairs

```
         -1
        :O:
         |
[ -1 :O—S—O: -1 ]2−
         |+2
        :O:
         -1
```

ox. #: S, +6; O, –2

8.52 (a) 16 e^-, 8 e^- pairs

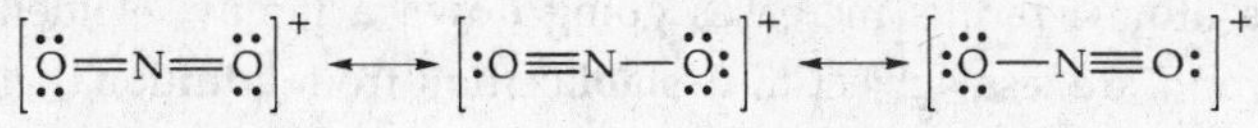

$$[\ddot{\underset{\cdot\cdot}{O}}=N=\ddot{\underset{\cdot\cdot}{O}}]^+ \longleftrightarrow [:O\equiv N-\ddot{\underset{\cdot\cdot}{O}}:]^+ \longleftrightarrow [:\ddot{\underset{\cdot\cdot}{O}}-N\equiv O:]^+$$

(b) More than one correct Lewis structure can be drawn, so resonance structures are needed to accurately describe the structure.

(c) NO_2^+ has 16 valence electrons. Consider other triatomic molecules involving second-row nonmetallic elements. O_3^{2+} or C_3^{4-} are not "common" (or stable). CO_2 is common and matches the description (as does N_3^-, azide ion).

8.54 The Lewis structures are as follows:

5 e^- pairs

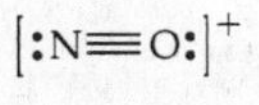

$$[:N\equiv O:]^+$$

9 e^- pairs

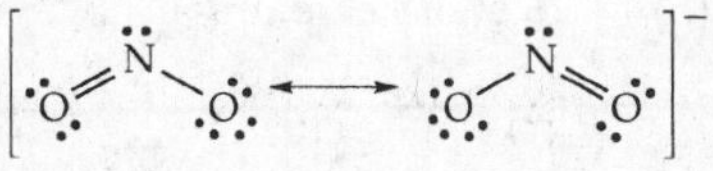

12 e^- pairs

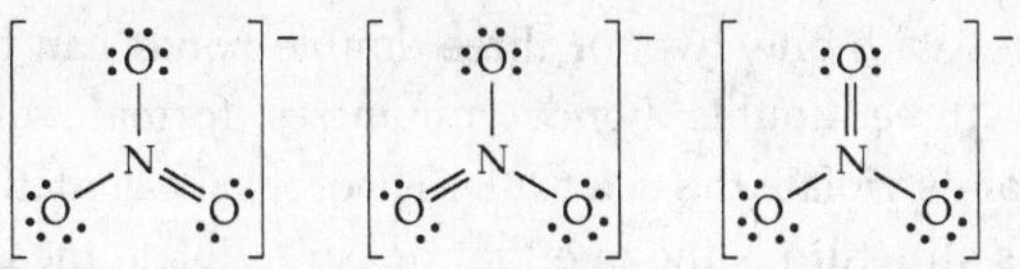

The average number of electron pairs in the N–O bond is 3.0 for NO^+, 1.5 for NO_2^-, and 1.33 for NO_3^-. The more electron pairs shared between two atoms, the shorter the bond. Thus the N–O bond lengths vary in the order $NO^+ < NO_2^- < NO_3^-$.

8.56 (a)

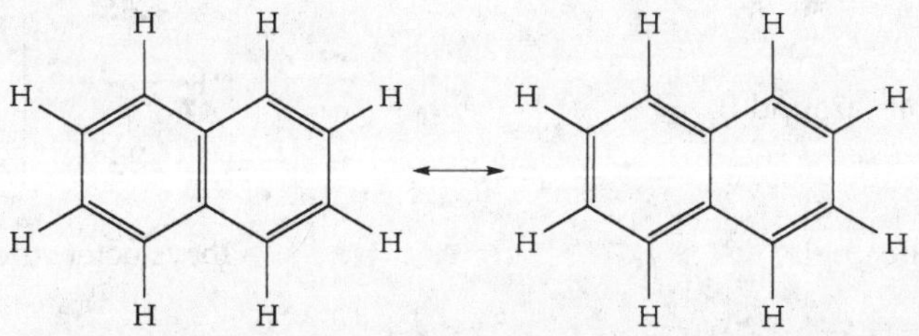

(b) The resonance model of this molecule has bonds that are neither single nor double, but somewhere in between. This results in bond lengths that are intermediate between C–C single and C=C double bond lengths.

(c)

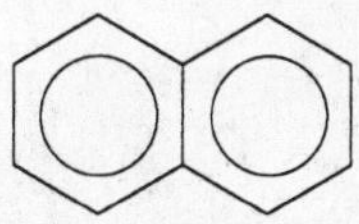

Exceptions to the Octet Rule

8.58 Carbon, in group 14, needs to form four single bonds to achieve an octet, as in CH_4. Nitrogen, in group 15, needs to form three, as in NH_3. If G = group number and n = the number of single bonds, G + n = 18 is a general relationship for the representative nonmetals.

Check: O as in H_2O (G = 16) + (n = 2 bonds) = 18

8.60 In the third period, atoms have the space and available orbitals to accommodate extra electrons. Since atomic radius increases going down a family, elements in the third period and beyond are less subject to destabilization from additional electron-electron repulsions. Also, the third shell contains d orbitals that are relatively close in energy to 3s and 3p orbitals (the ones that accommodate the octet) and provide an allowed energy state for the extra electrons.

8.62 (a) 16 e^-, 8 e^- pairs

:Ö=C=Ö: ⟷ :Ö—C≡O: ⟷ :O≡C—Ö:

Three resonance structures, all obey the octet rule. The left structure minimizes formal charge.

(b) 26 e^-, 13 e^- pairs

$[:\ddot{O}—\ddot{I}—\ddot{O}:]^-$ with a third :Ö: bonded to I

Other resonance structures with one, two, or three double bonds can be drawn. While a structure with three double bonds minimizes formal charges, all structures with double bonds violate the octet rule. Theoretical calculations show that the single best Lewis structure is the one that doesn't violate the octet rule. Such a structure is shown above.

(c) 6 e^-, 3 e^- pairs

H—B—H with a third H bonded to B

6 electrons around B.

(d) 32 e^-, 16 e^- pairs

$[BF_4]^-$: :F̤: bonded above, :F̤—B—F̤: , :F̤: bonded below

Obeys octet rule.

(e) 22 e^-, 11 e^- pair

:F̤—Xe—F̤: (three lone pairs on Xe)

Violates the octet rule; 10 e^- around central Xe.

8.64 (a) 19 e^-, 9.5 e^- pairs, odd electron molecule

:Ö—Cl—Ö: ⟷ :Ö—Cl—Ö· ⟷ ·Ö—Cl—Ö:

(b) None of the structures satisfies the octet rule. In each structure, one atom has only 7 e^- around it. If a molecule has an odd number of electrons in the valence shell, no Lewis structure can satisfy the octet rule.

(c) :Ö—Cl—Ö: ⟷ :Ö—Cl—Ö· ⟷ ·Ö—Cl—Ö:

Structure 1			Structure 2			Structure 3		
−1	+2	−1	−1	+1	0	0	+1	−1

Formal charge arguments predict that the two resonance structures with the odd electron on O are most important. This contradicts electronegativity arguments, which would predict that the less electronegative atom, Cl, would be more likely to have fewer than 8 e^- around it.

Bond Enthalpies

8.66 (a) $\Delta H = 3D(C\text{–}Br) + D(C\text{–}H) + D(Cl\text{–}Cl) - 3D(C\text{–}Br) - (C\text{–}Cl) - D(H\text{–}Cl)$

$= D(C\text{–}H) + D(Cl\text{–}Cl) - D(C\text{–}Cl) - D(H\text{–}Cl)$

$\Delta H = 413 + 242 - 328 - 431 = -104$ kJ

(b) $\Delta H = 4D(C\text{–}H) + 2D(C\text{–}S) + 2D(S\text{–}H) + D(C\text{–}C) + D(H\text{–}Br)$

$-4D(S\text{–}H) - D(C\text{–}C) - 2D(C\text{–}Br) - 4D(C\text{–}H)$

$= 2D(C\text{–}S) + D(H\text{–}Br) - 2D(S\text{–}H) - 2D(C\text{–}Br)$

$\Delta H = 2(259) + 366 - 2(339) - 2(276) = -346$ kJ

(c) $\Delta H = 4D(N\text{–}H) + D(N\text{–}N) + D(Cl\text{–}Cl) - 4D(N\text{–}H) - 2D(N\text{–}Cl)$

$= D(N\text{–}N) + D(Cl\text{–}Cl) - 2D(N\text{–}Cl)$

$\Delta H = 163 + 242 - 2(200) = 5$ kJ

8.68 *Plan.* Draw structural formulas so bonds can be visualized. *Solve.*

(a)

3 H—C(H)=C(H)—H ⟶ cyclohexane ring, each C bearing two H (H H, H H, H H, H H, H H, H H)

$\Delta H = 12D(C\text{–}H) + 3D(C{=}C) - 12D(C\text{–}H) - 6D(C\text{–}C)$

$= 3D(C{=}C) - 6D(C\text{–}C) = 3(614) - 6(348) = -246$ kJ

(b)

H—Si(Cl)(H)—H + 3 Cl—Cl ⟶ Cl—Si(Cl)(Cl)—Cl + 3 H—Cl

$\Delta H = 3D(Si\text{–}H) + D(Si\text{–}Cl) + 3D(Cl\text{–}Cl) - 4D(Si\text{–}Cl) - 3D(H\text{–}Cl)$

$= 3D(Si\text{–}H) + 3D(Cl\text{–}Cl) - 3D(Si\text{–}Cl) - 3D(H\text{–}Cl)$

$= 3(323) + 3(242) - 3(464) - 3(431) = -990$ kJ

(c) *Plan.* Use bond enthalpies to calculate ΔH for the reaction with $S_8(g)$ as a product. Then,

$8H_2S(g) \rightarrow 8H_2(g) + S_8(g)$	ΔH
$S_8(g) \rightarrow S_8(s)$	$-\Delta H_f^\circ$ for $S_8(g)$
$8H_2S(g) \rightarrow 8H_2(g) + S_8(s)$	$\Delta H_{rxn} = [\Delta H - \Delta H_f^\circ\ S_8(g)]$

$\Delta H = 16D(S\text{-}H) - 8(H\text{-}H) - 8(S\text{-}S)$

$= 16(339) - 8(436) - 8(266) = -192 \text{ kJ}$

$\Delta H_{rxn} = \Delta H - \Delta H_f^\circ\ S_8(g) = -192 \text{ kJ} - 102.3 \text{ kJ} = -294.3 = -294 \text{ kJ}$

8.70 (a) $H_2C{=}CH_2 + H{-}H \longrightarrow H_3C{-}CH_3$

$\Delta H = 4D(C\text{-}H) + D(C{=}C) + D(H\text{-}H) - 6D(C\text{-}H) - D(C\text{-}C)$

$= D(C{=}C) + D(H\text{-}H) - 2D(C\text{-}H) - D(C\text{-}C)$

$\Delta H = 614 + 436 - 2(413) - 348 = -124 \text{ kJ}$

(b) $\Delta H^\circ = \Delta H_f^\circ\ C_2H_6(g) - \Delta H_f^\circ\ C_2H_4(g) - \Delta H_f^\circ\ H_2(g)$

$= -84.68 - 52.30 - 0 = -136.98 \text{ kJ}$

The values of ΔH for the reaction differ because the bond enthalpies used in part (a) are average values that can differ from one compound to another. For example, the exact enthalpy of a C-H bond in C_2H_4 is probably not equal to the enthalpy of a C-H bond in C_2H_6. Thus, reaction enthalpies calculated from average bond enthalpies are estimates. On the other hand, standard enthalpies of formation are measured quantities and should lead to accurate reaction enthalpies. The advantage of average bond enthalpies is that they can be used for reactions where no measured enthalpies of formation are available.

8.72 (a) (i) $C + 2\ F{-}F \longrightarrow CF_4$

$\Delta H = 2D(F\text{-}F) - 4D(C\text{-}F) = 2(155) - 4(485) = -1630 \text{ kJ}$

(ii) $C{\equiv}O + 3\ F{-}F \longrightarrow CF_4 + F{-}O{-}F$

$\Delta H = D(C{\equiv}O) + 3D(F\text{-}F) - 4D(C\text{-}F) - 2D(D\text{-}F)$

$= 1072 + 3(155) - 4(485) - 2(190) = -783 \text{ kJ}$

(iii) $O{=}C{=}O + 4\ (F{-}F) \longrightarrow CF_4 + 2\ F{-}O{-}F$

$\Delta H = 2D(C{=}O) + 4D(F\text{-}F) - 4D(C\text{-}F) - 4D(O\text{-}F)$

$= 2(799) + 4(155) - 4(485) - 4(190) = -482 \text{ kJ}$

Reaction (i) is most exothermic.

(b) The more oxygen atoms bound to carbon, the less exothermic the reaction in this series.

Additional Exercises

8.74 (a) Lattice energy is proportional to Q_1Q_2/d. For each of these compounds, Q_1Q_2 is the same. The anion H^- is present in each compound, but the ionic radius of the cation increases going from Be to Ba. Thus, the value of d (the cation-anion separation) increases and the ratio Q_1Q_2/d decreases. This is reflected in the decrease in lattice energy going from BeH_2 to BaH_2.

(b) Again, Q_1Q_2 for ZnH_2 is the same as that for the other compounds in the series and the anion is H^-. The lattice energy of ZnH_2, 2870 kJ, is closest to that of MgH_2, 2791 kJ. The ionic radius of Zn^{2+} is similar to that of Mg^{2+}.

8.76 $$E = \frac{-8.99 \times 10^9\ J \bullet m}{C^2} \times \frac{4(1.60 \times 10^{-19}\ C)^2}{(0.99 + 1.40) \times 10^{-10}\ m} = -3.852 \times 10^{-18} = -3.85 \times 10^{-18}\ J$$

On a molar basis: $(-3.852 \times 10^{-18}\ J)(6.022 \times 10^{23}) = -2.319 \times 10^6\ J = -2320\ kJ$

Note that its absolute value is less than the lattice energy, 3414 kJ/mol. The difference represents the added energy of putting all the $Ca^{2+}O^{2-}$ ion pairs together in a three-dimensional array, similar to the one in Figure 8.3.

8.77 $E = Q_1Q_2/d$; $k = 8.99 \times 10^9\ J\ m/coul^2$

(a) $$Na^+, Br^-:\ E = \frac{-8.99 \times 10^9\ J \bullet m}{C^2} \times \frac{(1.60 \times 10^{-19}\ C)^2}{(0.97 + 1.96) \times 10^{-10}\ m} = -7.8547 \times 10^{-19}$$

$$= -7.85 \times 10^{-19}\ J$$

The sign of E is negative because one of the interacting ions is an anion; this is an attractive interaction.

On a molar basis: $-7.855 \times 10^{-19} \times 6.022 \times 10^{23} = -4.73 \times 10^5\ J = -473\ kJ$

(b) $$Rb^+, Br^-: E = \frac{-8.99 \times 10^9\ J \bullet m}{C^2} \times \frac{(1.60 \times 10^{-19}\ C)^2}{(1.47 + 1.96) \times 10^{-10}\ m} = -6.71 \times 10^{-19}\ J$$

On a molar basis: $-4.04 \times 10^5\ J = -404\ kJ$

(c) $$Sr^{2+}, S^{2-}: E = \frac{-8.99 \times 10^9\ J \bullet m}{C^2} \times \frac{(2 \times 1.60 \times 10^{-19}\ C)^2}{(1.13 + 1.84) \times 10^{-10}\ m} = -3.10 \times 10^{-18}\ J$$

On a molar basis: $-1.87 \times 10^6\ J = -1.87 \times 10^3\ kJ$

8.79 Molecule (b) H_2S and ion (c) NO_2^- contain polar bonds. The atoms that form the bonds (H-S) and N-O) have different electronegativity values.

8.80 (a) B-O. The most polar bond will be formed by the two elements with the greatest difference in electronegativity. Since electronegativity increases moving right and up on the periodic chart, the possibilities are B-O and Te-O. These two bonds are likely to have similar electronegativitiy differences (3 columns apart vs. 3 rows apart). Values from Figure 8.6 confirm the similarity, and show that B-O is slightly more polar.

(b) Te-I. Both are in the fifth row of the periodic chart and have the two largest covalent radii among this group of elements.

(c) TeI_2. Te needs to participate in two covalent bonds to satisfy the octet rule, and each I atom needs to participate in one bond, so by forming a TeI_2 molecule, the octet rule can be satisfied for all three atoms.

:Ï—Te—Ï:

(d) B_2O_3. Although this is probably not a purely ionic compound, it can be understood in terms of gaining and losing electrons to achieve a noble-gas configuration. If each B atom were to lose 3 e^- and each O atom were to gain 2 e^-, charge balance and the octet rule would be satisfied.

P_2O_3. Each P atom needs to share 3 e^- and each O atom 2 e^- to achieve an octet. Although the correct number of electrons seem to be available, a correct Lewis structure is difficult to imagine. In fact, phosphorus (III) oxide exists as P_4O_6 rather than P_2O_3 (Chapter 22).

8.82 Use the method detailed in Section 8.5, *A Closer Look*, to estimate partial charges from electronegativity values. From Figure 8.6, the electronegativity of Br is 2.8 and of Cl is 3.0.

Br has 2.8/(3.0 + 2.8) = 0.48 of the charge of the bonding e^- pair.

Cl has 3.0/(3.0 + 2.8) = 0.52 of the charge of the bonding e^- pair.

This amounts to $0.52 \times 2e = 1.04e$ on Cl or $0.04e$ more than a neutral Cl atom. This implies a –0.04 charge on Cl and +0.04 charge on Br.

From Figure 7.5, the covalent radius of Br is 1.14 Å and of Cl is 0.99 Å. The Br–Cl separation is 2.13 Å.

$$\mu = Qr = 0.04e \times \frac{1.60 \times 10^{-19}\ \text{C}}{e} \times 2.13\ \text{Å} \times \frac{1 \times 10^{-10}\ \text{m}}{\text{Å}} \times \frac{1\text{D}}{3.34 \times 10^{-30}\ \text{C}\bullet\text{m}} = 0.41\ \text{D}$$

Clearly, this method is approximate. The estimated dipole moment of 0.41 D is within 28% of the measured value of 0.57 D.

8.83 I_3^- has a Lewis structure with an expanded octet of electrons around the central I.

:Ï—Ï—Ï:

F cannot accommodate an expanded octet because it is too small and has no available d orbitals in its valence shell.

8.85 (a) 14e^-, 7 e^- pairs

$[:\ddot{\text{Cl}}—\ddot{\text{O}}:]^-$

FC on Cl = 7 - [6 + 1/2(2)] = 0

32 e^-, 16 e^- pairs

$[ClO_4]^-$: Cl bonded to four :Ö: atoms

FC on Cl = 7 - [0 + 1/2(8)] = +3

(b) The oxidation number of Cl is +1 in ClO^- and +7 in ClO_4^-.

(c) The definition of formal charge assumes that all bonding pairs of electrons are equally shared by the two bonded atoms, that all bonds are purely covalent. The definition of oxidation number assumes that the more electronegative element in the bond gets all of the bonding electrons, that the bonds are purely ionic. These two definitions represent the two extremes of how electron density is distributed between bonded atoms.

In ClO^- and ClO_4^-, Cl is the less electronegative element, so the oxidation numbers have a higher positive value than the formal charges. The true description of the electron density distribution is somewhere between the extremes indicated by formal charge and oxidation number.

8.86 (a) $:N\equiv N-\ddot{\underset{..}{O}}: \longleftrightarrow :\ddot{\underset{..}{N}}-N\equiv O: \longleftrightarrow :\ddot{N}=N=\ddot{O}:$

0 +1 −1 −2 +1 −1 −1 +1 0

In the leftmost structure, the more electronegative O atom has the negative formal charge, so this structure is likely to be most important.

(b) In general, the more shared pairs of electrons between two atoms, the shorter the bond, and vice versa. That the N-N bond length in N_2O is slightly longer than the typical N≡N indicates that the middle and right resonance structures where the N atoms share less than three electron pairs are contributors to the true structure. That the N-O bond length is slightly shorter than a typical N=O indicates that the middle structure, where N and O share more than two electron pairs, does contribute to the true structure. This physical data indicates that while formal charge can be used to predict which resonance form will be more important to the observed structure, the influence of minor contributors on the true structure cannot be ignored.

8.88 (a) $\Delta H = 5D(C\text{–}H) + D(C\text{–}C) + D(C\text{–}O) + D(O\text{–}H) - 6D(C\text{–}H) - 2D(C\text{–}O)$

$= D(C\text{–}C) + D(O\text{–}H) - D(C\text{–}H) - D(C\text{–}O)$

$= 348\text{ kJ} + 463\text{ kJ} - 413\text{ kJ} - 358\text{ kJ}$

$\Delta H = +40$ kJ; ethanol has the lower enthalpy

(b) $\Delta H = 4D(C\text{–}H) + D(C\text{–}C) + 2D(C\text{–}O) - 4D(C\text{–}H) - D(C\text{–}C) - D(C{=}O)$

$= 2D(C\text{–}O) - D(C{=}O)$

$= 2(358\text{ kJ}) - 799\text{ kJ}$

$\Delta H = -83$ kJ; acetaldehyde has the lower enthalpy

(c) $\Delta H = 8D(C\text{–}H) + 4D(C\text{–}C) + D(C{=}C) - 8D(C\text{–}H) - 2D(C\text{–}C) - 2D(C{=}C)$

$= 2D(C\text{–}C) - D(C{=}C)$

$= 2(348\text{ kJ}) - 614\text{ kJ}$

$\Delta H = +82$ kJ; cyclopentene has the lower enthalpy

(d) $\Delta H = 3D(C\text{–}H) + D(C\text{–}N) + D(C\equiv N) - 3D(C\text{–}H) - D(C\text{–}C) - D(C\equiv N)$

$= D(C\text{–}N) - D(C\text{–}C)$

$= 293\text{ kJ} - 348\text{ kJ}$

$\Delta H = -55$ kJ; acetonitrile has the lower enthalpy

8.89 (a) 4 [H—C(O—NO₂)(H)—C(O—NO₂)(H)—C(O—NO₂)(H)—H] → 6 N≡N + 12 O=C=O + 10 H—O—H + O=O

nitroglycerine

$$\Delta H = 20D(\text{C-H}) + 8D(\text{C-C}) + 12D(\text{C-O}) + 24D(\text{O-N}) + 12D(\text{N=O})$$
$$- [6D(\text{N} \equiv \text{N}) + 24D(\text{C=O}) + 20D(\text{H-O}) + D(\text{O=O})]$$

$$\Delta H = 20(413) + 8(348) + 12(358) + 24(201) + 12(607)$$
$$- [6(941) + 24(799) + 20(463) + 495]$$
$$= -7129 \text{ kJ}$$

$$1.00 \text{ g } C_3H_5N_3O_9 \times \frac{1 \text{ mol } C_3H_5N_3O_9}{227.1 \text{ g } C_3H_5N_3O_9} \times \frac{-7129 \text{ kJ}}{4 \text{ mol } C_3H_5N_3O_9} = 7.85 \text{ kJ/g } C_3H_5N_3O_9$$

(b) $4C_7H_5N_3O_6(s) \rightarrow 6N_2(g) + 7CO_2(g) + 10H_2O(g) + 21C(s)$

8.91 (a) $\Delta H = 2D(\text{A=A}) - 4D(\text{A-A})$

(b) For the reaction to be exothermic ($-\Delta H$), $D(\text{A=A}) < 2D(\text{A-A})$.

(c) If the reaction is exothermic, the second bond between atoms in A=A must be weaker than the first bond. (The first bond is A–A. If A=A < 2A–A, then the second bond has a smaller bond dissociation enthalpy than the first.)

8.92

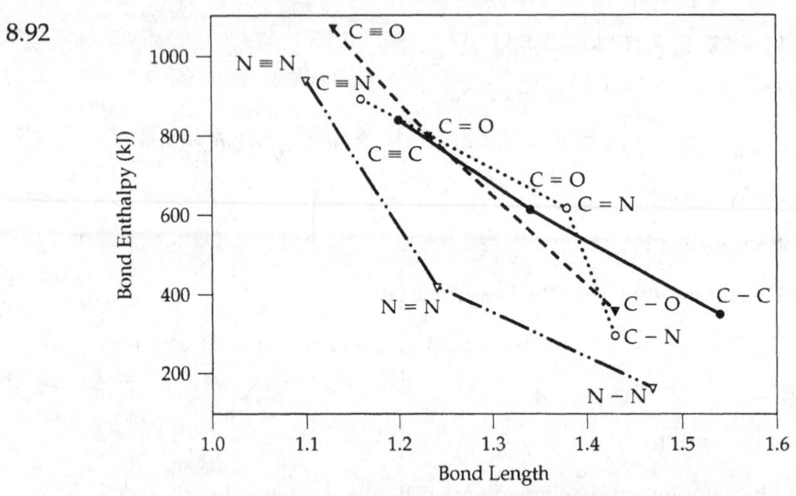

When comparing the same pair of bonded atoms (C-N vs. C=N vs. C≡N), the shorter the bond the greater the bond energy, but the two quantities are not necessarily directly proportional. The plot clearly shows that there are no simple length/strength correlations for single bonds alone, double bonds alone, triple bonds alone, or among different pairs of bonded atoms (all C-C bonds vs. all C-N bonds, etc.).

Integrative Exercises

8.95 (a)

$Sr(s) \rightarrow Sr(g)$	ΔH_f° Sr(g) [ΔH_{sub}° Sr(s)]
$Sr(g) \rightarrow Sr^+(g) + 1\,e^-$	I_1 Sr
$Sr^+(g) \rightarrow Sr^{2+}(g) + 1\,e^-$	I_2 Sr
$Cl_2(g) \rightarrow 2Cl(g)$	$2\,\Delta H_f^\circ$ Cl(g) [D(Cl_2)]
$2Cl(g) + 2\,e^- \rightarrow 2Cl^-(g)$	$2E_1$ Cl
$SrCl_2(s) \rightarrow Sr(s) + Cl_2(g)$	$-\Delta H_f^\circ$ $SrCl_2$
$SrCl_2(s) \rightarrow Sr^{2+}(g) + 2Cl^-(g)$	ΔH_{latt}

(b) $\Delta H_f^\circ SrCl_2(s) = \Delta H_f^\circ Sr(g) + I_1(Sr) + I_2(Sr) + 2\Delta H_f^\circ Cl(g) + 2E(Cl) - \Delta H_{latt} SrCl_2$

$\Delta H_f^\circ SrCl_2(s) = 164.4 \text{ kJ} + 549 \text{ kJ} + 1064 \text{ kJ} + 2(121.7) \text{ kJ} + 2(-349) \text{ kJ} - 2127 \text{ kJ}$

$= -804 \text{ kJ}$

8.96 The pathway to the formation of K_2O can be written:

$2K(s) \rightarrow 2K(g)$	$2\Delta H_f^\circ$ K(g)
$2K(g) \rightarrow 2K^+(g) + 2\,e^-$	$2\,I_1(K)$
$1/2\,O_2(g) \rightarrow O(g)$	ΔH_f° O(g)
$O(g) + 1\,e^- \rightarrow O^-(g)$	$E_1(O)$
$O^-(g) + 1\,e^- \rightarrow O^{2-}(g)$	$E_2(O)$
$2K^+(g) + O^{2-}(g) \rightarrow K_2O(s)$	$-\Delta H_{latt}\, K_2O(s)$
$2K(s) + 1/2\,O_2(g) \rightarrow K_2O(s)$	$\Delta H_f^\circ\, K_2O(s)$

$\Delta H_f^\circ K_2O(s) = 2\Delta H_f^\circ K(g) + 2\,I_1(K) + \Delta H_f^\circ O(g) + E_1(O) + E_2(O) - \Delta H_{latt}\, K_2O(s)$

$E_2(O) = \Delta H_f^\circ K_2O(s) + \Delta H_{latt}\, K_2O(s) - 2\Delta H_f^\circ K(g) - 2\,I_1(K) - \Delta H_f^\circ O(g) - E_1(O)$

$E_2(O) = -363.2 \text{ kJ} + 2238 \text{ kJ} - 2(89.99) \text{ kJ} - 2(419) \text{ kJ} - 247.5 \text{ kJ} - (-141) \text{ kJ}$

$= +750 \text{ kJ}$

8.98 (a) Even though Cl has the greater (more negative) electron affinity, F has a much larger ionization energy, so the electronegativity of F is greater.

F: k(IE-EA) = k(1681 - (-328)) = k(2009)

Cl: k(IE-EA) = k(1251 - (-349)) = k(1600)

(b) Electronegativiy is the ability of an atom in a molecule to attract electrons to itself. It can be thought of as the ability to hold its own electrons (as measured by ionization energy) and the capacity to attract the electrons of other atoms (as measured by electron affinity). Thus, both properties are relevant to the concept of electronegativity.

(c) EN = k(IE - EA). For F: 4.0 = k(2009), k = 4.0/2009 = 2.0×10^{-3}

(d) Cl: EN = 2.0×10^{-3} (1600) = 3.2

O: EN = 2.0×10^{-3} (1314 - (-141)) = 2.9

These values do not follow the trend on Figure 8.6. The Pauling scale on the figure shows O to be second only to F in electronegativity, more electronegative than Cl. The simple definition EN = k(IE - EA) that employs thermochemical properties of isolated, gas phase atoms does not take into account the complex bonding environment of molecules.

8.99 (a) Assume 100 g.

$$14.52 \text{ g C} \times \frac{1 \text{ mol}}{12.011 \text{ g C}} = 1.209 \text{ mol C}; 1.209/1.209 = 1$$

$$1.83 \text{ g H} \times \frac{1 \text{ mol}}{1.008 \text{ g H}} = 1.816 \text{ mol H}; 1.816/1.209 = 1.5$$

$$64.30 \text{ g Cl} \times \frac{1 \text{ mol}}{35.453 \text{ g Cl}} = 1.814 \text{ mol Cl}; 1.814/1.209 = 1.5$$

$$19.35 \text{ g O} \times \frac{1 \text{ mol}}{15.9994 \text{ g O}} = 1.209 \text{ mol O}; 1.209/1.209 = 1.0$$

Multiplying by 2 to obtain an integer ratio, the empirical formula is $C_2H_3Cl_3O_2$.

(b) The empirical formula weight is 2(12.0) + 3(1.0) + 3(35.5) + 2(16) = 165.5. The empirical formula is the molecular formula.

(c) 44 e^-, 22 e^- pairs

```
     ..    ..
    :Cl:  :O—H
     |     |
 ..         ..
:Cl—C——C—O—H
 ..         ..
     |     |
    :Cl:   H
     ..
```

8.101 (a) C_2H_2: 10 e^-, 5 e^- pair N_2: 10 e^-, 5 e^- pair

H—C≡C—H :N≡N:

(b) N_2 is an extremely stable, unreactive compound. Under appropriate conditions, it can be either oxidized (Section 22.7) or reduced (Sections 14.7 and 15.1). C_2H_2 is a reactive gas, used in combination with O_2 for welding and as starting material for organic synthesis (Section 25.4).

(c) $2N_2(g) + 5O_2(g) \rightarrow 2N_2O_5(g)$

$2C_2H_2(g) + 5O_2(g) \rightarrow 4CO_2(g) + 2H_2O(g)$

(d) $\Delta H^\circ_{rxn}(N_2) = 2\Delta H^\circ_f N_2O_5(g) - 2\Delta H^\circ_f N_2(g) - 5\Delta H^\circ_f O_2(g)$
$= 2(11.30) - 2(0) - 5(0) = 22.60$ kJ

$\Delta H^\circ_{ox} = 11.30$ kJ/mol N_2

$\Delta H^\circ_{rxn}(C_2H_2) = 4\Delta H^\circ_f CO_2(g) + 2\Delta H^\circ_f H_2O(g) - 2\Delta H^\circ_f C_2H_2(g) - 5\Delta H^\circ_f O_2(g)$
$= 4(-393.5 \text{ kJ}) + 2(-241.82 \text{ kJ}) - 2(226.7 \text{ kJ}) - 5(0)$
$= -2511.0$ kJ

$\Delta H^\circ_{ox}(C_2H_2) = -1255.5$ kJ/mol C_2H_2

The oxidation of C_2H_2 is highly exothermic, which means that the energy state of the combined products is much lower than that of the reactants. The reaction is "downhill" in an energy sense, and occurs readily. The oxidation of N_2 is mildly endothermic (energy of products higher than reactants) and the reaction does not readily occur. This is in agreement with the general reactivities from part (b).

Referring to bond enthalpies in Table 8.4, when the C–H bonds are taken into account, even more energy is required for bond breaking in the oxidation of C_2H_2 than in the oxidation of N_2. The difference seems to be in the enthalpies of formation of the products. $CO_2(g)$ and $H_2O(g)$ have extremely exothermic ΔH_f° values, which cause the oxidation of C_2H_2 to be energetically favorable. $N_2O_5(g)$ has an endothermic ΔH_f° value, which causes the oxidation of N_2 to be energetically unfavorable.

8.102 (a) Assume 100 g of compound

$$69.6 \text{ g S} \times \frac{1 \text{ mol S}}{32.07 \text{ g}} = 2.17 \text{ mol S}$$

$$30.4 \text{ g N} \times \frac{1 \text{ mol N}}{14.01 \text{ g}} = 2.17 \text{ mol N}$$

S and N are present in a 1:1 mol ratio, so the empirical formula is SN. The empirical formula weight is 46. MM/FW = 184.3/46 = 4 The molecular formula is S_4N_4.

(b) 44 e^-, 22 e^- pairs. Because of its small radius, N is unlikely to have an expanded octet. Begin with alternating S and N atoms in the ring. Try to satisfy the octet rule with single bonds and lone pairs. At least two double bonds somewhere in the ring are required.

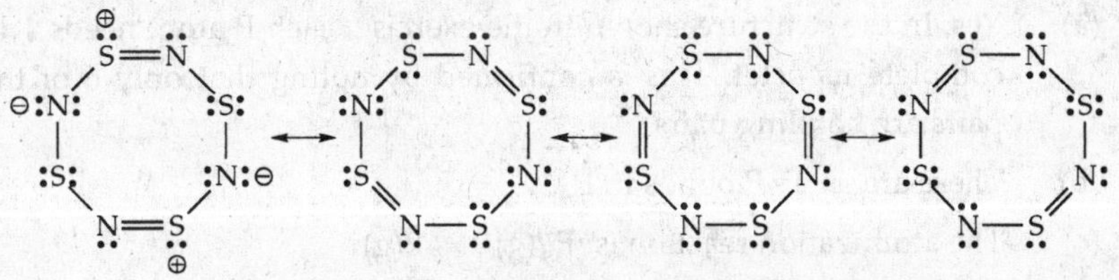

These structures carry formal charges on S and N atoms as shown. Other possibilities include:

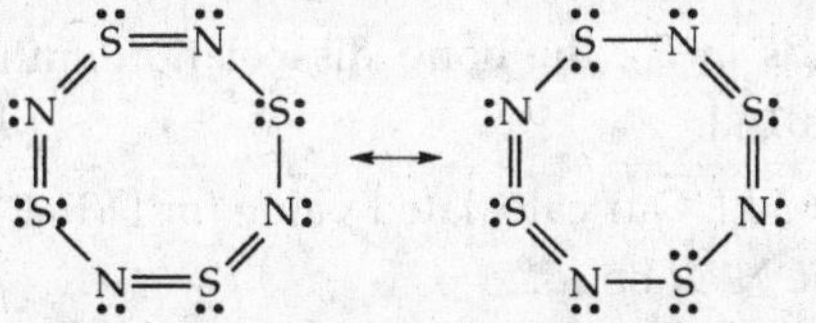

These structures have zero formal charges on all atoms and are likely to contribute to the true structure. Note that the S atoms that are shown with two double bonds are not necessarily linear, because S has an expanded octet. Other resonance structures with four double bonds are.

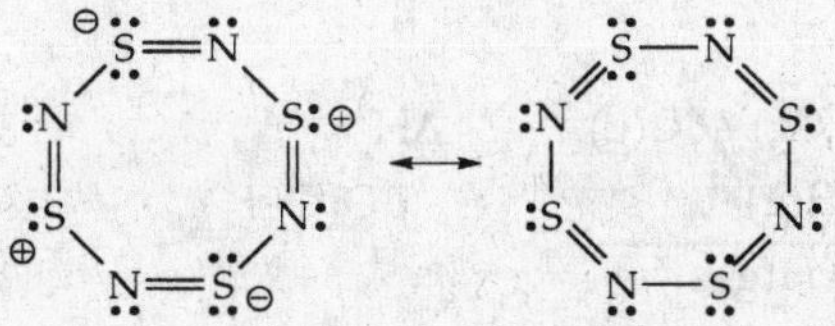

In either resonance structure, the two 'extra' electron pairs can be placed on any pair of S atoms in ring, leading to a total of 10 resonance structures. The sulfur atoms alternately carry formal charges of +1 and –1. Without further structural information, it is not possible to eliminate any of the above structures. Clearly, the S_4N_4 molecule stretches the limits of the Lewis model of chemical bonding.

(c) Each resonance structure has 8 total bonds and more than 8 but less than 16 bonding e^- pairs, so an "average" bond will be intermediate between a S–N single and double bond. We estimate an average S–N single bond length to be 1.77 Å (sum of bonding atomic radii from Figure 7.6). We do not have a direct value for a S–N double bond length. Comparing double and single bond lengths for C–C (1.34 Å, 1.54 Å), N–N (1.24 Å, 1.47 Å) and O–O (1.21 Å, 1.48 Å) bonds from Table 8.5, we see that, on average, a double bond is approximately 0.23 Å shorter than a single bond. Applying this difference to the S–N single bond length, we estimate the S–N double bond length as 1.54 Å. Finally, the intermediate S–N bond length in S_4N_4 should be between these two values, approximately 1.60–165 Å. (The measured bond length is 1.62 Å.)

(d) $S_4N_4 \rightarrow 4S(g) + 4N(g)$

$\Delta H = 4\Delta H_f^\circ\ S(g) + 4\Delta H_f^\circ\ N(g) - \Delta H_f^\circ\ S_4N_4$

$\Delta H = 4(222.8\ kJ) + 4(472.7\ kJ) - 480\ kJ = 2302\ kJ$

This energy, 2302 kJ, represents the dissociation of 8 S–N bonds in the molecule; the average dissociation energy of one S–N bond in S_4N_4 is then 2302 kJ/8 bonds = 287.8 kJ.

8.103 (a) Yes. In the structure shown in the exercise, each P atom needs 1 unshared pair to complete its octet. This is confirmed by noting that only 6 of the 10 valence e^- pairs are bonding pairs.

(b) There are six P–P bonds in P_4.

(c) The atomization reaction is: $P_4(g) \rightarrow 4P(g)$

$\Delta H_{atom} = 4\Delta H_f^\circ\ P(g) - \Delta H_f^\circ\ P_4(g)$

$= 4(316.4\ kJ) - 58.9\ kJ = 1206.7\ kJ$

(d) Since there are six P–P bonds in P_4, the bond dissociation enthalpy for an P–P bond, D(P–P) = 1206.7/6 = 201 kJ.

(e) From Table 8.4, D(N–N) = 163 kJ. Our calculated value for D(P–P) is 201 kJ, so the P–P bond is stronger than the N–N bond.

8.105 (a) $C_6H_6(g) \rightarrow 6H(g) + 6C(g)$

$\Delta H^\circ = 6\Delta H_f^\circ\ H(g) + 6\Delta H_f^\circ\ C(g) - \Delta H_f^\circ\ C_6H_6(g)$

$\Delta H^\circ = 6(217.94)\ kJ + 6(718.4)\ kJ - 82.9\ kJ = 5535\ kJ$

(b) $C_6H_6(g) \rightarrow 6CH(g)$

(c)

$C_6H_6(g) \rightarrow 6H(g) + 6C(g)$	ΔH°	5535 kJ
$6H(g) + 6C(g) \rightarrow 6CH(g)$	$-6D(C\text{–}H)$	$-6(413)$ kJ
$C_6H_6(g) \rightarrow 6CH(g)$		3057 kJ

3057 kJ is the energy required to break the six C-C bonds in $C_6H_6(g)$. The average bond dissociation energy for one carbon-carbon bond in $C_6H_6(g)$ is

$$\frac{3057\text{ kJ}}{6\text{C}-\text{C bonds}} = 509.5\text{ kJ}.$$

(d) The value of 509.5 kJ is between the average value for a C-C single bond (348 kJ) and a C=C double bond (614 kJ). It is somewhat greater than the average of these two values, indicating that the carbon-carbon bond in benzene is a bit stronger than we might expect.

8.106 (a) $Br_2(l) \rightarrow 2Br(g)$ $\Delta H^\circ = 2\Delta H_f^\circ\ Br(g) = 2(111.8)\text{ kJ} = 223.6\text{ kJ}$

(b) $CCl_4(l) \rightarrow C(g) + 4Cl(g)$

$\Delta H^\circ = \Delta H_f^\circ\ C(g) + 4\Delta H_f^\circ\ Cl(g) - \Delta H_f^\circ\ CCl_4(l)$

$= 718.4\text{ kJ} + 4(121.7)\text{ kJ} - (-139.3)\text{ kJ} = 1344.5$

$$\frac{1344.5\text{ kJ}}{4\text{ C}-\text{Cl bonds}} = 336.1\text{ kJ}$$

(c)

$H_2O_2(l) \rightarrow 2H(g) + 2O(g)$

$2H(g) + 2O(g) \rightarrow 2OH(g)$

$H_2O_2(l) \rightarrow 2OH(g)$

$D(O-O)(l) = 2\Delta H_f^\circ\ H(g) + 2\Delta H_f^\circ\ O(g) - \Delta H_f^\circ\ H_2O_2(l) - 2D(O-H)(g)$

$= 2(217.94)\text{ kJ} + 2(247.5)\text{ kJ} - (-187.8)\text{ kJ} - 2(463)\text{ kJ}$

$= 193\text{ kJ}$

(d) The data are listed below.

bond	D gas kJ/mol	D liquid kJ/mol
Br–Br	193	223.6
C-Cl	328	336.1
O-O	146	192.7

Breaking bonds in the liquid requires more energy than breaking bonds in the gas phase. For simple molecules, bond dissociation from the liquid phase can be thought of in two steps:

molecule (l) → molecule (g)

molecule (g) → atoms (g)

The first step is evaporation or vaporization of the liquid and the second is bond dissociation in the gas phase. Average bond enthalpy in the liquid phase is then the sum of the enthalpy of vaporization for the molecule and the gas phase bond dissociation enthalpies, divided by the number of bonds dissociated. This is greater than the gas phase bond dissociation enthalpy owing to the contribution from the enthalpy of vaporization.

9 Molecular Geometry and Bonding Theories

Visualizing Concepts

9.1 Removing an atom from the equatorial plane of trigonal bipyramid in Figure 9.3 creates a seesaw shape. It might appear that you could also obtain a seesaw by removing two atoms from the square plane of the octahedron. However, one of the B–A–B angles in the seesaw is 120°, so it must be derived from a trigonal bipyramid.

9.2 (a) 120°

(b) If the blue balloon expands, the angle between red and green balloons decreases.

(c) Nonbonding (lone) electron pairs exert greater repulsive forces than bonding pairs, resulting in compression of adjacent bond angles.

9.4 (a) 4 e^- domains

(b) The molecule has a non-zero dipole moment, because the C–H and C–F bond dipoles do not cancel each other.

(c) The dipole moment vector bisects the F–C–F and H–C–H angles, with the negative end of the vector toward the F atoms.

9.5 (a) The reference point for zero energy on the diagram corresponds to a state where the two Cl atoms are separate and not interfacing. This corresponds to an infinite Cl–Cl distance beyond the right extreme of the horizontal axis. The point near the left side of the plot where the curve intersects the x-axis at E = 0 has no special meaning.

(b) Energy decreases as atom separation decreases because the valence electrons of one atom come close enough to the other atom to be stabilized by both nuclei instead of just one nucleus.

(c) The Cl–Cl distance at the energy minimum on the plot is the Cl–Cl bond distance.

(d) At interatomic separations shorter than the bond distance, the two nuclei begin to repel each other, increasing the overall energy of the system.

9.7 (a) The ground state electron configuration of Si is [Ne]$3s^2 3p^2$. The left side of the orbital diagram corresponds to the configuration $3s^1 3p^3$, so one or more electrons has been promoted.

(b) The diagram illustrates mixing of a single s and three p atomic orbitals to form sp^3 hybrids.

9.8 (a) Recall that π bonds require p atomic orbitals, so the maximum hybridization of a C atom involved in a double bond is sp^2 and in a triple bond is sp. There are 6 C atoms in the molecule. Starting on the left, the hybridizations are: sp^2, sp^2, sp^3, sp, sp, sp^3.

(b) All single bonds are σ bonds. Double and triple bonds each contain 1 σ bond. This molecule has 8 C–H σ bonds and 5 C–C σ bonds, for a total of 13 σ bonds.

(c) Double bonds have 1 π bond and triple bonds have 2 π bonds. This molecule has a total of 3 π bonds.

9.10 (a) The diagram has five electrons in MOs formed by 2p atomic orbitals. C has two 2p electrons, so X must have three 2p electrons. X is N.

(b) The molecule has an unpaired electron, so it is paramagnetic.

(c) Atom X is N, which is more electronegative than C. The atomic orbitals of the more electronegative N are slightly lower in energy than those of C. The lower energy π_{2p} bonding molecular orbitals will have a greater contribution from the lower energy N atomic orbitals. (Higher energy π^*_{2p} MOs will have a greater contribution from higher energy C atomic orbitals.)

Molecular Shapes; the VSEPR Model

9.12 (a) In a symmetrical tetrahedron, the four bond angles are equal to each other, with values of 109.5°. The H–C–H angles in CH_4 and the O–Cl–O angles in ClO_4^- will have values close to 109.5°.

(b) Two, the H–N–H angle and the N–H distance, assuming that NH_3 is a symmetric trigonal pyramid with equal bond distances and angles. In a trigonal pyramidal molecule, there are three bonding and one nonbonding electron domains. Since a nonbonding electron domain compresses bond angles, the H–N–H angles will not be exactly 109.5°, the bond angles in a perfect tetrahedron. The bond angle must be specified. Also, N does not sit in the plane of the H atoms. The distance of N out of the plane is determined by the N–H distances, as well as the H–N–H angles. The N–H distance must also be specified.

9.14 (a) The number of electron domains in a molecule or ion is the number of bonds (double and triple bonds count as one domain) **plus** the number of nonbonding (lone) electron pairs.

(b) A *bonding electron domain* is a region between two bonded atoms that contains one or more pairs of bonding electrons. A *nonbonding electron domain* is localized on a single atom and contains one pair of nonbonding electrons (a lone pair).

9.16 (a) 3 (or 5 if 90° angles are also present)

(b) 2 (or 5 if 120° angles are also present, 6 if more than one 180° angle is present)

(c) 4 (d) 6 (or 5 if 120° angles are also present)

9.18 If the electron-domain geometry is trigonal bipyramidal, there are five total electron domains around the central atom. An AB_3 molecule has three bonding domains, so there must be two nonbonding domains on A.

9.20 (a) —X— trigonal planar trigonal planar

(b) tetrahedral trigonal pyramidal

(c) trigonal bipyramidal linear

9.22 bent (b), linear (l), octahedral (oh), seesaw (ss), square pyramidal (sp), square planar (spl), tetrahedral (td), trigonal bipyramidal (tbp), trigonal planar (tr), trigonal pyramidal (tp), T-shaped (T)

	Molecule or ion	Valence electrons	Lewis structure		Electron-domain geometry	Molecular geometry
(a)	HCN	10	:N≡C—H	N≡C—H	l	l
(b)	SO_3^{2-}	26	* $[:\ddot{O}—\ddot{S}—\ddot{O}:]^{2-}$ with :O: below S	$[S(O)_3]^{2-}$	td	tp
(c)	SF_4	34	:F—S—F: with :F: above and below S	SF_4 with lone pair on S	tbp	ss
(d)	PF_6^-	48	$[PF_6]^-$	$[PF_6]^-$	oh	oh
(e)	NH_3Cl^+	14	$[H—N—H$ with :Cl: above and H below$]^+$	$[NH_3Cl]^+$	tb	tb
(f)	N_3^-	16	* $[\ddot{N}—N≡N:]^-$	$[N—N≡N]^-$	l	l

*More than one resonance structure is possible. All equivalent resonance structures predict the same molecular geometry.

9.24 (a) Electron-domain geometries: i, octahedral; ii, tedrahedral; iii, trigonal bipyramial

(b) nonbonding electron domains: i, 2; ii, 0; iii, 1

(c) S or Se. Shape iii has five electron domains, so A must be in or below the third row of the periodic table. This eliminates Be and C. Assuming each F atom has three nonbonding electron domains and forms only single bonds with A, A must have six valence electrons to produce these electron-domain and molecular geometries.

(d) Xe. (See Table 9.3) Assuming F behaves typically, A must be in or below the third row and have eight valence electrons. Only Xe fits this description. (Noble gas elements above Xe have not been shown to form molecules of the type AF_4. See Section 7.8.)

9.26 (a) 1 - 109°, 2 - 120° (b) 3 - 109°, 4 - 120°

(c) 5 - 109°, 6 - 109° (d) 7 - 180°, 8 - 109°

9.28

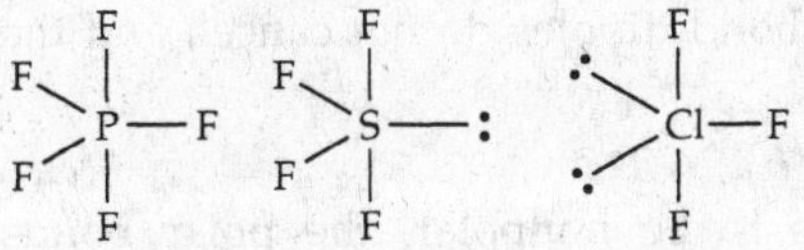

The three nonbonded electron pairs on each F atom have been omitted for clarity.

The F (axial) –A–F (equatorial) angle is largest in PF_5 and smallest in ClF_3. As the number of nonbonding domains in the equatorial plane increases, they push back the axial A–F bonds, decreasing the F (axial) –A–F (equatorial) bond angles.

9.30 (a) ClO_2^- 20 e^-, 10 e^- pr

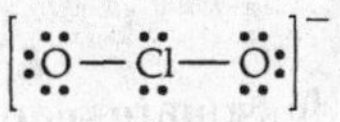

4 e^- domains around Cl, tetrahedral e^- domain geometry,

bent molecular geometry bond angle $\lesssim 109.5°$

NO_2^- 18 e^-, 9 e^- pr

3 e^- domains about N (both resonance structures), trigonal planar e^- domain geometry bent molecular geometry bond angle $\lesssim 120°$

Both molecular geometries are described as 'bent" because both molecules have two nonlinear bonding electron domains. The bond angles (the angle between the two bonding domains) in the two ions are different because the total number of electron domains, and thus the electron domain geometries are different.

(b) XeF_2 22 e^-, 11 e^- pr

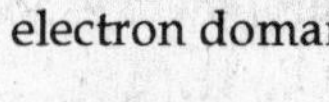

5 e^- domains around Xe, trigonal bipyramidal e^- domain geometry, linear molecular geometry

The question here really is: why do the three nonbonding domains all occupy the equatorial plane of the trigonal bipyramid? In a tbp, there are several different kinds of repulsions, bonding domain-bonding domain (bd-bd), bonding domain-nonbonding domain (bd-nd), and nonbonding domain-nonbonding domain (nd-nd). Each of these can have 90°, 120°, or 180° geometry. Since nonbonding domains occupy more space, 90° nd-nd repulsions are most significant and least desirable. The various electron domains arrange themselves to minimize these 90° nd-nd interactions. The arrangement shown above has no 90° nd-nd repulsions. An arrangement with one or two nonbonding domains in axial positions would lead to at least two 90° nd-nd repulsions, a less stable situation. (To convince yourself, tabulate the number and kinds of repulsions for each possible tbp arrangement of 2bd's and 3nd's.)

Polarity of Polyatomic Molecules

9.32 If PH_3 were planar, the PH_3 bond dipoles would cancel, and the molecule would be nonpolar. Since PH_3 is polar, the 3 P–H bond dipoles do not cancel, and the molecule can't be planar.

9.34 (a) For a molecule with polar bonds to be nonpolar, the polar bonds must be (symmetrically) arranged so that the bond dipoles cancel. In most cases, nonbonding e^- domains must be absent from the central atom. Square planar structures may not meet the second condition.

(b) AB_2: linear e^- domain geometry (edg), linear molecular geometry (mg), trigonal bipyramidal edg, linear mg

AB_3: trigonal planar edg, trigonal planar mg

AB_4: tetrahedral edg, tetrahedral mg; octahedral edg, square planar mg

9.36 (a) Polar, $\Delta EN > 0$

I–F

(b) Nonpolar, the molecule is linear and the bond dipoles cancel.

S=C=S

(c) Nonpolar, in a symmetrical trigonal planar structure, the bond dipoles cancel.

O
||
S
O O

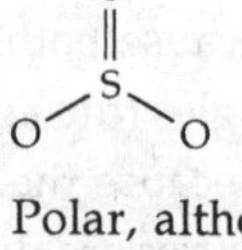

(d) Polar, although the bond dipoles are essentially zero, there is an unequal charge distribution due to the nonbonded electron pair on P.

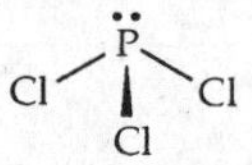

P
Cl Cl
Cl

(e) Nonpolar, symmetrical octahedron

(f) Polar, square pyramidal molecular geometry, bond dipoles do not cancel.

9.38 Each C–Cl bond is polar. The question is whether the vector sum of the C–Cl bond dipoles in each molecule will be nonzero. In the *ortho* and *meta* isomers, the C–Cl vectors are at 60° and 120° angles, respectively, and their resultant dipole moments are nonzero. In the *para* isomer, the C–Cl vectors are opposite, at an angle of 180°, with a resultant dipole moment of zero. The *ortho* and *meta* isomers are polar, the *para* isomer is nonpolar.

Orbital Overlap; Hybrid Orbitals

9.40 (a)

2s 2s

(b)

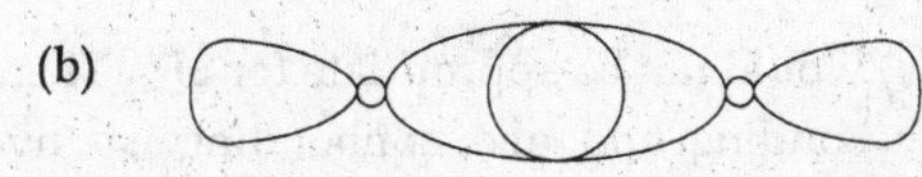

$2p_z$ $2p_z$

(c)

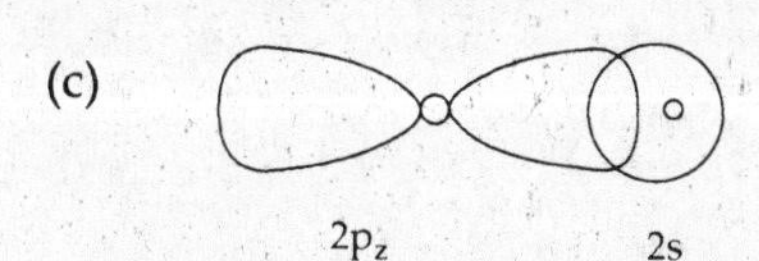

$2p_z$ 2s

9.42 (a) 8 valence e^-, 4 e^- pairs

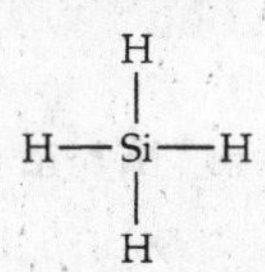

4 bonding e^- domains, tetrahedral domain and molecular geometry

(b) SiH_4 has four equivalent Si-H bonds. A free Si atom has the valence electron configuration $2s^22p^2$; the two 2p electrons are unpaired and available for bonding, but the 2s electrons are paired and unavailable. In order to have four unpaired electrons available for bonding, one of the 2s electrons must be promoted to the empty 2p orbital.

(c) The tetrahedral electron domain geometry of SiH_4 requires sp^3 hydridization.

(d) 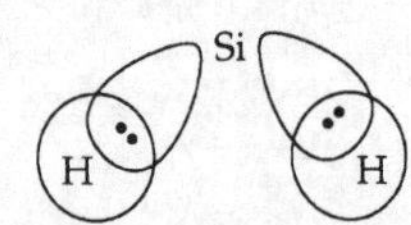

The other two Si–H bonds are in a plane perpendicular to the plane of the page, and pointing up. They form angles of 109.5° with each other and the two Si–H bonds drawn above.

9.44 (a) sp^2 - 120° angles in a plane

(b) sp^3d - 90°, 120° and 180° bond angles (trigonal bipyramid)

(c) sp^3d^2 - 90° and 180° bond angles (octahedron)

9.46 (a) S: $[Ne]3s^23p^4$

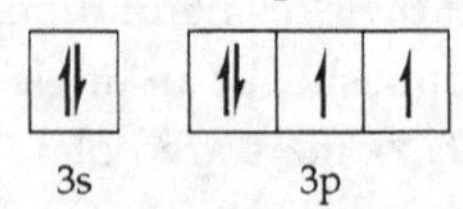

hybridize
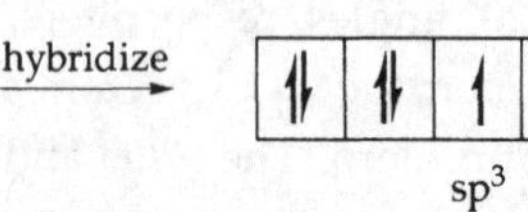

(b) The hybrid orbitals are called sp^3.

(c) 

(d) The hybrid orbitals formed in (a) would not be appropriate for SF_4. There are five electron domains in SF_4, four bonding and one nonbonding, so five hybrid orbitals are required. A set of four sp^3 hybrid orbitals could not accommodate all the electron pairs around S.

9.48 (a) 32 e^-, 16 e^- pairs

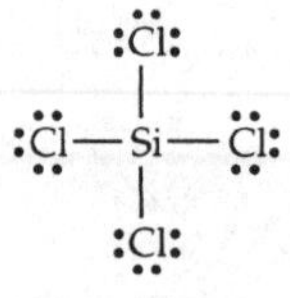

4 e^- pairs around Si, tetrahedral e^- domain geometry, sp^3 hybrid orbitals

(b) 10 e^-, 5 e^- pairs

H—C≡N:

2 e^- domains around C, linear e^- domain geometry, sp hybrid orbitals

(c) 24 e^-, 12 e^- pairs

:O—S—O:
‖
:O:

(other resonance structures are possible)

3 e^- domains around S, trigonal planar e^- domain geometry, sp^2 hybrid orbitals

(d) 22 e^-, 11 e^- pairs

:Cl:
I
:Cl:

5 e^- domains around I, trigonal bipyramidal e^- domain geometry, sp^3d hybrid orbitals (In a trigonal bipyramid, placing nonbonding e^- pairs in the equatorial position minimizes repulsion.)

(e) 36 e^-, 18 e^- pairs

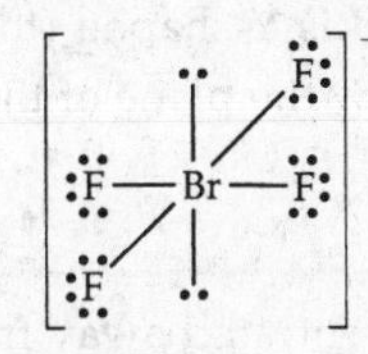

6 e^- domains around Br, octahedral e^- domain geometry, sp^3d^2 hybrid orbitals

Multiple Bonds

9.50 (a) Two unhybridized p orbitals remain, and the atom can form two pi bonds.

(b) A triple bond is composed of 1 σ and 2 π bonds.

(c) There is free rotation of attached groups around a σ bond, but not around a π bond. Rotation about a π bond would require that the bond be broken; the p orbitals would no longer be in the correct orientation for π overlap. The π overlap that is part of all multiple bonds introduces rigidity into molecules.

9.52 H—N—N—H :N≡N:
H H

The N atoms in N_2H_4 are sp^3 hybridized; there are no unhybridized p orbitals available for π bonding. In N_2, the N atoms are sp hybridized, with two unhybridized p orbitals on each N atom available to form the two π bonds in the N≡N triple bond.

9.54 (a) The C bound to O has three electron domains and is sp^2 hybridized; the other three C atoms are sp^3 hybridized.

(b) $C_4H_8O_2$ has 4(4) + 8(1) + 2(6) = 36 valence electrons.

(c) 13 pairs or 26 total valence electrons form σ bonds

(d) 1 pair or 2 total valence electrons form π bonds

(e) 4 pairs or 8 total valence electrons are nonbonding

9.56 (a) 1, 120°; 2, 120°; 3, 109°

(b) 1, sp^2; 2, sp^2; 3, sp^3

(c) 21 σ bonds

9.58 (a) 24 e^-, 12 e^- pairs (b)

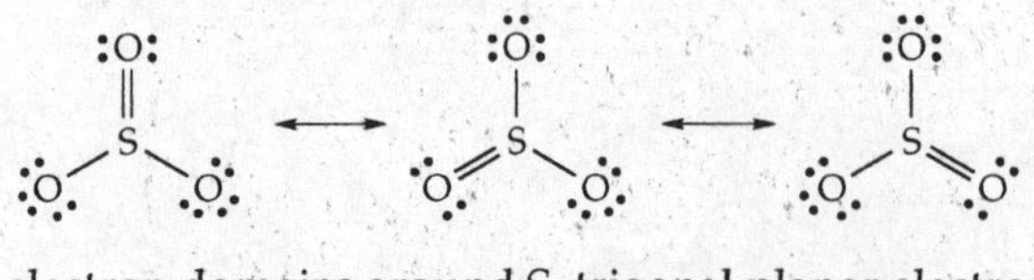

3 electron domains around S, trigonal planar electron-domain geometry, sp^2 hybrid orbitals

(c) The multiple resonance structures indicate delocalized π bonding. All four atoms lie in the trigonal plane of the sp^2 hybrid orbitals. On each atom there is a p atomic orbital perpendicular to this plane in the correct orientation for π overlap. The resulting delocalized π electron cloud is Y-shaped (the shape of the molecule) and has electron density above and below the plane of the molecule.

Molecular Orbitals

9.60 (a) In the σ anti-bonding MO, electron density is concentrated away from the nuclei; an electron in this orbital experiences less stabilization by the nucleus than an electron in an isolated atom.

(b) Yes. The Pauli principle, that no two electrons can have the same four quantum numbers, means that an orbital can hold at most two electrons. (Since n, l, and m_l are the same for a particular orbital and m_s has only two possible values, an orbital can hold at most two electrons). This is true for atomic and molecular orbitals.

(c) Four. When AOs combine to form MOs, the total number of orbitals is conserved. Combination of four AOs must result in formation of four MOs. Both can accommodate eight electrons.

9.62 (a)

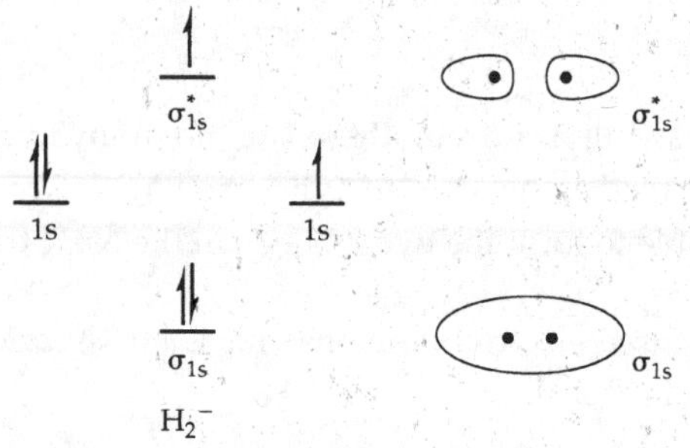

σ^*_{1s} σ^*_{1s}

1s 1s

σ_{1s} σ_{1s}

H_2^-

(b) σ^*_{1s}

σ_{1s}

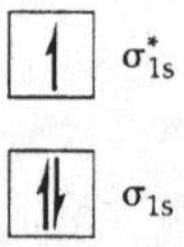

(c) Bond order = 1/2 (2-1) = 1/2

(d) If one electron moves from σ_{1s} to σ^*_{1s}, the bond order becomes -1/2. There is a net increase in energy relative to isolated H atoms, so the ion will decompose.

$$H_2^- \xrightarrow{h\nu} H + H^-$$

9.64 (a) Zero

(b) The two π_{2p} molecular orbitals are degenerate; they have the same energy, but they have different spatial orientations 90° apart.

(c) In the bonding MO the electrons are stabilized by both nuclei. In an antibonding MO, the electrons are directed away from the nuclei, so π_{2p} is lower in energy than π^*_{2p}.

9.66 (a) O_2^{2-} has a bond order of 1.0, while O_2^- has a bond order of 1.5. For the same bonded atoms, the greater the bond order the shorter the bond, so O_2^- has the shorter bond.

(b) The two possible orbital energy level diagrams are:

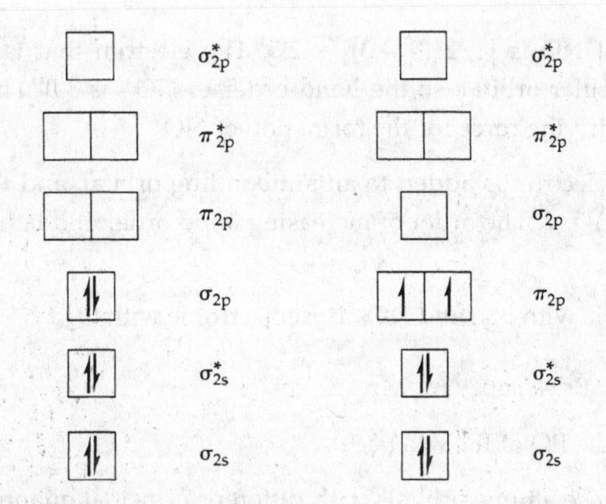

If the σ_{2p} molecular orbital is lower in energy than the π_{2p} orbitals, there are no unpaired electrons, and the molecule is diamagnetic. Switching the order of σ_{2p} and π_{2p} gives one unpaired electron in each degenerate π_{2p} orbital and explains the observed paramagnetism of B_2 (see Figure 9.45).

9.68 (a) Substances with unpaired electrons are attracted into a magnetic field. This property is called *paramagnetism*.

(b) Weigh the substance normally and in a magnetic field, as shown in Figure 9.46. Paramagnetic substances appear to have a larger mass when weighed in a magnetic field.

(c) See Figures 9.37 and 9.45. O_2^+, one unpaired electron; N_2^{2-}, two unpaired electrons; Li_2^+, one unpaired electron

9.70 Determine the number of "valence" (non-core) electrons in each molecule or ion. Use the homonuclear diatomic MO diagram from Figure 9.42 (shown below) to calculate bond order and magnetic properties of each species. The electronegativity difference between heteroatomics increases the energy difference between the 2S AO on one atom and the 2p AO on the other, rendering the "no interaction" MO diagram in Figure 9.42 appropriate.

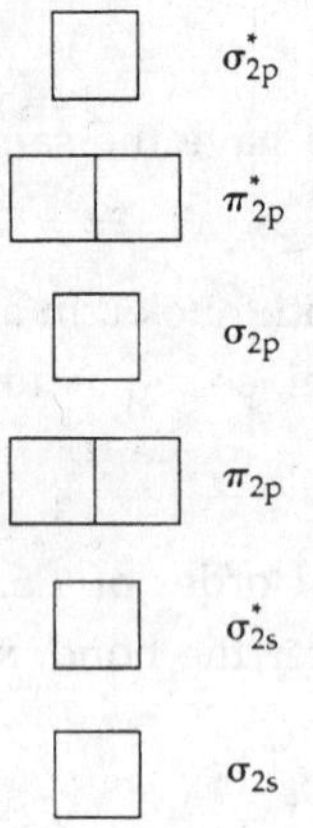

σ^*_{2p}

π^*_{2p}

σ_{2p}

π_{2p}

σ^*_{2s}

σ_{2s}

(a) CO^+: 9 e^-, B.O. = (7 – 2) / 2 = 2.5, paramagnetic

(b) NO^-: 12 e^-, B.O. = (8 – 4) / 2 = 2.0, paramagnetic

(c) OF^+: 12 e^-, B.O. = (8 – 4) / 2 = 2.0, paramagnetic

(d) NeF^+: 14 e^-, B.O. = (8 – 6) / 2 = 1.0, diamagnetic

9.72 (a) The bond order of NO is [1/2 (8 – 3)] = 2.5. The electron that is lost is in an antibonding molecular orbital, so the bond order in NO^+ is 3.0. The increase in bond order is the driving force for the formation of NO^+.

(b) To form NO^-, an electron is added to an antibonding orbital, and the new bond order is [1/2 (8 – 4)] = 2. The order of increasing bond order and bond strength is: $NO^- < NO < NO^+$.

(c) NO^+ is isoelectronic with N_2, and NO^- is isoelectronic with O_2.

9.74 (a) I: 5s, $5p_x$, $5p_y$, $5p_z$; Br: 4s, $4p_x$, $4p_y$, $4p_z$

(b) By analogy to F_2, the BO of IBr will be 1.

(c) I and Br have valence atomic orbitals with different principal quantum numbers. This means that the radial extensions (sizes) of the valence atomic orbital that contribute to the MO are different. The n = 5 valence AOs on I are larger than the n = 4 valence AOs on Br.

(d) σ^*_{np}

(e) None

Additional Exercises

9.75 (a) The physical basis of VSEPR is the electrostatic repulsion of like-charged particles, in this case groups or domains of electrons. That is, owing to electrostatic repulsion, electron domains will arrange themselves to be as far apart as possible.

(b) The σ-bond electrons are localized in the region along the internuclear axes. The positions of the atoms and geometry of the molecule are thus closely tied to the locations of these electron pairs. Because the π-bond electrons are distributed above and below the plane that contains the σ bonds, these electron pairs do not, in effect, influence the geometry of the molecule. Thus, all σ- and π-bond electrons localized between two atoms are located in the same electron domain.

9.77 For any triangle, the law of cosines gives the length of side c as $c^2 = a^2 + b^2 - 2ab\cos\theta$.

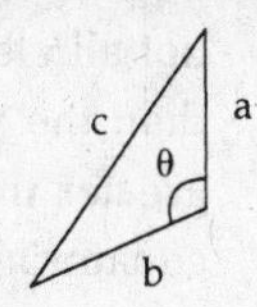

Let the edge length of the cube (uy = vy = vz) = X

The length of the face diagonal (uv) is

$(uv)^2 = (uy)^2 + (vy)^2 - 2(uy)(vy)\cos 90$

$(uv)^2 = X^2 + X^2 - 2(X)(X)\cos 90$

$(uv)^2 = 2X^2;\ uv = \sqrt{2}X$

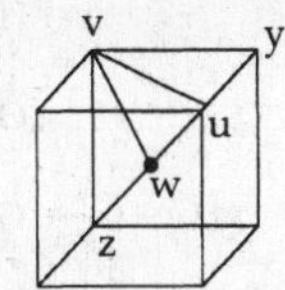

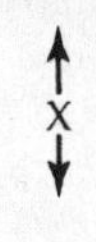

The length of the body diagonal (uz) is

$(uz)^2 = (vz^2) + (uv)^2 - 2(vz)(uv)\cos 90$

$(uz)^2 = X^2 + (\sqrt{2}X)^2 - 2(X)(\sqrt{2}X)\cos 90$

$(uz)^2 = 3X^2;\ uz = \sqrt{3}X$

For calculating the characteristic tetrahedral angle, the appropriate triangle has vertices u, v, and w. Theta, θ, is the angle formed by sides wu and wv and the hypotenuse is side uv.

$wu = wv = uz/2 = \sqrt{3}/2X;\ uv = \sqrt{2}X$

$(\sqrt{2}X)^2 = (\sqrt{3}/2X)^2 + (\sqrt{3}/2X)^2 - 2(\sqrt{3}/2X)(\sqrt{3}/2)\cos\theta$

$2X^2 = 3/4\,X^2 + 3/4\,X^2 - 3/2\,X^2\cos\theta$

$2X^2 = 3/2\,X^2 - 3/2\,X^2\cos\theta$

$1/2\,X^2 = -3/2\,X^2\cos\theta$

$\cos\theta = -(1/2\,X^2)\,/\,(3/2\,X^2) = -1/3 = -0.3333$

$\theta = 109.47°$

9.78 (a) 40 e^-, 20 e^- pairs

:Cl:
:F—P—F:
:F
:F:

5 e^- domains
trigonal pyramidal electron domain geometry

(b) The greater the electronegativity of the terminal atom, the larger the negative charge centered on the atom, the greater the effective size of the P–X electron domain. A P–F bond will produce a larger electron domain than a P–Cl bond.

(c) The molecular geometry (shape) is also trigonal bipyramidal, because all five electron domains are bonding domains. Because we predicted the P–F electron domain to be larger, the maximum number of three P–F bonds will occupy the equatorial plane of the molecule, minimizing the number of 90° P–F to P–F repulsions. This is the same argument that places a "larger" nonbonding domain in the equatorial position of a molecule like SF_4. The P–Cl bond is then axial, as shown in the Lewis structure.

(d) The molecular geometry is distorted from a perfect trigonal bipyramid because not all electron domains are alike. The 90° P–F to P–F repulsions will be greater than the 90° P–F to P–Cl repulsions, so the F(axial)–P–F angles will be slightly greater than 90°, and the Cl(axial)–P–F angles will be slightly less than 90°. The equatorial F–P–F angles of 120° are distorted little, if at all.

9.80 (a)

```
            H
            |
      H   :O:  :O:
      |    |    ||     ..
  H — C —  C —  C  —  O — H
      |    |           ..
      H    H
```

$3(4) + 3(6) + 6(1) = 36\ e^-, 18\ e^-$ pr

(b) There are 11 σ and 1 π bonds.

(c) The C=O on the right-hand C atom is shortest. For the same bonded atoms, in this case C and O, the greater the bond order, the shorter the bond.

(d, e) The right-most C has three e^- domains, so the hybridization is sp^2; bond angles about this C atom are approximately 120°. The middle and left-hand C atoms both have four e^- domains, are sp^3 hybridized, and have bond angles of approximately 109°.

9.81

```
     ..               F
     |                |
     P                B
   / | \            /   \
  F  |  F          F     F
     F
  μ = 1.03D         μ = 0
```

BF_3 is a trigonal planar molecule with the central B atom symmetrically surrounded by the three F atoms (Figure 9.13). The individual B–F bond dipoles cancel, and the molecule has a net dipole moment of zero. PF_3 has tetrahedral electron-domain geometry with one of the positions in the tetrahedron occupied by a nonbonding electron pair. The individual P–F bond dipoles do not cancel and the presence of a nonbonding electron pair ensures an asymmetrical electron distribution; the molecule is polar.

9.83

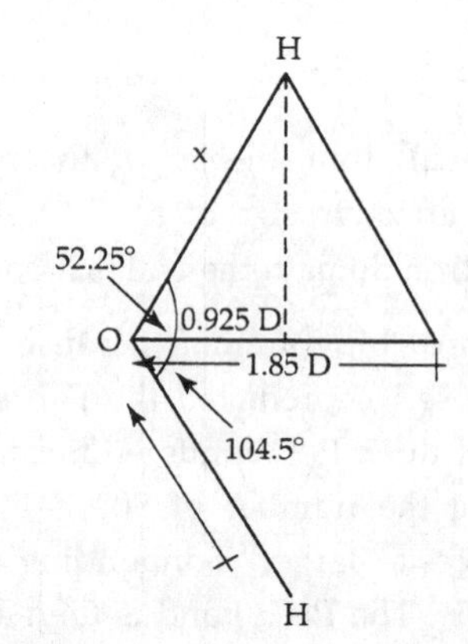

(a) The bond dipoles in H_2O lie along the O-H bonds with the positive end at H and the negative end at O. The dipole moment vector of the H_2O molecule is the resultant (vector sum) of the two bond dipoles. This vector bisects the H-O-H angle and has a magnitude of 1.85 D with the negative end pointing toward O.

(b) Since the dipole moment vector bisects the H-O-H bond angle, the angle between one H-O bond and the dipole moment vector is 1/2 the H-O-H bond angle, 52.25°. Dropping a perpendicular line from H to the dipole moment vector creates the right triangle pictured. If x = the magnitude of the O-H bond dipole, x cos (52.25) = 0.925 D. x = 1.51 D.

(c) The X-H bond dipoles (Table 8.3) and the electronegativity values of X (Figure 8.7) are

	Electronegativity	Bond dipole
F	4.0	1.82
O	3.5	1.51
Cl	3.0	1.08

Since the electronegativity of O is midway between the values for F and Cl, the O-H bond dipole should be approximately midway between the bond dipoles of HF and HCl. The value of the O-H bond dipole calculated in part (b) is consistent with this prediction.

9.84 (a) XeF_6 50 e^-, 25 e^- pairs

:F: :F:
:F—Xe—F:
:F: :F:

(b) There are seven electron domains around Xe, and the maximum number of e^- domains in Table 9.3 is six.

(c) Tie seven balloons together and see what arrangement they adopt (seriously! see Figure 9.5). Alternatively, go to the chemical literature where VSEPR was first proposed and see if there is a preferred orientation for seven e^- domains.

(d) Since the hybrid orbitals for five e^- domains involve one d orbital and for six pairs, two d orbitals, a reasonable suggestion would be sp^3d^3.

(e) One of the seven e^- domains is a nonbonded domain. The question is whether it occupies an axial or equatorial position. The equatorial plane of a pentagonal bipyramid has F-Xe-F angles of 72°. Placing the nonbonded domain in the equatorial plane would create severe repulsions between it and the adjacent bonded domains. Thus, the nonbonded domain will reside in the axial position. The molecular structure is a pentagonal pyramid.

9.86 (a) 16 e^-, 8 e^- pairs

$[N{=}N{=}N]^- \longleftrightarrow [:N{\equiv}N{-}N:]^- \longleftrightarrow [:N{-}N{\equiv}N:]^-$

(b) The observed bond length of 1.16 Å is intermediate between the values for N=N, 1.24 Å, and N ≡ N, 1.10 Å. This is consistent with the resonance structures, which indicate contribution from formally double and triple bonds to the true bonding picture in N_3^-.

(c) In each resonance structure, the central N has two electron domains, so it must be sp hybridized. It is difficult to predict the hybridization of terminal atoms in molecules where there are resonance structures because there are a different number of electron domains around the terminal atoms in each structure. Since the "true" electronic arrangement is a combination of all resonance structures, we will assume that the terminal N–N bonds have some triple bond character and that the terminal N atoms are sp hybridized. (There is no experimental measure of hybridization at terminal atoms, since there are no bond angles to observe.)

(d) In each resonance structure, N–N σ bonds are formed by sp hybrids and π bonds are formed by unhybridized p orbitals. Nonbonding e^- pairs can reside in sp hybrids or p atomic orbitals.

(e) Recall that electrons in 2s orbitals are on the average closer to the nucleus than electrons in 2p orbitals. Since sp hybrids have greater s orbital character, it is reasonable to expect the radial extension of sp orbitals to be smaller than that of sp^2 or sp^3 orbitals and σ bonds formed by sp orbitals to be slightly shorter than those formed by other hybrid orbitals, assuming the same bonded atoms.

There are no solitary σ bonds in N_3^-. That is, the two σ bonds in N_3^- are each accompanied by at least one π bond between the bonding pair of atoms. Sigma bonds that are part of a double or triple bond must be shorter so that the p orbitals can overlap enough for the π bond to form. Thus, the observation is not applicable to this molecule. (Comparison of C–H bond lengths in C_2H_2, C_2H_4, C_2H_6 and related molecules would confirm or deny the observation.)

9.87 (a) $\ddot{\underset{..}{O}}=\ddot{O}-\ddot{\underset{..}{O}}:$ ⟷ $:\ddot{\underset{..}{O}}-\ddot{O}=\ddot{\underset{..}{O}}$

To accommodate the π bonding by all 3 O atoms indicated in the resonance structures above, all O atoms are sp^2 hybridized.

(b) For the first resonance structure, both sigma bonds are formed by overlap of sp^2 hybrid orbitals, the π bond is formed by overlap of atomic p orbitals, one of the nonbonded pairs on the right terminal O atom is in a p atomic orbital, and the remaining five nonbonded pairs are in sp^2 hybrid orbitals.

(c) Only unhybridized p atomic orbitals can be used to form a delocalized π system.

(d) The unhybridized p orbital on each O atom is used to form the delocalized π system, and in both resonance structures one nonbonded electron pair resides in a p atomic orbital. The delocalized π system then contains four electrons, two from the π bond and two from the nonbonded pair in the p orbital.

9.88 (a) Each C atom is surrounded by three electron domains (two single bonds and one double bond), so bond angles at each C atom will be approximately 120°.

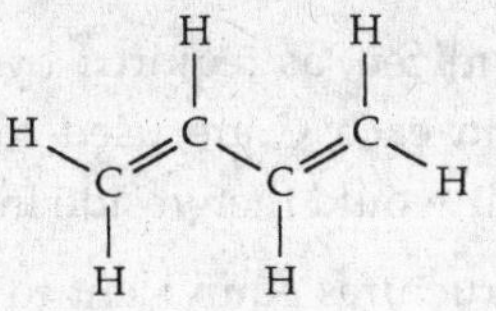

Since there is free rotation around the central C–C single bond, other conformations are possible.

(b) According to Table 8.5, the average C–C length is 1.54 Å, and the average C=C length is 1.34 Å. While the C=C bonds in butadiene appear "normal," the central C–C is significantly shorter than average. Examination of the bonding in butadiene reveals that each C atom is sp^2 hybridized and the π bonds are formed by the remaining unhybridized 2p orbital on each atom. If the central C–C bond is rotated so that all four C atoms are coplanar, the four 2p orbitals are parallel, and some delocalization of the π electrons occurs.

9.89 (a) The diagram shows two s atomic orbitals with opposite phases. Because they are spherically symmetric, the interaction of s orbitals can only produce a σ molecular orbital. Because the two orbitals in the diagram have opposite phases, the interaction excludes electron density from the region between the nuclei. The resulting MO has a node between the two nuclei. Relative to Figure 9.42, the MO is a σ^*_{2s}. The principal quantum number designation is arbitrary, because it defines only the size of the pertinent AOs and MOs. Shapes and phases of MOs depend only on these same characteristics of the interacting AOs.

(b) The diagram shows two p atomic orbitals with oppositely phased lobes pointing at each other. End-to-end overlap produces a σ-type MO; opposite phases mean a node between the nuclei and an antibonding MO. The interaction results in a σ^*_{2p} MO.

(c) The diagram shows parallel p atomic orbitals with like-phased lobes aligned. Side-to-side overlap produces a π-type MO; overlap of like-phased lobes concentrates electron density between the nuclei and a bonding MO. The interaction results in a π_{2p} MO.

9.90 (a) $C_5H_5^-$, 26 e^-, 13 e^- pair

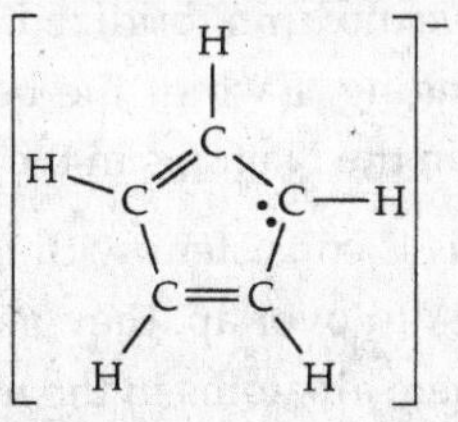

No. According to the single Lewis structure above, the four C atoms involved in double bonds would be sp^2 hybridized, but the C atom with the lone pair would be sp^3 hybridized. Not all C atoms would have the same hybridization.

(b) Equivalent sp^2 hydridization at all C atoms is consistent with the planar structure of the ion. The VSEPR model applied to the single Lewis structure above does not predict uniform hybridization or planarity of the ion. The VSEPR/Lewis model is consistent with the known structural features of the ion only if the other resonance structures are considered.

(c) If all C atoms are sp^2 hybridized, as required by the planar structure of the ion, the three hybrid orbitals on each C are used to form the σ framework of the molecule. The unshared pair would then reside in an unhybridized p orbital.

(d) Yes, there are resonance structures equivalent to the Lewis structure above. It is possible to place the unshared pair on any of the five C atoms, resulting in five equivalent resonance structures (four plus the one above).

(e) The five resonance structures indicate that there is delocalization of the π electron density in the molecule. The interior circle conveys uniform delocalization over the entire ring, as in benzene.

(f) In a rigid planar structure with each C atom sp^2 hybridized, the unhybridized p orbitals are aligned parallel, facilitating overlap to form a delocalized π network above and below the plane of the molecule. Since the unshared pair occupies an unhybridized p orbital (part c), it is part of the delocalized π network, along with the four electrons involved in π bonds; the delocalized system contains a total of six electrons.

9.92 Paramagnetic materials appear to weigh more when the mass measurement is made in the presence of a magnetic field. Air contains $O_2(g)$, which is paramagnetic because of its two unpaired electrons in a σ^*_{2p} molecular orbital (Figure 9.45). In the presence of a magnetic field, mass measurements made in air would be skewed by the paramagnetism of O_2, and lead to inaccurate results.

9.93 We will refer to azo benzene (on the left) as A and hydrazobenzene (on the right) as H.

(a) A: sp^2; H: sp^3

(b) A: Each N and C atom has one unhybridized p orbital. H: Each C atom has one unhybridized p orbital, but the N atoms have no unhybridized p orbitals.

(c) A: 120°; H: 109°

(d) Since all C and N atoms in A have unhybridized p orbitals, all can participate in delocalized π bonding. The delocalized π system extends over the entire molecule, including both benzene rings and the azo "bridge." In H, the N atoms have no unhybridized p orbitals, so they cannot participate in delocalized π bonding. Each of the benzene rings in H is delocalized, but the network cannot span the N atoms in the bridge.

(e) This is consistent with the answer to (d). In order for the unhybridized p orbitals in A to overlap, they must be parallel. This requires a planar σ bond framework where all atoms in the molecule are coplanar.

(f) For a substance to appear colored, it must absorb light in the visible region of the electromagnetic spectrum. This requires a HOMO-LUMO energy gap in the visible region. For organic molecules, the size of the gap is related to the number of conjugated π bonds; the more conjugated π bonds, the smaller the gap and the more likely the molecule is to be colored. Azobenzene has seven conjugated π bonds (π network delocalized over entire molecule) while hydrazobenzene has only three (π network on benzene rings only). Thus, the size of the HOMO-LUMO energy gap is smaller in A, and A is more likely than H to appear colored.

9.94 (a) H: $1s^1$; F: $[He]2s^22p^5$

When molecular orbitals are formed from atomic orbitals, the total number of orbitals is conserved. Since H and F have a total of five valence AOs ($H_{1s} + F_{2s} + 3F_{2p}$), the MO diagram for HF has five MOs.

(b) H and F have a total of eight valence electrons. Since each MO can hold a maximum of two electrons, four of the five MOs would be occupied.

(c) No. The MO diagram for NO in Figure 9.48 has eight MOs, and the diagram for HF has only five. The number, type, and energy spacing of the MOs in HF will be different.

Integrative Exercises

9.97 (a) $2SF_4(g) + O_2(g) \rightarrow 2OSF_4(g)$

(b) 40 e^-, 20 e^- pairs

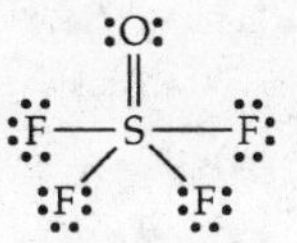

:O:
‖
:F—S—F:
:F: :F:

There must be a double bond drawn between O and S in order for their formal charges to be zero.

(c) $\Delta H = 8D(S\text{–}F) + D(O{=}O) - 8D(S\text{–}F) - 2D(S{=}O)$

$\Delta H = D(O{=}O) - 2D(S{=}O) = 495 - 2(523) = -551$ kJ, exothermic

(d) trigonal bipyramidal electron-domain geometry

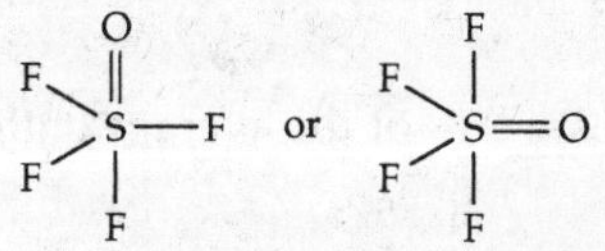

O, F, F, S, F, F (O axial) or F, F, F, S=O, F (O equatorial)

(e) Because F is more electronegative than O, the structure that minimizes S–F repulsions is more likely (see Solution 9.98). That is, the structure with fewer 90° F–S–F angles and more 120°F–S–F angles is favored. The structure on the left, with O in the axial position is more likely. Note that a double bond involving an atom with an expanded octet of electrons, such as the S=O in this molecule, does not have the same geometric implications as a double bond to a first row element.

9.98 (a) PX_3, 26 valence e^-, 13^- pairs

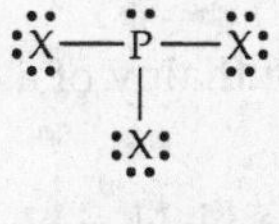

:X—P—X:
|
:X:

4 electron domains around P, tetrahedral e^- domain geometry, 107° bond angles (less than 109° due to repulsion with nonbonded pair)

(b) As electronegativity increases (I < Br < Cl < F), the X–P–X angles decreases.

(c) For P–I, ΔEN is (2.5 – 2.1) = 0.4 and the bond dipole is small. For P–F, ΔEN is (4.0 – 2.1) = 1.9 and the bond dipole is large. The greater the ΔEN and bond dipole, the larger the magnitude of negative charge centered on X. The more negative charge centered on X, the greater the repulsion between X and the nonbonding electron pair on P and the smaller the bond angle. (Since electrostatic attractions and repulsions vary as 1/r, the shorter P–F bond may also contribute to this effect.)

(d) $PBrCl_4$, 40 valence electrons, 20 e^- pairs. The molecule will have trigonal bipyramidal electron-domain geometry (similar to PCl_5 in Table 9.3.) Based on the argument in part (c), the P–Br bond will have smaller repulsions with P–Cl bonds than P–Cl bonds have with each other. Therefore, the Br will occupy an axial position in the trigonal bipyramid, so that the more unfavorable P–Cl to P–Cl repulsions can be situated at larger angles in the equatorial plane.

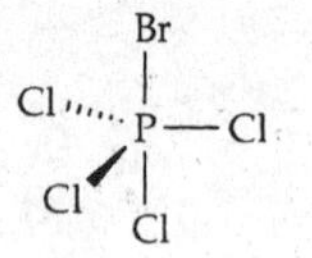

Br
Cl.... P—Cl
Cl
Cl

9.100 (a) C ≡ C 839 kJ/mol (1 σ, 2 π)

C=C 614 kJ/mol (1 σ, 1 π)

C–C 348 kJ/mol (1 σ)

The contribution from 1 π bond would be (614–348) 266 kJ/mol. From a second π bond, (839 – 614), 225 kJ/mol. An average π bond contribution would be (266 + 225)/2 = 246 kJ/mol.

This is $\dfrac{246 \text{ kJ}/\pi \text{ bond}}{348 \text{ kJ}/\sigma \text{ bond}} \times 100 = 71\%$ of the average enthalpy of a σ bond.

(b) N ≡ N 941 kJ/mol

N=N 418 kJ/mol

N–N 163 kJ/mol

first π = (418 – 163) = 255 kJ/mol

second π = (941 – 418) = 523 kJ/mol

average π bond enthalpy = (255 + 523)/2 = 389 kJ/mol

This is $\dfrac{389 \text{ kJ}/\pi \text{ bond}}{163 \text{ kJ}/\sigma \text{ bond}} \times 100 = 240\%$ of the average enthalpy of a σ bond.

N–N σ bonds are weaker than C–C σ bonds, while N–N π bonds are stronger than C–C π bonds. The relative energies of C–C σ and π bonds are similar, while N–N π bonds are much stronger than N–N σ bonds.

(c) N_2H_4, 14 valence e^-, 7 e^- pairs

$$\mathrm{H{-}\ddot{N}{-}\ddot{N}{-}H}$$ (each N also bonded to H)

4 electron domains around N, sp^3 hybridization

N_2H_2, 12 valence e^-, 6 e^- pairs

$$\mathrm{H{-}\ddot{N}{=}\ddot{N}{-}H}$$

3 electron domains around N, sp^2 hybridization

N_2, 10 valence e^-, 5 e^- pairs

$$\mathrm{:N{\equiv}N:}$$

2 electron domains around N, sp hybridization

(d) In the three types of N–N bonds, each N atom has a nonbonding or lone pair of electrons. The lone pair to bond pair repulsions are minimized going from 109° to 120° to 180° bond angles, making the π bonds stronger relative to the σ bond. In the three types of C–C bonds, no lone-pair to bond-pair repulsions exist, and the σ and π bonds have more similar energies.

9.102 (a) 1 eV = 96.485 kJ/mol

$$H_2\text{: } 15.4\text{ eV} \times \frac{96.485\text{ kJ/mol}}{1\text{ eV}} = 1486 = 1.49 \times 10^3\text{ kJ/mol}$$

$$N_2\text{: } 15.6\text{ eV} \times \frac{96.485\text{ kJ/mol}}{1\text{ eV}} = 1505 = 1.51 \times 10^3\text{ kJ/mol}$$

$$O_2\text{: } 12.1\text{ eV} \times \frac{96.485\text{ kJ/mol}}{1\text{ eV}} = 1167 = 1.17 \times 10^3\text{ kJ/mol}$$

$$F_2\text{: } 15.7\text{ eV} \times \frac{96.485\text{ kJ/mol}}{1\text{ eV}} = 1515 = 1.52 \times 10^3\text{ kJ/mol}$$

(b)

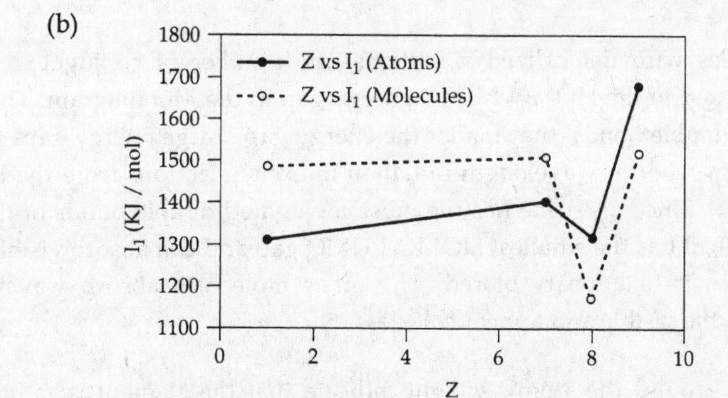

(c) In general, I_1 for atoms and molecules increases going across a row of the periodic chart. In both cases, there is a discontinuity at oxygen. The details of the trends are different. The deviation at O is larger for the molecules than the atoms, while the increase at F is much greater for the atoms than the molecules.

(d) According to Figures 9.35 and 9.45, H_2, N_2, and F_2 are diamagnetic and O_2 is paramagnetic. That is, ionization in H_2, N_2, and F_2 has to overcome spin-pairing energy, while ionization of O_2 removes an already unpaired electron. Thus, the ionization energy of O_2 is much less than I_1 for H_2, N_2, and F_2.

Despite differences in bond order, bond length, and the bonding or antibonding nature of the HOMO in H_2, N_2, and F_2, the ionization energies for these molecules are very similar.

9.104 (a) The molecular and empirical formulas of the four molecules are:

benzene: molecular, C_6H_6; empirical, CH

napthalene: molecular, $C_{10}H_8$; empirical, C_5H_4

anthracene: molecular, $C_{14}H_{10}$; empirical, C_7H_5

tetracene: molecular, $C_{18}H_{12}$, empirical, C_3H_2

(b) Yes. Since the compounds all have different empirical formulas, combustion analysis could in principle be used to distinguish them. In practice, the mass % of C in the four compounds is not very different, so the data would have to be precise to at least 3 decimal places and 4 would be better.

(c) $C_{10}H_8(s) + 12O_2(g) \rightarrow 10CO_2(g) + 4H_2O(l)$

(d) $\Delta H_{comb} = 5D(C{=}C) + 5D(C{-}C) + 8D(C{-}H) + 12D(O{=}O) - 20D(C{=}O) - 8D(O{-}H)$

$= 5(614) + 5(348) + 8(413) + 12(495) - 20(799) - 8(463)$

$= -5630$ kJ/mol $C_{10}H_8$

(e) Yes.

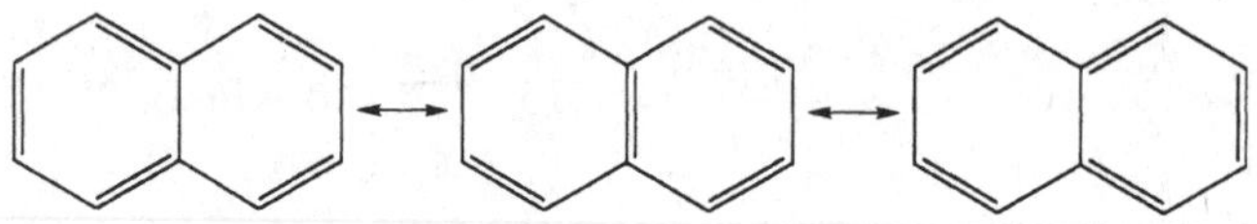

(f) For molecules with delocalized π systems, the number of conjugated double bonds is related to the HOMO-LUMO energy gap in the MO diagram. The more conjugated double bonds, the smaller the energy gap. Large energy gaps require higher energy, shorter wavelength radiation to excite electrons from the HOMO to the LUMO. Since tetracene has the most conjugated double bonds of the four molecules (9), it has the smallest HOMO-LUMO gap and can absorb visible light, which causes it to appear colored. The other molecules absorb wavelengths shorter than the visible and appear colorless.

9.105 (a) Parentheses around the atomic weight indicate that the element is radioactive. The nuclei of radioactive elements spontaneously decay to form other, lighter nuclei. The difficulty with studying At is that it decays before a reliable set of physical and chemical properties can be measured.

(b) $1s^2 2s^2 2p^6 3s^2 3p^6 4s^2 3d^{10} 4p^6 5s^2 4d^{10} 5p^6 6s^2 4f^{14} 5d^{10} 6p^5$

(c) AtI is expected to have a slightly polar covalent bond, similar to other interhalogen molecules. Since At and I are adjacent in the halogen family, their electronegativities will be very similar, but not equal. When two atoms with similar electronegativies combine, the result is a covalent molecule with a small net dipole moment.

(d) 22 e^-, 11 e^- pairs;

$$\left[:\ddot{\underset{..}{I}}-\ddot{\underset{..}{At}}-\ddot{\underset{..}{I}}:\right]^-$$

trigonal bipyramidal electron-domain geometry

linear molecular geometry

(In a trigonal bipyramid, placing nonbonding electron pairs in the equatorial plane minimizes repulsions.)

(e) By analogy to F_2 in Figure 9.45, the bond order should be 1 and the HOMO is π^*_{6p}.

10 Gases

Visualizing Concepts

10.2 At constant temperature and volume, pressure depends on total number of particles (Charles' Law). In order to reduce the pressure by a factor of 2, the number of particles must be reduced by a factor of 2. At the lower pressure, the container would have half as many particles as at the higher pressure.

10.3. (a) At constant pressure and temperature, the container volume is directly proportional to the number of particles present (Avogadro's Law). As the reaction proceeds, 3 gas molecules are converted to 2 gas molecules, so the container volume decreases. If the reaction goes to completion, the final volume would be 2/3 of the initial volume.

(b) At constant volume, pressure is directly proportional to the number of particles (Charles' Law). Since the number of molecules decreases as the reaction proceeds, the pressure also decreases. At completion, the final pressure would be 2/3 the initial pressure.

10.5 (a) Partial pressure depends on the number of particles of each gas present. Red has the fewest particles, then yellow, then blue. $P_{red} < P_{yellow} < P_{blue}$

(b) $P_{gas} = \chi_{gas} P_t$. Calculate the mole fraction, χ_{gas} = [mol gas / total moles] or [particles gas / total particles]. This is true because Avogadro's number is a counting number, and mole ratios are also particle ratios.

χ_{red} = 2 red atoms / 10 total atoms = 0.2; P_{red} = 0.2(0.90 atm) = 0.18 atm

χ_{yellow} = 3 yellow atoms / 10 total atoms = 0.3; P_{yellow} = 0.3(0.90 atm) = 0.27 atm

χ_{blue} = 5 blue atoms / 10 total atoms = 0.5; P_{blue} = 0.5(0.90 atm) = 0.45 atm

10.6

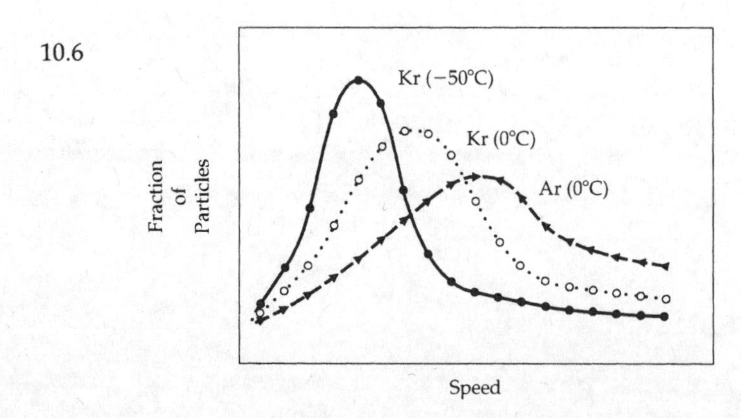

10.8 (a) Total pressure is directly related to total number of particles (or total mol particles). P(ii) < P(i) = P(iii)

(b) Partial pressure of He is directly related to number of He atoms (yellow) or mol He atoms. $P_{He}(iii) < P_{He}(ii) < P_{He}(i)$

(c) Density is total mass of gas per unit volume. We can use the atomic or molar masses of He (4) and N_2(28), as relative masses of the particles.

$$\text{mass(i)} = 5(4) + 2(28) = 76$$

$$\text{mass(ii)} = 3(4) + 1(28) = 40$$

$$\text{mass(iii)} = 2(4) + 5(28) = 148$$

Since the container volumes are equal, d(ii) < d(i) < d(iii).

(d) At the same temperature, all gases have the same "avg" kinetic energy. The average kinetic energies of the particles in the three containers are equal.

Gas Characteristics; Pressure

10.10 (a) Because gas molecules are far apart and in constant motion, the gas expands to fill the container. Attractive forces hold liquid molecules together and the volume of the liquid does not change.

(b) H_2O and CCl_4 molecules are too dissimilar to displace each other and mix in the liquid state. All mixtures of gases are homogeneous. (See Solution 10.9 (c)).

(c) Because gas molecules are far apart, the mass present in 1 mL of a gas is very small. The mass of a gas present in 1 L is on the same order of magnitude as the mass of a liquid present in 1 mL.

10.12 $P = m \times a/A$; $1\ \text{Pa} = 1\ \text{kg/m} \bullet \text{s}^2$; $A = 3.0\ \text{cm} \times 4.1\ \text{cm} \times 4 = 49.2 = 49\ \text{cm}^2$

$$\frac{262\ \text{kg}}{49.2\ \text{cm}^2} \times \frac{9.81\ \text{m}}{\text{s}^2} \times \frac{(100)^2\ \text{cm}^2}{1\ \text{m}^2} = 5.224 \times 10^5 \frac{\text{kg}}{\text{m} \bullet \text{s}^2} = 5.2 \times 10^5\ \text{Pa}$$

$$P_{total} = P_{atm} + P_{H_2O} = 0.95\ \text{atm} + 1.061\ \text{atm} = 2.011 = 2.0\ \text{atm}$$

10.14 Using the relationship derived in Solution 10.13 for two liquids under the influence of gravity, $(d \times h)_{lid} = (d \times h)_{Hg}$. At 752 torr, the height of an Hg barometer is 752 mm.

$$\frac{1.20\ \text{g}}{1\ \text{mL}} \times h_{lid} = \frac{13.6\ \text{g}}{1\ \text{mL}} \times 760\ \text{mm};\ h_{lid} = \frac{13.6\ \text{g/mL} \times 752\ \text{mm}}{1.20\ \text{g/mL}} = 8.52 \times 10^3\ \text{mm} = 8.52\ \text{m}$$

10.16 The mercury would fill the tube completely; there would be no vacuum at the closed end. This is because atmospheric pressure will support a mercury column higher than 70 cm, while our tube is only 50 cm. No mercury flows from the tube into the dish and no vacuum forms at the top of the tube.

10.18 (a) $0.850\ \text{atm} \times \frac{760\ \text{torr}}{1\ \text{atm}} = 646\ \text{torr}$

(b) $785\ \text{torr} \times \frac{101.325\ \text{kPa}}{760\ \text{torr}} = 105\ \text{kPa}$

(c) $655 \text{ mm Hg} \times \dfrac{1 \text{ atm}}{760 \text{ mm Hg}} = 0.862 \text{ atm}$

(d) $1.323 \times 10^5 \text{ Pa} \times \dfrac{1 \text{ atm}}{1.01325 \times 10^5 \text{ Pa}} = 1.3057 = 1.306 \text{ atm}$

(e) $2.50 \text{ atm} \times \dfrac{1.01325 \times 10^5 \text{ Pa}}{1 \text{ atm}} \times \dfrac{1 \text{ bar}}{1 \times 10^5 \text{ Pa}} = 2.53 \text{ bar}$

10.20 (a) $\dfrac{1.63105 \text{ Pa}}{1 \text{ Titan atm}} \times \dfrac{1 \text{ Earth atm}}{101{,}325 \text{ Pa}} = \dfrac{1.60972 \times 10^{-5} \text{ Earth atm}}{1 \text{ Titan atm}}$

(b) $\dfrac{90 \text{ Earth atm}}{1 \text{ Venus atm}} \times \dfrac{101.3 \text{ kPa}}{1 \text{ Earth atm}} = \dfrac{9.1 \times 10^3 \text{ kPa}}{1 \text{ Venus atm}}$

$P_{gas} = 10.3 \text{ cm} \times \dfrac{1 \text{ atm}}{76 \text{ cm}} = 0.136 \text{ atm}$

10.22 (a) The atmosphere is exerting 10.7 cm = 107 mm Hg (torr) more pressure than the gas.

$$P_{gas} = P_{atm} - 107 \text{ torr} = \left(0.977 \text{ atm} \times \frac{760 \text{ torr}}{1 \text{ atm}}\right) - 107 \text{ torr} = 636 \text{ torr}$$

(b) The gas is exerting 9.5 mm Hg (torr) more pressure than the atmosphere.

$$P_{gas} = P_{atm} + 9.5 \text{ torr} = \left(1.02 \text{ atm} \times \frac{760 \text{ torr}}{1 \text{ atm}}\right) + 9.5 \text{ torr} = 785 \text{ torr}$$

The Gas Laws

10.24 *Analyze.* Given: initial P, V, T. Find: final values of P, V, T for certain changes of condition. *Plan.* Select the appropriate gas law relationships from Section 10.3; solve for final conditions, paying attention to units. *Solve.*

(a) $P_1V_1 = P_2V_2$; the proportionality holds true for any pressure or volume units.

$P_1 = 735 \text{ torr}, V_1 = 5.22 \text{ L}, P_2 = 1.88 \text{ atm}$

$$V_2 = \frac{P_1V_1}{P_2} = \frac{735 \text{ torr} \times 5.22 \text{ L}}{1.88 \text{ atm}} \times \frac{1 \text{ atm}}{760 \text{ torr}} = 2.69 \text{ L}$$

Check. As pressure increases, volume should decrease; our result agrees with this.

(b) $V_1/T_1 = V_2/T_2$; T must be in Kelvins for the relationship to be true.

$V_1 = 5.22 \text{ L}, T_1 = 23°\text{C} = 296 \text{ K}, T_2 = 165°\text{C} = 438 \text{ K}$

$$V_2 = \frac{V_1T_2}{T_1} = \frac{5.22 \text{ L} \times 438 \text{ K}}{296 \text{ K}} = 7.72 \text{ L}$$

Check. As temperature increases, volume should increase; our result is consistent with this.

10.26 According to Avogadro's hypothesis, the mole ratios in the chemical equation will be volume ratios for the gases if they are at the same temperature and pressure.

$$N_2(g) + 3H_2(g) \rightarrow 2NH_3(g)$$

The volumes of H_2 and N_2 are in a stoichiometric $\frac{3.6\ L}{1.2\ L}$ or $\frac{3\ vol\ H_2}{1\ vol\ N_2}$ ratio, so either can be used to determine the volume of $NH_3(g)$ produced.

$$1.2\ L\ N_2 \times \frac{2\ mol\ NH_3}{1\ mol\ N_2} = 2.4\ L\ NH_3(g)\ \text{produced.}$$

The Ideal-Gas Equation

(In *Solutions to Exercises*, the symbol for molar mass is MM.)

10.28 (a) STP stands for standard temperature, 0°C (or 273 K), and standard pressure, 1 atm.

(b) $V = \frac{nRT}{P}$; $V = 1\ mol \times \frac{0.08206\ L\bullet atm}{K\bullet mol} \times \frac{273\ K}{1\ atm}$

$V = 22.4$ L for 1 mole of gas at STP

(c) 25°C + 273 = 298 K

$V = \frac{nRT}{P}$; $V = 1\ mol \times \frac{0.08206\ L\bullet atm}{K\bullet mol} \times \frac{298\ K}{1\ atm}$

$V = 24.5$ L for 1 mol of gas at 1 atm and 25°C

10.30 $n = g/MM$; $PV = nRT = gRT/MM$; $MM = gRT/PV$.

2-L flask: $MM = 4.8\ RT/2.0(X) = 2.4\ RT/X$

3-L flask: $MM = 0.36\ RT/3.0\ (0.1\ X) = 1.2\ RT/X$

The molar masses of the two gases are not equal. The gas in the 2-L flask has a molar mass that is twice as large as the gas in the 3-L flask.

10.32 *Analyze/Plan.* Follow the strategy for calculations involving many variables given in Section 10.4. *Solve.*

(a) n = 1.75 mol, P = 0.985 atm, T = –6°C = 267 K

$$V = \frac{nRT}{P} = 1.75\ mol \times \frac{0.08206\ L\bullet atm}{K\bullet mol} \times \frac{267\ K}{0.985\ atm} = 38.9\ L$$

(b) $n = 3.33 \times 10^{-3}$ mol, V = 255 mL = 0.255 L

$$P = 720\ torr \times \frac{1\ atm}{760\ torr} = 0.9474 = 0.947\ atm$$

$$T = \frac{PV}{nR} = 0.9474\ atm \times \frac{0.255\ L}{3.33 \times 10^{-3}\ mol} \times \frac{1\ K\bullet mol}{0.08206\ L\bullet atm} = 884\ K$$

(c) $n = 4.67 \times 10^{-2}$ mol, V = 413 mL = 0.413 L, T = 122°C = 395 K

$$P = \frac{nRT}{V}; = 0.00467\ mol \times \frac{0.08206\ L\bullet atm}{K\bullet mol} \times \frac{395\ K}{0.413\ L} = 3.67\ atm$$

(d) V = 67.5 L, T = 54°C = 327 K,

$$P = 11.25\ kPa \times \frac{1\ atm}{101.325\ kPa} = 0.11103 = 0.1110\ atm$$

$$n = \frac{PV}{RT} = 0.11103\ atm \times \frac{K\bullet mol}{0.08206\ L\bullet atm} \times \frac{67.5\ L}{327\ K} = 0.279\ mol$$

10.34 Find the volume of the tube in cm^3; $1\ cm^3 = 1\ mL$.

$r = d/2 = 2.5\ cm/2 = 1.25 = 1.3\ cm;\ h = 5.5\ m = 5.5 \times 10^2\ cm$

$V = \pi r^2 h = 3.14159 \times (1.25\ cm)^2 \times (5.5 \times 10^2\ cm) = 2.700 \times 10^3\ cm^3 = 2.7\ L$

$$PV = \frac{g}{MM}RT;\ g = \frac{MM \times PV}{RT};\ P = 1.78\ torr \times \frac{1\ atm}{760\ torr} = 2.342 \times 10^{-3} = 2.34 \times 10^{-3}\ atm$$

$$g = \frac{20.18\ g\ Ne}{1\ mol\ Ne} \times \frac{K \bullet mol}{0.08206\ L \bullet atm} \times \frac{2.342 \times 10^{-3}\ atm \times 2.700\ L}{308\ K} = 5.049 \times 10^{-3} = 5.0 \times 10^{-3}\ g\ Ne$$

10.36 $P_{O_3} = 3.0 \times 10^{-3}\ atm;\ T = 250\ K;\ V = 1\ L\ (exact)$

$$\#\ of\ O_3\ molecules = \frac{PV}{RT} \times 6.022 \times 10^{23}$$

$$\# = \frac{3.0 \times 10^{-3}\ atm \times 1\ L}{250\ K} \times \frac{K \bullet mol}{0.08206\ L \bullet atm} \times \frac{6.022 \times 10^{23}\ molecules}{mol}$$

$$= 8.8 \times 10^{19}\ O_3\ molecules$$

10.38 (a) $V = 0.250\ L,\ T = 23°C = 296\ K,\ n = 2.30\ g\ C_3H_8 \times \frac{1\ mol\ C_3H_8}{44.1\ g\ C_3H_8} = 0.052154$

$$= 0.0522\ mol$$

$$P = \frac{nRT}{V} = 0.052154\ mol \times \frac{0.08206\ L \bullet atm}{K \bullet mol} \times \frac{296\ K}{0.250\ L} = 5.07\ atm$$

(b) STP = 1.00 atm, 273 K

$$V = \frac{nRT}{P} = 0.052154\ mol \times \frac{0.08206\ L \bullet atm}{K \bullet mol} \times \frac{273\ K}{1.00\ atm} = 1.1684 = 1.17\ L$$

(c) $°C = 5/9\ (°F - 32°);\ K = °C + 273.15 = 5/9\ (130°F - 32°) + 273.15 = 327.59 = 328\ K$

$$P = \frac{nRT}{V} = 0.052154\ mol \times \frac{0.08206\ L \bullet atm}{K \bullet mol} \times \frac{327.59\ K}{0.250\ L} = 5.608 = 5.61\ atm$$

10.40 $V = 65.0\ L,\ T = 23°C = 296\ K,\ P = 16{,}500\ kPa \times \frac{1\ atm}{101.325} = 162.84 = 163\ atm$

(a) $$g = \frac{MM \times PV}{RT};\ g = \frac{32.0\ g\ O_2}{1\ mol\ O_2} \times \frac{K \bullet mol}{0.08206\ L \bullet atm} \times \frac{162.84\ atm}{296\ K} \times 65.0\ L$$

$$= 1.39446 \times 10^4\ g\ O_2 = 13.9\ kg\ O_2$$

(b) $$V_2 = \frac{P_1V_1T_2}{T_1P_2} = \frac{16{,}500\ kPa \times 65.0\ L \times 273\ K}{296\ K \times 101.325\ kPa} = 9.76 \times 10^3\ L$$

(c) $$T_2 = \frac{P_2T_1}{P_1} = \frac{150.0\ atm \times 296\ K}{16{,}500\ kPa} \times \frac{101.325\ kPa}{1\ atm} = 272.7 = 2.73\ K$$

(d) $$P_2 = \frac{P_1V_1T_2}{V_2T_1} = \frac{16{,}500\ kPa \times 65.0\ L \times 297\ K}{55.0\ L \times 296\ K} = 19{,}566 = 1.96 \times 10^4\ kPa$$

10.42 $mass = 1800 \times 10^{-9}\ g = 1.8 \times 10^{-6}\ g;\ V = 1\ m^3 = 1 \times 10^3\ L;\ T = 273 + 10°\ C = 283\ K$

(a) $$P = \frac{g\ RT}{MM \times V};\ P = \frac{1.8 \times 10^{-6}\ g\ Hg \times 1\ mol\ Hg}{200.6\ g\ Hg} \times \frac{0.08206\ L \bullet atm}{K \bullet mol} \times \frac{283\ K}{1 \times 10^3\ L}$$

$$= 2.1 \times 10^{-10}\ atm$$

(b) $$\frac{1.8 \times 10^{-6} \text{ g Hg}}{1 \text{ m}^3} \times \frac{1 \text{ mol Hg}}{200.6 \text{ g Hg}} \times \frac{6.022 \times 10^{23} \text{ Hg atoms}}{1 \text{ mol Hg}} = 5.4 \times 10^{15} \text{ Hg atoms/m}^3$$

(c) $$1600 \text{ km}^3 \times \frac{1000^3 \text{ m}^3}{1 \text{ km}^3} \times \frac{1.8 \times 10^{-6} \text{ g Hg}}{1 \text{ m}^3} = 2.9 \times 10^6 \text{ g Hg/day}$$

Further Applications of the Ideal-Gas Equation

10.44 (c) $CO_2(g)$ is least dense. For gases at the same conditions, density is directly proportional to molar mass, and CO_2 has the smallest molar mass.

10.46 (b) Xe atoms have a higher mass than N_2 molecules. Because both gases at STP have the same number of molecules per unit volume, the Xe gas must be denser.

10.48 (a) $d = \frac{MM \times P}{RT}$; MM = 146.1 g/mol, T = 28°C = 301 K, P = 678 torr

$$d = \frac{146.1 \text{ g}}{1 \text{ mol}} \times \frac{\text{K} \bullet \text{mol}}{0.08206 \text{ L} \bullet \text{atm}} \times \frac{678 \text{ torr}}{301 \text{ K}} \times \frac{1 \text{ atm}}{760 \text{ torr}} = 5.28 \text{ g/L}$$

(b) $$MM = \frac{dRT}{P} = \frac{7.135 \text{ g}}{1 \text{ L}} \times \frac{0.08206 \text{ L} \bullet \text{atm}}{\text{K} \bullet \text{mol}} \times \frac{285 \text{ K}}{743 \text{ torr}} \times \frac{760 \text{ torr}}{1 \text{ atm}} = 171 \text{ g/mol}$$

10.50 $$MM = \frac{gRT}{PV} = \frac{0.846 \text{ g}}{0.354 \text{ L}} \times \frac{0.08206 \text{ L} \bullet \text{atm}}{\text{K} \bullet \text{atm}} \times \frac{373 \text{ K}}{752 \text{ torr}} \times \frac{760 \text{ torr}}{1 \text{ atm}} = 73.9 \text{ g/mol}$$

10.52 $n_{H_2} = \frac{P_{H_2} V}{RT}$; $P = 814 \text{ torr} \times \frac{1 \text{ atm}}{760 \text{ torr}} = 1.071 = 1.07 \text{ atm}$; T = 273 + 21°C = 294 K

$$n_{H_2} = 1.071 \text{ atm} \times \frac{\text{K} \bullet \text{mol}}{0.08206 \text{ L} \bullet \text{atm}} \times \frac{53.5 \text{ L}}{294 \text{ K}} = 2.3751 = 2.38 \text{ mol } H_2$$

$$2.3751 \text{ mol } H_2 \times \frac{1 \text{ mol } CaH_2}{2 \text{ mol } H_2} \times \frac{42.10 \text{ g } CaH_2}{1 \text{ mol } CaH_2} = 50.0 \text{ g } CaH_2$$

10.54 Balance the chemical equation.

Then, g $C_5H_{12} \rightarrow$ mol $C_5H_{12} \rightarrow$ mol $O_2 \rightarrow$ V O_2 at given P, T.

$C_5H_{12}(l) + 8O_2(g) \rightarrow 5CO_2(g) + 6H_2O(g)$

$$2.50 \text{ g } C_5H_{12} \times \frac{1 \text{ mol } C_5H_{12}}{72.15 \text{ g } C_5H_{12}} \times \frac{8 \text{ mol } O_2}{1 \text{ mol } C_5H_{12}} = 0.2772 = 0.277 \text{ mol } O_2$$

$$V = \frac{nRT}{P} = 0.2772 \text{ mol } O_2 \times \frac{296 \text{ K}}{0.980 \text{ atm}} \times \frac{0.08206 \text{ L} \bullet \text{atm}}{\text{K} \bullet \text{mol}} = 6.87054 = 6.87 \text{ L } O_2$$

According to Avogadro's law, mole ratios are volume ratios for gases at the same pressure and temperature. Use mole ratios from the chemical equation to calculate volumes of $CO_2(g)$ and $H_2O(g)$.

$$6.87054 \text{ L } O_2 \times \frac{5 \, CO_2}{8 \, O_2} = 4.29 \text{ L } CO_2; \; 6.87054 \text{ L } O_2 \times \frac{6 \, H_2O}{8 \, O_2} = 5.15 \text{ L } H_2O$$

10.56 The gas sample is a mixture of $C_2H_2(g)$ and $H_2O(g)$. Find the partial pressure of C_2H_2, then moles CaC_2 and C_2H_2.

$P_t = 726 \text{ torr} = P_{C_2H_2} + P_{H_2O}$. P_{H_2O} at 26° C = 25.21 torr

$$P_{C_2H_2} = (726 \text{ torr} - 25.21 \text{ torr}) \times \frac{1 \text{ atm}}{760 \text{ torr}} = 0.9221 = 0.922 \text{ atm}$$

$$0.887 \text{ g } CaC_2 \times \frac{1 \text{ mol } CaC_2}{64.10 \text{ g}} \times \frac{1 \text{ mol } C_2H_2}{1 \text{ mol } CaC_2} = 0.013838 = 0.0138 \text{ mol } C_2H_2$$

$$V = 0.013838 \text{ mol} \times \frac{0.08206 \text{ L} \bullet \text{atm}}{\text{K} \bullet \text{mol}} \times \frac{299 \text{ K}}{0.9221 \text{ atm}} = 0.368 \text{ L } C_2H_2$$

Partial Pressures

10.58 (a) The partial pressure of gas A is **not affected** by the addition of gas C. The partial pressure of A depends only on moles of A, volume of container, and conditions; none of these factors changes when gas C is added.

(b) The total pressure in the vessel **increases** when gas C is added, because the total number of moles of gas increases.

(c) The mole fraction of gas B **decreases** when gas C is added. The moles of gas B stay the same, but the total moles increase, so the mole fraction of B (n_B/n_t) decreases.

10.60 (a) $$2.50 \text{ g } CH_4 \times \frac{1 \text{ mol } CH_4}{16.04 \text{ g } CH_4} = 0.15586 = 0.156 \text{ mol } CH_4$$

$$P_{CH_4} = \frac{nRT}{V} = 0.15586 \text{ mol} \times \frac{0.08206 \text{ L} \bullet \text{atm}}{\text{K} \bullet \text{mol}} \times \frac{288 \text{ K}}{2.00 \text{ L}} = 1.842 = 1.84 \text{ atm}$$

$$2.50 \text{ g } C_2H_4 \times \frac{1 \text{ mol } C_2H_4}{28.05 \text{ g } C_2H_4} = 0.08913 = 0.0891 \text{ mol } C_2H_4$$

$$P_{C_2H_4} = \frac{nRT}{V} = 0.08913 \text{ mol } C_2H_4 \times \frac{0.08206 \text{ L} \bullet \text{atm}}{\text{K} \bullet \text{mol}} \times \frac{288 \text{ K}}{2.00 \text{ L}} = 1.053 = 1.05 \text{ atm}$$

$$2.50 \text{ g } C_4H_{10} \times \frac{1 \text{ mol } C_4H_{10}}{58.12 \text{ g } C_4H_{10}} = 0.04301 = 0.0430 \text{ mol } C_4H_{10}$$

$$P_{C_4H_{10}} = \frac{nRT}{V} = 0.04301 \text{ mol} \times \frac{0.08206 \text{ L} \bullet \text{atm}}{\text{K} \bullet \text{mol}} \times \frac{288 \text{ K}}{2.00 \text{ L}} = 0.5083 = 0.508 \text{ atm}$$

(b) P_t = 1.842 atm + 1.053 atm + 0.508 atm = 3.403 = 3.40 atm

10.62 $$V\ C_4H_{10}O(l) \xrightarrow{\text{density}} \text{mass } C_4H_{10}O \rightarrow \text{mol } C_4H_{10}O \rightarrow P_{C_4H_{10}O} \text{ at given V, T.}$$

$P_t = P_{N_2} + P_{O_2} + P_{C_4H_{10}O}$; T = 273.15 + 35.0°C = 308.15 = 308.2 K

(a) $$4.00 \text{ mL } C_4H_{10}O \times \frac{0.7134 \text{ g } C_4H_{10}O}{\text{mL}} \times \frac{1 \text{ mol } C_4H_{10}O}{74.12 \text{ g } C_4H_{10}O} = 0.0385 \text{ mol } C_4H_{10}O$$

$$P = \frac{nRT}{V} = 0.03850 \text{ mol} \times \frac{308.15}{5.00 \text{ L}} \times \frac{0.08206 \text{ L} \bullet \text{atm}}{\text{K} \bullet \text{mol}} = 0.1947 = 0.195 \text{ atm}$$

(b) $P_t = P_{N_2} + P_{O_2} + P_{C_4H_{10}O} = 0.751 \text{ atm} + 0.208 \text{ atm} + 0.195 \text{ atm} = 1.154 \text{ atm}$

10.64 $n_{N_2} = 10.25\text{ g } N_2 \times \frac{1\text{ mol}}{28.02\text{ g}} = 0.3658\text{ mol}$; $n_{H_2} = 2.05\text{ g } H_2 \times \frac{1\text{ mol}}{2.016\text{ g}} = 1.0169 = 1.02\text{ mol}$

$$n_{NH_3} = 7.63\text{ g } NH_3 \times \frac{1\text{ mol}}{17.03\text{ g}} = 0.448\text{ mol};\ n_t = 0.3658 + 1.0169 + 0.4480 = 1.8307$$

$$= 1.83\text{ mol}$$

$$P_{N_2} = \frac{n_{N_2}}{n_t} \times P_t = \frac{0.3658}{1.8307} \times 2.35\text{ atm} = 0.470\text{ atm}$$

$$P_{H_2} = \frac{1.0169}{1.8307} \times 2.35\text{ atm} = 1.31\text{ atm};\ P_{NH_3} = \frac{0.4480}{1.8307} \times 2.35\text{ atm} = 0.575\text{ atm}$$

10.66 (a) $n_{O_2} = 5.08\text{ g } O_2 \times \frac{1\text{ mol}}{32.00\text{ g}} = 0.159\text{ mol}$; $n_{N_2} = 7.17\text{ g } N_2 \times \frac{1\text{ mol}}{28.02\text{ g}} = 0.256\text{ mol}$

$$n_{H_2} = 1.32\text{ g } H_2 \times \frac{1\text{ mol}}{2.016\text{ g}} = 0.655\text{ mol};\ n_t = 0.159 + 0.256 + 0.655 = 1.070\text{ mol}$$

$$\chi_{O_2} = \frac{n_{O_2}}{n_t} = \frac{0.159}{1.07} = 0.149;\ \chi_{N_2} = \frac{n_{N_2}}{n_t} = \frac{0.256}{1.07} = 0.239$$

$$\chi_{H_2} = \frac{0.655}{1.07} = 0.612$$

(b) $P_{O_2} = n \times \frac{RT}{V}$; $P_{O_2} = 0.159\text{ mol} \times \frac{0.08206\text{ L} \bullet \text{atm}}{\text{K} \bullet \text{mol}} \times \frac{288\text{ K}}{12.40\text{ L}} = 0.303\text{ atm}$

$$P_{N_2} = 0.256\text{ mol} \times \frac{0.08206\text{ L} \bullet \text{atm}}{\text{K} \bullet \text{mol}} \times \frac{288\text{ K}}{12.40\text{ L}} = 0.488\text{ atm}$$

$$P_{H_2} = 0.655\text{ mol} \times \frac{0.08206\text{ L} \bullet \text{atm}}{\text{K} \bullet \text{mol}} \times \frac{288\text{ K}}{12.40\text{ L}} = 1.25\text{ atm}$$

10.68 Calculate the pressure of the gas in the second vessel directly from mass and conditions using the ideal-gas equation.

(a) $P_{SO_2} = \frac{gRT}{MV} = \frac{4.00\text{ g } SO_2}{64.07\text{ g } SO_2/\text{mol}} \times \frac{0.08206\text{ L} \bullet \text{atm}}{\text{K} \bullet \text{mol}} \times \frac{298\text{ K}}{10.0\text{ L}} = 0.15267$

$$= 0.153\text{ atm}$$

(b) $P_{N_2} = \frac{gRT}{MV} = \frac{2.35\text{ g } N_2}{28.01\text{ g } N_2/\text{mol}} \times \frac{0.08206\text{ L} \bullet \text{atm}}{\text{K} \bullet \text{mol}} \times \frac{298\text{ K}}{10.0\text{ L}} = 0.20516$

$$= 0.205\text{ atm}$$

(c) $P_t = P_{SO_2} + P_{N_2} = 0.15267\text{ atm} + 0.20516\text{ atm} = 0.358\text{ atm}$

Kinetic-Molecular Theory; Graham's Law

10.70 (a) False. The average kinetic energy per molecule in a collection of gas molecules is the same for all gases at the same temperature.

b) True.

(c) False. The molecules in a gas sample at a given temperature exhibit a distribution of kinetic energies.

(d) True.

10.72 Newton's model provides no explanation of the effect of a change in temperature on the pressure of a gas at constant volume or on the volume of a gas at constant pressure. On the other hand, the assumption that the average kinetic energy of gas molecules increases with increasing temperature explains Charles' Law, that an increase in temperature requires an increase in volume to maintain constant pressure.

10.74 (a) They have the same number of molecules (equal volumes of gases at the same temperature and pressure contain equal numbers of molecules).

(b) N_2 is more dense because it has the larger molar mass. Since the volumes of the samples and the number of molecules are equal, the gas with the larger molar mass will have the greater density.

(c) The average kinetic energies are equal (statement 5, section 10.7).

(d) CH_4 will effuse faster. The lighter the gas molecules, the faster they will effuse (Graham's Law).

10.76 (a) *Plan.* The greater the molecular (and molar) mass, the smaller the rms speed of the molecules. Calculate the molar mass of each gas, and place them in decreasing order of mass and increasing order of rms speed. *Solve.*

CO = 28 g/mol; SF_6 = 146 g/mol; H_2S = 34 g/mol; Cl_2 = 71 g/mol; HBr = 81 g/mol. In order of increasing speed (and decreasing molar mass):

$SF_6 < HBr < Cl_2 < H_2S < CO$

(b) *Plan.* Follow the logic of Sample Exercise 10.14. *Solve.*

$$u_{CO} = \sqrt{\frac{3RT}{M}} = \left(\frac{3 \times 8.314 \text{ kg} \bullet \text{m}^2/\text{s}^2 \bullet \text{K} \bullet \text{mol} \times 300 \text{ K}}{28.0 \times 10^{-3} \text{ kg/mol}}\right)^{1/2} = 5.17 \times 10^2 \text{ m/s}$$

$$u_{Cl_2} = \left(\frac{3 \times 8.314 \text{ kg} \bullet \text{m}^2/\text{s}^2 \bullet \text{K} \bullet \text{mol} \times 300 \text{ K}}{70.9 \times 10^{-3} \text{ kg/mol}}\right)^{1/2} = 3.25 \times 10^2 \text{ m/s}$$

As expected, the lighter molecule moves at the greater speed.

10.78 $$\frac{\text{rate } ^{235}\text{U}}{\text{rate } ^{238}\text{U}} = \sqrt{\frac{238.05}{235.04}} = \sqrt{1.0128} = 1.0064$$

There is a slightly greater rate enhancement for $^{235}U(g)$ atoms than $^{235}UF_6(g)$ molecules (1.0043), because ^{235}U is a greater percentage (100%) of the mass of the diffusing particles than in $^{235}UF_6$ molecules. The masses of the isotopes were taken from *The Handbook of Chemistry and Physics.*

10.80 The time required is proportional to the reciprocal of the effusion rate.

$$\frac{\text{rate (X)}}{\text{rate (O}_2)} = \frac{105 \text{ s}}{31 \text{ s}} = \left[\frac{32 \text{ g O}_2}{\text{MM}_x}\right]^{1/2}; \text{MM}_x = 32 \text{ g O}_2 \times \left[\frac{105}{31}\right]^2 = 370 \text{g/mol (two sig figs)}$$

Nonideal-Gas Behavior

10.82 Ideal-gas behavior is most likely to occur at high temperature and low pressure, so the atmosphere on Mercury is more likely to obey the ideal-gas law. The higher temperature on Mercury means that the kinetic energies of the molecules will be larger relative to intermolecular attractive forces. Further, the gravitational attractive forces on Mercury are lower because the planet has a much smaller mass. This means that for the same column mass of gas (Figure 10.1), atmospheric pressure on Mercury will be lower.

10.84 The constant *a* is a measure of the strength of intermolecular attractions among gas molecules; *b* is a measure of molecular volume. Both increase with increasing molecular mass and structural complexity.

10.86 (a) At STP, the molar volume $= 1 \text{ mol} \times \frac{0.08206 \text{ L} \bullet \text{atm}}{\text{K} \bullet \text{mol}} \times \frac{273 \text{ K}}{1 \text{ atm}} = 22.4 \text{ L}$

Dividing the value for *b*, 0.0322 L/mol, by 4, we obtain 0.0080 L. Thus, the volume of the Ar atoms is (0.0080/22.4)100 = 0.036% of the total volume.

(b) At 100 atm pressure (and 0°C, standard temperature) the molar volume is 0.224 L, and the volume of the Ar atoms is 3.6% of the total volume.

Additional Exercises

10.88 Only item (b) is satisfactory. Item (c) would not have supported a column of Hg because it is open at both ends. The atmosphere would exert pressure on the top of the column as well as on the reservoir; the column would only be as high as the reservoir and the height would not change with changing pressure. Item (d) is not tall enough to support a nearly 760 mm Hg column. Items (a) and (e) are inappropriate for the same reason: they don't have a uniform cross-sectional area. The height of the Hg column is a direct measure of atmospheric pressure only if the cross-sectional area is constant over the entire tube.

10.89 $P_1V_1 = P_2V_2$; $V_2 = P_1V_1/P_2$

$$V_2 = \frac{3.0 \text{ atm} \times 1.0 \text{ mm}^3}{695 \text{ torr}} \times \frac{760 \text{ torr}}{1 \text{ atm}} = 3.3 \text{ mm}^3$$

10.91 $P = \frac{nRT}{V}$; $n = 1.4 \times 10^{-5}$ mol, V = 0.600 L, T = 23°C = 296 K

$$P = 1.4 \times 10^{-5} \text{ mol} \times \frac{0.08206 \text{ L} \bullet \text{atm}}{\text{K} \bullet \text{mol}} \times \frac{296 \text{ K}}{0.600 \text{ L}} = 5.7 \times 10^{-4} \text{ atm} = 0.43 \text{ mm Hg}$$

10.92 (a) $n = \frac{PV}{RT} = 3.00 \text{ atm} \times \frac{\text{K} \bullet \text{mol}}{0.08206 \text{ L} \bullet \text{atm}} \times \frac{110 \text{ L}}{300 \text{ K}} = 13.4 \text{ mol } C_3H_8(g)$

(b) $\frac{0.590 \text{ g } C_3H_8 \text{ (l)}}{1 \text{ mL}} \times 110 \times 10^3 \text{ mL} \times \frac{1 \text{ mol } C_3H_8}{44.094 \text{ g}} = 1.47 \times 10^3 \text{ mol } C_3H_8 \text{ (l)}$

(c) Using C_3H_8 in a 110 L container as an example, the ratio of moles liquid to moles gas that can be stored in a certain volume is $\frac{1.47 \times 10^3 \text{ mol liquid}}{13.4 \text{ mol gas}} = 110.$

A container with a fixed volume holds many more moles (molecules) of C_3H_8(l) because in the liquid phase the molecules are touching. In the gas phase, the molecules are far apart (statement 2, section 10.7), and many fewer molecules will fit in the container.

10.94 Volume of laboratory $= 54\ m^2 \times 3.1\ m \times \dfrac{1000\ L}{1\ m^3} = 1.674 \times 10^5 = 1.7 \times 10^5\ L$

Calculate the **total** moles of gas in the laboratory at the conditions given.

$$n_t = \frac{PV}{RT} = 1.00\ atm \times \frac{K \bullet mol}{0.08206\ L \bullet atm} \times \frac{1.674 \times 10^5\ L}{297\ K} = 6.869 \times 10^3 = 6.9 \times 10^3\ mol\ gas$$

An $Ni(CO)_4$ concentration of 1 part in 10^9 means 1 mol $Ni(CO)_4$ in 1×10^9 total moles of gas.

$$\frac{x\ mol\ Ni(CO)_4}{6.869 \times 10^3\ mol\ gas} = \frac{1}{10^9} = 6.869 \times 10^{-6}\ mol\ Ni(CO)_4$$

$$6.869 \times 10^{-6}\ mol\ Ni(CO)_4 \times \frac{170.74 g\ Ni(CO)_4}{1\ mol\ Ni(CO)_4} = 1.2 \times 10^{-3}\ g\ Ni(CO)_4$$

10.95 It is simplest to calculate the partial pressure of each gas as it expands into the total volume, then sum the partial pressures.

$P_2 = P_1V_1/V_2;\ P_{N_2} = 265\ torr\ (1.0\ L/2.5\ L) = 106 = 1.1 \times 10^2\ torr$

$P_{Ne} = 800\ torr\ (1.0\ L/2.5\ L) = 320 = 3.2 \times 10^2\ torr;\ P_{H_2} = 532\ torr\ (0.5\ L/2.5\ L)$
$= 106 = 1.1 \times 10^2\ torr$

$P_t = P_{N_2} + P_{Ne} + P_{H_2} = (106 + 320 + 106)\ torr = 532 = 5.3 \times 10^2\ torr$

10.97 (a) $5.00\ g\ HCl \times \dfrac{1\ mol\ HCl}{36.46\ g\ HCl} = 0.1371 = 0.137\ mol\ HCl$

$5.00\ g\ NH_3 \times \dfrac{1\ mol\ NH_3}{17.03\ g\ NH_3} = 0.2936 = 0.294\ mol\ NH_3$

The gases react in a 1:1 mole ratio, HCl is the limiting reactant and is completely consumed. (0.2936 mol - 0.1371 mol) = 0.1565 = 0.157 mol NH_3 remain in the system. $NH_3(g)$ is the only gas remaining after reaction. $V_t = 4.00\ L$

(b) $P = \dfrac{nRT}{V} = 0.1565\ mol \times \dfrac{0.08206 L \bullet atm}{K \bullet mol} \times \dfrac{298\ K}{4.00 L} = 0.957\ atm$

10.98 V and T are the same for He and O_2.

$P_{He}V = n_{He}RT,\ \ P_{He} / n_{He} = RT/V;\ \ P_{O_2} / n_{O_2} = RT/V$

$$\frac{P_{He}}{n_{He}} = \frac{P_{O_2}}{n_{O_2}} = n_{O_2} = \frac{P_{O_2} \times n_{He}}{P_{He}};\ n_{He} = 1.42\ g\ He \times \frac{1\ mol\ He}{4.003\ g\ He} = 0.3547 = 0.355\ mol\ He$$

$$n_{O_2} = \frac{158\ torr}{42.5\ torr} \times 0.355\ mol = 1.3188 = 1.32\ mol\ O_2;\ 1.3188\ mol\ O_2 \times \frac{32.00\ g\ O_2}{1 mol\ O_2} = 42.2 g\ O_2$$

10.100 Calculate the number of moles of Ar in the vessel:

n = (339.854 - 337.428)/39.948 = 0.060729 = 0.06073 mol

The total number of moles of the mixed gas is the same (Avogadro's Law). Thus, the average atomic weight is (339.076 - 337.428)/0.060729 = 27.137 = 27.14. Let the mole fraction of Ne be χ. Then,

χ (20.183) + (1 - χ) (39.948) = 27.137; 12.811 = 19.765 χ; χ = 0.6482

Neon is thus 64.82 mole percent of the mixture.

10.101 (a) The quantity $d/P = MM/RT$ should be a constant at all pressures for an ideal gas. It is not, however, because of nonideal behavior. If we graph d/P vs P, the ratio should approach ideal behavior at low P. At P = 0, $d/P = 2.2525$. Using this value in the formula $MM = d/P \times RT$, MM = 50.46 g/mol.

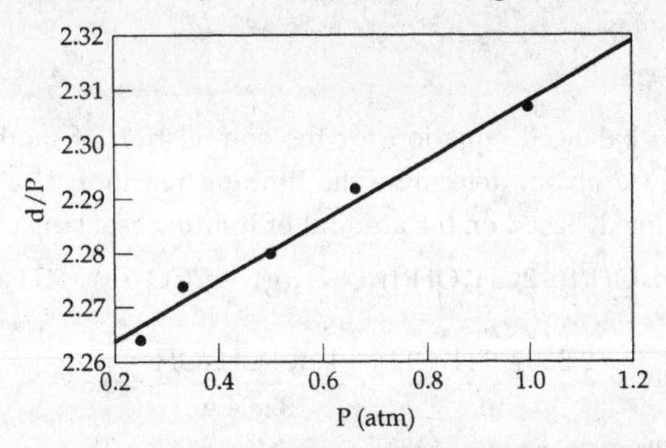

(b) The ratio d/P varies with pressure because of the finite volumes of gas molecules and attractive intermolecular forces.

10.103 (a) The initial drop in the value of PV/RT is due to attractive forces between molecules. These intermolecular forces cause the molecules to "stick" together and behave as if there are fewer net particles in the sample. At lower pressures, this is the dominant effect. At the same time, the real volume of gas molecules causes the amount of free space in the gas sample to be less than the container volume. Using the container volume to calculate PV/RT gives a value larger than that for an ideal gas (which assumes that the total container volume is free space). At higher pressures, this effect more than compensates for molecular attraction, and PV/RT is greater than 1.

(b) As the temperature of a gas increases, the average kinetic energy of the particles increases. The increased kinetic energy overcomes the attractive forces between molecules and keeps them separate.

10.104 (a) $$120.00 \text{ kg } N_2(g) \times \frac{1000 \text{ g}}{1 \text{ kg}} \times \frac{1 \text{ mol } N_2}{28.0135 \text{ g N}} = 4283.6 \text{ mol } N_2$$

$$P = \frac{nRT}{V} = 4283.6 \text{ mol} \times \frac{0.08206 \text{ L} \bullet \text{atm}}{\text{K} \bullet \text{mol}} \times \frac{553 \text{ K}}{1100.0 \text{ L}} = 176.72 = 177 \text{ atm}$$

(b) According to Equation [10.26], $P = \dfrac{nRT}{V - nb} - \dfrac{n^2 a}{V^2}$

$$P = \frac{(4283.6 \text{ mol})(0.08206 \text{ L} \bullet \text{atm/K} \bullet \text{mol})(553 \text{ K})}{1100.0 \text{ L} - (4283.6 \text{ mol})(0.0391 \text{ L/mol})} - \frac{(4283.6 \text{ mol})^2 (1.39 \text{ L}^2 \bullet \text{atm/mol}^2)}{(1100.0 \text{ L})^2}$$

$$P = \frac{194{,}388 \text{ L} \bullet \text{atm}}{1100.0 \text{ L} - 167.5 \text{ L}} - 21.1 \text{ atm} = 208.5 \text{ atm} - 21.1 \text{ atm} = 187.4 \text{ atm}$$

(c) The pressure corrected for the real volume of the N_2 molecules is 208.5 atm, 31.8 atm higher than the ideal pressure of 176.7 atm. The 21.1 atm correction for intermolecular forces reduces the calculated pressure somewhat, but the "real" pressure is still higher than the ideal pressure. The correction for the real volume of molecules dominates. Even though the value of b is small, the number of moles of N_2 is large enough so that the molecular volume correction is larger than the attractive forces correction.

Integrative Exercises

10.106 *Plan.* Write the balanced equation for the combustion of methanol. Since amounts of both reactants are given, determine the limiting reactant. Use mole ratios to calculate mol H_2O produced, based on the amount of limiting reactant. *Solve.*

methanol = $CH_3OH(l)$. $2CH_3OH(l) + 3O_2(g) \rightarrow 2CO_2(g) + 4H_2O(g)$

$$25.0\,\text{mL}\,CH_3OH \times \frac{0.850\,\text{g}\,CH_3OH}{\text{mL}} \times \frac{1\,\text{mol}\,CH_3OH}{32.04\,\text{g}} = 0.6632 = 0.663\,\text{mol}\,CH_3OH$$

$$\text{mol}\,O_2 = n = \frac{PV}{RT} = 1.00\,\text{atm} \times \frac{12.5\,\text{L}}{273\,\text{K}} \times \frac{\text{K}\bullet\text{mol}}{0.08206\,\text{L}\bullet\text{atm}} = 0.5580 = 0.558\,\text{mol}\,O_2$$

$$0.558\,\text{mol}\,O_2 \times \frac{2\,\text{mol}\,CH_3OH}{3\,\text{mol}\,O_2} = 0.372\,\text{mol}\,CH_3OH$$

0.558 mol O_2 can react with only 0.372 mol CH_3OH, so O_2 is the limiting reactant. Note that a large volume of $O_2(g)$ is required to completely react with a relatively small volume of $CH_3OH(l)$.

$$0.558\,\text{mol}\,O_2 \times \frac{4\,\text{mol}\,H_2O}{3\,\text{mol}\,O_2} = 0.744\,\text{mol}\,H_2O$$

10.107 (a) *Plan.* Use the ideal-gas law to calculate the moles CO_2 that react.

Solve. P(reacted) = P(initial) - P(final), at constant V, T. Since both CaO and BaO react with CO_2 in a 1:1 mole ratio, mol CaO + mol BaO = mol CO_2. Use molar masses to calculate % CaO in sample.

$$P(\text{reacted}) = 730\,\text{torr} - 150\,\text{torr} = 580\,\text{torr};\ 580\,\text{torr} \times \frac{1\,\text{atm}}{760\,\text{torr}} = 0.76316 = 0.763\,\text{atm}$$

$$n = \frac{PV}{RT} = 0.76316\,\text{atm} \times \frac{1.0\,\text{L}}{298\,\text{K}} \times \frac{\text{K}\bullet\text{mol}}{0.08206\,\text{L}\bullet\text{atm}} = 0.03121 = 0.0312\,\text{mol}\,CO_2$$

(b) *Plan.* Use the stoichiometry of the reaction and definition of moles to calculate the mass and Mass % of CaO.

Solve. $CaO(s) + CO_2(s) \rightarrow CaCO_3(s)$. $BaO(s) + CO_2(g) \rightarrow BaCO_3(s)$

mol CO_2 reacted = mol CaO + mol BaO

Let x = g CaO, 4.00 – x = g BaO

$$0.03121 = \frac{x}{56.08} + \frac{4.00 - x}{153.3}$$

$0.03121(56.08)(153.3) = 153.3x + 56.08(4.00 - x)$

$268.3 = (153.3x - 56.08x) + 224.3$

$43.98 = 97.22x$, $x = 0.452 = 0.45$ g CaO

$$\frac{0.452 \text{ g CaO}}{4.00 \text{ g sample}} \times 100 = 11.3 = 11\% \text{ CaO}$$

(By strict sig fig rules, the result has 2 sig figs, because 268 – 224 = 44 has 0 decimal places and 2 sig figs.)

10.109 (a) *Analyze/Plan.* $AgF + S_8 \xrightarrow{\Delta} \text{unknown gas}$

The gas probably contains the elements S and F in an unknown mole ratio. (Most compounds containing Ag are ionic and therefore solids.) Calculate the molar mass of the gas from its density at the given conditions. Determine the relative amount of F from the data on the reaction of the gas with water to produce HF.

Solve.

$$MM = \frac{dRT}{P} = \frac{0.803 \text{ g}}{L} \times \frac{0.08206 \text{ L} \bullet \text{atm}}{\text{K} \bullet \text{mol}} \times \frac{305 \text{K}}{150 \text{mm}} \times \frac{760 \text{mm}}{1 \text{atm}} = 101.83$$

$$= 102 \text{ g/mol}$$

mol F in 480 mL sample: $M \times L = \text{mol}$

$0.081\ M \text{ HF} \times 0.080 \text{ L} = 0.00648 = 0.0065 \text{ mol HF} = 0.0065 \text{ mol F}$ in the sample total mol gas in 480 mL sample: $n = PV/RT$

$$n = 126 \text{ mm} \times \frac{1 \text{ atm}}{760 \text{ mm}} \times \frac{0.480 \text{ L}}{301 \text{ K}} \times \frac{\text{K} \bullet \text{mol}}{0.08206 \text{ L} \bullet \text{atm}} = 3.222 \times 10^{-3}$$

$$= 3.22 \times 10^{-3} \text{ mol}$$

mole ratio of S and F:

total g gas $= 3.222 \times 10^{-3}$ mol gas $\times$ 101.83 g/mol; = 0.32808 = 0.328 g gas

g F $= 6.48 \times 10^{-3}$ mol F $\times$ 18.998 g/mol F = 0.12311 = 0.123 g F

g S = 0.32808 – 0.12311 = 0.20497 = 0.205 g S

mol S = 0.205 g S/32.07 g/mol = 0.0639 mol S

0.00639 mol S/0.00648 mol F = 1:1 mole ratio of S:F

The empirical formula is SF; empirical FW = 32.07 + 19.00 = 51.07;

since MM = 102, the empirical formula is S_2F_2

Check. The empirical and molecular formula weights are in an integer ratio, so the result is reasonable.

(b) 26 valence e^-, 13 e^- pairs

:F̤—S̤—S̤—F̤:   :F̤—S(=S:)—F̤: ⟷ :F̤—S(—S:)—F̤:

(c) According to VSEPR the electron domain geometry about S in each of the molecules will be tetrahedral, with bond angles of 109° or less. The left molecule will have a structure similar to hydrogen peroxide, H_2O_2; the 4 atoms are not necessarily coplanar, and in fact we expect a dihedral angle of approximately 110°. Since the right molecule has a single central atom, we can describe the molecular geometry as trigonal pyramidal.

From the given bond distances, we expect the single-bond covalent radii to be S = 1.02 Å, F = 0.72 Å. The simple conclusion is that the S-S distances will be ~2.04 Å and the S-F distances ~1.74 Å. In fact, the S-S distance in the left compound is 1.89 Å and in the right it is 1.86 Å. Clearly each of these bonds has some double bond character, as indicated by one of the resonance structures for the right molecule. The actual S-F distances are 1.63 Å (left) and 1.60 Å (right), also shorter than the predicted S-F single bond distance. One possible conclusion is that each of the bonds has some double bond character, that some of the "nonbonding" electron density is incorporated into a delocalized π-bonding network in the molecules.

10.110 (a) 19 e^-, 9.5 e^- pairs

:Ö—Ċl—Ö:

Resonance structures can be drawn with the odd electron on O, but electronegativity considerations predict that it will be on Cl for most of the time.

(b) ClO_2 is very reactive because it is an odd-electron molecule. Adding an electron (reduction) both pairs the odd electron and completes the octet of Cl. Thus, ClO_2 has a strong tendency to gain an electron and be reduced.

(c) ClO_2^-, 20 e^-, 10 e^- pairs

[:Ö—C̈l—Ö:]$^-$

(d) 4 e^- domains around Cl, O-Cl-O bond angle ~109°

(e) Calculate mol Cl_2 from ideal-gas equation; determine limiting reactant; mass ClO_2 via mol ratios.

$$\text{mol } Cl_2 = \frac{PV}{RT} = 1.50 \text{ atm} \times \frac{2.00 \text{ L}}{294 \text{ K}} \times \frac{\text{K} \bullet \text{mol}}{0.08206 \text{ L} \bullet \text{atm}} = 0.1243 = 0.124 \text{ mol } Cl_2$$

$$10.0 \text{ g NaClO}_2 \times \frac{1 \text{ mol NaClO}_2}{90.44 \text{ g}} = 0.1106 = 0.111 \text{ mol NaClO}_2$$

2 mol $NaClO_2$ are required for 1 mol Cl_2, so $NaClO_2$ is the limiting reactant. For every 2 mol $NaClO_2$ reacted, 2 mol ClO_2 are produced, so mol ClO_2 = mol $NaClO_2$.

$$0.1106 \text{ mol ClO}_2 \times \frac{67.45 \text{ g ClO}_2}{\text{mol}} = 7.46 \text{ g ClO}_2$$

10.111 (a) ft^3 CH_4 → L CH_4 → mol CH_4 → mol CH_3OH → g CH_3OH → L CH_3OH

$$10.7 \times 10^9 \text{ ft CH}_4 \times \frac{1 \text{ yd}^3}{3^3 \text{ ft}^3} \times \frac{1 \text{m}^3}{(1.0936)^3 \text{ yd}^3} \times \frac{1 \text{ L}}{1 \times 10^{-3} \text{ m}^3} = 3.03001 \times 10^{11}$$

$$= 3.03 \times 10^{11} \text{ L CH}_4$$

$$n = \frac{PV}{RT} = \frac{3.03 \times 10^{11} \text{ L} \times 1.00 \text{atm}}{298 \text{ K}} \times \frac{\text{K} \bullet \text{mol}}{0.08206 \text{ L} \bullet \text{atm}} = 1.2391 \times 10^{10}$$

$$= 1.24 \times 10^{10} \text{ mol CH}_4$$

1 mol CH_4 = 1 mol CH_3OH

$$1.2391 \times 10^{10} \text{ mol } CH_3OH \times \frac{32.04 \text{ g } CH_3OH}{\text{mol } CH_3OH} \times \frac{1 \text{ mL } CH_3OH}{0.791 \text{ g}} \times \frac{1 \text{ L}}{1000 \text{ mL}}$$

$$= 5.0189 \times 10^8 = 5.02 \times 10^8 \text{ L } CH_3OH$$

(b) $CH_4(g) + 2O_2(g) \rightarrow CO_2(g) + 2H_2O(l)$ $\Delta H° = -890.4$ kJ (see Solution 10.108)

$$1.2391 \times 10^{10} \text{ mol } CH_4 \times \frac{-890.4 \text{ kJ}}{1 \text{ mol } CH_4} = -1.10 \times 10^{13} \text{ kJ}$$

$CH_3OH(l) + 3/2\, O_2(g) \rightarrow CO_2(g) + 2H_2O(l)$ $\Delta H° = -726.6$ kJ

$$\Delta H° = \Delta H_f^\circ\, CO_2(g) + 2\Delta H_f^\circ\, H_2O(l) - \Delta H_f^\circ\, CH_3OH(l) - 3/2\, \Delta H_f^\circ\, O_2(g)$$

$$= -393.5 \text{ kJ} + 2(-285.83 \text{ kJ}) - (-238.6 \text{ kJ}) - 0 = -726.6 \text{ kJ}$$

$$1.2391 \times 10^{10} \text{ mol } CH_3OH \times \frac{-726.6 \text{ kJ}}{1 \text{ mol } CH_3OH} = -9.00 \times 10^{12} \text{ kJ}$$

(c) Assume a volume of 1.00 L of each liquid.

$$1.00 \text{ L } CH_4(l) \times \frac{466 \text{ g}}{1 \text{ L}} \times \frac{1 \text{ mol}}{16.04 \text{ g}} \times \frac{-890.4 \text{ kJ}}{\text{mol } CH_4} = -2.59 \times 10^4 \text{ kJ/L } CH_4$$

$$1.00 \text{ L } CH_3OH \times \frac{791 \text{ g}}{1 \text{ L}} \times \frac{1 \text{ mol}}{32.04 \text{ g}} \times \frac{-726.6 \text{ kJ}}{\text{mol } CH_3OH} = -1.79 \times 10^4 \text{ kJ/L } CH_3OH$$

Clearly $CH_4(l)$ has the higher enthalpy of combustion per unit volume.

10.113 (a) $MgCO_3(s) + 2HCl(aq) \rightarrow MgCl_2(aq) + H_2O(l) + CO_2(g)$

$CaCO_3(s) + 2HCl(aq) \rightarrow CaCl_2(aq) + H_2O(l) + CO_2(g)$

(b) $$n = \frac{PV}{RT} = 743 \text{ torr} \times \frac{1 \text{ atm}}{760 \text{ torr}} \times \frac{\text{K} \bullet \text{mol}}{0.08206 \text{ L} \bullet \text{atm}} \times \frac{1.72 \text{ L}}{301 \text{ K}}$$

$$= 0.06808 = 0.0681 \text{ mol } CO_2$$

(c) $x = \text{g } MgCO_3$, $y = \text{g } CaCO_3$, $x + y = 6.53$ g

mol $MgCO_3$ + mol $CaCO_3$ = mol CO_2 total

$$\frac{x}{84.32} + \frac{y}{100.09} = 0.06808;\ y = 6.53 - x$$

$$\frac{x}{84.32} + \frac{6.53 - x}{100.09} = 0.06808$$

$$100.09x - 84.32x + 84.32(6.53) = 0.06808\,(84.32)(100.09)$$

$$15.77x + 550.610 = 574.549;\ x = 1.52 \text{ g } MgCO_3$$

$$\text{mass \% } MgCO_3 = \frac{1.52 \text{ g } MgCO_3}{6.53 \text{ g sample}} \times 100 = 23.3\%$$

[By strict sig fig rules, the answer has 2 sig figs: 15.77x + 551 (3 digits from 6.53) = 575; 575 – 551 = 24 (no decimal places, 2 sig figs) leads to 1.5 g $MgCO_3$.]

11 Intermolecular Forces, Liquids and Solids

Visualizing Concepts

11.2 (a) Hydrogen bonding; H–F interactions qualify for this narrowly defined interaction.

(b) London dispersion forces, the only intermolecular forces between nonpolar F_2 molecules.

(c) Ion-dipole forces between Na^+ cation and the negative end of a polar covalent water molecule.

(d) Dipole-dipole forces between oppositely charged portions of two polar covalent SO_2 molecules.

11.3 The viscosity of glycerol will be greater than that of 1-propanol. Viscosity is the resistance of a substance to flow. The stronger the intermolecular forces in a liquid, the greater its viscosity. Hydrogen bonding is the predominant force for both molecules. Glycerol has three times as many O–H groups and many more H-bonding interactions than 1-propanol, so it experiences stronger intermolecular forces and greater viscosity. (Both molecules have the same carbon-chain length, so dispersion forces are similar.)

11.5 The stronger the intermolecular forces, the greater the average kinetic energy required to escape these forces, and the higher the boiling point. $CH_3CH_2CH_2OH$ has hydrogen bonding, by virtue of its -OH group, so it has the higher boiling point. Dispersion forces are similar because molar masses are the same for both molecules.

11.6 (a) 360 K, the normal boiling point; 265 K, normal freezing point. The left-most line is the freezing/melting curve, the right-most line is the condensation/boiling curve. The normal boiling and freezing points are the temperatures of boiling and freezing at 1 atm pressure.

(b) The material is solid in the white zone, liquid in the blue zone, and gas in the yellow zone. (i) gas (ii) solid (iii) liquid

11.8 (a) Clearly, the structure is close-packed. The question is: cubic or hexagonal? Without looking deeper into the layers of oranges, one cannot distinguish whether the layer structure is cubic (ABCABC) or hexagonal (ABABAB) close-packed.

(b) CN = 12, regardless of whether the structure is hexagonal or cubic close packed.

(c) Molecular. There are no strong bonds between particles.

Kinetic-Molecular Theory

11.10 (a) In solids, particles are in essentially fixed positions relative to each other, so the average energy of attraction is stronger than average kinetic energy. In liquids, particles are close together but moving relative to each other. The average attractive energy and average kinetic energy are approximately balanced. In gases, particles are far apart and in constant, random motion. Average kinetic energy is much greater than average energy of attraction.

(b) As the temperature of a substance is increased, the average kinetic energy of the particles increases. In a collection of particles (molecules), the state is determined by the strength of interparticle forces relative to the average kinetic energy of the particles. As the average kinetic energy increases, more particles are able to overcome intermolecular attractive forces and move to a less ordered state, from solid to liquid to gas.

(c) At constant temperature, the average kinetic energy of a collection of particles is constant. Compression brings particles closer together and increases the number of particle-particle collisions. With more collisions, the likelihood of intermolecular attractions causing the particles to coalesce (liquefy) is greater.

11.12 (a) The average distance between molecules is greater in the liquid state. Density is the ratio of the mass of a substance to the volume it occupies. For the same substance in different states, mass will be the same. The smaller the density, the greater the volume occupied, and the greater the distance between molecules. The liquid at 130° has the lower density (1.08 g/cm^3), so the average distance between molecules is greater.

(b) As the temperature of a substance increases, the average kinetic energy and speed of the molecules increases. At the melting point the molecules, on average, have enough kinetic energy to break away from the very orderly array that was present in the solid. As the translational motion of the molecules increases, the occupied volume increases and the density decreases. Thus, the solid density, 1.266 g/cm^3 at 15°C, is greater than the liquid density, 1.08 g/cm^3 at 130°C.

Intermolecular Forces

11.14 (a) London-dispersion forces

(b) dipole-dipole and London-dispersion forces

(c) hydrogen bonding (dominates properties of these small molecules) and London-dispersion forces (less significant)

11.16 (a) CH_3OH experiences hydrogen bonding, but CH_3SH does not.

(b) Both gases are influenced by London-dispersion forces. The heavier the gas particles, the stronger the London-dispersion forces. The heavier Xe is a liquid at the specified conditions, while the lighter Ar is a gas.

(c) Both gases are influenced by London-dispersion forces. The larger, diatomic Cl_2 molecules are more polarizable, experience stronger dispersion forces, and have the higher boiling point.

(d) Acetone and 2-methylpropane are molecules with similar molar masses and London-dispersion forces. Acetone also experiences dipole-dipole forces and has the higher boiling point.

11.18 (a) A more polarizable molecule can develop a larger transient dipole, increasing the strength of electrostatic attractions among polarized molecules.

(b) The noble gases are all monoatomic. Going down the column, the atomic radius and the size of the electron cloud increase. The larger the electron cloud, the more polarizable the atom, the stronger the London-dispersion forces and the higher the boiling point.

(c) It is generally true that the greater the molecular weight, the stronger the dispersion forces experienced by a molecule. This is true because trends in molecular size and molecular weight are usually parallel.

(d) It is usually true that as the number of electrons in a molecule increases, the size of the molecule increases. Larger molecules tend to have diffuse electron clouds, which lead to greater polarizability. Thus, the statement that more electrons lead to increased dispersion forces (and greater polarizability) is correct.

11.20 For molecules with similar structures, the strength of dispersion forces increases with molecular size (molecular weight and number of electrons in the molecule).

(a) Br_2

(b) $CH_3CH_2CH_2SH$

(c) $CH_3CH_2CH_2Cl$. These two molecules have the same molecular formula and molecular weight (C_3H_7Cl, molecular weight = 78.5 amu), so the shapes of the molecules determine which has the stronger dispersion forces. According to Figure 11.6, the cylindrical (not branched) molecule will have stronger dispersion forces.

11.22 Both molecules experience hydrogen bonding through their –OH groups and dispersion forces between their hydrocarbon portions. The position of the –OH group in isopropyl alcohol shields it somewhat from approach by other molecules and slightly decreases the extent of hydrogen bonding. Also, isopropyl alcohol is less rod-like (it has a shorter chain) than propyl alcohol, so dispersion forces are weaker. Since hydrogen bonding and dispersion forces are weaker in isopropyl alcohol, it has the lower boiling point.

11.24 (a) Replacing a hydroxyl hydrogen with a CH_3 group eliminates hydrogen bonding in that part of the molecule. This reduces the strength of intermolecular forces and leads to a (much) lower boiling point.

(b) $CH_3OCH_2CH_2OCH_3$ is a larger, more polarizable molecule with stronger London-dispersion forces and thus a higher boiling point.

11.26 (a) C_6H_{14}, dispersion; C_8H_{18}, dispersion. C_8H_{18} has the higher boiling point due to greater molar mass and similar strength of forces.

(b) C_3H_8, dispersion; CH_3OCH_3, dipole-dipole, and dispersion. CH_3OCH_3 has the higher boiling point due to stronger intermolecular forces and similar molar mass.

(c) HOOH, hydrogen bonding, dipole-dipole, and dispersion; HSSH, dipole-dipole, and dispersion. HOOH has the higher boiling point due to the influence of hydrogen bonding (Figure 11.7).

(d) NH_2NH_2, hydrogen bonding, dipole-dipole, and dispersion; CH_3CH_3, dispersion. NH_2NH_2 has the higher boiling point due to much stronger intermolecular forces.

11.28 (a) In the solid state, NH_3 molecules are arranged so as to form the maximum number of hydrogen bonds. At the melting point, the average kinetic energy of the molecules is large enough so that they are free to move relative to each other. As they move, old hydrogen bonds break and new ones form, but the strict relative order required for maximum hydrogen bonding is no longer present.

(b) In the liquid state, molecules are moving relative to one another while touching, which makes some hydrogen bonding possible. When molecules achieve enough kinetic energy to vaporize, the distance between them increases beyond the point where hydrogen bonds can form.

Viscosity and Surface Tension

11.30 (a) *Cohesive* forces bind molecules to each other.
Adhesive forces bind molecules to surfaces.

(b) The cohesive forces are hydrogen bonding among water molecules and also hydrogen bonding among cellulose molecules in the paper towel. Adhesive forces are any attractive forces between water and cellulose (the paper towel), likely also hydrogen bonding. If adhesive forces between cellulose and water weren't significant, paper towels wouldn't absorb water.

(c) The shape of a meniscus depends on the strength of the cohesive forces within a liquid relative to the adhesive forces between the walls of the tube and the liquid. Adhesive forces between polar water molecules and silicates in glass (Figure 11.16) are even stronger than cohesive hydrogen-bonding forces among water molecules, so the meniscus is U-shaped (concave-upward).

11.32 (a)

```
H—N̈—N̈—H      H—Ö—Ö—H      H—Ö—H
  |  |           ¨  ¨           ¨
  H  H
```

(b) All have bonds (N–H or O–H, respectively) capable of forming hydrogen bonds. Hydrogen bonding is the strongest intermolecular interaction between neutral molecules and leads to very strong cohesive forces in liquids. The stronger the cohesive forces in a liquid, the greater the surface tension.

Changes of State

11.34 (a) condensation, exothermic

(b) sublimation, endothermic

(c) vaporization (evaporation), endothermic

(d) freezing, exothermic

11.36 (a) Liquid ethyl chloride at room temperature is far above its boiling point. When the liquid contacts the metal surface, heat sufficient to vaporize the liquid is transferred from the metal to the ethyl chloride, and the heat content of the molecules increases. At constant atmospheric pressure, $\Delta H = q$, so the heat content and the enthalpy content of $C_2H_5Cl(g)$ is higher than that of $C_2H_5Cl(l)$.

(b) Attractive intermolecular forces hold the C_2H_5Cl molecules in close contact in the liquid phase. In order to overcome these attractive forces and maintain separation in the gas phase, the enthalpy content of the C_2H_5Cl molecules must increase when they change from the liquid to the gaseous state.

11.38 Energy released when 100 g of H_2O is cooled from 18°C to 0°C:

$$\frac{4.184 \text{ J}}{\text{g} \bullet \text{K}} \times 100 \text{ g } H_2O \times 18°\text{C} = 7.53 \times 10^3 \text{ J} = 7.5 \text{ kJ}$$

Energy released when 100 g of H_2O is frozen (there is no change in temperature during a change of state):

$$\frac{334 \text{ J}}{\text{g}} \times 100 \text{ g } H_2O = 3.34 \times 10^4 \text{ J} = 33.4 \text{ kJ}$$

Total energy released = 7.53 kJ + 33.4 kJ = 40.93 = 40.9 kJ

Mass of freon that will absorb 40.9 kJ when vaporized:

$$40.93 \text{ kJ} \times \frac{1 \times 10^3 \text{ J}}{1 \text{ kJ}} \times \frac{1 \text{ g } CCl_2F_2}{289 \text{ J}} = 142 \text{ g } CCl_2F_2$$

11.40 Consider the process in steps, using the appropriate thermochemical constant.

Heat the liquid from 5.00°C to 47.6°C (278.00 K to 320.6 K), using the specific heat of the liquid.

$$25.0 \text{ g } C_2Cl_3F_3 \times \frac{0.91 \text{ J}}{\text{g} \bullet \text{K}} \times 42.6 \text{ K} \times \frac{1 \text{ kJ}}{1000 \text{ J}} = 0.969 = 0.97 \text{ kJ}$$

Boil the liquid at 47.6°C (320.6 K), using the enthalpy of vaporization.

$$25.0 \text{ g } C_2Cl_3F_3 \times \frac{1 \text{ mol } C_2Cl_3F_3}{187.4 \text{ g } C_2Cl_3F_3} \times \frac{27.49 \text{ kJ}}{\text{mol}} = 3.667 = 3.67 \text{ kJ}$$

Heat the gas from 47.6°C to 82.00°C (320.6 K to 355.00 K), using the specific heat of the gas.

$$25.0 \text{ g } C_2Cl_3F_3 \times \frac{0.67 \text{ J}}{\text{g} \bullet \text{K}} \times 34.4 \text{ K} \times \frac{1 \text{ kJ}}{1000 \text{ J}} = 0.576 = 0.58 \text{ kJ}$$

The total energy required is 0.969 kJ + 3.667 kJ + 0.576 kJ = 5.21 kJ.

11.42 (a) According to Solution 11.41(b), the higher the critical temperature, the stronger the intermolecular forces of a substance. Therefore, the strength of intermolecular forces decreases moving from left to right across the series and as molecular weight decreases.

(b) The molecules in this series experience London-dispersion forces and, except for CF_4, dipole-dipole forces. We expect the strength of dispersion forces to increase with increasing molecular weight, which agrees with the trends in critical temperature and pressure.

Vapor Pressure and Boiling Point

11.44 (a) The pressure difference on the manometer is 130 mm Hg and the gas in the vessel is essentially 100% molecules of the substance in the vapor phase. When the vessel is evacuated, virtually all air is removed. As the frozen liquid warms, it establishes a vapor pressure of 130 mm Hg. This is the pressure difference on the manometer.

(b) The pressure difference is 1 atm. The gas is 130 mm Hg of the molecular vapor and the rest is air. The liquid vaporizes in contact with the atmosphere, so atmospheric pressure is maintained above the liquid, but the equilibrium gas composition reflects the amount of vapor necessary to maintain 130 mm pressure, plus enough air to maintain a total pressure of 1 atm.

(c) The pressure difference is 890 mm Hg (1 atm + 130 mm Hg) and the gas is a mixture of 130 mm vapor and 1 atm air. The initial air pressure in the flask is 1 atm and no air is allowed to escape. The gas in the flask is not in equilibrium with the atmosphere and the final pressure in the flask does not equal atmospheric pressure. After a most of the liquid vaporizes, the total gas pressure is the result of 130 mm vapor and 1 atm air.

11.46 Both molecules are pyramidal, with a nonbonding electron pair on the central atom. Even though the molecules are polar covalent, differences in their intermolecular forces and physical properties are likely to be dominated by differences in dispersion forces.

(a) PCl_3

(b) $AsCl_3$

(c) At the same temperature, the average kinetic energy of molecules of the two substances is equal.

(d) The strength of dispersion forces increases with increasing molecular weight, so $AsCl_3$ will experience stronger intermolecular forces. This is the basis of predictions in parts (a) and (b) above.

11.48 (a) On a humid day, there are more gaseous water molecules in the air and more are recaptured by the surface of the liquid, making evaporation slower.

(b) At high altitude, atmospheric pressure is lower and water boils at a lower temperature. The eggs must be cooked longer at the lower temperature.

11.50 (a) The boiling point of a liquid is the temperature at which its vapor pressure equals atmospheric pressure. According to Appendix B, the vapor pressure of water is 680 torr at approximately 97°C.

(b) The temperature at which the vapor pressure of water is 752 mm Hg is almost 100°C, greater than the boiling temperature at 680 mm Hg. The average kinetic energy of the H_2O molecules at the boiling temperature in Chicago is greater than the average kinetic energy of the molecules at the boiling temperature in Reno. A liquid boils when its vapor pressure equals the external pressure acting on the liquid. If the external pressure acting on the liquid molecules is smaller (as is the atmospheric pressure in Reno), a smaller average kinetic energy is required for boiling (bubble formation within the liquid).

Phase Diagrams

11.52 (a) The *triple point* on a phase diagram represents the temperature and pressure at which the gas, liquid, and solid phases are in equilibrium.

(b) No. A phase diagram represents a closed system, one where no matter can escape and no substance other than the one under consideration is present; air cannot be present in the system. Even if air is excluded, at 1 atm of external pressure, the triple point of water is inaccessible, regardless of temperature [see Sample Exercise 11.6(b)].

11.54 (a) Solid CO_2 sublimes to form $CO_2(g)$ at a temperature of about -60°C.

(b) Solid CO_2 melts to form $CO_2(l)$ at a temperature of about -50°C. The $CO_2(l)$ boils when the temperature reaches approximately -40°C.

11.56 (a)

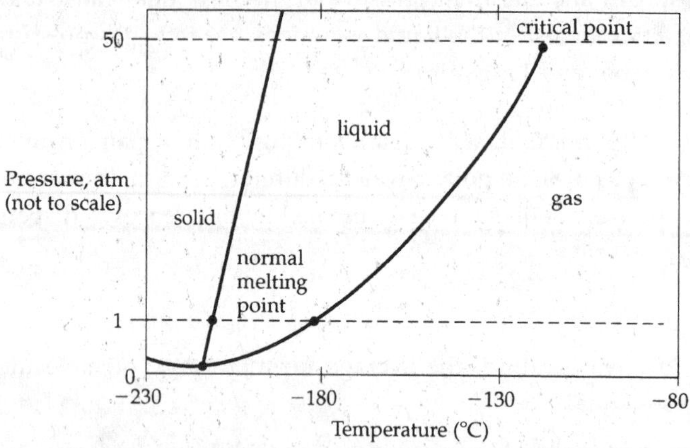

On the diagram above, the triple point is at the dot closest to the x-axis. The normal melting point is at the intersection of the solid-liquid line and the 1-atm dashed line. The normal boiling point is at the intersection of the liquid-gas line and the 1-atm line. The critical point is clearly marked.

(b) $O_2(s)$ will not float on $O_2(l)$. $O_2(s)$ is denser than $O_2(l)$ because the solid-liquid line on the phase diagram is normal. That is, as pressure increases, the melting temperature increases. [Note that the solid-liquid line for O_2 is nearly vertical, indicating a small difference in the densities of $O_2(s)$ and $O_2(l)$].

(c) $O_2(s)$ will melt when heated at a pressure of 1 atm, since this is a much greater pressure than the pressure at the triple point.

Structures of Solids

11.58 In amorphous silica (SiO_2) the regular structure of quartz is disrupted; the loose, disordered structure, Figure 11.30(b), has many vacant "pockets" throughout. There are fewer SiO_2 groups per volume in the amorphous solid; the packing is less efficient and less dense.

11.60 (a) Ti: 8 corners × 1/8 sphere/corner + [1 center × 1 sphere/center] = 2 Ti atoms

O: 4 faces × 1/2 sphere/face + [2 interior × 1 sphere/interior] = 4 O atoms

Formula: TiO_2

(b) Rutile is an ionic solid; ion-ion forces among Ti^{4+} cations and O^{2-} anions are quite strong, owing to the magnitudes of the charges, and lead to the ordered structure.

11.62 (a) 8 corners × 1/8 atom/corner + 6 faces × ½ atom/face = 4 atoms

(b) Each aluminum atom is in contact with 12 nearest neighbors, 6 in one plane, 3 above that plane, and 3 below. Its coordination number is thus 12.

(c) The length of the face diagonal of a face-centered cubic unit cell is four times the radius of the metal and $\sqrt{2}$ times the unit cell dimension (usually designated *a* for cubic cells).

$$4 \times 1.43\ \text{Å} = \sqrt{2} \times a;\ a = \frac{4 \times 1.43\ \text{Å}}{\sqrt{2}} = 4.0447 = 4.04\ \text{Å} = 4.04 \times 10^{-8}\ \text{cm}$$

(d) The density of the metal is the mass of the unit cell contents divided by the volume of the unit cell.

$$\text{density} = \frac{4\ \text{Al atoms}}{(4.0447 \times 10^{-8}\ \text{cm})^3} \times \frac{26.98\ \text{g Al}}{6.022 \times 10^{23}\ \text{Al atoms}} = 2.71\ \text{g/cm}^3$$

11.64 Avogadro's number is the number of KCl formula units in 74.55 g of KCl.

$$74.55\ \text{g KCl} \times \frac{1\ \text{cm}^3}{1.984\ \text{g}} \times \frac{(1 \times 10^{10}\ \text{pm})^3}{1\ \text{cm}^3} \times \frac{4\ \text{KCl units}}{628^3\ \text{pm}^3} = 6.07 \times 10^{23}\ \text{KCl formula units}$$

11.66 (a) Na^+, 6 (b) Zn^{2+}, 4 (c) Ca^{2+}, 8

11.68 In the face-centered cubic structure, there are four NiO units in the unit cell. Density is the mass of the unit cell contents divided by the unit cell volume (a^3).

$$\text{density} = \frac{4\ \text{NiO units}}{(4.18\ \text{Å})^3} \times \frac{74.69\ \text{g NiO}}{6.022 \times 10^{23}\ \text{NiO units}} \times \left(\frac{1\ \text{Å}}{1 \times 10^{-8}\ \text{cm}}\right)^3 = 6.79\ \text{g/cm}^3$$

11.70 (a) According to Figure 11.42(b), there are 4 HgS units in a unit cell with the zinc blende structure. [4 complete Hg^{2+} ions, 6(1/2) + 8(1/8) S^{2-} ions]

$$\text{density} = \frac{4\ \text{HgS units}}{(5.852\ \text{Å})^3} \times \frac{232.655\ \text{g}}{6.022 \times 10^{23}\ \text{HgS units}} \times \left(\frac{1\ \text{Å}}{1 \times 10^{-8}\ \text{cm}}\right)^3 = 7.711\ \text{g/cm}^3$$

(b) We expect Se^{2-} to have a larger ionic radius than S^{2-}, since Se is below S in the chalcogen family and both ions have the same charge. Thus, HgSe will occupy a larger volume and the unit cell edge will be longer.

(c) For HgSe:

$$\text{density} = \frac{4 \text{ HgSe units}}{(6.085 \text{ Å})^3} \times \frac{279.55 \text{ g HgSe}}{6.022 \times 10^{23} \text{ HgSe units}} \times \left(\frac{1 \text{ Å}}{1 \times 10^{-8} \text{ cm}}\right)^3 = 8.241 \text{ g/cm}^3$$

Even though HgSe has a larger unit cell volume than HgS, it also has a larger molar mass. The mass of Se is more than twice that of S, while the radius of Se^{2-} is only slightly larger than that of S (Figure 7.7). The greater mass of Se accounts for the greater density of HgSe.

Bonding in Solids

11.72 (a) ionic (b) metallic

(c) ionic (somewhat borderline, could be modeled as ionic with some covalent character to the bonds, in keeping with the high oxidation state of Zr, or as a network solid with ionic character to the bonding, in keeping with the electronegativity difference between Zr and O.)

(d) molecular (e) molecular (f) molecular

11.74 (a) metallic

(b) molecular or metallic (physical properties of metals vary widely)

(c) covalent-network or ionic (d) covalent-network (e) ionic

11.76 According to Table 11.7, the solid could be either ionic with low water solubility or network covalent. Due to the extremely high sublimation temperature, it is probably covalent-network.

11.78 (a) HF – hydrogen bonding versus dipole-dipole for HCl

(b) C(graphite) – covalent-network bonding versus London dispersion forces for CH_4

(c) KBr – ionic versus dispersion forces for nonpolar Br_2

(d) MgF_2 – higher charge on Mg^{2+} than Li^+

Additional Exercises

11.80 (a) Correct.

(b) The lower boiling liquid must experience less total intermolecular forces.

(c) If both liquids are structurally similar nonpolar molecules, the lower boiling liquid has a lower molecular weight than the higher boiling liquid.

(d) Correct.

(e) At their boiling points, both liquids have vapor pressures of 760 mm Hg.

11.81 (a) The *cis* isomer has stronger dipole-dipole forces; the *trans* isomer is nonpolar. The higher boiling point of the *cis* isomer supports this conclusion.

(b) While boiling points are primarily a measure of strength of intermolecular forces, melting points are influenced by crystal packing efficiency as well as intermolecular forces. Since the nonpolar *trans* isomer with weaker intermolecular forces has the higher melting point, it must pack more efficiently.

11.83 When a halogen atom (Cl or Br) is substituted for H in benzene, the molecule becomes polar. These molecules experience dispersion forces similar to those in benzene plus dipole-dipole forces, so they have higher boiling points than benzene. C_6H_5Br has a higher molar mass and is more polarizable than C_6H_5Cl, so it has the higher boiling point. C_6H_5OH experiences hydrogen bonding, the strongest force between neutral molecules, so it has the highest boiling point.

11.84 (a) Propylamine experiences hydrogen bonding interactions while trimethylamine, with no N-H bonds, does not. Also, the rod-like shape of propylamine (see Solution 11.21) leads to stronger dispersion forces than in pyramidal trimethylamine. The stronger intermolecular forces in propylamine lead to the higher boiling point.

(b) Propylamine is most soluble in water by virtue of its $-NH_2$ group, which can act as both a donor and acceptor in hydrogen bonding. Trimethylamine is much more soluble than the structurally similar isobutane, because the pyramidal $\ddot{N}$ atom can act as a hydrogen bond acceptor. Trimethylamine is less soluble than propylamine because its participation in hydrogen bonding is less extensive.

11.86 The more carbon atoms in the hydrocarbon, the longer the chain, the more polarizable the electron cloud, the higher the boiling point. A plot of the number of carbon atoms versus boiling point is shown below. For 8 C atoms, C_8H_{18}, the boiling point is approximately 130°C.

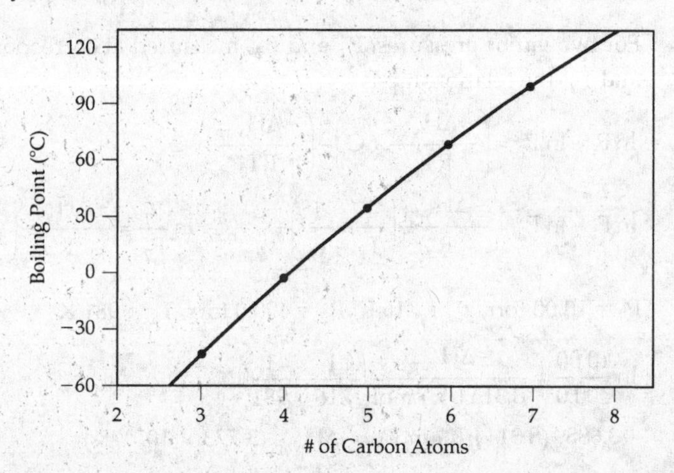

11.87 According to Figure 11.20, as the pressure of a gas above its critical temperature increases beyond critical pressure, the solubility of a solute increases. The solubility of the solute is essentially zero below critical pressure. Supercritical CO_2 at very high pressure dissolves caffeine, and the solution leaves the extractor. The pressure reduction value reduces the pressure of CO_2 enough so that the caffeine becomes insoluble. The solid caffeine is deposited in the separator, and the low pressure CO_2 gas is recycled.

11.89 (a) If the Clausius-Clapeyron equation is obeyed, a graph of ln P vs 1/T(K) should be linear. Here are the data in a form form for graphing.

T(K)	1/T	P(torr)	ln P
280.0	3.571×10^{-3}	32.42	3.479
300.0	3.333×10^{-3}	92.47	4.527
320.0	3.125×10^{-3}	225.1	5.417
330.0	3.030×10^{-3}	334.4	5.812
340.0	2.941×10^{-3}	482.9	6.180

According to the graph, the Clausius-Clapeyron equation is obeyed, to a first approximation.

$$\Delta H_{vap} = -\text{slope} \times R;\ \text{slope} = \frac{3.479 - 6.180}{(3.571 - 2.941) \times 10^{-3}} = -\frac{2.701}{0.630 \times 10^{-3}} = -4.29 \times 10^{3}$$

$$\Delta H_{vap} = -(-4.29 \times 10^{3}) \times 8.314\ \text{J/K} \bullet \text{mol} = 35.7\ \text{kJ/mol}$$

(b) The normal boiling point is the temperature at which the vapor pressure of the liquid equals atmospheric pressure, 760 torr. From the graph,

ln 760 = 6.63, 1/T for this vapor pressure = 2.828×10^{-3}; T = 353.6 K

11.90 (a) The Clausius-Clapeyron equation is $\ln P = \frac{-\Delta H_{vap}}{RT} + C$.

For two vapor pressures, P_1 and P_2, measured at corresponding temperatures T_1 and T_2, the relationship is

$$\ln P_1 - \ln P_2 = \left(\frac{-\Delta H_{vap}}{RT_1} + C\right) - \left(\frac{-\Delta H_{vap}}{RT_2} + C\right)$$

$$\ln P_1 - \ln P_2 = \frac{-\Delta H_{vap}}{R}\left(\frac{1}{T_1} - \frac{1}{T_2}\right) + C - C;\ \ln\frac{P_1}{P_2} = \frac{-\Delta H_{vap}}{R}\left(\frac{1}{T_1} - \frac{1}{T_2}\right)$$

(b) $P_1 = 10.00$ torr, $T_1 = 716$ K; $P_2 = 400.0$ torr, $T_2 = 981$ K

$$\ln\frac{10.00}{400.0} = \frac{-\Delta H_{vap}}{8.314\ \text{J/K} \bullet \text{mol}}\left(\frac{1}{716} - \frac{1}{981}\right)$$

$$-3.6889\,(8.314\ \text{J/K} \bullet \text{mol}) = -\Delta H_{vap}\,(3.773 \times 10^{-4}\ /\text{K})$$

$$\Delta H_{vap} = 8.129 \times 10^{4} = 8.1 \times 10^{4}\ \text{J/mol} = 81\ \text{kJ/mol}$$

$[(1/T_1) - (1/T_2)]$ has 2 sig figs and so does the result.

(c) The normal boiling point of a liquid is the temperature at which the vapor pressure of the liquid is 760 torr.

$P_1 = 400.0$ torr, $T_1 = 981$ K; $P_2 = 760$ torr, T_2 = b.p. of potassium

$$\ln\left(\frac{400.0}{760.0}\right) = \frac{-8.129 \times 10^4 \text{ J/mol}}{8.314 \text{ J/K} \bullet \text{mol}}\left(\frac{1}{981 \text{ K}} - \frac{1}{T_2}\right)$$

$$\frac{-0.64185}{-9.7775 \times 10^3} = 1.0194 \times 10^{-3} - \frac{1}{T_2}; \ \frac{1}{T_2} = 1.0194 \times 10^{-3} - 6.565 \times 10^{-5}$$

$$\frac{1}{T_2} = 9.5375 \times 10^{-4} = 9.5 \times 10^{-4}; \ T_2 = 1048 = 1.0 \times 10^3 \text{ K } (8 \times 10^2 \text{ °C})$$

(d) P_1 = VP of K(l) at 100°C, $T_1 = 373$ K; $P_2 = 10.00$ torr, $T_2 = 716$ K

$$\ln\frac{P_1}{10.00 \text{ torr}} = \frac{-8.129 \times 10^4 \text{ J/mol}}{8.314 \text{ J/K} \bullet \text{mol}}\left(\frac{1}{373} - \frac{1}{716}\right)$$

$$\ln\frac{P_1}{10.00 \text{ torr}} = \frac{-8.129 \times 10^4 \text{ J/mol}}{8.314 \text{ J/K} \bullet \text{mol}} \times 1.284 \times 10^{-3} = -12.5543$$

$$\frac{P_1}{10.00 \text{ torr}} = e^{-12.5543} = 3.530 \times 10^{-6}; \ P_1 = 3.5 \times 10^{-5} \text{ torr}$$

11.91 Physical data for the two compounds from the *Handbook of Chemistry and Physics*:

	MM	dipole moment	boiling point
CH_2Cl_2	85 g/mol	1.60 D	40.0°C
CH_3I	142 g/mol	1.62 D	42.4°C

(a) The two substances have very similar molecular structures; each is an unsymmetrical tetrahedron with a single central carbon atom and no hydrogen bonding. Since the structures are very similar, the magnitudes of the dipole-dipole forces should be similar. This is verified by their very similar dipole moments. The heavier compound, CH_3I, will have slightly stronger London dispersion forces. Since the nature and magnitude of the intermolecular forces in the two compounds are nearly the same, it is very difficult to predict which will be more volatile [or which will have the higher boiling point as in part (b)].

(b) Given the structural similarities discussed in part (a), one would expect the boiling points to be very similar, and they are. Based on its larger molar mass (and dipole-dipole forces being essentially equal) one might predict that CH_3I would have a slightly higher boiling point; this is verified by the known boiling points.

(c) According to Equation 11.1, $\ln P = \frac{-\Delta H_{vap}}{RT} + C$

A plot of ln P vs. 1/T for each compound is linear. Since the order of volatility changes with temperature for the two compounds, the two lines must cross at some temperature; the slopes of the two lines, ΔH_{vap} for the two compounds, and the y-intercepts, C, must be different.

(d) CH_2Cl_2

ln P	T(K)	1/T
2.303	229.9	4.351×10^{-3}
3.689	250.9	3.986×10^{-3}
4.605	266.9	3.747×10^{-3}
5.991	297.3	3.364×10^{-3}

CH_3I

ln P	T(K)	1/T
2.303	227.4	4.398×10^{-3}
3.689	249.0	4.016×10^{-3}
4.605	266.2	3.757×10^{-3}
5.991	298.5	3.350×10^{-3}

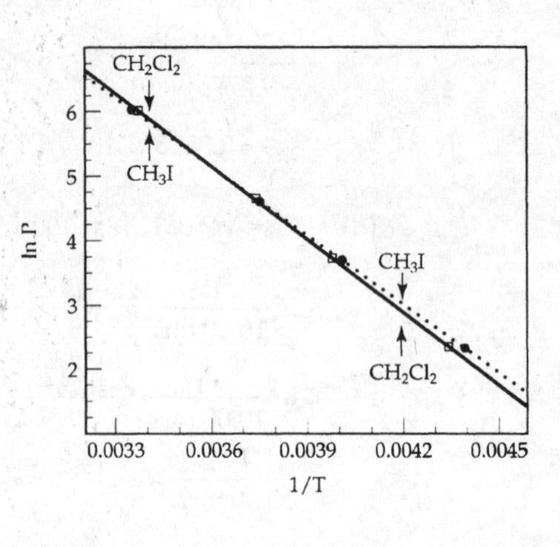

$$\text{For } CH_2Cl_2, -\Delta H_{vap}/R = \text{slope} = \frac{(5.991 - 2.303)}{(3.364 \times 10^{-3} - 4.350 \times 10^{-3})} = \frac{-3.688}{0.987 \times 10^{-3}}$$

$$= -3.74 \times 10^3 = -\Delta H_{vap}/R$$

$$\Delta H_{vap} = 8.314\,(3.74 \times 10^3) = 3.107 \times 10^4 \text{ J/mol} = 31.1 \text{ kJ/mol}$$

$$\text{For } CH_3I, -\Delta H_{vap}/R = \text{slope} = \frac{(5.991 - 2.303)}{(3.350 \times 10^{-3} - 4.398 \times 10^{-3})} = \frac{-3.688}{1.048 \times 10^{-3}} = -3.519 \times 10^3$$

$$= -\Delta H_{vap}/R$$

$$\Delta H_{vap} = 8.314\,(3.519 \times 10^3) = 2.926 \times 10^4 \text{ J/mol} = 29.3 \text{ kJ/mol}$$

11.93 (a) The face diagonal of a face-centered cubic unit cell has length $a\sqrt{2}$ and also 4 r, where a is the cubic cell dimension and r is the atomic radius.

$$4\,r = a\sqrt{2};\ r = (a\sqrt{2})/4 = (4.078 \text{ Å})(\sqrt{2})/4 = 1.44179 = 1.442 \text{ Å}$$

(b) Density is the mass of the unit cell contents divided by the unit cell volume (a^3). In a face-centered cubic unit cell, there are 4 Au atoms.

$$\text{density} = \frac{4 \text{ Au atoms}}{(4.078 \text{ Å})^3} \times \frac{196.97 \text{ g}}{6.0221 \times 10^{23} \text{ Au atoms}} \times \left(\frac{1 \text{ Å}}{1 \times 10^{-8} \text{ cm}}\right)^3 = 19.29 \text{ g/cm}^3$$

11.94 (a) (i) 8 corners × 1/8 atom/corner = 1 atom

(ii) 8 corners × 1/8 atom/corner + 1 center × 1 atom/center = 2 atoms

(iii) 8 corners × 1/8 atom/corner + 6 faces × 1/2 atom/face = 4 atoms

(b) Fundamentally, $CaCl_2$ must have a different crystal structure than NaCl because the two formulas are different. Na^+ has an ionic radius of 0.97 Å and Ca^{2+}, 0.99 Å; these values are very similar. So, ion size does not prohibit Ca^{2+} from replacing Na^+ in an ionic lattice, but electrostatic effects do. In $CaCl_2$, there are two Cl^- anions for every Ca^{2+} cation.

Use Figure 11.35(b) structure to visualize the contents of one face-centered cubic unit cell of the NaCl structure.

Na^+ (purple): (8 corners × 1/8 Na^+/corner) + (6 faces × 1/2 Na^+/face) = 4 Na^+

Cl^- (green): (12 edges × 1/4 Cl^-/edge) + (1 center × 1 Cl^-/center) = 4 Cl^-

In the NaCl structure, there is one anion site for every cation. $CaCl_2$ requires two anion sites for each cation, so it cannot have the same crystal structure as NaCl. (In fact, the unit cell of $CaCl_2$ has three unequal edge lengths and is not cubic; it is much different than the NaCl unit cell.)

11.96 $n\lambda = 2d \sin\theta$; n = 1, λ = 1.54 Å, θ = 14.22 °; calculate d.

$$d = \frac{n\lambda}{2 \sin \theta} = \frac{1 \times 1.54 \text{ Å}}{2 \sin(14.22)} = 3.1346 = 3.13 \text{ Å}$$

11.97 (a) Both diamond (d = 3.5 g/cm^3) and graphite (d = 2.3 g/cm^3) are covalent-network solids with efficient packing arrangements in the solid state; there is relatively little empty space in their respective crystal lattices. Diamond, with bonded C–C distances of 1.54 Å in all directions, is more dense than graphite, with shorter C–C distances within carbon sheets but longer 3.41 Å separations between sheets (Figure 11.41). Buckminsterfullerene has much more empty space, both inside each C_{60} "ball" and between balls, than either diamond or graphite, so its density will be considerably less than 2.3 g/cm^3.

(b) In a face-centered cubic unit cell, there are 4 complete C_{60} units.

$$\frac{4 \text{ C}_{60} \text{ units}}{(14.2 \text{ Å})^3} \times \frac{720.66 \text{ g}}{6.022 \times 10^{23} \text{ C}_{60} \text{ units}} \times \left(\frac{1 \text{ Å}}{1 \times 10^{-8} \text{ cm}}\right)^3 = 1.67 \text{ g/cm}^3$$

(1.67 g/cm^3 is the smallest density of the three allotropes, diamond, graphite, and buckminsterfullerene.)

Integrative Exercises

11.99 (a) In Table 11.4, viscosity increases as the length of the carbon chain increases. Longer molecular chains become increasingly entangled, increasing resistance to flow.

(b) Whereas viscosity depends on molecular chain length in a critical way, surface tension depends on the strengths of intermolecular interactions between molecules. These dispersion forces do not increase as rapidly with increasing chain length and molecular weight as viscosity does.

(c) The –OH group in n-octyl alcohol gives rise to hydrogen bonding among molecules, which increases molecular entanglement and leads to greater viscosity and higher boiling point.

11.100 (a) 24 valence e^-, 12 e^- pairs

```
   H  :O:  H
   |   ||  |
H—C——C——C—H
   |       |
   H       H
```

The geometry around the central C atom is trigonal planar, and around the two terminal C atoms, tetrahedral.

(b) Polar. The C=O bond is quite polar and the dipoles in the trigonal plane around the central C atom do not cancel.

(c) Dipole-dipole and London-dispersion forces

(d) Since the molecular weights of acetone and 1-propanol are similar, the strength of the London-dispersion forces in the two compounds is also similar. The big difference is that 1-propanol has hydrogen bonding, while acetone does not. These relatively strong attractive forces lead to the higher boiling point for 1-propanol.

11.102 (a) In order for butane to be stored as a liquid at temperatures above its boiling point (-5°C), the pressure in the tank must be greater than atmospheric pressure. In terms of the phase diagram of butane, the pressure must be high enough so that, at tank conditions, the butane is "above" the gas-liquid line and in the liquid region of the diagram.

The pressure of a gas is described by the ideal gas law as P = nRT/V; pressure is directly proportional to moles of gas. The more moles of gas present in the tank the greater the pressure, until sufficient pressure is achieved for the gas to liquify. At the point where liquid and gas are in equilibrium and temperature is constant, liquid will vaporize or condense to maintain the equilibrium vapor pressure. That is, as long as some liquid is present, the gas pressure in the tank will be constant.

(b) If butane gas escapes the tank, butane liquid will vaporize (evaporate) to maintain the equilibrium vapor pressure. Vaporization is an endothermic process, so the butane will absorb heat from the surroundings. The temperature of the tank and the liquid butane will decrease.

(c) $$155\text{ g } C_4H_{10} \times \frac{1\text{ mol } C_4H_{10}}{58.12\text{ g } C_4H_{10}} \times \frac{21.3\text{ kJ}}{\text{mol}} = 56.8\text{ kJ}$$

$$V = \frac{nRT}{P} = 155\text{ g} \times \frac{1\text{ mol}}{58.12\text{ g}} \times \frac{0.08206\text{ L} \bullet \text{atm}}{\text{mol} \bullet \text{K}} \times \frac{308\text{ K}}{755\text{ torr}} \times \frac{760\text{ torr}}{1\text{ atm}} = 67.851 = 67.9\text{ L}$$

11.103 *Plan:*

(i) Using thermochemical data from Appendix B, calculate the energy (enthalpy) required to melt and heat the H_2O.

(ii) Using Hess's Law, calculate the enthalpy of combustion, ΔH_{comb}, for C_3H_8.

(iii) Solve the stoichiometry problem.

Solve.

(i) Heat $H_2O(s)$ from $-14.0°C$ to $0.0°C$; $2500 \text{ g } H_2O \times \frac{2.092 \text{ J}}{\text{g} \bullet °C} \times 14.0°C = 73.22$

$= 73.2 \text{ kJ}$

Melt $H_2O(s)$; $2500 \text{ g } H_2O \times \frac{6.008 \text{ kJ}}{\text{mol } H_2O} \times \frac{1 \text{ mol } H_2O}{18.02 \text{ g } H_2O} = 833.52 = 834 \text{ kJ}$

Heat $H_2O(l)$ from $0.0°C$ to $60.0°C$; $2500 \text{ g } H_2O \times \frac{4.184 \text{ J}}{\text{g} \bullet °C} \times 60.0°C = 627.6 = 628 \text{ kJ}$

Total energy $= 73.22 \text{ kJ} + 833.52 \text{ kJ} + 627.6 \text{ kJ} = 1534.34 = 1.534 \times 10^3 \text{ kJ}$

(The result has zero decimal places because 628 kJ has zero decimal places.)

(ii) $C_3H_8(g) + 5O_2(g) \rightarrow 3CO_2(g) + 4H_2O(l)$

Assume that one product is $H_2O(l)$, since this leads to a more negative ΔH_{comb} and fewer grams of $C_3H_8(g)$ required.

$$\Delta H_{comb} = 3\Delta H_f^\circ CO_2(g) + 4\Delta H_f^\circ H_2O(l) - \Delta H_f^\circ C_3H_8(g) - 5\Delta H_f^\circ O_2(g)$$

$$= 3(-393.5 \text{ kJ}) + 4(-285.83 \text{ kJ}) - (-103.85 \text{ kJ}) - 5(0) = -2219.97 = -2220 \text{ kJ}$$

(iii) $1.53434 \times 10^3 \text{ kJ required} \times \frac{1 \text{ mol } C_3H_8}{2219.97 \text{ kJ}} \times \frac{44.096 \text{ g } C_3H_8}{1 \text{ mol } C_3H_8} = 30.48 \text{ g } C_3H_8$

(1.534×10^3 kJ required has 4 sig figs and so does the result)

11.104 (a) Low viscosity, low surface tension, and, especially, high thermal conductivity owing to metallic properties.

(b) Metallic bonding between sodium atoms, which persists in the liquid state, is probably the main inhibitor of movement of atoms relative to one another. As temperature increases, thermal motion of the atoms increases and the liquid expands, weakening bonding interactions relative to thermal energies.

11.106 *Plan.* relative humidity and v.p. of H_2O at $23°C \rightarrow P_{H_2O} \rightarrow$ ideal-gas law $\rightarrow$ mol $H_2O(g) \rightarrow H_2O$ molecules.

Solve. r.h. $= (P_{H_2O}$ in air / v.p. of $H_2O) \times 100$

From Appendix B, v.p. of H_2O at $23°C = 21.07$ torr

P_{H_2O} in air $=$ r.h. $\times$ v.p. of $H_2O/100 = 45 \times 21.07 \text{ torr}/100 = 9.4815 = 9.5$ torr

$n = PV/RT$; $V = 14 \text{ m} \times 9.0 \text{ m} \times 8.6 \text{ m} \times \frac{1 \text{ dm}^3}{(0.1)^3 \text{ m}^3} \times \frac{1 \text{ L}}{\text{dm}^3} = 1.0836 \times 10^6 = 1.1 \times 10^6 \text{ L}$

$n = 9.4815 \text{ torr} \times \frac{1 \text{ atm}}{760 \text{ torr}} \times \frac{\text{mol} \bullet \text{K}}{0.08206 \text{ L} \bullet \text{atm}} \times \frac{1.0836 \times 10^6 \text{ L}}{296 \text{ K}} = 556.6 = 5.6 \times 10^2 \text{ mol } H_2O$

$556.6 \text{ mol } H_2O \times \frac{6.022 \times 10^{23} \text{ molecules}}{1 \text{ mol}} = 3.4 \times 10^{26} \text{ } H_2O \text{ molecules}$

11.107 Data are taken from the 74th edition of the *Handbook of Chemistry and Physics.* T_m = melting point, T_b = boiling point

(a) W: $T_m = 3410°C$, $T_b = 5660\ °C$; WF_6: $T_m = 2.5°C$, $T_b = 17.5°C$

W is a metal, with strong metallic bonding, and very high T_m and T_b. WF_6 is an octahedral, nonpolar molecule. Even though it has high molar mass, the spherical shape of the molecule prevents extensive molecular contacts. The resulting London-dispersion forces are very weak, which leads to the low T_m and T_b.

(b) SO_2: $T_m = -72.7°C$, $T_b = -10°C$; SF_4: $T_m = -124°C$, $T_b = -40°C$ (sublimes)

Both SO_2 and SF_4 are polar covalent molecules with a nonbonding electron pair on the central S atom. The electron-domain geometry in SO_2 is trigonal planar and the molecule shape is bent. The electron-domain geometry in SF_4 is trigonal bipyramidal and the molecular shape is see-saw. Both are gases at ambient temperature and pressure. SF_4 has higher molar mass but lower melting and boiling points than SO_2. This indicates that dipole-dipole forces are more influential on the properties of these molecules and that SF_4 has a smaller dipole moment than SO_2.

(c) SiO_2: $T_m = 1723°C$, $T_b = 2230°C$, MM = 60 g/mol

$SiCl_4$: $T_m = -70°C$, $T_b = 57.57°C$, MM = 170 g/mol

SiO_2 is a covalent-network substance, with high T_m and T_b. Covalent bonds hold SiO_2 units in a rigid lattice, and high energy is required to break these bonds and melt or boil the substance. $SiCl_4$ is a tetrahedral, nonpolar molecule. Weak dispersion forces between the approximately spherical molecules result in predictably low T_m and T_b.

12 Modern Materials

Visualizing Concepts

12.2 Material Y, with the smaller band gap, is more suitable for solar energy conversion. According to Section 12.4, semiconductors with small band gaps are most appropriate, because they are able to capture a wider range of solar wavelengths. Semiconductors with larger band gaps capture only higher energy, shorter wavelength light.

12.3 The polymer chains in cartoon (a) are much more tightly wound than those in (b). This indicates that there are stronger intermolecular forces attracting polymer (a) chains to each other. Large empty spaces in polymer (b) suggest only weak attractive forces among chains.

12.5 (a) Addition polymers are formed from monomers with C–C double or triple bonds, but not by aromatic compounds derived from benzene. Molecule (i), with a terminal C=C, is the only monomer shown that is capable of addition polymerization.

(b) Condensation is the combination of two or more molecules to form one larger molecule, in this case a polymer, and a small molecule, usually H_2O or NH_3. Monomers for condensation polymerization contain a carboxyl (–COOH) group and either an alcohol (–OH) or amine (–NH_2) group. Molecule (iii) contains both a carboxyl group and an amine group; it can "condense" with like monomers to form a polymer and NH_3.

(c) Liquid crystals are typically formed by columnar molecules that contain functional groups capable of strong intermolecular interactions that promote long-range order. Molecule (ii) fits this description (and is the only remaining choice). The –CN group experiences dipole-dipole and London dispersion forces. The rod-like 5 C chain and the planar benzene-like (phenyl) ring both have shapes the encourage strong dispersion forces. Taken together, these intermolecular forces are likely to encourage the long range order required to form a liquid crystal.

12.6 An integrated circuit with 20 billion components has 20,000,000,000 or 20×10^9 components. The log of this number is 10.30. By finding this number on the y-axis of the graph, following the value across to the red line, and reading the corresponding year from the x-axis, we see that the year is ~2015. That is, Moore's law predicts that current chip technology will reach its complexity (and performance) limit in 2015.

Classes of Materials

12.8 (a) InAs – semiconductor, Gp 3A + Gp 5A

(b) MgO – insulator; ionic compound

(c) HgS – semiconductor, like CdS, average 4 e^- per atom

(d) Sn – metal; p-block metal left of metalloid line

12.10 p-type semiconductors have a slight e^- deficit. If the dopant replaces As, Group 5A, it should have fewer than five valence electrons. The dopant will be a 4A element, again probably Si or Ge. Si would be closer to As in bonding atomic radius.

12.12 (a) True. 400 kJ/mol is sufficient energy to break many chemical bonds. It is an energy barrier large enough to prohibit any charge mobility.

(b) True. The energy difference between valence and conduction can be small as in the case of metals, but the conduction band is always at the higher energy. Otherwise all materials would be conductors.

(c) False. Electrons can conduct only if they are in a partially filled band. If electrons in filled bands could conduct, there would be no insulators.

(d) False. "Holes" are empty electron sites. They are the absence of an electron where one would normally exist, leading to a region of at least partial positive charge.

12.14 A conductive metal such as Ag conducts electricity with a characteristic resistance to the flow of electrons given by Ohm's law, E = IR. A superconducting substance such as Nb_3Sn below its transition temperature conducts electricity with no resistance to the flow of electrons. Such a superconductor can transfer energy with no net loss, while a metallic conductor cannot be 100% efficient.

12.16 (a) The superconducting transition temperature, T_c, is the temperature at which a material loses all resistance to the flow of electrical current, the temperature below which the material becomes superconducting.

(b) The temperature 77 K is significant because that is the temperature of liquid nitrogen, a readily available, inexpensive, and safe coolant. Materials with T_c temperatures above 77 K produce more financially viable devices than materials which must be cooled with liquid helium below 77 K to achieve superconductivity.

12.18 The phenomenon that superconductors exclude all magnetic fields from their volume is the Meisner effect (Figure 12.4). This can be used to levitate trains, by having either the tracks or the train wheels made from a magnetic material, and the other from a superconductor. The superconductor would need to be cooled below its transition temperature. There are some practical advantages to having the superconductor aboard the train and the magnetic field along the tracks. The train components require less of the costly superconductor and can be cooled more efficiently than the tracks. This is the reverse of the orientation shown in Figure 12.4, where the magnet floats above the cooled superconductor.

Materials for Structure

12.20 Monomers are small molecules with low molecular mass that are joined together to form polymers. They are the repeating units of a polymer. Three (of the many) monomers mentioned in this chapter are

$CH_2{=}CH{-}CH_3$ propylene (propene)

$CH_2{=}CH{-}C_6H_5$ styrene (phenyl ethene)

$CH_2{=}C(CH_3){-}CH{=}CH_2$ isoprene (2-methyl-1,3-butadiene)

12.22 $n\ HO{-}C(=O){-}CH_2{-}CH_2{-}C(=O){-}O{-}H + n\ H{-}N(H){-}CH_2{-}CH_2{-}NH_2 \rightarrow$

$$\left[{-}C(=O){-}CH_2{-}CH_2{-}C(=O){-}NH{-}CH_2{-}CH_2{-}NH{-}\right] + 2n\ H_2O$$

12.24 (a) By analogy to polyisoprene, Equation [12.4],

$$n\ CH_2{=}CH{-}C(Cl){=}CH_2 \longrightarrow \left[{-}CH_2{-}CH{=}C(Cl){-}CH_2{-}\right]_n$$

(b) $$n\ CH_2{=}CH(CN) \longrightarrow \left[{-}CH_2{-}CH(CN){-}\right]_n$$

12.26 $H_2N{-}CH(R){-}C(=O){-}O{-}H + H{-}N(H){-}CH(R){-}C(=O){-}O{-}H \longrightarrow$

$$\left[{-}NH{-}CH(R){-}C(=O){-}NH{-}CH(R){-}C(=O){-}\right]_n + nH_2O$$

12.28 $HOOC{-}C_6H_4{-}COOH$ diacid terephthalic acid; $H_2N{-}C_6H_4{-}NH_2$ diamine *p*-diaminobenzene

Note that these monomers are the same as those in Nomex (Solution 12.25) except for the orientation of the functional groups on the benzene rings. Clearly monomer structure strongly impacts polymer structure.

12.30 At the molecular level, the longer, unbranched chains of HDPE fit closer together and have more crystalline (ordered, aligned) regions than the shorter, branched chains of LDPE. Closer packing leads to higher density.

12.32 (a) An elastomer is a polymer material that recovers its shape when released from a distorting force. A typical elastomeric polymer can be stretched to at least twice its original length and return to its original dimensions upon release.

(b) A thermoplastic material can be shaped and reshaped by application of heat and/or pressure.

(c) A thermosetting plastic can be shaped once, through chemical reaction in the shape-forming process, but cannot easily be reshaped, due to the presence of chemical bonds that cross-link the polymer chains.

(d) A plasticizer is a substance of relatively low molar mass added to a polymer material to soften it.

12.34 (a) The object that shatters is ceramic; the one that dents is metal.

(b) The two materials differ in their behavior because of their different solid state bonding characteristics. Ceramics are formed from inorganic materials linked by ionic or highly polar covalent bonds into three-dimensional bonding networks. During catastrophic failure (dropping 10 feet onto cement), the network structure prevents atoms from sliding over one another and the ceramic shatters. A series of bond ruptures occurs, often along planes in the three-dimensional structure, leading to fragments with sharp edges.

Metallic bonding is characterized by delocalization of loosely held valence electrons among metal atoms. Unlike the covalent or ionic network bonding in ceramics, metallic bonding is multidirectional. This allows metal atoms to slide over each other during deformation, resulting in dents and stress cracks rather than shattering.

12.36 Since Zr and Ti are in the same family, assume that the stoichiometry of the compounds in a sol-gel process will be the same for the two metals.

i. Alkoxide formation: oxidation-reduction reaction

$$Zr(s) + 4CH_3CH_2OH(l) \rightarrow \underset{\text{alkoxide}}{Zr(OCH_2CH_3)_4(s)} + 2H_2(g)$$

ii. Sol formation: metathesis reaction

$$Zr(OCH_2CH_3)_4(soln) + 4H_2O(l) \rightarrow \underset{\substack{\text{"precipitate"} \\ \text{sol}}}{Zr(OH)_4(s)} + \underset{\text{nonelectrolyte}}{4CH_3CH_2OH(l)}$$

$Zr(OCH_2CH_3)_4(s)$ is dissolved in an alcohol solvent and then reacted with water. In general, reaction with water is called *hydrolysis*. The alkoxide anions ($CH_3CH_2O^-$) combine with H^+ from H_2O to form the nonelectrolyte $CH_3CH_2OH(l)$, and Zr^{2+} cations combine with OH^- to form the $Zr(OH)_4$ solid. The product $Zr(OH)_4(s)$ is not a traditional coagulated precipitate, but a finely divided evenly dispersed collection of particles called a sol.

iii. Gel formation: condensation reaction

$$(OH)_3Zr{-}O{-}H(s) + H{-}O{-}Zr(OH)_3(s) \rightarrow \underset{\text{gel}}{(HO)_3Zr{-}O{-}Zr(OH)_3(s)} + H_2O(l)$$

Adjusting the acidity of the $Zr(OH)_4$ sol initiates condensation, the splitting-out of $H_2O(l)$ and formation of a zirconium-oxide network solid. The solid remains suspended in the solvent mixture and is called a gel.

iv. Processing: physical changes

The gel is heated to drive off solvent and the resulting solid consists of dry, uniform and finely divided ZrO_2 particles.

12.38 By analogy to the ZnS structure, the C atoms form a face-centered cubic array with Si atoms occupying alternate tetrahedral holes in the lattice. This means that the coordination numbers of both Si and C are 4; each Si is bound to four C atoms in a tetrahedral arrangement, and each C is bound to four Si atoms in a tetrahedral arrangement, producing an extended three-dimensional network. ZnS, an ionic solid, sublimes at 1185° and 1 atm pressure and melts at 1850° and 150 atm pressure. The considerably higher melting point of SiC, 2800° at 1 atm, indicates that SiC is probably not a purely ionic solid and that the Si–C bonding network has significant covalent character. This is reasonable, since the electronegativities of Si and C are similar (Figure 8.7). SiC is high-melting because a great deal of chemical energy is stored in the covalent Si–C bonds, and it is hard because the three-dimensional lattice resists any change that would weaken the Si–C bonding network.

12.40 Correlation and learning vary with each learner. The interpretations below are one of many possibilities, with a lettered statement for each verse of the poem.

(a) The bulk properties of a substance depend on its component particles. A metal composed of Pb atoms will have the properties of lead, not those of gold. The properties of materials depend on their component atoms and molecules, and the nature of the bonding between particles.

(b) Silicon and graphite, a form of carbon, are semiconductors because of their ordered covalent-network structure and the resulting bands of energy states. These solid-state structures were determined by X-ray crystallography (Chapter 11).

(c) Many of the properties of metals, including ductility and conductivity, are due to the delocalized electrons characteristic of metallic bonding. The electrons are delocalized because of the nearly continuous valence and conduction bonds (Figure 12.2) in metals.

(d) Ceramics are inorganic solids with localized electrons and a large band gap; "no free electrons form a lubricating tide."

(e) Glass is one of the only recyclable ceramics. It is an amorphous solid, lacking the three-dimensional order of most ceramics. It can be melted and reformed, because its properties do not depend on a perfectly ordered structure.

(f) Polymers are very long-chain, high molecular weight substances, formed by covalent bonding among one or more types of monomer units. Intermolecular forces between chains are typically weak, so the materials are flexible. Those that can be formed into shapes are called plastics. Rigidity can be increased by cross-linking, chemical bonding between chains. Proteins and nucleic acids, "cross-linked helixes unknown to *Robert Hooke*," are copolymers formed by bonding several different monomers.

(g) The melting points of all materials depend on their structure.

(h) Electrical conductivity depends on the size of the band gap in a material. Doping with very small amounts of appropriate impurities increases the conductivity of semiconductors.

(i) Doped semiconductors and ceramic superconductors are examples of nonstoichiometric solids discussed in this chapter. p-type semiconductors have "strange holes" that "wander loose," leading to their semiconductivity.

(j) Semiconductors can emit light if electrons have been excited across the band gap by an applied voltage or light of energy equal to or higher than the band gap. (Figure 12.41)

Nano particles of the semiconductor Cd_3P_2 absorb different colors of visible light and appear different colors, depending on particle size. "Each element absorbs its signature."

(k) Superconductors have no resistance to current flow. They exclude all magnetic fields from their volume (Meisner effect). While the mechanism of superconductivity in ceramics is not well understood, "Diffuse material becomes magnetic when another Field aligns domains."

Materials for Medicine

12.42 One structural characteristic of polymers that forms effective interfaces with biological systems is the presence of polar functional groups in the polymer backbone or as substituents. Polystyrene is a hydrocarbon; it has no polar functional groups and is a nonpolar substance. Polyurethane has polar carbon–oxygen, carbon–nitrogen, and nitrogen–hydrogen functional groups. The N-H groups mean that it can act as a hydrogen-bond donor as well as an acceptor. In fact, the polyurethane backbone is very similar to the protein backbone shown in this section. We expect polyurethane to be the superior biointerface.

12.44 Surface roughness in synthetic heart valves causes hemolysis, the breakdown of red blood cells. The surface of the valve implant was probably not smooth enough.

12.46 Polystyrene is an essentially nonpolar hydrocarbon, while polyethyleneterephthalate (PET) contains polar ester groups, as well as nonpolar hydrocarbon portions. PET is more appropriate, because it provides polar ester groups $\left(-\overset{\overset{\large O}{\|}}{C}-O-C- \right)$ with hydrogen bonding capabilities where the cells can attach. Also, the ester linkages are susceptible to hydrolysis (the reverse of condensation); this renders the synthetic matrix biodegradable when employed in the body.

Materials for Electronics

12.48 Impurities can be thought of as dopants. Since dopants at concentrations of parts per million (ppm) can change the conductivity of Si, it needs to be more pure than the ppm level to ensure known and constant conductivity. Purity at ppm level would be 99.9999% pure. Purity of 99.999999999% is pure at nearly the parts per trillion level.

12.50 "Plastic" semiconductors would be shapable as well as flexible. This would be very desirable for medical applications such as prosthetic hands or feet, where the flexibility would render the device more natural and more durable to constant movements. Flexible semiconductors would also be useful in forming irregularly shaped components for a wide variety of devices.

12.52 From Table 12.1, the band gap energy for TiO_2 is 3.0 eV.

$$\lambda = 6.626 \times 10^{-34}\ \text{J}\bullet\text{s} \times \frac{3.00 \times 10^{8}\ \text{m}}{\text{s}} \times \frac{1}{3.0\ \text{eV}} \times \frac{1\ \text{eV}}{1.602 \times 10^{-19}\ \text{J}} = 4.136 \times 10^{-7}$$

$$= 4.1 \times 10^{-7}\ \text{m}$$

TiO_2 absorbs 4.1×10^{-7} to 2.0×10^{-7}, a span of 2.1×10^{-7} m.

$\frac{2.1 \times 10^{-7}}{28 \times 10^{-7}} \times 100 = 7.5\%$ of the wavelengths in the solar spectrum.

This is a much smaller portion of the spectrum, and of the total flux, than Si absorbs.

Materials for Optics

12.54 In an ordinary liquid, molecules are oriented randomly and their relative orientations are continuously changing. In liquid crystals, the molecules are aligned in at least one dimension. The relative orientations in the other two dimensions may change, but alignment in the oriented direction is maintained.

12.56 Reinitzer observed that cholesteryl benzoate has a phase that exhibits properties intermediate between those of the solid and liquid phases. This "liquid-crystalline" phase, formed by melting at 145°C, is opaque, changes color as the temperature increases, and becomes clear at 179°C.

12.58 Because order is maintained in at least one dimension, the molecules in a liquid-crystalline phase are not totally free to change orientation. This makes the liquid-crystalline phase more resistant to flow, more viscous, than the isotropic liquid.

12.60 The "LCD molecule" is long relative to its thickness. It has C=C and C≡N groups that promote rigidity and polarizability along the length of the molecule. The C≡N group also provides dipole-dipole interactions that encourage alignment. Unlike the molecules in Figure 12.34, which contain planar phenyl rings, the LCD molecule contains nonaromatic, nonplanar six-membered rings. These rings could contribute to specific physical properties such as the liquid crystal temperature range that make this molecule particularly functional in LCD displays.

12.62 As the temperature of a substance increases, the average kinetic energy of the molecules increases. More molecules have sufficient kinetic energy to overcome intermolecular attractive forces, so overall ordering of the molecules decreases as temperature increases. Melting provides kinetic energy sufficient to disrupt alignment in one dimension in the solid, producing a smectic phase with ordering in two dimensions. Additional heating of the smectic phase provides kinetic energy sufficient to disrupt alignment in another dimension, producing a nematic phase with one-dimensional order.

12.64 From Table 12.1, E_g for GaAs (x = 0) is 1.4 and for GaP (x = 1) is 2.2. If E_g varies linearly with x, the band gap for x = 0.5 should be approximately the average of the two extreme values: (1.4 + 2.2)/2 = 1.8 eV.

$$\lambda = hc/E = 6.626 \times 10^{-34}\ \text{J}\bullet\text{s} \times \frac{3.00 \times 10^{8}\ \text{m}}{\text{s}} \times \frac{1}{1.8\ \text{eV}} \times \frac{1\ \text{eV}}{1.602 \times 10^{-19}\ \text{J}} = 6.89 \times 10^{-7}$$

$$= 6.9 \times 10^{-7}\ \text{m} = 690\ \text{nm}$$

Materials for Nanotechnology

12.66 *Analyze.* Given: 1/16 inch diameter pin, Encyclopedia Britannica, EB (over 20 volumes), line width = 32 atoms. Find: Can we print EB on the head of a pin, using a 32 atom line width?

Plan. There are many ways to approach this problem. Some data about EB must be discovered, via either a trip to the library or the internet, or both. Then, use this data about EB, along with that given in the problem, to test the validity of Feynman's assertion. The exact approach depends on the EB data discovered. It will probably include calculating the area of a pinhead using πr^2, and estimating the diameter of an "average" atom.

Solve. In the exercise, we are given information about the width of a line and the total printing area for the nano printing of EB. Calculate these dimensions explicitly, using any necessary assumptions.

Line width: If each line is 32 atoms wide, what is the diameter of an 'atom'? Use Figure 7.6 to calculate the radius and diameter of an average atom, ignoring H (because it is an outlier) and the noble gases (because they don't typically form bonds). The smallest atom is F and the largest is Rb. The average of their diameters is [2(0.71) + 2(2.11)]/2 = 2.82 Å. (This is a generous estimate; Feynman used 2.5 Å.)

$$\frac{32\ \text{atoms}}{\text{line}} \times \frac{2.82\ \text{Å}}{\text{atom}} \times \frac{1 \times 10^{-8}\ \text{cm}}{\text{Å}} = 9.024 \times 10^{-7} = 9.0 \times 10^{-7}\ \text{cm}$$

Area of pinhead: πr^2, d = 1/16 inch, r = 1/32 inch

$$\frac{1}{32}\ \text{inch} \times 2.54\ \text{cm} = 0.07938 = 0.079\ \text{cm}$$

$$\text{area} = 3.14159\ (0.07938\ \text{cm})^2 = 0.01979 = 0.020\ \text{cm}^2$$

The question is, how do we relate the one-dimensional line-width to a two-dimensional area? The smallest area we can print is a square (9.0×10^{-7} cm) by (9.0×10^{-7} cm) = $8.1 \times 10^{-13}\ \text{cm}^2$; define this area as one pixel. Each character (letter) is a combination of printed lines and white space. Assume that an area 9 × 9 pixels will be sufficient to print any character, including all lines and white space. This accommodates any character and generalizes our solution to languages other than English This is the method used by dot-matrix printers. A word or a sentence is then a string of these boxes. The number of characters we can print on the pinhead is then

$$\frac{0.020\ \text{cm}^2}{\text{pinhead}} \times \frac{1\ \text{character}}{8.1 \times 10^{-13}\ \text{cm}^2} = 2.5 \times 10^{10}\ \text{characters.}$$

Is this enough to print EB? Here's where research comes in. According to the EB website, the modern encyclopedia has 32 volumes and 44 million (44×10^6 or 4.4×10^7) words.

$$\frac{2.5 \times 10^{10}\ \text{characters}}{4.4 \times 10^7\ \text{words}} = 561\ \text{characters/word.}$$

Since no word, let alone an "average" one, contains 561 characters, it's safe to say that Feynman's assertion is correct, with room to spare.

12.68 True. Blue light has short wavelengths, corresponding to a relatively large band gap. As particle size decreases, band gap increases and wavelength decreases. We could begin with a semiconductor with a smaller band gap and make it a nanoparticle to increase E_g and decrease wavelength. (Nanoparticle size becomes one more way to tune the properties of semiconductors.)

12.70 Vol of unit cell = $(4.08\ \text{Å})^3 = 67.9173 = 67.9\ \text{Å}^3$

$$22\ \text{nm diameter} = 11\ \text{nm radius};\ 11\ \text{nm} \times \frac{10\ \text{Å}}{1\ \text{nm}} = 110\ \text{Å radius}$$

vol of sphere = $4/3 \times 3.14159 \times (110\ \text{Å})^3 = 5.5753 \times 10^6 = 5.58 \times 10^6\ \text{Å}^3$

$$\frac{4\ \text{Au atoms}}{67.9173\ \text{Å}^3} = \frac{x\ \text{Au atoms}}{5.5753 \times 10^6\ \text{Å}^3};\ x = 3.2836 \times 10^5 = 3.28 \times 10^5\ \text{Au atoms}$$

Additional Exercises

12.71 Semiconductors have a filled valence band and an empty conduction bond, separated by a characteristic difference in energy, the band gap, E_g. When a semiconductor is heated, more electrons have sufficient energy to jump the band gap, and conductivity increases. Metals have a partially-filled continuous energy band. Heating a metal increases the average kinetic energy of the metal atoms, usually through increased vibrations within the lattice. The greater vibrational energy of the atoms leads to imperfections in the lattice and discontinuities in the energy band. Thermal vibrations create barriers to electron delocalization and reduce the conductivity of the metal.

12.73 $\left[-CF_2-CF_2- \right]_n$ or [zigzag chain of C atoms, each bearing two F]

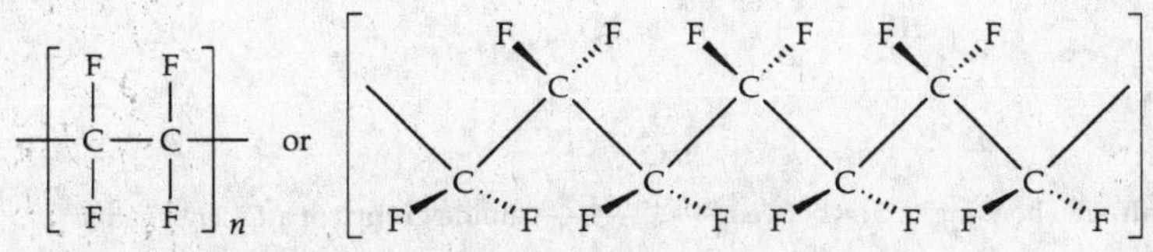

Teflon™ is formed by addition polymerization.

12.74 (a) polymer (b) ceramic (c) ceramic (d) polymer

(e) liquid crystal (an organic molecule with a characteristic long axis and the kinds of functional groups often found in compounds with liquid-crystalline phases (Figure 12.34); not enough repeating units to be a polymer)

12.75 Ceramics are usually three-dimensional network solids, whereas plastics most often consist of large, chain-like molecules (the chain may be branched) held loosely together by relatively weak van der Waals forces. Ceramics are rigid precisely because of the many strong bonding interactions intrinsic to the network. Once a crack forms, atoms near the defect are subject to great stress, and the crack is propagated. They are stable to high temperatures because tremendous kinetic energy (temperature) is required for an atom to break free from the bonding network. On the other hand, plastics are flexible because the molecules themselves are flexible (free rotation around the sigma bonds in the polymer chain), and it is easy for the molecules to move relative to one another (weak intermolecular forces). [However, recall that rigidity of the plastic can be increased in a number of ways, including cross-linking (Figure 12.15), or reinforcement with a second polymer (Figure 12.17).] Plastics are not thermally stable because their largely organic molecules are subject to oxidation and/or bond breaking at high temperatures.

12.76 In a liquid crystal display (Figure 12.37), the molecules must be free to rotate by 90°. The long directions of molecules remain aligned but any attractive forces between the ends of molecules are disrupted. At low Antarctic temperatures, the liquid crystalline phase is closer to its freezing point. The molecules have less kinetic energy due to temperature and the applied voltage may not be sufficient to overcome orienting forces among the ends of molecules. If some or all of the molecules do not rotate when the voltage is applied, the display will not function properly.

12.78 This phenomenon is similar to supercooling, Section 11.4. When the isotropic liquid is cooled below the liquid crystal-liquid transition temperature, the kinetic energy of the molecules has been decreased enough so that formation of the liquid crystalline phase is energetically favorable. However, the molecules may not be correctly organized so that long range ordering can take place.

12.79

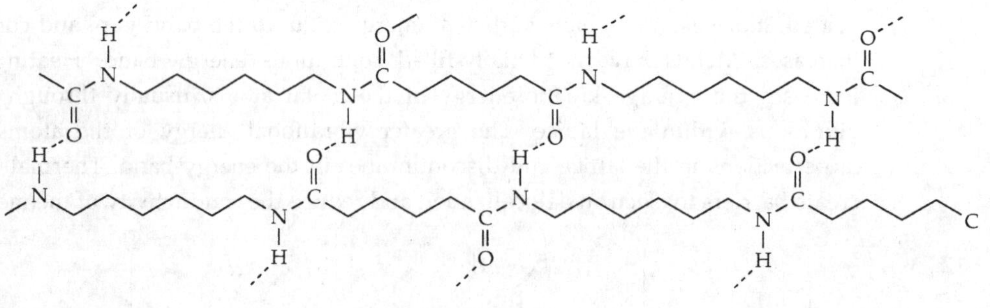

Hydrogen bonding occurs between $-\mathrm{C}(=\mathrm{O})-\mathrm{N}(\mathrm{H})-$ amide groups of adjacent chains.

12.80

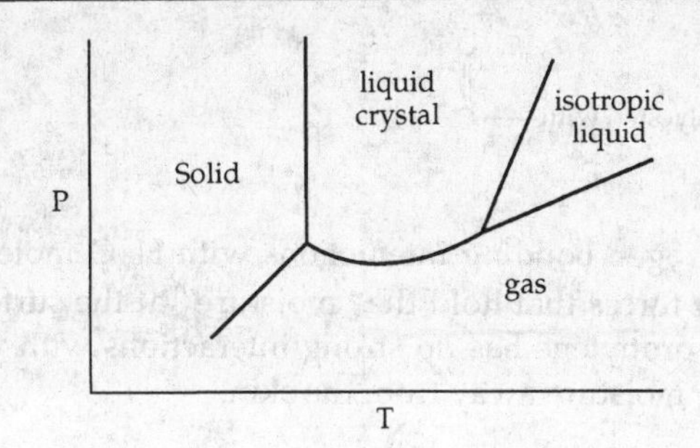

12.82 Solar energy conversion is the process of changing light energy from the sun into useful electrical energy, electricity. Semiconductors are appropriate for this purpose because sunlight excites electrons from the valence to the conduction band, rendering them useful. In order to absorb the full range of solar wavelengths, the semiconductor should have a relatively small band gap, ~50–150 kJ/mol. In order to efficiently translate excited electrons into useful current, the material should have a high conductivity. Practical considerations are, as usual, to be cheap, available, and nontoxic.

Integrative Exercises

12.84 (a) sp^3 hybrid orbitals at C, 109° bond angles around C

(b)

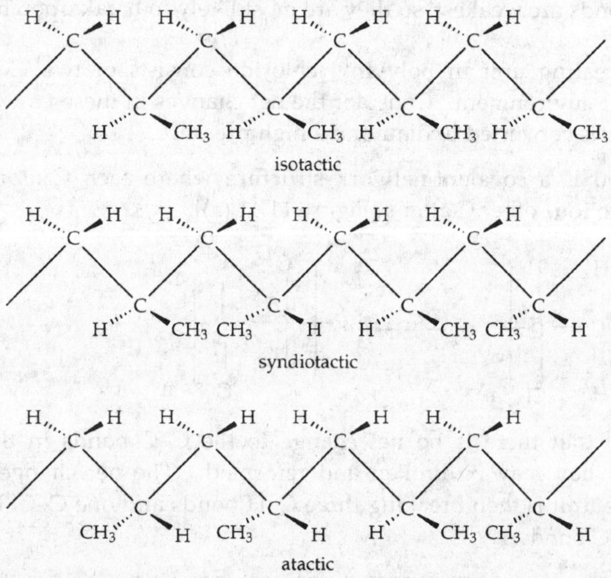

isotactic

syndiotactic

atactic

Isotactic polypropylene has the highest degree of crystallinity and highest melting point. The regular shape of the polymer backbone allows for close, orderly (almost zipper-like) contact between chains. This maximizes dispersion forces between chains and produces higher order (crystallinity) and melting point. Atactic polypropylene has the least order and the lowest melting point.

(c) Cotton, with —C(OH)— groups and polyester, with —C(=O)—O—C groups, both participate in hydrogen bonding interactions with H_2O molecules. These are strong intermolecular forces that hold the "moisture" at the surface of the fabric next to the skin. Polypropylene has no strong interactions with water, and capillary action "wicks" the moisture away from the skin.

12.85 If the expected oxidation states on Y and Ba are +3 and +2, respectively, the average oxidation state of Cu is +2 1/3. That is, two Cu ions are in the +2 state and one is in the +3 state. Y^{3+} and Ba^{2+} have the stable electron configurations of their nearest noble gases, while Cu has an incomplete d orbital set. Cu(II) is d^9, Cu(III) is d^8 and both have unpaired electrons. Although the mechanism by which copper 3d electrons interact through bridging oxygen atoms to form a superconducting state is still not clear, it is evident that the electronic structure of copper ions is essential to the observed superconductivity.

12.87 (a) $\left[CH_2 — CH(Cl) \right]_n$

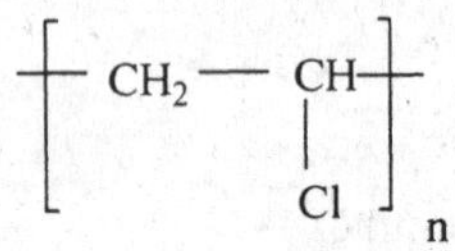

C—Cl	328 kJ/mol	← lowest
C—C	348 kJ/mol	
C—H	413 kJ/mol	

(b) C-Cl bonds are weakest, so they are most likely to break upon heating.

(c) The repeating unit in polyvinyl chloride consists of two C atoms, each in a different environment. Consider the net changes in these two C atoms when the polymer is converted to diamond a high pressure.

Diamond is a covalent-network structure where each C atom is tetrahedrally bound to four other C atoms [Figure 11.41 (a)].

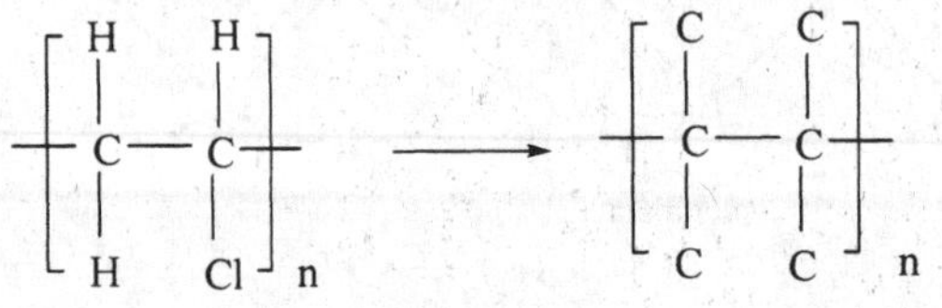

$$\left[\begin{matrix} H & H \\ | & | \\ —C— & C— \\ | & | \\ H & Cl \end{matrix} \right]_n \longrightarrow \left[\begin{matrix} C & C \\ | & | \\ —C— & C— \\ | & | \\ C & C \end{matrix} \right]_n$$

Assume that there is no net change to the C–C bonds in the structure, even though they may be broken and reformed. The net change to the 2-C vinyl chloride unit is then breaking three C–H bonds and one C–Cl bond, and making four C–C bonds.

ΔH = D(C-H) - D(C-l) - 4D(C-) = 3(413) + 328 – 4(348) =

523 kJ/vinyl choride unit

12.88 (a) $CH_3—\ddot{O}—C_6H_4—N(\ddot{O}:)=N—C_6H_4—\ddot{O}—CH_3$

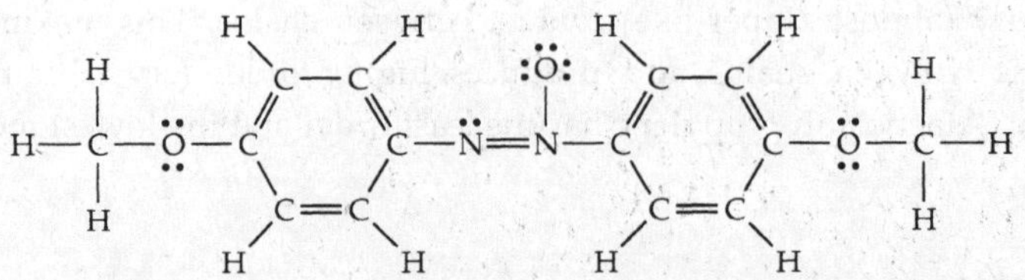

There are several other resonance structures involving alternate placement of the double bonds in the benzene rings.

(b) Both N atoms are surrounded by 3 VSEPR electron domains, so the hybridization at both atoms is sp^2. The bond angles around the N attached to O will be approximately 120°.

(c) When the $-OCH_3$ group is replaced by the $-CH_2CH_2CH_2CH_3$ group, a small rather compact group with some polarity is replaced by a larger, more flexible, nonpolar group. Thus, the molecules don't line up as well in the solid, and the melting point and liquid crystal temperature range are lower.

(d) The density decreases going from solid to nematic liquid crystal to isotropic liquid. In the nematic liquid crystal, most of the long range order of the solid state is lost. The molecules are moving relative to one another with their long axes more or less aligned. The result is more empty space and a lower density than the solid. There is a further small decrease in density when the last degree of order is lost and the substance becomes an isotropic liquid.

12.89 (a) Follow the logic outlined in Solution 12.69.

vol. of unit cell = $(5.43\ \text{Å})^3 = 160.1030 = 1.60 \times 10^2\ \text{Å}^3$

$$1\ \text{cm}^3 \times \frac{(1)^3\ \text{Å}^3}{(1 \times 10^{-8})^3\ \text{cm}^3} = 1 \times 10^{24}\ \text{Å}^3 \text{ (volume of material)}$$

$$\frac{4\ \text{Si atoms}}{160.103\ \text{Å}^3} = \frac{x\ \text{Si atoms}}{1 \times 10^{24}\ \text{Å}^3};\ x = 2.4984 \times 10^{22} = 2.50 \times 10^{22}\ \text{Si atoms}$$

(To 1 sig fig, the result is 2×10^{22} Si atoms.)

(b) 1 ppm phosphorus = 1 P atom per 1×10^6 Si atoms

$$\frac{1\ \text{P atom}}{1 \times 10^6\ \text{Si atoms}} = \frac{x\ \text{P atoms}}{2.4984 \times 10^{22}\ \text{Si Atoms}};\ x = 2.4984 \times 10^{22}$$

$$= 2.50 \times 10^{16}\ \text{P atoms}$$

$$2.4984 \times 10^{16}\ \text{P atom} \times \frac{1\ \text{mol}}{6.022 \times 10^{23}\ \text{atoms}} \times \frac{30.97376\ \text{g P}}{\text{mol}} \times \frac{1\ \text{mg}}{1 \times 10^{-3}\ \text{g}}$$

$$= 1.29 \times 10^{-3}\ \text{mg P } (1.29\ \mu\text{g})$$

12.90 (a) According to Table 12.1, the values of E_g for the Group 4A elements with the diamond structure are: C(diamond), 5.5 eV; Si, 1.1 eV; Ge, 0.67 eV; Sn(gray), 0.08 eV. Going down Group 4A, covalent radius and bond length increases, while E_g decreases.

(b) For the III–V semiconductors, E_g values are: GaP, 2.2 eV; GaAs, 1.43 eV. For the II–VI semiconductors: CdS, 2.4 eV; CdSe, 1.7 eV; CdTe, 1.44 eV. In both sets of mixed semiconductors, holding the element with fewer valence electrons constant, the value of E_g decreases moving down the group of elements with more valence electrons.

(c) I_1 Values from Chapter 7.

element	E_g, eV	I_1, kJ/mol
C(dia)	5.5	1086
Si	1.1	786
Ge	0.67	762
Sn(gray)	0.08	709

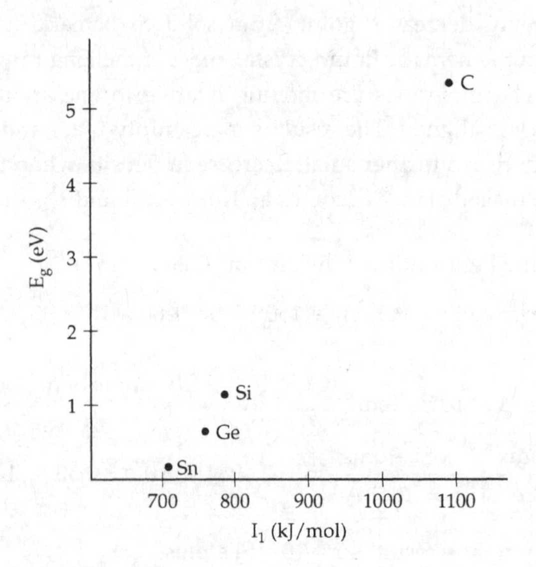

The plot shows a strong relationship between I_1, and E_g; as I_1 increases, E_g increases. I_1 is the energy required to completely remove an electron from a gas-phase atom of the element. E_g is the energy required to excite an electron from the valence band to the conduction band in a bulk sample of the element. I_1 is largely determined by the attraction of an electron for the nucleus. Going down a family, valence electrons are farther from the nucleus and shielded from the full nuclear charge by an increasing electron core, so effective nuclear charge and ionization energy decrease. This relationship carries over to properties of the bulk elements like metallic character and E_g. As electrons are less strongly attracted to nuclei, they are more easily delocalized and the energy required to excite them to the conduction band decreases.

13 Properties of Solutions

Visualizing Concepts

13.2 ΔH_3 contains the interaction of a cation with the solvent. In the figure for Exercise 13.1, we see a single Na^+ cation separated from the bulk sample, and solvent molecules interacting with each other as well as the cation. The diagram shows attractive solvent-solvent and ion-solvent interactions, which contribute to an overall negative (-) ΔH. In Figure 13.4, only ΔH_3 is negative.

13.3 The pink solid is hydrated $CoCl_2$, $CoCl_2 \bullet xH_2O$, where x is a specific integer. The waters of hydration are either associated with Co^{2+}, Cl^-, or sit in specific sites in the crystal lattice. When heated in an oven, the water molecules incorporated into the crystal lattice gradually gain kinetic energy and vaporize. The blue solid is anhydrous $CoCl_2$, absent the waters of hydration and with a different solid-state structure than the pink hydrate.

13.4 Diagram (b) is the best representation of a saturated solution. There is some undissolved solid with particles that are close together and ordered, in contact with a solution containing mobile, separated solute particles. As much solute has dissolved as can dissolve, leaving some undissolved solid in contact with the saturated solution.

13.5 Solubility increases in the order Ar, 1.50×10^{-3} *M* < Kr, 2.79×10^{-3} *M* < Xe, 5×10^{-5} *M*, the order of increasing polarizability. As the molar mass of the ideal gas increases, atomic size increases and the electron cloud is less tightly held by the nucleus, causing the cloud to be more polarizable. The greater the polarizability, the stronger the dispersion forces between the gas atoms and water, the more likely the gas atom is to stay dissolved rather than escape the solution, the greater the solubility of the gas.

13.7 (a) Yes, the *molarity* changes with a change in temperature. Molarity is defined as moles solute per unit volume of solution. If solution volume is different, molarity is different.

(b) No, *molality* does not change with change in temperature. Molality is defined as moles solute per kilogram of solvent. Even though the volume of solution has changed due to increased kinetic energy, the mass of solute and solvent have not changed, and the molality stays the same.

13.8 Ideally, 0.50 L. If the volume outside the balloon is very large compared to 0.25 L, solvent will flow across the semipermeable membrane until the molarities of the inner and outer solutions are equal, 0.10 *M*. This requires an "inner" solution volume twice as large as the initial volume, or 0.50 L. (In reality, osmosis across the balloon membrane is not perfect. The solution concentration inside the balloon will be slightly greater than 0.10 *M* and the volume of the balloon will be slightly less than 0.50 L.)

13.10 According to Figure 13.18, the solubility of CO at 25°C and 1 atm pressure is approximately 0.96 mM. By Henry's Law, $S_g = k P_g$. At the same temperature and pressure, k will be the same, so $S_1/P_1 = S_2/P_2$.

$$\frac{0.96 \text{ m}M}{1 \text{ atm}} = \frac{2.5 \text{ m}M}{\text{x atm}}; \text{x} = \frac{2.5 \text{ m}M \times 1 \text{ atm}}{0.96 \text{ atm}} = 2.6 \text{ atm}$$

The Solution Process

13.12 (a) For the same solute, NaCl, in different solvents, solute-solute interactions (ΔH_1) are the same. Because water experiences hydrogen bonding while benzene has only dispersion forces, solvent-solvent interactions (ΔH_2) are greater for water. On the other hand, solute-solvent interactions (ΔH_3) are much weaker between ionic NaCl and nonpolar benzene than between ionic NaCl and polar water. It is the large difference in ΔH_3 that causes NaCl to be soluble in water but not in benzene.

(b) Lattice energy is the main component of ΔH_1, the enthalpy required to separate solute particles. If ΔH_1 is too large, the dissolving process is prohibitively endothermic, and the substance is not very soluble.

(c) Ion-dipole forces between cations and water molecules and relatively small lattice energies (ion-ion forces between cations and anions) lead to strongly hydrated cations.

13.14 From weakest to strongest solvent-solute interactions:

(b), dispersion forces < (c), hydrogen bonding < (a), ion-dipole

13.16 Separation of solvent molecules, ΔH_2, will be smallest in this case, because hydrogen bonding is the weakest of the intermolecular forces involved. ΔH_1 involves breaking ionic bonds, and ΔH_3 involves formation of ion-dipole interactions, both stronger forces than hydrogen bonding.

13.18 KBr is quite soluble in water because of the sizeable increase in disorder of the system (ordered KBr lattice → freely moving hydrated ions) associated with the dissolving process. An increase in disorder or randomness in a process tends to make that process spontaneous.

Saturated Solutions; Factors Affecting Solubility

13.20 (a) $$\frac{1.22 \text{ mol } MnSO_4 \bullet H_2O}{1 \text{ L soln}} \times \frac{169.0 \text{ g } MnSO_4 \bullet H_2O}{1 \text{ mol}} \times 0.100 \text{ L}$$

$$= 20.6 \text{ g } MnSO_4 \bullet H_2O/100 \text{ mL}$$

The 1.22 *M* solution is unsaturated.

(b) Add a known mass, say 5.0 g, of $MnSO_4 \bullet H_2O$, to the unknown solution. If the solid dissolves, the solution is unsaturated. If there is undissolved $MnSO_4 \bullet H_2O$, filter the solution and weigh the solid. If there is less than 5.0 g of solid, some of the added $MnSO_4 \bullet H_2O$ dissolved and the unknown solution is unsaturated. If there is exactly 5.0 g, no additional solid dissolved and the unknown is saturated. If there is more than 5.0 g, excess solute has precipitated and the solution is supersaturated.

13.22 (a) at 30°C, $\frac{10 \text{ g KClO}_3}{100 \text{ g H}_2\text{O}} \times 250 \text{ g H}_2\text{O} = 25 \text{ g KClO}_3$

(b) $\frac{66 \text{ g Pb(NO}_3)_2}{100 \text{ g H}_2\text{O}} \times 250 \text{ g H}_2\text{O} = 165 = 1.7 \times 10^2 \text{ g Pb(NO}_3)_2$

(c) $\frac{3 \text{ g Ce}_2(\text{SO}_4)_3}{100 \text{ g H}_2\text{O}} \times 250 \text{ g H}_2\text{O} = 7.5 = 8 \text{ g Ce}_2(\text{SO}_4)_3$

13.24 Immiscible means that oil and water do not mix homogeneously; they do not dissolve. Many substances are called "oil," but they are typically nonpolar carbon-based molecules with fairly high molecular weights. As such, there are fairly strong dispersion forces among oil molecules. The properties of water are dominated by its strong hydrogen bonding. The dispersion-dipole interactions between water and oil are likely to be weak. Thus, ΔH_1 and ΔH_2 are large and positive, while ΔH_3 is small and negative. The net ΔH_{soln} is large and positive, and mixing does not occur.

13.26 For small *n* values, the dominant interactions among acid molecules will be hydrogen-bonding. As *n* increases, dispersion forces between carbon chains become more important and eventually dominate. Thus, as *n* increases, water solubility decreases and hexane solubility increases.

13.28 *Analyze/Plan.* Water, H_2O, is a polar solvent that forms hydrogen bonds with other H_2O molecules. The more soluble solute in each case will have intermolecular interactions that are most similar to the hydrogen bonding in H_2O. *Solve.*

(a) Glucose, $C_6H_{12}O_6$, is more soluble because it is capable of hydrogen bonding (Figure 13.12). Nonpolar C_6H_{12} is capable only of dispersion interactions and does not have strong intermolecular interactions with polar (hydrogen bonding) H_2O.

(b) Ionic sodium propionate, CH_3CH_2COONa, is more soluble. Sodium propionate is a crystalline solid, while propionic acid is a liquid. The increase in disorder or entropy when an ionic solid dissolves leads to significant water solubility, despite the strong ion-ion forces (large ΔH_1) present in the solute (see Solution 13.18).

(c) HCl is more soluble because it is a strong electrolyte and completely ionized in water. Ionization leads to ion-dipole solute-solvent interactions, and an increase in disorder. CH_3CH_2Cl is a molecular solute capable of relatively weak dipole-dipole solute-solvent interactions and is much less soluble in water.

13.30 Pressure has an effect on O_2 solubility in water because, at constant temperature and volume, pressure is directly related to the amount of O_2 available to dissolve. The greater the partial pressure of Oabove water, the more O_2 molecules are available for dissolution, and the more molecules that strike the surface of the liquid.

Pressure does not affect the amount or physical properties of NaCl, or ionic solids in general, so it has little influence on the dissolving of NaCl in water.

13.32 $665 \text{ torr} \times \frac{1 \text{ atm}}{760 \text{ torr}} = 0.875 \text{ atm}; P_{O_2} = \chi_{O_2}(P_t) = 0.21(0.875 \text{ atm}) = 0.1838 = 0.18 \text{ atm}$

$$S_{O_2} = kP_{O_2} = \frac{1.38 \times 10^{-3} \text{ mol}}{\text{L} \bullet \text{atm}} \times 0.1838 \text{ atm} = 2.5 \times 10^{-4}\ M$$

Concentrations of Solutions

13.34 (a) $\text{mass \%} = \frac{\text{mass solute}}{\text{total mass solution}} \times 100$

$$\text{mass solute} = 0.045 \text{ mol } I_2 \times \frac{253.8 \text{ g } I_2}{1 \text{ mol } I_2} = 11.421 = 11 \text{ g } I_2$$

$$\text{mass \% } I_2 = \frac{11.421 \text{ g } I_2}{11.421 \text{ g } I_2 + 115 \text{ g } CCl_4} \times 100 = 9.034 = 9.0\% \ I_2$$

(b) $\text{ppm} = \frac{\text{mass solute}}{\text{total mass solution}} \times 10^6 = \frac{0.0079 \text{ g } Sr^{2+}}{1 \times 10^3 \text{ g } H_2O} \times 10^6 = 7.9 \text{ ppm } Sr^{2+}$

13.36 (a) $\frac{25.5 \text{ g } C_6H_5OH}{94.11 \text{ g/mol}} = 0.2710 = 0.271 \text{ mol } C_6H_5OH$

$$\frac{495 \text{ g } CH_3CH_2OH}{46.07 \text{ g/mol}} = 10.7445 = 10.7 \text{ mol } CH_3CH_2OH$$

$$\chi_{C_6H_5OH} = \frac{0.2710}{0.2710 + 10.7445} = 0.02460 = 0.0246$$

(b) $\text{mass \%} = \frac{25.5 \text{ g } C_6H_5OH}{25.5 \text{ g } C_6H_5OH + 495 \text{ g } CH_3CH_2OH} \times 100 = 4.90\% \ C_6H_5OH$

(c) $\text{m} = \frac{0.2710 \text{ mol } C_6H_5OH}{0.495 \text{ kg } CH_3CH_2OH} = 0.54747 = 0.547 \text{ m } C_6H_5OH$

13.38 (a) $M = \frac{\text{mol solute}}{\text{L soln}}; \frac{25.0 \text{ g } Al_2(SO_4)_3}{0.350 \text{ L soln}} \times \frac{1 \text{ mol } Al_2(SO_4)_3}{342.2 \text{ g } Al_2(SO_4)_3} = 0.209\ M \ AL_2(SO_4)_3$

(b) $\frac{5.25 \text{ g } Mn(NO_3)_2 \bullet 2H_2O}{0.175 \text{ L soln}} \times \frac{1 \text{ mol } Mn(NO_3)_2 \bullet 2H_2O}{215.0 \text{ g } Mn(NO_3)_2 \bullet 2H_2O} = 0.140\ M \ Mn(NO_3)_2$

(c) $M_c \times L_c = M_d \times L_d$; $9.00\ M \ H_2SO_4 \times 0.0350 \text{ L} = ?M \ H_2SO_4 \times 0.500 \text{ L}$

500 mL of 0.630 M H_2SO_4

13.40 (a) $16.0 \text{ mol } H_2O \times \frac{18.02 \text{ g } H_2O}{1 \text{ mol } H_2O} = 288.3 \text{ g } H_2O = 0.288 \text{ kg } H_2O$

$$m = \frac{1.50 \text{ mol KCl}}{0.2883 \text{ kg } H_2O} = 5.2026 = 5.20\ m \text{ KCl}$$

(b) $m = \frac{\text{mol solute}}{\text{kg solute}}$; $\text{mol } S_8 = m \times \text{kg } C_{10}H_8 = 0.12\, m \times 0.1000 \text{ kg } C_{10}H_8 = 0.012 \text{ mol}$

$$0.012 \text{ mol } S_8 \times \frac{256.5 \text{ g } S_8}{1 \text{ mol } S_8} = 3.078 = 3.1 \text{ g } S_8$$

13.42 (a) $\text{mass \%} = \frac{\text{mass } C_6H_8O_6}{\text{total mass solution}} \times 100$;

$$\frac{80.5 \text{ g } C_6H_8O_6}{80.5 \text{ g } C_6H_8O_6 + 210 \text{ g } H_2O} \times 100 = 27.71 = 27.7\%\ C_6H_8O_6$$

(b) $\text{mol } C_6H_8O_6 = \frac{80.5 \text{ g } C_6H_8O_6}{176.1 \text{ g/mol}} = 0.4571 = 0.457 \text{ mol } C_6H_8O_6$

$$\text{mol } H_2O = \frac{210 \text{ g } H_2O}{18.02 \text{ g/mol}} = 11.654 = 11.7 \text{ mol } H_2O$$

$$\chi_{C_6H_8O_6} = \frac{0.4571 \text{ mol } C_6H_8O_6}{0.4571 \text{ mol } C_6H_8O_6 + 11.654 \text{ mol } H_2O} = 0.0377$$

(c) $m = \frac{0.4571 \text{ mol } C_6H_8O_6}{0.210 \text{ kg } H_2O} = 2.18\, m\ C_6H_8O_6$

(d) $M = \frac{\text{mol } C_6H_8O_6}{\text{L solution}}$; $290.5 \text{ g soln} \times \frac{1 \text{ mL}}{1.22 \text{ g}} \times \frac{1 \text{ L}}{1000 \text{ mL}} = 0.2381 = 0.238 \text{ L}$

$$M = \frac{0.4571 \text{ mol } C_6H_8O_6}{0.2381 \text{ L soln}} = 1.92\, M\ C_6H_8O_6$$

13.44 Given: $10.0 \text{ g } C_4H_4S$, 1.065 g/mL; $250.0 \text{ mL } C_7H_8$, 0.867 g/mL

(a) $\text{mol } C_4H_4S = 10.0 \text{ g } C_4H_4S \times \frac{1 \text{ mol } C_4H_4S}{84.15 \text{ g } C_4H_4S} = 0.1188 = 0.119 \text{ mol } C_4H_4S$

$$\text{mol } C_7H_8 = \frac{0.867 \text{ g}}{1 \text{ mL}} \times 250.0 \text{ mL} \times \frac{1 \text{ mol } C_7H_8}{92.14 \text{ g } C_7H_8} = 2.352 = 2.35 \text{ mol}$$

$$\chi_{C_4H_4S} = \frac{0.1188 \text{ mol } C_4H_4S}{0.1188 \text{ mol } C_4H_4S + 2.352 \text{ mol } C_7H_8} = 0.04809 = 0.0481$$

(b) $m_{C_4H_4S} = \frac{\text{mol } C_4H_4S}{\text{kg } C_7H_8}$; $250.0 \text{ mL} \times \frac{0.867 \text{ g}}{1 \text{ mL}} \times \frac{1 \text{ kg}}{1000 \text{ g}} = 0.2168 = 0.217 \text{ kg } C_7H_8$

$$m_{C_4H_4S} = \frac{0.1188 \text{ mol } C_4H_4S}{0.2168 \text{ kg } C_7H_8} = 0.548\, m\ C_4H_4S$$

(c) $10.0 \text{ g } C_4H_4S \times \frac{1 \text{ mL}}{1.065 \text{ g}} = 9.390 = 9.39 \text{ mL } C_4H_4S$;

$$V_{soln} = 9.39 \text{ mL } C_4H_4S + 250.0 \text{ mL } C_7H_8 = 259.4 \text{ mL}$$

$$M_{C_4H_4S} = \frac{0.1188 \text{ mol } C_4H_4S}{0.2594 \text{ L soln}} = 0.458\, M\ C_4H_4S$$

13.46 (a) $\frac{1.50 \text{ mol } HNO_3}{1 \text{ L soln}} \times 0.245 \text{ L} = 0.3675 = 0.368 \text{ mol } HNO_3$

(b) Assume that for dilute aqueous solutions, the mass of the solvent is the mass of solution.

$$\frac{1.25 \text{ mol NaCl}}{1 \text{ kg H}_2\text{O}} \times \frac{\text{x mol}}{50.0 \times 10^{-6} \text{ kg}}; \text{x} = 6.25 \times 10^{-5} \text{ mol NaCl}$$

(c) $$\frac{1.50 \text{ g C}_{12}\text{H}_{22}\text{O}_{11}}{100 \text{ g soln}} = \frac{\text{x g C}_{12}\text{H}_{22}\text{O}_{11}}{124.0 \text{ g soln}}; \text{x} = 1.125 = 1.13 \text{ g C}_{12}\text{H}_{22}\text{O}_{11}$$

$$1.125 \text{ g C}_{12}\text{H}_{22}\text{O}_{11} \times \frac{1 \text{ mol C}_{12}\text{H}_{22}\text{O}_{11}}{342.3 \text{ g C}_{12}\text{H}_{22}\text{O}_{11}} = 3.287 \times 10^{-3} = 3.29 \times 10^{-3} \text{ mol C}_{12}\text{H}_{22}\text{O}_{11}$$

13.48 (a) $$\frac{0.110 \text{ mol (NH}_4)_2\text{SO}_4}{1 \text{ L soln}} \times 1.50 \text{ L} \times \frac{132.2 \text{ g (NH}_4)_2\text{SO}_4}{1 \text{ mol (NH}_4)_2\text{SO}_4} = 21.81 = 21.8 \text{ g (NH}_4)_2\text{SO}_4$$

Weigh 21.8 g $(NH_4)_2SO_4$, dissolve in a small amount of water, continue adding water with thorough mixing up to a total solution volume of 1.50 L.

(b) Determine the mass fraction of Na_2CO_3 in the solution:

$$\frac{0.65 \text{ mol Na}_2\text{CO}_3}{1000 \text{ g H}_2\text{O}} \times \frac{106.0 \text{ g Na}_2\text{CO}_3}{1 \text{ mol Na}_2\text{CO}_3} = 68.9 \text{ g} = \frac{69 \text{ g Na}_2\text{CO}_3}{1000 \text{ g H}_2\text{O}}$$

$$\text{mass fraction} = \frac{68.9 \text{ g Na}_2\text{CO}_3}{1000 \text{ g H}_2\text{O} + 68.9 \text{ g Na}_2\text{CO}_3} = 0.06446 = 0.064$$

In 120 g of solution, there are 0.06446(120) = 7.735 = 7.7 g Na_2CO_3.

Weigh out 7.7 g Na_2CO_3 and dissolve it in 120 - 7.7 = 112.3 g H_2O to make exactly 120 g of solution.

(112.3 g H_2O/0.997 g H_2O/mL @ 25° = 112.6 mL H_2O)

(c) $$1.20 \text{ L} \times \frac{1000 \text{ mL}}{1 \text{ L}} \times \frac{1.16 \text{ g}}{1 \text{ mL}} = 1392 \text{ g solution; } 0.150(1392 \text{ g soln}) = 209 \text{ g Pb(NO}_3)_2$$

Weigh 209 g $Pb(NO_3)$ and add (1392 – 209) = 1183 g H_2O to make exactly (1392 = 1.39×10^3) g or 1.20 L of solution.

(1183 g H_2O/0.997 g/mL @ 25°C = 1187 mL H_2O)

(d) Calculate the mol HCl necessary to neutralize 5.5 g $Ba(OH)_2$.

$Ba(OH)_2(s) + 2HCl(aq) \rightarrow BaCl_2(aq) + 2H_2O(l)$

$$5.5 \text{ g Ba(OH)}_2 + \frac{1 \text{ mol Ba(OH)}_2}{171 \text{ g Ba(OH)}_2} \times \frac{2 \text{ mol HCl}}{1 \text{ mol Ba(OH)}_2} = 0.0643 = 0.064 \text{ mol HCl}$$

$$M = \frac{\text{mol}}{\text{L}}; \text{L} = \frac{\text{mol}}{M} = \frac{0.0643 \text{ mol HCl}}{0.50\ M \text{ HCl}} = 0.1287 = 0.13 \text{ L} = 130 \text{ mL}$$

130 mL of 0.50 M HCl are needed.

$M_c \times L_c = M_d \times L_d$; $6.0\ M \times L_c = 0.50\ M \times 0.1287$ L; $L_c = 0.01072$ L = 11 mL

Using a pipette, measure exactly 11 mL of 6.0 M HCl and dilute with water to a total volume of 130 mL.

13.50 *Analyze/Plan.* Assume 1.00 L of solution. Calculate mass of 1 L of solution using density. Calculate mass of NH_3 using mass %, then mol NH_3 in 1.00 L. *Solve.*

$$1.00 \text{ L soln} \times \frac{1000 \text{ mL}}{1 \text{ L}} \times \frac{0.90 \text{ g soln}}{1 \text{ mL soln}} = 9.0 \times 10^2 \text{ g soln/L}$$

$$\frac{900 \text{ g soln}}{1.00 \text{ L soln}} \times \frac{28 \text{ g } NH_3}{100 \text{ g soln}} \times \frac{1 \text{ mol } NH_3}{17.03 \text{ g } NH_3} = 14.80 = 15 \text{ mol } NH_3\text{/L soln} = 15\ M\ NH_3$$

13.52 (a) $$\frac{0.0750 \text{ mol } C_8H_{10}N_4O_2}{1 \text{ kg } CHCl_3} \times \frac{194.2 \text{ g } C_8H_{10}N_4O_2}{1 \text{ mol } C_8H_{10}N_4O_2} = 14.565$$

$$= 14.6 \text{ g } C_8H_{10}N_4O_2\text{/kg } CHCl_3$$

$$\frac{14.565 \text{ g } C_8H_{10}N_4O_2}{14.565 \text{ g } C_8H_{10}N_4O_2 + 1000.00 \text{ g } CHCl_3} \times 100 = 1.436 = 1.44\%\ C_8H_{10}N_4O_2 \text{ by}$$
mass

(b) $$1000 \text{ g } CHCl_3 \times \frac{1 \text{ mol } CHCl_3}{119.4\ CHCl_3} = 8.375 = 8.38 \text{ mol } CHCl_3$$

$$\chi_{C_8H_{10}N_4O_2} = \frac{0.0750}{0.0750 + 8.375} = 0.00888$$

13.54 (a) For gases at the same temperature and pressure, volume % = mol %. The volume and mol % of CO_2 in this breathing air is 4.0%.

(b) $$P_{CO_2} = \chi_{CO_2} \times P_t = 0.040\,(1 \text{ atm}) = 0.040 \text{ atm}$$

$$M_{CO_2} = \frac{P_{CO_2}}{RT} = \frac{0.040 \text{ atm}}{310 \text{ K}} \times \frac{\text{K} \bullet \text{mol}}{0.08206 \text{ L} \bullet \text{atm}} = 1.6 \times 10^{-3}\ M$$

Colligative Properties

13.56 (a) decrease (b) decrease
(c) increase (d) increase

13.58 (a) An *ideal solution* is a solution that obeys Raoult's Law.

(b) *Analyze/Plan.* Calculate the vapor pressure predicted by Raoult's law and compare it to the experimental vapor pressure. Assume ethylene glycol (eg) is the solute. *Solve.*

$$\chi_{H_2O} = \chi_{eg} = 0.500;\ P_A = \chi_A P_A{}^\circ = 0.500(149) \text{ mm Hg} = 74.5 \text{ mm Hg}$$

The experimental vapor pressure (P_A), 67 mm Hg, is less than the value predicted by Raoult's law for an ideal solution. The solution is not ideal.

Check. An ethylene glycol-water solution has extensive hydrogen bonding, which causes deviation from ideal behavior. We expect the experimental vapor pressure to be less than the ideal value and it is.

13.60 (a) H_2O vapor pressure will be determined by the mole fraction of H_2O in the solution. The vapor pressure of pure H_2O at 343 K (70°C) = 233.7 torr.

$$\frac{35.0 \text{ g } C_3H_8O_3}{92.10 \text{ g/mol}} = 0.3800 = 0.380 \text{ mol}; \quad \frac{125 \text{ g } H_2O}{18.02 \text{ g/mol}} = 6.937 = 6.94 \text{ mol}$$

$$P_{H_2O} = \frac{6.937 \text{ mol } H_2O}{6.937 + 0.380} \times 233.7 \text{ torr} = 221.6 = 222 \text{ torr}$$

(b) Calculate χ_B by vapor pressure lowering; $\chi_B = \Delta P_A / P_A°$ (see Solution 13.59(b)). Given moles solvent, calculate moles solute from the definition of mole fraction.

$$\chi_{C_2H_6O_2} = \frac{10.0 \text{ torr}}{100 \text{ torr}} = 0.100$$

$$\frac{1.00 \times 10^3 \text{ g } C_2H_5OH}{46.07 \text{ g/mol}} = 21.71 = 21.7 \text{ mol } C_2H_5OH; \text{ let y = mol } C_2H_6O_2$$

$$\chi_{C_2H_6O_2} = \frac{\text{y mol } C_2H_6O_2}{\text{y mol } C_2H_6O_2 + 21.71 \text{ mol } C_2H_5OH} = 0.100 = \frac{y}{y + 21.71}$$

$0.100 \text{ y} + 2.171 = \text{y}; 0.900 \text{ y} = 2.171; \text{y} = 2.412 = 2.41 \text{ mol } C_2H_6O_2$

$$2.412 \text{ mol } C_2H_6O_2 \times \frac{62.07 \text{ g}}{1 \text{ mol}} = 150 \text{ g } C_2H_6O_2$$

13.62 (a) Since C_6H_6 and C_7H_8 form an ideal solution, we can use Raoult's Law. Since both components are volatile, both contribute to the total vapor pressure of 35 torr.

$$P_t = P_{C_6H_6} + P_{C_7H_8}; P_{C_6H_6} = \chi_{C_6H_6} P^o_{C_6H_6}; P_{C_7H_8} = \chi_{C_7H_8} P^o_{C_7H_8}$$

$$\chi_{C_7H_8} = 1 - \chi_{C_6H_6}; P_T = \chi_{C_6H_6} P^o_{C_6H_6} + (1 - \chi_{C_6H_6}) P^o_{C_7H_8}$$

$$35 \text{ torr} = \chi_{C_6H_6}(75 \text{ torr}) + (1 - \chi_{C_6H_6}) 22 \text{ torr}$$

$$13 \text{ torr} = 53 \text{ torr}(\chi_{C_6H_6}); \chi_{C_6H_6} = \frac{13 \text{ torr}}{53 \text{ torr}} = 0.2453 = 0.25; \quad \chi_{C_7H_8} = 0.7547 = 0.75$$

(b) $P_{C_6H_6} = 0.2453(75 \text{ torr}) = 18.4 \text{ torr}; \; P_{C_7H_8} = 0.7547(22 \text{ torr}) = 16.6 \text{ torr}$

In the vapor, $\chi_{C_6H_6} = \frac{P_{C_6H_6}}{P_t} = \frac{18.4 \text{ torr}}{35 \text{ torr}} = 0.53; \quad \chi_{C_7H_8} = 0.47$

13.64 *Analyze/Plan.* ΔT_b depends on mol dissolved particles. Assume 100 g of each solution, calculate mol solute and mol dissolved particles. Glucose and sucrose are molecular solutes, but $NaNO_3$ dissociates into 2 mol particles per mol solute. *Solve.*

10% by mass means 10 g solute in 100 g solution. If we have 10 g of each solute, the one with the smallest molar mass will have the largest mol solute. The molar masses are: glucose, 180.2 g/mol; sucrose, 342.3 g/mol; $NaNO_3$, 85.0 g/mol. $NaNO_3$ has most mol solute, and twice as many dissolved particles, so it will have the highest boiling point. Sucrose has least mol solute and lowest boiling point. Glucose is intermediate.

In order of increasing boiling point: 10% sucrose < 10% glucose < 10% $NaNO_3$.

13.66 0.030 *m* phenol > 0.040 *m* glycerin = 0.020 *m* KBr. Phenol is very slightly ionized in water, but not enough to match the number of particles in a 0.040 *m* glycerin solution. The KBr solution is 0.040 *m* in particles, so it has the same freezing point as 0.040 *m* glycerin, which is a nonelectrolyte.

13.68 $\Delta T = K(m)$; first calculate the **molality** of the solute particles.

(a) 0.40 m

(b) $$\frac{20.0\ g\ C_{10}H_{22}}{0.455\ kg\ CHCl_3} \times \frac{1\ mol\ C_{10}H_{22}}{142.3\ g\ C_{10}H_{22}} = 0.3089 = 0.309\ m$$

(c) $$m = \frac{0.45\ mol\ eg + 2(0.15)\ mol\ KBr}{0.150\ kg\ H_2O} = \frac{0.75\ mol\ particles}{0.150\ kg\ H_2O} = 5.0\ m$$

Then, f.p. = $T_f - K_f(m)$; b.p. = $T_b + K_b(m)$; T in °C

	m	T_f	$-K_f(m)$	f.p.	T_b	$+K_b(m)$	b.p.
(a)	0.40	−114.6	−1.99(0.40) = −0.80	−115.4	78.4	1.22(0.40) = 0.49	78.9
(b)	3.09	−63.5	−4.68(3.09) = −14.5	−78.0	61.2	3.63(3.09) = 11.2	72.4
(c)	5.0	0.0	−1.86(5.0) = −9.3	−9.3	100.0	0.51(5.0) = 2.6	102.6

13.70 $\pi = MRT$; T = 20°C + 273 = 293 K

$$M\ (\text{of ions}) = \frac{mol\ NaCl \times 2}{L\ soln} = \frac{3.4\ g\ NaCl}{1\ L\ soln} \times \frac{1\ mol\ NaCl}{58.4\ g\ NaCl} \times \frac{2\ mol\ ions}{1\ mol\ NaCl} = 0.116 = 0.12\ M$$

$$\pi = \frac{0.116\ mol}{L} \times \frac{0.08206\ L \bullet atm}{K \bullet mol} \times 293\ K = 2.8\ atm$$

13.72 $$\Delta T_f = 5.5 - 4.1 = 1.4;\quad m = \frac{\Delta T_f}{K_f} = \frac{1.4}{5.12} = 0.273 = 0.27\ m$$

$$\text{MM lauryl alcohol} = \frac{g\ lauryl\ alcohol}{m \times kg\ C_6H_6} = \frac{5.00\ g\ lauryl\ alcohol}{0.273 \times 0.100\ kg\ C_6H_6}$$
$$= 1.8 \times 10^2\ g/mol\ lauryl\ alcohol$$

13.74 $$M = \pi/RT = \frac{0.605\ atm}{298\ K} \times \frac{mol \bullet K}{0.08206\ L \bullet atm} = 0.02474 = 0.0247\ M$$

$$MM = \frac{g}{M \times L} = \frac{2.35\ g}{0.02474\ M \times 0.250\ L} = 380\ g/mol$$

13.76 If these were ideal solutions, they would have equal ion concentrations and equal ΔT_f values. Data in Table 13.5 indicates that the van't Hoff factors (i) for both salts are less than the ideal values. For 0.030 m NaCl, i is between 1.87 and 1.94, about 1.92. For 0.020 m K_2SO_4, i is between 2.32 and 2.70, about 2.62. From Equation 13.14,

ΔT_f (measured) = $i \times \Delta T_f$ (calculated for nonelectrolyte)

NaCl: ΔT_f (measured) = $1.92 \times 0.030\ m \times 1.86\ °C/m = 0.11\ °C$

K_2SO_4: ΔT_f (measured) = $2.62 \times 0.020\ m \times 1.86°C/m = 0.097\ °C$

0.030 m NaCl would have the larger ΔT_f.

The deviations from ideal behavior are due to ion-pairing in the two electrolyte solutions. K_2SO_4 has more extensive ion-pairing and a larger deviation from ideality because of the higher charge on SO_4^{2-} relative to Cl^-.

Colloids

13.78 (a) Suspensions are classified as solutions or colloids according to the size of the dispersed particles. Solute particles have diameters less than 10 Å. Clearly a protein with a molecular mass of 30,000 amu will be longer than 10 Å. The aqueous suspensions are colloids because of the size of protein molecules.

(b) Emulsion. An emulsifying agent is one that aids in the formation of an emulsion. It usually has a polar part and a nonpolar part, to facilitate mixing of immiscible liquids with very different molecular polarities.

13.80 (a) When the colloid *particle mass* becomes large enough so that gravitational and interparticle attractive forces are greater than the kinetic energies of the particles, settling and aggregation can occur.

(b) *Hydrophobic* colloids do not attract a sheath of water molecules around them and thus tend to aggregate from aqueous solution. They can be stabilized as colloids by adsorbing charges on their surfaces. The charged particles interact with solvent water, stabilizing the colloid.

(c) *Charges on colloid particles* can stabilize them against aggregation. Particles carrying like charges repel one another and are thus prevented from aggregating and settling out.

13.82 (a) The nonpolar hydrophobic tails of soap particles (the hydrocarbon chain of stearate ions) establish attractive intermolecular dispersion forces with the nonpolar oil molecules, while the charged hydrophilic head of the soap particles interacts with H_2O to keep the oil molecules suspended. (This is the mechanism by which laundry detergents remove greasy dirt from clothes.)

(b) Electrolytes from the acid neutralize surface charges of the suspended particles in milk, causing the colloid to coagulate.

Additional Exercises

13.84 In this equilibrium system, molecules move from the surface of the solid into solution, while molecules in solution are deposited on the surface of the solid. As molecules leave the surface of the small particles of powder, the reverse process preferentially deposits other molecules on the surface of a single crystal. Eventually, all molecules that were present in the 50 g of powder are deposited on the surface of a 50 g crystal; this can only happen if the dissolution and deposition processes are ongoing.

13.86 (a) $C_{Rn} = kP_{Rn}; k = C_{Rn}/P_{Rn} = 7.27 \times 10^{-3}$ M/1 atm $= 7.27 \times 10^{-3}$ mol/L•atm

(b) $P_{Rn} = \chi_{Rn}P_{total}; P_{Rn} = 3.5 \times 10^{-6}$ (32 atm) $= 1.12 \times 10^{-4} = 1.1 \times 10^{-4}$ atm

$$S_{Rn} = k\,P_{Rn}; S_{Rn} = \frac{7.27 \times 10^{-3} \text{ mol}}{\text{L} \bullet \text{atm}} \times 1.12 \times 10^{-4} \text{ atm} = 8.1 \times 10^{-7}\, M$$

13.87 0.10% by mass means 0.10 g glucose/100 g blood.

(a) $$\text{ppm glucose} = \frac{\text{g glucose}}{\text{g solution}} \times 10^6 = \frac{0.10 \text{ g glucose}}{100 \text{ g blood}} \times 10^6 = 1000 \text{ ppm glucose}$$

(b) m = mol glucose/kg solvent. Assume that the mixture of nonglucose components is the 'solvent'.

mass solvent = 100 g blood – 0.10 g glucose = 99.9 g solvent = 0.0999 kg solvent

$$\text{mol glucose} = 0.10\,\text{g} \times \frac{1\,\text{mol}}{180.2\,\text{g}\,C_6H_{12}O_6} = 5.55 \times 10^{-4} = 5.6 \times 10^{-4}\ \text{mol glucose}$$

$$m = \frac{5.55 \times 10^{-4}\ \text{mol glucose}}{0.0999\ \text{kg solvent}} = 5.6 \times 10^{-3}\ m\ \text{glucose}$$

In order to calculate molarity, solution volume must be known. The density of blood is needed to relate mass and volume.

13.89 Both solutions have 15 g alcohol per 100 g solution. If the densities are equal, then equal volumes of the two solutions contain equal masses of solute.

(a) M = mol solute/L solution. The molar mass of ethanol, CH_3CH_2OH, is less than the molar mass of propanol, $CH_3CH_2CH_2OH$. In equal volumes of the two solutions with equal masses of solute, the ethanol solution will contain more moles of solute particles and have the greater molarity.

(b) m = mol solute/kg solvent. For equal solution mass and equal solute mass, solvent mass must also be equal. According to part (a), there are more mol solute in the ethanol solution. So it also has the greater molality.

(c) χ = mol solute/total mol. We can express the two ratios as:

$$\chi_E = \frac{\text{mol E}}{\text{mol E} + \text{mol solv}}; \chi_P = \frac{\text{mol P}}{\text{mol P} + \text{mol solv}}$$

We have established that mol E > mol P, and that mol solv is the same in both solutions. The question is, are the mole fraction ratios equal?

No, because mol solv is constant in both denominators. Even though the larger (mol E) appears in both numerator and denominator of χ_E, the value of (mol solv) is not proportionally larger, and the value of χ_E is higher.

13.90 (a) $$\frac{1.80\ \text{mol LiBr}}{1\ \text{L soln}} \times \frac{86.85\ \text{g LiBr}}{1\ \text{mol LiBr}} = 156.3 = 156\ \text{g LiBr}$$

1 L soln = 826 g soln; g CH_3CN = 826 – 156.3 = 669.7 = 670 g CH_3CN

$$m\ \text{LiBr} = \frac{1.80\ \text{mol LiBr}}{0.6697\ \text{kg}\ CH_3CN} = 2.69\ m$$

(b) $$\frac{669.7\ \text{g}\ CH_3CN}{41.05\ \text{g/mol}} = 16.31 = 16.3\ \text{mol}\ CH_3CN;\ \chi_{LiBr} = \frac{1.80}{1.80 + 16.31} = 0.0994$$

(c) $$\text{mass \%} = \frac{669.7\ \text{g}\ CH_3CN}{826\ \text{g soln}} \times 100 = 81.1\%\ CH_3CN$$

13.92 Mole fraction ethyl alcohol, $$\chi_{C_2H_5OH} = \frac{P_{C_2H_5OH}}{P^\circ_{C_2H_5OH}} = \frac{8\ \text{torr}}{100\ \text{torr}} = 0.08$$

$$\frac{620 \times 10^3 \text{ g } C_{24}H_{50}}{338.6 \text{ g/mol}} = 1.83 \times 10^3 \text{ mol } C_{24}H_{50}; \quad \text{let y = mol } C_2H_5OH$$

$$\chi_{C_2H_5OH} = 0.08 = \frac{y}{y + 1.83 \times 10^3}; \; 0.92\,y = 146.4; \; y = 1.6 \times 10^2 \text{ mol } C_2H_5OH$$

(Strictly speaking, y should have 1 sig fig because 0.08 has 1 sig fig, but this severely limits the calculation.)

$$1.6 \times 10^2 \text{ mol } C_2H_5OH \times \frac{46 \text{ g } C_2H_5OH}{1 \text{ mol}} = 7.4 \times 10^3 \text{ g or } 7.4 \text{ kg } C_2H_5OH$$

13.93 (a) The solvent vapor pressure over each solution is determined by the total particle concentrations present in the solutions. When the particle concentrations are equal, the vapor pressures will be equal and equilibrium established. The particle concentration of the nonelectrolyte is just 0.050 *M*, the ion concentration of the NaCl is 2 × 0.035 *M* = 0.070 *M*. Solvent will diffuse from the less concentrated nonelectrolyte solution. The level of the NaCl solution will rise, and the level of the nonelectrolyte solution will fall.

(b) Let x = volume of solvent transferred

$$\frac{0.050\,M \times 30.0 \text{ mL}}{(30.0 - x) \text{ mL}} = \frac{0.070\,M \times 30.0 \text{ mL}}{(30.0 + x) \text{ mL}}; \; 1.5(30.0 + x) = 2.1(30.0 - x)$$

45 + 1.5 x = 63 - 2.1 x; 3.6 x = 18; x = 5.0 = 5 mL transferred

The volume in the nonelectrolyte beaker is (30.0 – 5.0) = 25.0 mL; in the NaCl beaker (30.0 + 5.0) = 35.0 mL.

13.94 (a) 0.100 *m* K_2SO_4 is 0.300 *m* in particles. H_2O is the solvent.

$\Delta T_f = K_f m = -1.86(0.300) = -0.558$; $T_f = 0.0 - 0.558 = -0.558°C = -0.6°C$

(b) ΔT_f (nonelectrolyte) = -1.86(0.100) = -0.186; $T_f = 0.0 - 0.186 = -0.186°C = -0.2°C$

T_f (measured) = $i \times T_f$ (nonelectrolyte)

From Table 13.5, *i* for 0.100 *m* K_2SO_4 = 2.32

T_f (measured) = 2.32(–0.186°C) = –0.432°C = –0.4°C

13.96 (a) $K_b = \frac{\Delta T_b}{m}$; $\Delta T_b = 47.46°C - 46.30°C = 1.16°C$

$$m = \frac{\text{mol solute}}{\text{kg } CS_2} = \frac{0.250 \text{ mol}}{400.0 \text{ mL } CS_2} \times \frac{1 \text{ mL } CS_2}{1.261 \text{ g } CS_2} \times \frac{1000 \text{ g}}{1 \text{ kg}} = 0.4956 = 0.496\,m$$

$$K_b = \frac{1.16°C}{0.4956\,m} = 2.34°C/m$$

(b) $$m = \frac{\Delta T_b}{K_b} = \frac{(47.08 - 46.30)°C}{2.34°C/m} = 0.333 = 0.33\,m$$

$$m = \frac{\text{mol unknown}}{\text{kg } CS_2}; \quad m \times \text{kg } CS_2 = \frac{\text{g unknown}}{\text{MM unknown}}; \quad \text{MM} = \frac{\text{g unknown}}{m \times \text{kg } CS_2}$$

$$50.0\ \text{mL } CS_2 \times \frac{1.261\ \text{g } CS_2}{1\ \text{mL}} \times \frac{1\ \text{kg}}{1000\ \text{g}} = 0.06305 = 0.0631\ \text{kg } CS_2$$

$$MM = \frac{5.39\ \text{g unknown}}{0.333\ m \times 0.06305\ \text{kg } CS_2} = 257 = 2.6 \times 10^2\ \text{g/mol}$$

13.97 (a) Assume 1000 g of solution. $1000\ \text{g soln} \times \frac{1\ \text{mL}}{1.22\ \text{g}} = 819.7 = 820\ \text{mL}$

$$1000\ \text{g soln} \times \frac{40.0\ \text{g KSCN}}{100\ \text{g soln}} = 400\ \text{g KSCN};\ 1000\ \text{g soln} - 400\ \text{g KSCN} = 600\ \text{g } H_2O$$

$$\frac{400\ \text{g KSCN}}{97.19\ \text{g/mol}} = 4.116 = 4.12\ \text{mol KSCN};\ \frac{600\ \text{g } H_2O}{18.02\ \text{g/mol}} = 33.30 = 33.3\ \text{mol } H_2O$$

$$\chi_{KSCN} = \frac{4.116}{4.116 + 33.30} = 0.110;\ m = \frac{4.116\ \text{mol KSCN}}{0.600\ \text{kg } H_2O} = 6.86\ m$$

$$M = \frac{4.116\ \text{mol KSCN}}{0.8197\ \text{L}} = 5.02\ M$$

(b) If there are 4.12 mol of KSCN, there are 8.24 moles of ions. There are then 33.3 mol H_2O/8.24 mol ions ≈ 4 mol H_2O for each mol of ions, or 4 water molecules for each ion. This is too few water molecules to completely hydrate the anions and cations in the solution.

For a solution that is this concentrated, one would expect significant ion-pairing, because the ions are not completely surrounded and separated by H_2O molecules. Because of ion-pairing, the effective number of particles will be less than that indicated by m and M, so the observed colligative properties will be significantly different from those predicted by formulas for ideal solutions. The observed freezing point will be higher, the boiling point lower, and the osmotic pressure lower than predicted.

Integrative Exercises

13.100 $$\frac{0.015\ \text{g } N_2}{1\ \text{L blood}} \times \frac{1\ \text{mol } N_2}{28.01\ \text{g } N_2} = 5.355 \times 10^{-4} = 5.4 \times 10^{-4}\ \text{mol } N_2\text{/L blood}$$

At 100 ft, the partial pressure of N_2 in air is 0.78 (4.0 atm) = 3.12 atm. This is just four times the partial pressure of N_2 at 1.0 atm air pressure. According to Henry's law, $S_g = kP_g$, a 4-fold increase in P_g results n a 4-fold increase in S_g, the solubility of the gas. Thus, the solubility of N_2 at 100 ft is $4(5.355 \times 10^{-4}\ M) = 2.142 \times 10^{-3} = 2.1 \times 10^{-3}\ M$. If the diver suddenly surfaces, the amount of N_2/L blood released is the difference in the solubilities at the two depths: $(2.142 \times 10^{-3}\ \text{mol/L} - 5.355 \times 10^{-4}\ \text{mol/L}) = 1.607 \times 10^{-3} = 1.6 \times 10^{-3}\ \text{mol } N_2\text{/L blood}$.

At surface conditions of 1.0 atm external pressure and 37°C = 310 K,

$$V = \frac{nRT}{P} = 1.607 \times 10^{-3}\ \text{mol} \times \frac{310\ \text{K}}{1.0\ \text{atm}} \times \frac{0.08206\ \text{L} \bullet \text{atm}}{\text{mol} \bullet \text{K}} = 0.041\ \text{L}$$

That is, 41 mL of tiny N_2 bubbles are released from each L of blood.

13.101 The stronger the intermolecular forces, the higher the heat (enthalpy) of vaporization.

(a) None of the substances are capable of hydrogen bonding in the pure liquid, and they have similar molar masses. All intermolecular forces are van der Waals forces, dipole-dipole, and dispersion forces. In decreasing order of strength of forces:

acetone > acetaldehyde > ethylene oxide > cyclopropane

The first three compounds have dipole-dipole and dispersion forces, the last only dispersion forces.

(b) The order of solubility in hexane should be the reverse of the order above. The least polar substance, propane, will be most soluble in hexane. Ethanol, CH_3CH_2OH, is capable of hydrogen bonding with the three polar compounds. Thus, acetaldehyde, acetone, and ethylene oxide should be more soluble than cyclopropane, but without further information we cannot distinguish among the polar molecules.

13.102 For ionic solids, the exothermic part of the solution process is step (3), surrounding the separated ions by solvent molecules. The released energy comes from the attractive interaction of the solvent with the separated ions. In hydrates, one or more water molecules are already associated with the ions, reducing the total energy released during solvation.

13.104 (a) $Zn(s) + H_2SO_4(aq) \rightarrow ZnSO_4(aq) + H_2(g)$

$$2.050 \text{ g Zn} \times \frac{1 \text{ mol Zn}}{65.39 \text{ g Zn}} = 0.03135 \text{ mol Zn}$$

$$1.00\ M\ H_2SO_4 \times 0.0150 \text{ L} = 0.0150 \text{ mol } H_2SO_4$$

Since Zn and H_2SO_4 react in a 1:1 mole ratio, H_2SO_4 is the limiting reactant; 0.0150 mol of $H_2(g)$ are produced.

(b) $$P = \frac{nRT}{V} = \frac{0.0150 \text{ mol}}{0.122 \text{ L}} \times \frac{0.08206 \text{ L} \bullet \text{atm}}{\text{mol} \bullet \text{K}} \times 298 \text{ K} = 3.0066 = 3.01 \text{ atm}$$

(c) $$S_{H_2} = kP_{H_2} = \frac{7.8 \times 10^{-4} \text{ mol}}{\text{L} \bullet \text{atm}} \times 3.0066 \text{ atm} = 0.002345 = 2.3 \times 10^{-3}\ M$$

$$\frac{0.002345 \text{ mol } H_2}{\text{L soln}} \times 0.0150 \text{ L} = 3.518 \times 10^{-5} = 3.5 \times 10^{-5} \text{ mol dissolved } H_2$$

$$\frac{3.5 \times 10^{-5} \text{ mol dissolved } H_2}{0.0150 \text{ mol } H_2 \text{ produced}} \times 100 = 0.23\% \text{ dissolved } H_2$$

This is approximately 2.3 ppt; for every 10,000 H_2 molecules, 23 are dissolved. It was reasonable to ignore dissolved $H_2(g)$ in part (b).

13.105 (a) $$\frac{1.3 \times 10^{-3} \text{ mol } CH_4}{\text{L soln}} \times 4.0 \text{ L} = 5.2 \times 10^{-3} \text{ mol } CH_4$$

$$V = \frac{nRT}{P} = \frac{5.2 \times 10^{-3} \text{ mol} \times 273 \text{ K}}{1.0 \text{ atm}} \times \frac{0.08206 \text{ L} \bullet \text{atm}}{\text{K} \bullet \text{mol}} = 0.12 \text{ L}$$

(b) All three hydrocarbons are nonpolar; they have zero net dipole moment. In CH_4 and C_2H_6, the C atoms are tetrahedral and all bonds are σ bonds. C_2H_6 has a higher molar mass than CH_4, which leads to stronger dispersion forces and greater water solubility. In C_2H_4, the C atoms are trigonal planar and the π electron cloud is symmetric above and below the plane that contains all the atoms. The π cloud in C_2H_4 is an area of concentrated electron density that experiences attractive forces with the positive ends of H_2O molecules. These forces increase the solubility of C_2H_4 relative to the other hydrocarbons.

(c) The molecules have similar molar masses. NO is most soluble because it is polar. The triple bond in N_2 is shorter than the double bond in O_2. It is more difficult for H_2O molecules to surround the smaller N_2 molecules, so they are less soluble than O_2 molecules.

(d) H_2S and SO_2 are polar molecules capable of hydrogen bonding with water. Hydrogen bonding is the strongest force between neutral molecules and causes the much greater solubility. H_2S is weakly acidic in water. SO_2 reacts with water to form H_2SO_3, a weak acid. The large solubility of SO_2 is a sure sign that a chemical process has occurred.

(e) N_2 and C_2H_4. N_2 is too small to be easily hydrated, so C_2H_4 is more soluble in H_2O.

NO (31) and C_2H_6 (30). The structures of these two molecules are very different, yet they have similar solubilities. NO is slightly polar, but too small to be easily hydrated. The larger C_2H_6 is nonpolar, but more polarizable (stronger dispersion forces).

NO (31) and O_2 (32). The slightly polar NO is more soluble than the slightly larger (longer O=O bond than $N \equiv O$ bond) but nonpolar O_2.

13.107 (a)

$$\Delta T_f = K_f m = K_f \times \frac{\text{mol } C_7H_6O_2}{\text{kg } C_6H_6} = K_f \times \frac{\text{g } C_7H_6O_2}{\text{kg } C_6H_6 \times \text{M } C_7H_6O_2}$$

$$MM = \frac{K_f \times \text{g } C_7H_6O_2}{\Delta T_f \times \text{kg } C_6H_6} = \frac{5.12 \times 0.55}{0.360 \times 0.032} = 2444.4 = 2.4 \times 10^2 \text{ g/mol}$$

(b) The formula weight of $C_7H_6O_2$ is 122 g/mol. The experimental molar mass is twice this value, indicating that benzoic acid is associated into dimers in benzene solution. This is reasonable, since the carboxyl group, -COOH, is capable of strong hydrogen bonding with itself. Many carboxylic acids exist as dimers in solution.

The structure of benzoic acid dimer in benzene solution is:

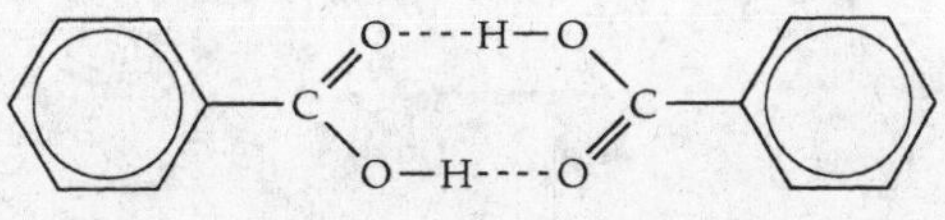

O----H—O
C C
O—H----O

13.108 $\chi_{CHCl_3} = \chi_{C_3H_6O} = 0.500$

(a) For an ideal solution, Raoult's Law is obeyed.

$P_t = P_{CHCl_3} + P_{C_3H_6O}$; $P_{CHCl_3} = 0.5(300 \text{ torr}) = 150 \text{ torr}$

$P_{C_3H_6O} = 0.5(360 \text{ torr}) = 180 \text{ torr}$; $P_t = 150 \text{ torr} + 180 \text{ torr} = 330 \text{ torr}$

(b) The real solution has a lower vapor pressure, 250 torr, than an ideal solution of the same composition, 330 torr. Thus, fewer molecules escape to the vapor phase from the liquid. This means that fewer molecules have sufficient kinetic energy to overcome intermolecular attractions. Clearly, even weak hydrogen bonds such as this one are stronger attractive forces than dipole-dipole or dispersion forces. These hydrogen bonds prevent molecules from escaping to the vapor phase and result in a lower than ideal vapor pressure for the solution. There is essentially no hydrogen bonding in the individual liquids.

(c) According to Coulomb's law, electrostatic attractive forces lead to an overall lowering of the energy of the system. Thus, when the two liquids mix and hydrogen bonds are formed, the energy of the system is decreased and $\Delta H_{soln} < 0$; the solution process is exothermic.

14 Chemical Kinetics

Visualizing Concepts

14.2 Chemical equation (d), $B \rightarrow 2A$, is consistent with the data. The concentration of A increases with time, and concentration B decreases with time, so B must be a reactant and A must be a product. The ending concentration of A is approximately twice as large as the starting concentration of B, so mole ratio of A:B is 2:1. The reaction is $B \rightarrow 2A$.

14.4 *Plan.* For a first-order reaction, a plot of ln[A] vs. time is linear, as shown in the diagram. The slope is $-k$, and the intercept is $[A]_0$. According to the Arrhenius equation [14.19], k increases with increasing temperature. *Solve.*

(a) Lines 1 and 2 have the same slope, and thus the same rate constant, k. These experiments are done at the same temperature. The y-intercepts of the two lines are different; the experiments had different initial concentrations of A.

(b) Lines 2 and 3 have the same y-intercept and thus the same starting concentration of A. The slopes of the two lines are different, so their rate constants are different and they occur at different temperatures. Line 3, with the smaller slope and k value will occur at the lower temperature.

14.6 On a plot of ln k vs. 1/T, the slope is $-E_a$ and the y-intercept is ln A, where E_a is activation energy and A is the frequency factor.

(a) If $E_a(2) > E_a(1)$ and $A_1 = A_2$, the lines will have the same y-intercept, negative slope direction, and the slope of line 2 will be steeper than the slope of line 1.

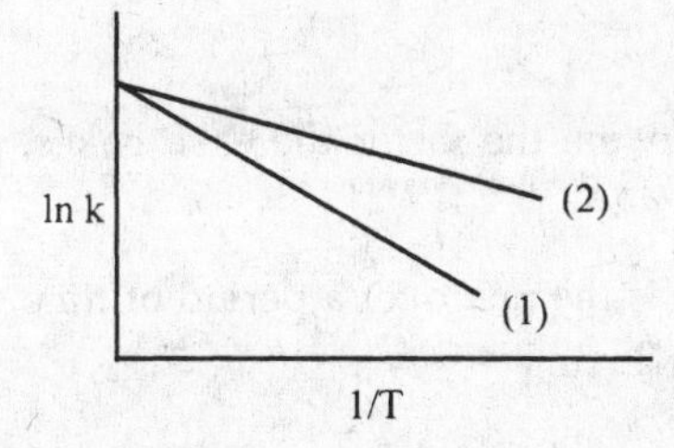

(b) If $A_2 > A_1$ and $E_a(1) = E_a(2)$, the lines will be parallel with the same negative slopes and different y-intercepts.

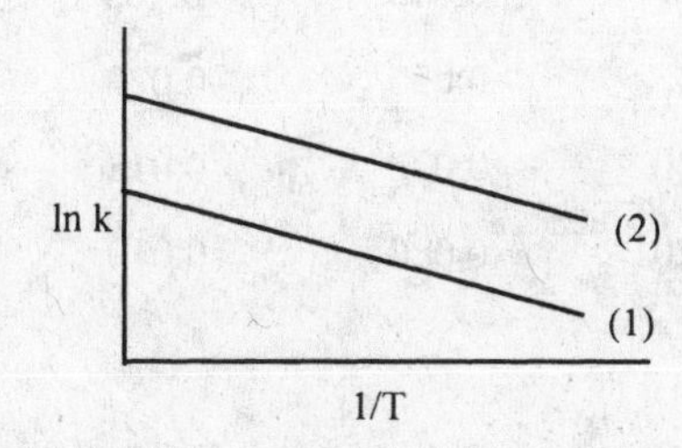

14.7 (a) $NO_2 + F_2 \rightarrow NO_2F + F$

$NO_2 + F \rightarrow NO_2F$

(b) $2NO_2 + F_2 \rightarrow 2NO_2F$

(c) F is an intermediate, because it is produced and then consumed during the reaction.

(d) rate = $k[NO_2][F_2]$

14.9 The most likely transition state shows the relative geometry of both reactants and products. It is reasonable to assume that multiple bonds, with greater total bond energy, reain intact at the expense of single bonds. In the black-and-white diagram below, open circles represent the red balls and closed circles represent the blue.

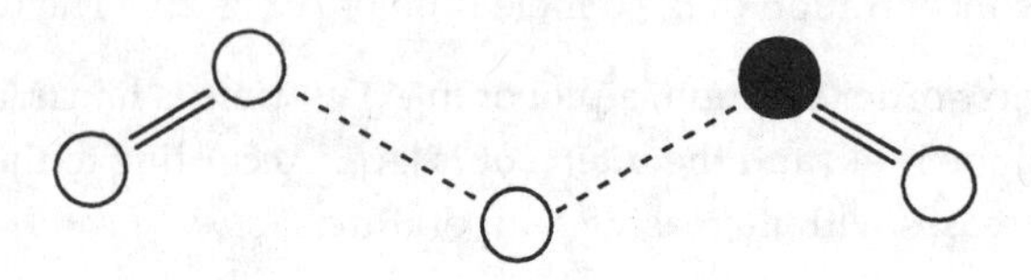

14.10 (a) $A_2 + AB + AC \rightarrow BA_2 + A + AC$

$BA_2 + A + AC \rightarrow A_2 + BA_2 + C$

net: $AB + AC \rightarrow BA_2 + C$

(b) A is the intermediate; it is produced and consumed.

(c) A_2 is the catalyst; it is consumed and reproduced.

Reaction Rates

14.12 (a) M/s

(b) The hotter the oven, the faster the cake bakes. Milk sours faster in hot weather than cool weather.

(c) The *average rate* is the rate over a period of time, while the *instantaneous rate* is the rate at a particular time.

14.14

Time(s)	Mol A	(a) Mol B	Δ Mol A	(b) Rate −(Δ mol A/s)
0	0.100	0.000		
40	0.067	0.033	−0.033	8.3×10^{-4}
80	0.045	0.055	−0.022	5.5×10^{-4}
120	0.030	0.070	−0.015	3.8×10^{-4}
160	0.020	0.080	−0.010	2.5×10^{-4}

(c) The volume of the container must be known to report the rate in units of concentration (mol/L) per time.

14.16 (a)

Time (min)	Time Interval (min)	Concentration (M)	ΔM	Rate (M/s)
0.0		1.85		
54.0	54.0	1.58	-0.27	8.3×10^{-5}
107.0	53.0	1.36	-0.22	6.9×10^{-5}
215.0	108	1.02	-0.34	5.2×10^{-5}
430.0	215	0.580	-0.44	3.4×10^{-5}

(b) From the slopes of the lines in the figure at the right, the rates are: at 75.0 min, 4.2×10^{-3} M/min, or 7.0×10^{-5} M/s; at 250 min, 2.1×10^{-3} M/min or 3.5×10^{-5} M/s

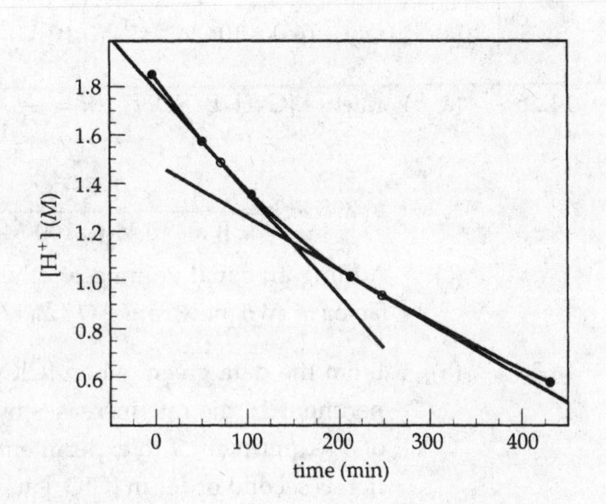

14.18 (a) rate = $-\Delta[HBr]/2\Delta t = \Delta[H_2]/\Delta t = \Delta[Br_2]/\Delta t$

(b) rate = $-\Delta[SO_2]/2\Delta t = -\Delta[O_2]/\Delta t = \Delta[SO_3]/2\Delta t$

(c) rate = $-\Delta[NO]/2\Delta t = -\Delta[H_2]/2\Delta t = \Delta[N_2]/\Delta t = \Delta[H_2O]/2\Delta t$

14.20 (a) $-\Delta[C_2H_4]/\Delta t = \Delta[CO_2]/2\Delta t = \Delta[H_2O]/2\Delta t$

$-2\Delta[C_2H_4]/\Delta t = \Delta[CO_2]/\Delta t = \Delta[H_2O]/\Delta t$

C_2H_4 is burning, $-\Delta[C_2H_4]/\Delta t = 0.37$ M/s

CO_2 and H_2O are produced, at twice the rate that C_2H_4 is consumed.

$\Delta[CO_2]/\Delta t = \Delta[H_2O]/\Delta t = 2(0.37)$ M/s = 0.74 M/s

(b) In this reaction, pressure is a measure of concentration.

$-\Delta[N_2H_4]/\Delta t = -\Delta[H_2]/\Delta t = \Delta[NH_3]/2\Delta t$

N_2H_4 is consumed, $-\Delta[N_2H_4]/\Delta t = 63$ torr/hr

H_2 is consumed, $-\Delta[H_2]/\Delta t = 63$ torr/hr

NH_3 is produced at twice the rate that N_2H_4 and H_2 are consumed,

$\Delta[NH_3]/\Delta t = -2\Delta[N_2H_4]/\Delta t = 2(63)$ torr/hr = 126 torr/hr

$\Delta P_T/\Delta t = (+126$ torr/hr $- 63$ torr/hr $- 63$ torr/hr$) = 0$ torr/hr

Rate Laws

14.22 (a) If [A] doubles, the rate will increase by a factor of four; the rate constant, k, is unchanged. Rate is proportional to $[A]^2$, so when the value of [A] doubles, rate changes by 2^2 or 4. The rate constant, k, is the proportionality constant that does not change (unless the temperature changes).

(b) The reaction is second order in A, first order in B, and third order overall.

(c) Units of $k = \frac{M/s}{M^3} = M^{-2}\,s^{-1}$

14.24 (a) rate = $k[H_2][NO]^2$

(b) rate = $(6.0 \times 10^4\ M^{-2}\,s^{-1})(0.025\ M)^2(0.015\ M) = 0.56\ M/s$

(c) rate = $(6.0 \times 10^4\ M^{-2}\,s^{-1})\ (0.10\ M)^2\ (0.010\ M) = 6.0\ M/s$

14.26 (a, b) rate = $k[C_2H_5Br][OH^-]$; $k = \frac{\text{rate}}{[C_2H_5Br][OH^-]}$

at 298 K, $k = \frac{1.7 \times 10^{-7}\ M/s}{[0.0477\ M][0.100\ M]} = 3.6 \times 10^{-5}\ M^{-1}s^{-1}$

(c) Adding an equal volume of ethyl alcohol reduces both $[C_2H_5Br]$ and $[OH^-]$ by a factor of two. new rate = (1/2)(1/2) = 1/4 of old rate

14.28 (a) From the data given, when $[ClO_2]$ increases by a factor of 3 (experiment 2 to experiment 1), the rate increases by a factor of 9. When $[OH^-]$ increases by a factor of 3 (experiment 2 to experiment 3), the rate increases by a factor of 3. The reaction is second order in $[ClO_2]$ and first order in $[OH^-]$. rate = $k[ClO_2]^2[OH^-]$.

(b) Using data from Expt 2:

$$k = \frac{\text{rate}}{[ClO_2]^2\,[OH^-]} = \frac{0.00276\ M/s}{(0.020\ M)^2\ (0.030\ M)} = 2.3 \times 10^2\ M^{-2}s^{-1}$$

(c) rate = $2.3 \times 10^2\ M^{-2}\,s^{-1}\ (0.010\ M)^2(0.025\ M) = 5.75 \times 10^{-4} = 5.8 \times 10^{-4}\ M/s$

14.30 *Analyze/Plan.* Follow the logic in Sample Exercise 14.6 to deduce the rate law. Rearrange the rate law to solve for k and deduce units. Calculate a k value for each set of concentrations and then average the three values. *Solve.*

(a) Doubling [NO] while holding $[O_2]$ constant increases the rate by a factor of 4 (experiments 1 and 3). Reducing $[O_2]$ by a factor of 2 while holding [NO] constant reduces the rate by a factor of 2 (experiments 2 and 3). The rate is second order in [NO] and first order in $[O_2]$. rate = $k[NO]^2[O_2]$

(b, c) From experiment 1: $k_1 = \frac{1.41 \times 10^{-2}\ M/s}{(0.0126\ M)^2\ (0.0125\ M)} = 7105 = 7.11 \times 10^3\ M^{-2}s^{-1}$

$k_2 = 0.113/(0.0252)^2(0.0250) = 7118 = 7.12 \times 10^3\ M^{-2}\,s^{-1}$

$k_3 = 5.64 \times 10^{-2}/(0.0252)^2(0.125) = 7105 = 7.11 \times 10^3\ M^{-2}\,s^{-1}$

$k_{avg} = (7105 + 7118 + 7105)/3 = 7109 = 7.11 \times 10^3\ M^{-2}\,s^{-1}$

(d) rate = $7.109 \times 10^3\ M^{-2}s^{-1}\ (0.100\ M)^2(0.0200\ M) = 1.422 = 1.42\ M/s$

(e) The data are given in terms of the disappearance of NO. Use Equation 14.4 to relate the disappearance of NO to the disappearance of O_2.

$-\Delta[NO]/2\Delta t = -[O_2]/\Delta t$

For the concentrations given in part (d), $\Delta[NO]/\Delta t = 1.42\ M/s$.

$\Delta[O_2]/\Delta t = \Delta[NO]/2\Delta t = 1.42\ M/s/2 = 0.711\ M/s$

14.32 (a) Increasing $[S_2O_8^{2-}]$ by a factor of 1.5 while holding $[I^-]$ constant increases the rate by a factor of 1.5 (Experiments 1 and 2). Doubling $[S_2O_8^{2-}]$ and increasing $[I^-]$ by a factor of 1.5 triples the rate ($2 \times 1.5 = 3$, experiments 1 and 3). Thus the reaction is first order in both $[S_2O_8^{2-}]$ and $[I^-]$; rate = k $[S_2O_8^{2-}]$ $[I^-]$.

(b) $k = \text{rate}/[S_2O_8^{2-}]\ [I^-]$

$k_1 = 2.6 \times 10^{-6}\ M/s/(0.018\ M)(0.036\ M) = 4.01 \times 10^{-3} = 4.0 \times 10^{-3}\ M^{-1}s^{-1}$

$k_2 = 3.9 \times 10^{-6}\ /(0.027)(0.036) = 4.01 \times 10^{-3} = 4.01 \times 10^{-3} = 4.0 \times 10^{-3}\ M^{-1}s^{-1}$

$k_3 = 7.8 \times 10^{-6}\ /(0.036)(0.054) = 4.01 \times 10^{-3} = 4.01 \times 10^{-3} = 4.0 \times 10^{-3}\ M^{-1}s^{-1}$

$k_4 = 1.4 \times 10^{-5}\ /(0.050)(0.072) = 3.89 \times 10^{-3} = 3.9 \times 10^{-3}\ M^{-1}s^{-1}$

$k_{avg} = 3.98 \times 10^{-3} = 4.0 \times 10^{-3}\ M^{-1}s^{-1}$

(c) $-\Delta[S_2O_8^{2-}]/\Delta t = -\Delta[I^-]/3\Delta t$; the rate of disappearance of $S_2O_8^{2-}$ is one-third the rate of disappearance of I^-.

(d) Note that the data are given in terms of disappearance of $S_2O_8^{2-}$.

$$\frac{-\Delta[I^-]}{\Delta t} = \frac{-3\Delta[S_2O_8^{2-}]}{\Delta t} = 3(3.98 \times 10^{-3}\ M^{-1}s^{-1})(0.015\ M)(0.040\ M) = 7.2 \times 10^{-6}\ M/s$$

Change of Concentration with Time

14.34 (a) A graph of $1/[A]$ vs time yields a straight line for a second-order reaction.

(b) The half-life of a first-order reaction is independent of $[A]_0$, $t_{1/2} = 0.693/k$. Whereas, the half-life of a second-order reaction does depend on $[A]_0$, $t_{1/2} = 1/k[A]_0$.

14.36 *Analyze.* Given rate constants for the decay of two radioisotopes, determine half-lives, decay rates, and amount remaining after three half-lives. *Plan.* Determine reaction order. Based on reaction-order, select the appropriate relationships for (a) rate constant and half-life and (c) rate-constant, time and concentration. In this example, mass is a measure of concentration.

Solve. Decay of radioiosotopes is a first-order process, since only one species is involved and the decay is not initiated by collision.

(a) For a first-order process, $t_{1/2} = 0.693/k$.

^{241}Am: $t_{1/2} = 0.693/1.6 \times 10^{-3}\ yr^{-1} = 433.1 = 4.3 \times 10^2\ yr$

^{125}I: $t_{1/2} = 0.693/0.011\ day^{-1} = 63.00 = 63\ days$

(b) For a given sample size, half of the ^{241}Am sample decays in 433 years, whereas half of the ^{125}I sample decays in 63 days. ^{125}I decays at a much faster rate.

(c) For a first order process, $\ln[A]_t - \ln[A]_0 = -kt$. $\ln[A]_t = -kt + \ln[A]_0$.

$[A]_0 = 1.0$ mg; $t = 3\ t_{1/2}$.

^{241}Am: $t = 3\ t_{1/2} = 3(433.1\ yr) = 1.299 \times 10^3 = 1.3 \times 10^3$ yr

$\ln [Am]_t = -1.6 \times 10^{-3}$ yr $(1.299 \times 10^3 - \ln(1.0) = -2.079 - 0 = -2.08$

$[Am]_t = 0.125 = 0.13$ mg

or, mass ^{241}Am remaining $= 1.0\ mg/2^3 = 0.125 = 0.13$ mg

^{125}I: For the same size starting sample and number of elapsed half-lives, the same mass, 0.13 mg ^{125}I, will remain. (The difference is that the elapsed time of 3 half-lives for ^{125}I is 3(63) = 189 days = 0.52 yr, vs. 433 yr for ^{241}Am.)

14.38 (a) Using Equation [14.13] for a first order reaction: $\ln[A]_t = -kt + \ln[A]_0$

2.5 min = 150 s; $[N_2O_5]_0 = (0.0250\ mol/2.0\ L) = 0.0125 = 0.013\ M$

$\ln[N_2O_5]_{150} = -(6.82 \times 10^{-3}\ s^{-1})(150\ s) + \ln(0.0125)$

$\ln[N_2O_5]_{150} = -1.0230 + (-4.3820) = -5.4050 = -5.41$

$[N_2O_5]_{150} = 4.494 \times 10^{-3} = 4.5 \times 10^{-3}\ M$; mol $N_2O_5 = 4.494 \times 10^{-3}\ M \times 2.0$ L

$= 9.0 \times 10^{-3}$ mol

(b) $[N_2O_5]_t = 0.010\ mol/2.0\ L = 0.0050\ M$; $[N_2O_5]_0 = 0.0125\ M$

$\ln(0.0050) = -(6.82 \times 10^{-3}\ s^{-1})(t) + \ln(0.0125)$

$$t = \frac{-[\ln(0.0050) - \ln(0.0125)]}{(6.82 \times 10^{-3}\ s^{-1})} = 134.35 = 1.3 \times 10^2\ s \times \frac{1\ min}{60\ s} = 2.24 = 2.2\ min$$

(c) $t_{1/2} = 0.693/k = 0.693/6.82 \times 10^{-3}\ s^{-1} = 101.6 = 102$ s or 1.69 min

14.40

t(s)	P_{CH_3NC}	$\ln P_{CH_3NC}$
0	502	6.219
2000	335	5.814
5000	180	5.193
8000	95.5	4.559
12000	41.7	3.731
15000	22.4	3.109

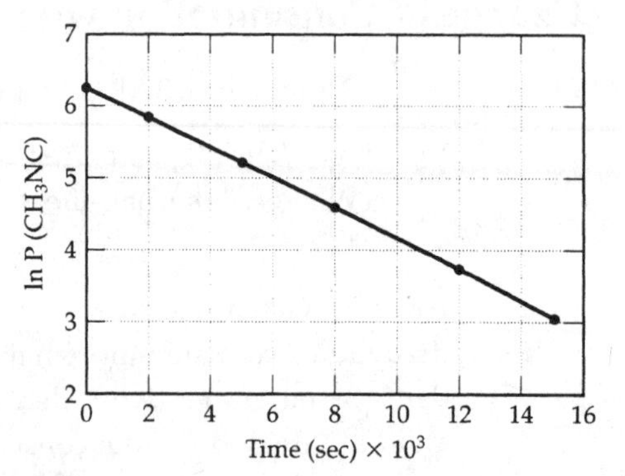

A graph of ln P vs t is linear with a slope of $-2.08 \times 10^{-4}\ s^{-1}$. The rate constant $k = -$slope $= 2.08 \times 10^{-4}\ s^{-1}$. Half-life $= t_{1/2} = 0.693/k = 3.33 \times 10^3$ s.

14.42 (a) Make both first- and second-order plots to see which is linear. Moles is a satisfactory concentration unit, since volume is constant.

time(s)	mol A	ln (mol A)	1/mol A
0	0.1000	-2.303	10.00
40	0.067	-2.70	14.9
80	0.045	-3.10	22.2
120	0.030	-3.51	33.3
160	0.020	-3.91	50.0

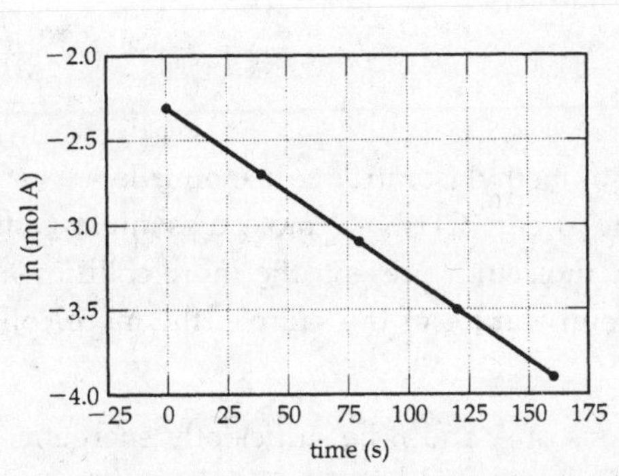

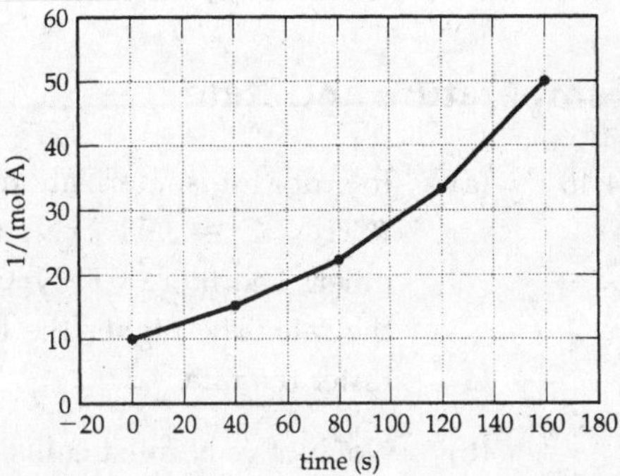

The plot of ln (mol A) vs time is linear, so the reaction is first-order in A.

(b) $k = -\text{slope} = -[-3.91 - (-2.70)]/120 = 0.010083 = 0.0101\ s^{-1}$
(The best fit to this line yields the same value for the slope, $0.01006 = 0.0101\ s^{-1}$)

(c) $t_{1/2} = 0.693/k = 0.693/0.010083\ s^{-1} = 68.7\ s$

14.44 (a) Make both first- and second-order plots to see which is linear.

time(min)	$[C_{12}H_{22}O_{11}]$(M)	$\ln[C_{12}H_{22}O_{11}]$	$1/[C_{12}H_{22}O_{11}]$
0	0.316	-1.152	3.16
39	0.274	-1.295	3.65
80	0.238	-1.435	4.20
140	0.190	-1.661	5.26
210	0.146	-1.924	6.85

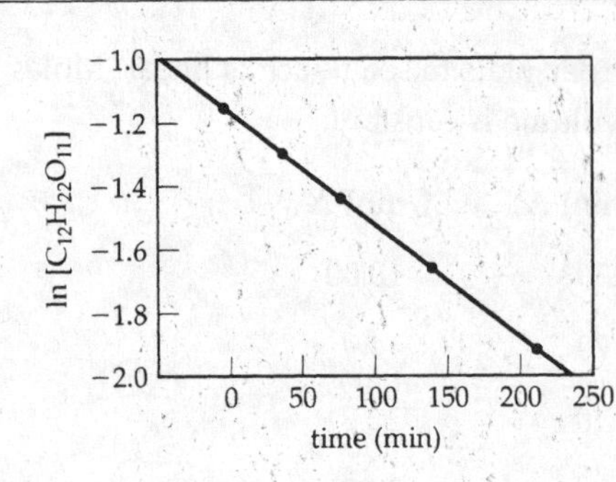

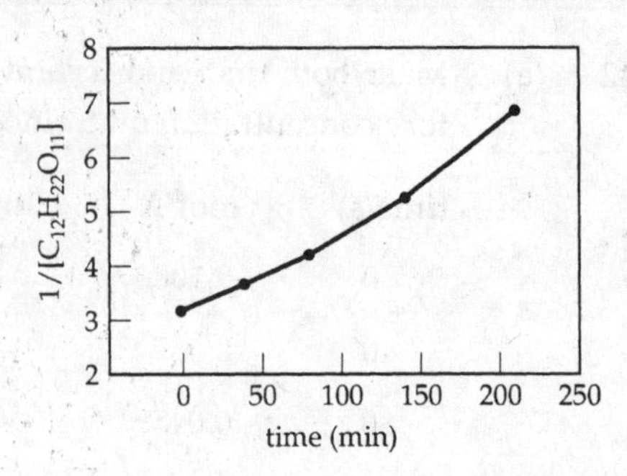

The plot of ln [$C_{12}H_{22}O_{11}$] is linear, so the reaction is first order in $C_{12}H_{22}O_{11}$.

(b) $k = -\text{slope} = -[-1.924 - (-1.295)] / 171 \text{ min} = 3.68 \times 10^{-3} \text{ min}^{-1}$

(The slope of the best-fit line is $-3.67 \times 10^{-3} \text{ min}^{-1}$.)

Temperature and Rate

14.46 (a) In order for isomerization of methyl isonitrile to acetonitrile ($CH_3N{\equiv}C \rightarrow CH_3C{\equiv}N$) to occur, $CH_3N{\equiv}C$ molecules must collide with each other. The more $CH_3N{\equiv}C$ molecules present, the more collisions and the faster the rate. The higher the temperature of the sample, the more collisions and the faster the rate.

(b) No. Not only must collisions of A and B be sufficiently energetic, A and B must collide in the correct orientation for the activated complex to form.

(c) The kinetic-molecular theory tells us that at some temperature T, there will be a distribution of molecular speeds and kinetic energies, and that the average kinetic energy of the sample is proportional to temperature. That is, as temperature of the sample increases, the average speed and kinetic energy of the molecules increases. At higher temperatures, there will be more molecular collisions (owing to greater speeds) and more energetic collisions (owing to greater kinetic energies). Overall there will be more collisions that have sufficient energy to form an activated complex, and the reaction rate will be greater.

14.48 (a) $f = e^{-E_a/RT}$ $E_a = 160 \text{ kJ/mol} = 1.60 \times 10^5 \text{ J/mol}, T = 500 \text{ K}$

$$-E_a/RT = -\frac{1.60 \times 10^5 \text{ J/mol}}{500 \text{ K}} \times \frac{\text{mol} \bullet \text{K}}{8.314 \text{ J}} = -38.489 = -38.5$$

$$f = e^{-38.489} = 1.924 \times 10^{-17} = 2 \times 10^{-17}$$

(b) $$-E_a/RT = -\frac{1.60 \times 10^5 \text{ J/mol}}{510 \text{ K}} \times \frac{\text{mol} \bullet \text{K}}{8.314 \text{ J}} = -37.735 = -37.7$$

$$f = e^{-37.735} = 4.093 \times 10^{-17} = 4.09 \times 10^{-17}$$

$$\frac{f \text{ at } 510 \text{ K}}{f \text{ at } 500 \text{ K}} = \frac{4.09 \times 10^{-17}}{1.92 \times 10^{-17}} = 2.13$$

An increase of 10 K means that 2.13 times more molecules have this energy.

14.50 *Analyze/Plan*. Use the definitions of activation energy ($E_{max} - E_{react}$) and ΔE ($E_{prod} - E_{react}$) to sketch the graph and calculate E_a for the reverse reaction. *Solve.*

(a) 

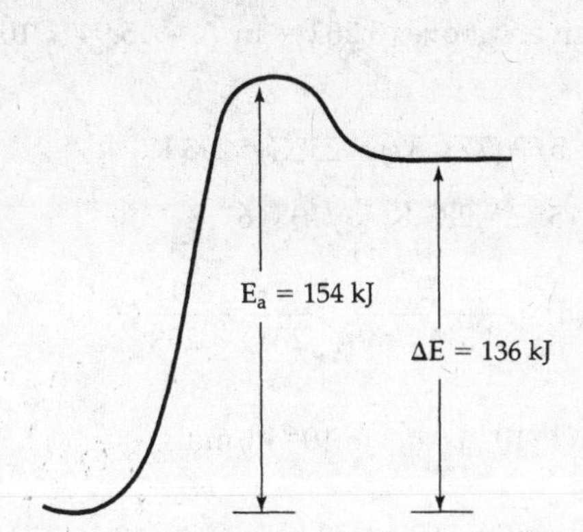

(b) E_a(reverse) = 18 kJ

14.52 E_a for the reverse reaction is:

(a) 45 – (–25) = 70 kJ (b) 35 – (–10) = 45 kJ (c) 55 – 10 = 45 kJ

Based on the magnitude of E_a, the reverse of reactions (b) and (c) occur at the same rate, which is faster than the reverse of reaction (a).

14.54 $T_1 = 737°C + 273 = 1010\ K$, $k_1 = 0.0796\ M^{-1}s^{-1}$;

$T_2 = 947°C + 273 = 1220\ K$, $k_2 = 0.0815\ M^{-1}s^{-1}$

$$\ln\left(\frac{k_1}{k_2}\right) = \frac{E_a}{R}\left(\frac{1}{T_2} - \frac{1}{T_1}\right)$$

$$\ln\left(\frac{0.0796}{0.0815}\right) = \frac{E_a}{8.314\ \text{J/mol}}\left(\frac{1}{1220} - \frac{1}{1010}\right)$$

$$-0.023589 = \frac{E_a\ (-1.704 \times 10^{-4})}{8.314\ \text{J/mol}}$$

$$E_a = \frac{8.314\,(-0.023589)\ \text{J/mol}}{(-1.704 \times 10^{-4})} = 1.151 \times 10^3\ \text{J/mol} = 1.15\ \text{kJ/mol}$$

14.56

k	ln k	T(K)	1/T(× 10^3)
0.028	-3.58	600	1.67
0.22	-1.51	650	1.54
1.3	0.26	700	1.43
6.0	1.79	750	1.33
23	3.14	800	1.25

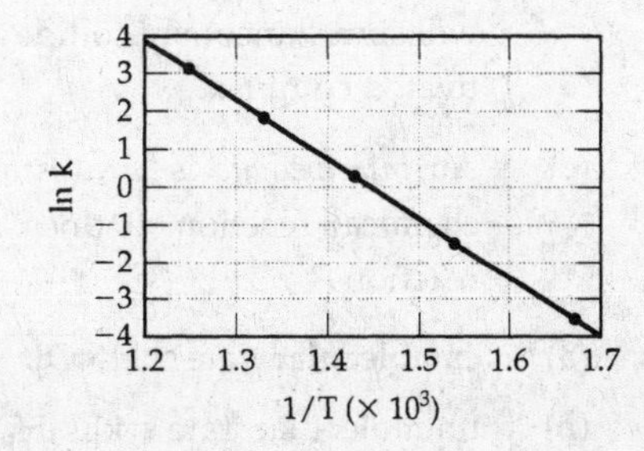

Using the relationship $\ln k = \ln A - E_a/RT$, the slope, $-15.94 \times 10^3 = -16 \times 10^3$, is $-E_a/R$. $E_a = 15.94 \times 10^3 \times 8.314$ J/mol $= 1.3 \times 10^2$ kJ/mol. To calculate A, we will use the rate data at 700 K. From the equation given above, $0.262 = \ln A - 15.94 \times 10^3/700$; $\ln A = 0.262 + 22.771$. $A = 1.0 \times 10^{10}$.

14.58 (a) $T_1 = 77°F$; $°C = 5/9\,(°F - 32) = 5/9\,(77 - 32) = 25°C = 298$ K

$T_2 = 59°F$; $°C = 5/9\,(59–32) = 15°C = 288$ K; $k_1/k_2 = 6$

$$\ln\left(\frac{k_1}{k_2}\right) = \frac{E_a}{R}\left[\frac{1}{T_2} - \frac{1}{T_1}\right];\ \ln(6) = \frac{E_a}{8.314\ \text{J/mol}}\left[\frac{1}{288} - \frac{1}{298}\right]$$

$$E_a = \frac{\ln(6)(8.314\ \text{J/mol})}{1.165 \times 10^{-4}} = 1.28 \times 10^5\ \text{J} = 1.3 \times 10^2\ \text{kJ/mol}$$

$T_1 = 77°F = 25°C = 298$ K; $T_2 = 41°F = 5°C = 278$ K, $k_1/k_2 = 40$

$$\ln(40) = \frac{E_a}{8.314\ \text{J/mol}}\left[\frac{1}{278} - \frac{1}{296}\right];\ E_a = \frac{\ln(40)(8.314\ \text{J/mol})}{2.414 \times 10^{-4}}$$

$E_a = 1.27 \times 10^5$ J $= 1.3 \times 10^2$ kJ/mol

The values are amazingly consistent, considering the precision of the data.

(b) For a first order reaction, $t_{1/2} = 0.693/k$, $k = 0.693/t_{1/2}$

k_1 at 298 K $= 0.693/2.7$ yr $= 0.257 = 0.26$ yr^{-1}

$T_1 = 298$ K, $T_2 = 273 - 15°C = 258$ K

$$\ln\left(\frac{0.257}{k_2}\right) = \frac{1.27 \times 10^5\ \text{J}}{8.314\ \text{J/mol}}\left[\frac{1}{258} - \frac{1}{298}\right] = 7.9497$$

$0.257/k_2 = e^{7.9497} = 2.835 \times 10^3$; $k_2 = 0.257/2.835 \times 10^3 = 9.066 \times 10^{-5} = 9.1 \times 10^{-5}$ yr^{-1}

$t_{1/2} = 0.693/k = 0.693/9.066 \times 10^{-5} = 7.64 \times 10^3$ yr $= 7.6 \times 10^3$ yr

Reaction Mechanisms

14.60 (a) The *molecularity* of a process indicates the number of molecules that participate as reactants in the process. A unimolecular process has one reactant molecule, a bimolecular process has two reactant molecules and a termolecular process has three reactant molecules.

(b) Termolecular processes are rare because it is highly unlikely that three molecules will simultaneously collide with the correct energy and orientation to form an activated complex.

(c) An *intermediate* is a substance that is produced and then consumed during a chemical reaction. It does not appear in the balanced equation for the overall reaction.

14.62 (a) bimolecular, rate = $k[NO]^2$

(b) unimolecular, rate = $k[C_3H_6]$

(c) unimolecular, rate = $k[SO_3]$

14.64 (a) Two elementary reactions; two energy maxima

(b) One intermediate; one energy minimum between reactants and products

(c) The second step is rate-limiting; second energy maximum and E_a is larger.

(d) Overall reaction is exothermic; energy of products is lower than energy of reactants.

14.66 *Analyze/Plan.* Follow the logic in Sample Exercise 14.14. *Solve.*

(a) First step: rate = $k_1[H_2O_2]\,[I^-]$; second step: rate = $k_2[IO^-]\,[H_2O_2]$

(b) $2H_2O_2(aq) \rightarrow 2H_2O(l) + O_2(g)$

(c) $IO^-(aq)$ is the intermediate.

(d) rate = $k[H_2O_2]\,[I^-]$

14.68 (a)

i. $HBr + O_2 \rightarrow HOOBr$

ii. $HOOBr + HBr \rightarrow 2HOBr$

iii. $2HOBr + 2HBr \rightarrow 2H_2O + 2Hr_2$

$$4HBr + O_2 \rightarrow 2H_2O + 2Br_2$$

(b) The observed rate law is: rate = $k[HBr][O_2]$, the rate law for the first elementary step. The first step must be rate-determining.

(c) HOOBr and HOBr are both intermediates; HOOBr is produced in i and consumed in ii and HOBr is produced in ii and consumed in iii.

(d) Since the first step is rate-determining, it is possible that neither of the intermediates accumulates enough to be detected. This does not disprove the mechanism, but indicates that steps ii and iii are very fast, relative to step i.

Catalysis

14.70 (a) The smaller the particle size of a solid catalyst, the greater the surface area. The greater the surface area, the more active sites and the greater the increase in reaction rate.

(b) Adsorption is the binding of reactants onto the surface of the heterogeneous catalyst. It is usually the first step in the catalyzed reaction.

14.72 (a)

$$2[NO(g) + N_2O(g) \rightarrow N_2(g) + NO_2(g)]$$
$$2NO_2(g) \rightarrow 2NO(g) + O_2(g)$$
$$2N_2O(g) \rightarrow 2N_2(g) + O_2(g)$$

(b) An intermediate is produced and then consumed during the course of the reaction. A catalyst is consumed and then reproduced. In other words, the catalyst is present when the reaction sequence begins and after the last step is completed. In this reaction, NO is the catalyst and NO_2 is an intermediate.

(c) No. The proposed mechanism cannot be ruled out, based on the behavior of NO_2. NO_2 functions as an intermediate; it is produced and then consumed during the reaction. That there is no measurable build-up of NO_2 indicates the first step is slow relative to the second; as soon as NO_2 is produced by the slow first step, it is consumed by the faster second step.

14.74 (a) Catalytic converters are heterogeneous catalysts that adsorb gaseous CO and hydrocarbons and speed up their oxidation to $CO_2(g)$ and $H_2O(g)$. They also adsorb nitrogen oxides, NO_x, and speed up their reduction to $N_2(g)$ and $O_2(g)$. If a catalytic converter is working effectively, the exhaust gas should have very small amounts of the undesirable gases CO, $(NO)_x$ and hydrocarbons.

(b) The high temperatures could increase the rate of the desired catalytic reactions given in part (a). It could also increase the rate of undesirable reactions such as corrosion, which decrease the lifetime of the catalytic converter.

(c) The rate of flow of exhaust gases over the converter will determine the rate of adsorption of CO, $(NO)_x$ and hydrocarbons onto the catalyst and thus the rate of conversion to desired products. Too fast an exhaust flow leads to less than maximum adsorption. A very slow flow leads to back pressure and potential damage to the exhaust system. Clearly the flow rate must be adjusted to balance chemical and mechanical efficiency of the catalytic converter.

14.76 Just as the π electrons in C_2H_4 are attracted to the surface of a hydrogenation catalyst, the nonbonding electron density on S causes compounds of S to be attracted to these same surfaces. Strong interactions could cause the sulfur compounds to be permanently attached to the surface, blocking active sites and reducing adsorption of alkenes for hydrogenation.

14.78 The individual structure of each enzyme molecule leads to a unique coiling and folding pattern. The resulting shape and electronic properties of the active site in each enzyme leads to its substrate specificity.

14.80 Let k and E_a equal the rate constant and activation energy for the uncatalyzed reaction. Let k_c and E_{ac} equal the rate constant and activation energy of the catalyzed reaction. A is the same for the uncatalyzed and catalyzed reactions. $k_c/k = 1 \times 10^5$, T = 37°C = 310 K.

According to Equation [14.20], ln k = E_a/RT + ln A. Subtracting ln k from ln k_c

$$\ln k_c - \ln k = \left[\frac{-E_{ac}}{RT}\right] + \ln A - \left[\frac{-E_a}{RT}\right] - \ln A$$

$$\ln (k_c/k) = \frac{E_a - E_{ac}}{RT};\ E_a - E_{ac} = RT \ln (k_c/k)$$

$$E_a - E_{ac} = \frac{8.314\ J}{K \bullet mol} \times 310\ K \times \ln (1 \times 10^5) = 2.966 \times 10^4\ J = 29.66\ kJ = 3 \times 10^1\ kJ$$

The enzyme must lower the activation energy by 30 kJ in order to achieve a 1×10^5-fold increase in reaction rate.

Additional Exercises

14.82 (a) $\text{rate} = \frac{-\Delta[NO]}{2\Delta t} = \frac{-\Delta[O_2]}{\Delta t} = \frac{9.3 \times 10^{-5}\ M/s}{2} = 4.7 \times 10^{-5}\ M/s$

(b,c) $\text{rate} = k[NO]^2[O_2]$; $k = \text{rate}/[NO]^2[O_2]$

$$k = \frac{4.7 \times 10^{-5}\ M/s}{(0.040\ M)^2\ (0.035\ M)} = 0.8393 = 0.84\ M^{-2}\ s^{-1}$$

(d) Since the reaction is second order in NO, if the [NO] is increased by a factor of 1.8, the rate would increase by a factor of 1.8^2, or (3.24) = 3.2.

14.83 (a) $\text{rate} = k[I^-]\ [OCl^-]/[OH^-]$

(b) Since the reaction is first order in $[I^-]$, tripling $[I^-]$ triples the rate.

(c) Since the rate is inversely proportional to $[OH^-]$, doubling $[OH^-]$ cuts the rate in half.

14.85 The units of rate are M/s. The reaction must be second order overall if the units of the rate constant are $M^{-1}\ s^{-1}$. If rate = $k[NO_2]^x$, then the cumulative units of $[NO_2]^x$ must be M^2, and x = 2.

If $[NO_2]_0 = 0.100\ M$ and $[NO_2]_t = 0.025\ M$, use the integrated form of the second order rate equation, $\frac{1}{[A]_t} = kt + \frac{1}{[A]_o}$, Equation [14.14], to solve for t.

$$\frac{1}{0.025\ M} = 0.63\ M^{-1}s^{-1}\ (t) + \frac{1}{0.100\ M};\quad \frac{(40 - 10)\ M^{-1}}{0.63\ M^{-1}s^{-1}} = t = 47.62 = 48\ s.$$

14.86 (a) $t_{1/2} = 0.693/k = 0.693/7.0 \times 10^{-4}\ s^{-1} = 990 = 9.9 \times 10^2\ s$

(b) $k = \frac{0.693}{t_{1/2}} = \frac{0.693}{56.3\ \text{min}} \times \frac{1\ \text{min}}{60\ s} = 2.05 \times 10^{-4}\ s^{-1}$

14.88 (a) A = abc, Equation [14.5]. A = 0.605, a = $5.60 \times 10^3\ cm^{-1}\ M^{-1}$, b = 1.00 cm

$$c = \frac{A}{ab} = \frac{0.605}{(5.60 \times 10^3\ cm^{-1}\ M^{-1})(1.00\ cm)} = 1.080 \times 10^{-4} = 1.08 \times 10^{-4}\ M$$

(b) Calculate $[c]_t$ using Beer's law. We calculated $[c]_0$ in part (a). Use Equation [14.13] to calculate k.

$$A_{30} = abc_{30};\ c_{30} = \frac{A_{30}}{ab} = \frac{0.250}{(5.60 \times 10^3\ cm^{-1}\ M^{-1})(1.00\ cm)} = 4.464 \times 10^{-5}\ M$$

$$\ln[c]_t = -kt + \ln[c]_0;\quad \frac{\ln[c]_0 - \ln[c]_t}{t} = k;\ t = 30\ \text{min} \times \frac{60\ s}{\text{min}} = 1800\ s$$

$$k = \ln(1.080 \times 10^{-4}) - \ln(4.464 \times 10^{-5})\ /\ 1800\ s = 4.910 \times 10^{-4} = 4.91 \times 10^{-4}\ s^{-1}$$

(c) For a first order reaction, $t_{1/2} = 0.693/k$.

$t_{1/2} = 0.693/4.910 \times 10^{-4}\ s^{-1} = 1.411 \times 10^3 = 1.41 \times 10^3\ s = 23.5\ \text{min}$

(d) $A_t = 0.100$; calculate c_t using Beer's law, then t from the first order integrated rate equation.

$$c_t = \frac{A}{ab} = \frac{0.100}{(5.60 \times 10^3 \text{ cm}^{-1} M^{-1})(1.00 \text{ cm})} = 1.786 \times 10^{-5} = 1.79 \times 10^{-5} M$$

$$t = \frac{\ln[c]_0 - \ln[c]_t}{k} = \frac{\ln(1.080 \times 10^{-4}) - \ln(1.786 \times 10^{-5})}{4.910 \times 10^{-4} \text{ s}^{-1}}$$

$t = 3.666 \times 10^3 = 3.67 \times 10^3 \text{ s} = 61.1 \text{ min}$

14.89

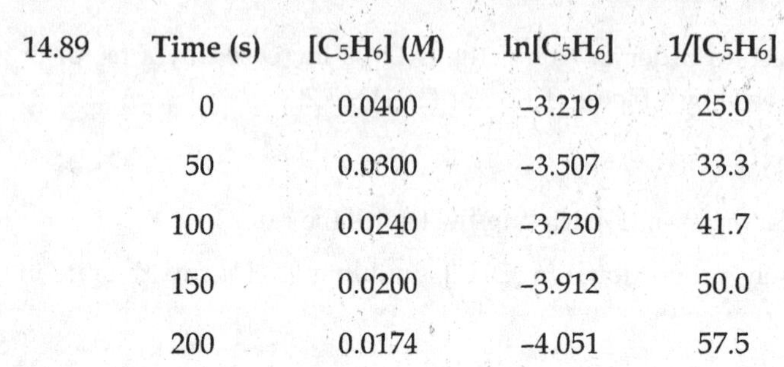

Time (s)	$[C_5H_6]$ (M)	$\ln[C_5H_6]$	$1/[C_5H_6]$
0	0.0400	-3.219	25.0
50	0.0300	-3.507	33.3
100	0.0240	-3.730	41.7
150	0.0200	-3.912	50.0
200	0.0174	-4.051	57.5

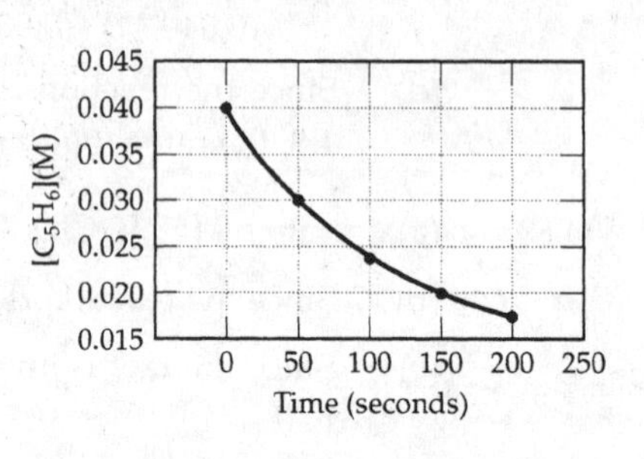

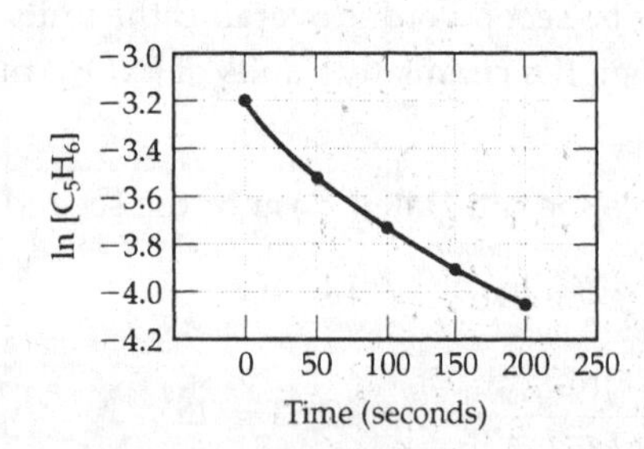

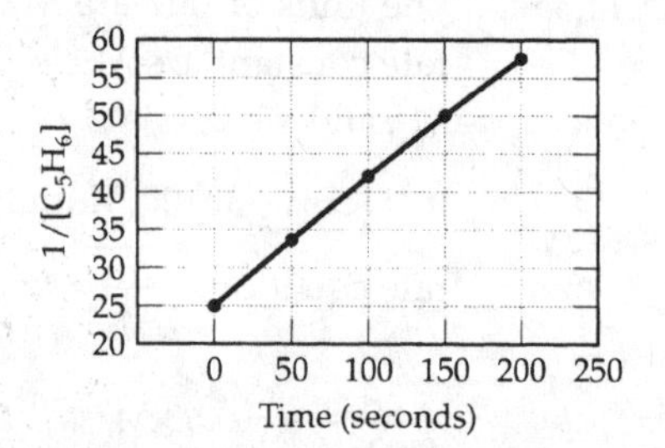

The plot of $1/[C_5H_6]$ vs time is linear and the reaction is second order.

The slope of this line is k. k = slope = $(50.0 - 25.0)\ M^{-1}/(150\text{–}0)\text{s} = 0.167\ M^{-1}\ \text{s}^{-1}$

(The best-fit slope and k value is $0.163\ M^{-1}\ \text{s}^{-1}$.)

14.90 (a) No. The value of A, which is related to frequency and effectiveness of collisions, can be different for each reaction and k is proportional to A.

(b) From Equation [14.21], reactions with different variations of k with respect to temperature have different activation energies, E_a. The fact that k for the two reactions is the same at a certain temperature is coincidental. The reaction with the higher rate at 35°C has the larger activation energy, because it was able to use the increase in energy more effectively.

14.92 (a)

$$\begin{array}{c} NO(g) + NO(g) \rightarrow N_2O_2(g) \\ N_2O_2(g) + H_2(g) \rightarrow N_2O(g) + H_2O(g) \\ \hline 2NO(g) + N_2O_2(g) + H_2(g) \rightarrow N_2O_2(g) + N_2O(g) + H_2O(g) \\ 2NO(g) + H_2(g) \rightarrow N_2O(g) + H_2O(g) \end{array}$$

(b) First reaction: $-\Delta[NO]/\Delta t = k[NO][NO] = k[NO]^2$

Second reaction: $-\Delta[H_2]/\Delta t = k[H_2][N_2O_2]$

(c) N_2O_2 is the intermediate: it is produced in the first step and consumed in the second.

(d) Since $[H_2]$ appears in the rate law, the second step must be slow relative to the first.

14.93 (a)

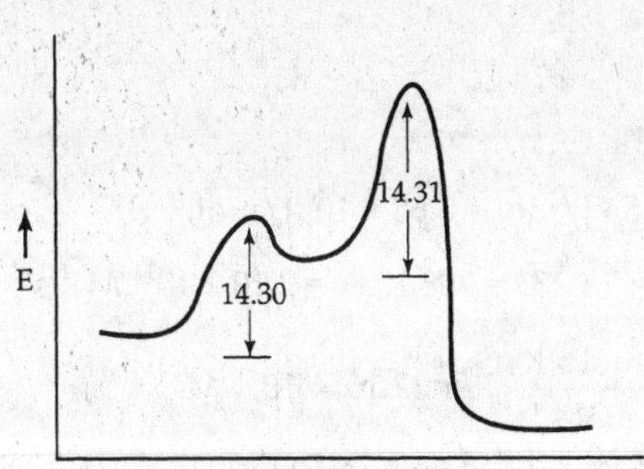

14.30 = E_a for reaction [14.30]

14.31 = E_a for reaction [14.31]

(b) The fact that Br_2 builds up during the reaction tells us that the appearance of Br_2 (reaction [14.30]) is faster than the disappearance of Br_2 (reaction [14.31]). This is the reason that E_a of [14.31] in the energy profile above is larger than E_a of [14.30].

14.95 (a) $(CH_3)_3AuPH_3 \rightarrow C_2H_6 + (CH_3)AuPH_3$

(b) $(CH_3)_3Au$, $(CH_3)Au$ and PH_3 are intermediates.

(c) Reaction 1 is unimolecular, Reaction 2 is unimolecular, Reaction 3 is bimolecular.

(d) Reaction 2, the slow one, is rate determining.

(e) If Reaction 2 is rate determining, rate = $k_2[(CH_3)_3Au]$.

$(CH_3)_3Au$ is an intermediate formed in Reaction 1, an equilibrium. By definition, the rates of the forward and reverse processes in Reaction 1 are equal:

$k_1[(CH_3)_3 AuPH_3] = k_{-1}[(CH_3)_3Au][PH_3]$; solving for $[(CH_3)_3Au]$,

$$[(CH_3)_3\ Au] = \frac{k_1[(CH_3)_3\ AuPH_3]}{k_{-1}[PH_3]}$$

Substituting into the rate law

$$\text{rate} = \left(\frac{k_2\,k_1}{k_{-1}}\right)\frac{[(CH_3)_3AuPH_3]}{[PH_3]} = \frac{k[(CH_3)_3AuPH_3]}{[PH_3]}$$

(f) The rate is inversely proportional to $[PH_3]$, so adding PH_3 to the $(CH_3)_3AuPH_3$ solution would decrease the rate of the reaction.

14.97 (a) The fact that the rate doubles with a doubling of the concentration of sugar tells us that the fraction of enzyme tied up in the form of an enzyme-substrate complex is small. A doubling of the substrate concentration leads to a doubling of the concentration of enzyme-substrate complex, because most of the enzyme molecules are available to bind substrates.

(b) The behavior of inositol suggests that it acts as a competitor with sucrose for binding at the active sites of the enzyme system. Such a competition results in a lower effective concentration of active sites for binding of sucrose, and thus results in a lower reaction rate.

Integrative Exercises

14.99 (a) $\ln k = -E_a/RT + \ln A$; $E_a = 86.8$ kJ/mol $= 8.68 \times 10^4$ J/mol;

$T = 35°C + 273 = 308$ K; $A = 2.10 \times 10^{11}\ M^{-1}\ s^{-1}$

$$\ln k = \frac{-8.68 \times 10^4 \text{ J/mol}}{308 \text{ K}} \times \frac{\text{mol} \bullet \text{K}}{8.314 \text{ J}} + \ln(2.10 \times 10^{11}\ M^{-1}\ s^{-1})$$

$\ln k = -33.8968 + 26.0704 = -7.8264$; $k = 3.99 \times 10^{-4}\ M^{-1}\ s^{-1}$

(b) $$\frac{0.335 \text{ g KOH}}{0.250 \text{ L soln}} \times \frac{1 \text{ mol KOH}}{56.1 \text{ g KOH}} = 0.02389 = 0.0239\ M \text{ KOH}$$

$$\frac{1.453 \text{ g } C_2H_5I}{0.250 \text{ L soln}} \times \frac{1 \text{ mol } C_2H_5I}{156.0 \text{ g } C_2H_5I} = 0.03726 = 0.0373\ M\ C_2H_5I$$

If equal volumes of the two solutions are mixed, the initial concentrations in the reaction mixture are 0.01194 *M* KOH and 0.01863 *M* C_2H_5I. Assuming the reaction is first order in each reactant:

rate = $k[C_2H_5I][OH^-] = 3.99 \times 10^{-4}\ M^{-1}s^{-1}\ (0.01194\ M)(0.01863\ M) = 8.88 \times 10^{-8}\ M/s$

(c) Since C_2H_5I and OH^- react in a 1 : 1 mole ratio and equal volumes of the solutions are mixed, the reactant with the smaller concentration, KOH, is the limiting reactant.

14.101 (a) $\ln k = -E_a/RT + \ln A$, Equation [14.20]. $E_a = 6.3$ kJ/mol $= 6.3 \times 10^3$ J/mol

$T = 100°C + 273 = 373$ K

$$\ln k = \frac{-6.3 \times 10^3 \text{ J/mol}}{8.314 \text{ J/K} \bullet \text{mol} \times 373 \text{ K}} + \ln(6.0 \times 10^8\ M^{-1}s^{-1})$$

$\ln k = -2.032 + 20.212 = 18.181 = 18.2$; $k = 7.87 \times 10^7 = 8 \times 10^7\ M^{-1}s^{-1}$

(b) NO, 11 valence e^-, 5.5 e^- pair (Assume the less electronegative N atom will be electron deficient.)

:Ṅ=Ö:

ONF, 18 valence e^-, 9 e^- pr

:Ö=N̈—F̤: ⟷ (:Ö̤—N̈=F̤)

The resonance form on the right is a very minor contributor to the true bonding picture, due to high formal charges and the unlikely double bond involving F.

(c) ONF has trigonal planar electron domain geometry, which leads to a "bent" structure with a bond angle of approximately 120°.

(d) [O═N ··· F—F]

(e) The electron deficient NO molecule is attracted to electron-rich F_2, so the driving force for formation of the transition state is greater than simple random collisions.

14.102 (a) $\Delta H^\circ_{rxn} = 2\Delta H^\circ_f\, H_2O(g) + 2\Delta H^\circ_f\, Br_2(g) - 4\Delta H^\circ_f\, HBr(g) - \Delta H^\circ_f\, O_2(g)$

$\Delta H^\circ_{rxn} = 2(-241.82) + 2(30.71) - 4(-36.23) - (0) = -277.30$ kJ

(b) Since the rate of the uncatalyzed reaction is very slow at room temperature, the magnitude of the activation energy for the rate-determining first step must be quite large. At room temperature, the reactant molecules have a distribution of kinetic energies (Chapter 10), but very few molecules even at the high end of the distribution have sufficient energy to form an activated complex. E_a for this step must be much greater than 3/2 RT, the average kinetic energy of the sample.

(c) 20 e⁻, 10 e⁻ pr

H—Ö—Ö—Br:

The intermediate resembles hydrogen peroxide, H_2O_2.

14.104 (a) If the reaction proceeds in a single elementary step, the coefficients in the balanced equation are the reaction orders for the respective reactants.

rate = $k[Ce^{4+}]^2\,[Tl^+]$

(b) If the uncatalyzed reaction occurs in a single step, it is **termolecular**. The activated complex requires collision of three particles with the correct energy and orientation for reaction. The probability of an effective three-particle collision is low and the rate is slow.

(c) The first step is rate-determining.

(d) The ability of Mn to adopt every oxidation state from +2 to +7 makes it especially suitable to catalyze this (and many other) reactions.

14.105 (a) D(Cl–Cl) = 242 kJ/mol Cl_2

$$\frac{242\text{ kJ}}{\text{mol }Cl_2} \times \frac{1000\text{ J}}{\text{kJ}} \times \frac{1\text{ mol}}{6.022 \times 10^{23}\text{ molecules}} = 4.019 \times 10^{-19} = 4.02 \times 10^{-19}\text{ J}$$

$$\lambda = hc/E = \frac{6.626 \times 10^{-34}\text{ J}\bullet\text{s} \times 2.998 \times 10^{8}\text{ m/s}}{4.019 \times 10^{-19}\text{ J}} = 4.94 \times 10^{-7}\text{ m}$$

This wavelength, 494 nm, is in the visible portion of the spectrum.

(b)

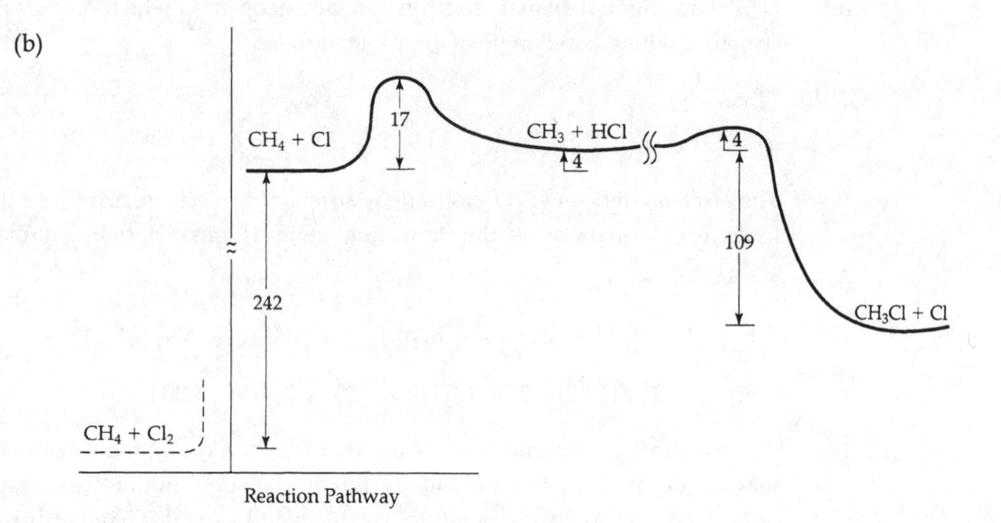

(c) Since D(Cl–Cl) is 242 kJ/mol, $CH_4(g) + Cl_2(g)$ should be about 242 kJ below the starting point on the diagram. For the reaction $CH_4(g) + Cl_2(g) \rightarrow CH_3(g) + HCl(g) + Cl(g)$, E_a is 242 + 17 = 259 kJ. (From bond dissociation enthalpies, ΔH for the overall reaction $CH_4(g) + Cl_2(g) \rightarrow CH_3Cl(g) + Cl(g)$ is –104 kJ, so the graph above is simply a sketch of the relative energies of some of the steps in the process.)

(d) CH_3, 7 valence e^-, odd electron species

$$\begin{array}{c} H-\dot{C}-H \\ | \\ H \end{array}$$

(e) This sequence is called a chain reaction because Cl• radicals are regenerated in Reaction 4, perpetuating the reaction. Absence of Cl• terminates the reaction, so $Cl\bullet + Cl\bullet \rightarrow Cl_2$ is a termination step.

15 Chemical Equilibrium

Visualizing Concepts

15.2 Yes. The first box is pure reactant A. As the reaction proceeds, some A changes to B. In the fourth and fifth boxes, the relative amounts (concentrations) of A and B are constant. Although the reaction is ongoing the rates of $A \rightarrow B$ and $B \rightarrow A$ are equal, and the relative amounts of A and B are constant.

15.3 *Analyze.* Given box diagram and reaction type, determine whether $K > 1$ for the equilibrium mixture depicted in the box.

Plan. Assign species in the box to reactants and products. Write an equilibrium expression in terms of concentrations. Find the relationship between numbers of molecules and concentration. Calculate K.

Solve. Let red = A, blue = X, red and blue pairs = AX. (The colors of A and X are arbitrary.) There are 3A, 2B, and 8AX in the box.

M = mol/L. Since moles is a counting unit for particles, mol ratios and particle ratios are equivalent. We can use numbers of particles in place of moles in the molarity formula. V = 1 L, so in this case, [A] = number of A particles.

$$K = \frac{[AX]}{[A][X]}; \quad [AX] = 8/V = 8; [A] = 3/V = 3; [X] = 2/V = 2$$

$$K = \frac{8}{[3][2]} = \frac{8}{6} = 1.33$$

15.5 *Analyze.* Given box diagrams, reaction type, and value of K_c, determine whether each reaction mixture is at equilibrium.

Plan. Analyze the contents of each box, express them as concentrations (see Solution 5.3). Write the equilibrium expression, calculate Q for each mixture, and compare it to K_c. If Q = K, the mixture is at equilibrium. If Q < K, the reaction shifts right (more product). If Q > K, the reaction shifts left (more reactant).

Solve. $K_c = \frac{[AB]^2}{[A_2][B_2]}$.

For this reaction, $\Delta n = 0$, so the volume terms cancel in the equilibrium expression. In this case, the number of each kind of particle can be used as a representation of moles (see Solution 5.3) and molarity.

(a) Mixture 1: $1A_2$, $1B_2$, 6AB; $Q = \frac{6^2}{(1)(1)} = 36$

$Q > K_c$, the mixture is not at equilibrium.

Mixture 2: $3A_2, 2B_2, 3AB$; $Q = \frac{3^2}{(3)(2)} = 1.5$

$Q = K_c$, the mixture is at equilibrium.

Mixture 3: $3A_2, 3B_2, 2AB$; $Q = \frac{2^2}{(3)(3)} = 0.44$

$Q < K_c$, the mixture is not at equilibrium.

(b) Mixture 1 proceeds toward reactants.

Mixture 3 proceeds toward products.

15.6 For the reaction $A_2(g) + B(g) \rightleftharpoons A(g) + AB(g)$, $\Delta n = 0$ and $K_p = K_c$. We can evaluate the equilibrium expression in terms of concentration. Also since $\Delta n = 0$, the volume terms in the expression cancel and we can use number of particles as a measure of moles and molarity. The mixture contains 2A, 4AB and $2A_2$.

$$K_c = \frac{[A][AB]}{[A_2][B]} = \frac{(2)(4)}{(2)(B)} = 2;\ B = 2$$

2 B atoms should be added to the diagram.

15.8 If temperature increases, K of an endothermic reaction increases and K of an exothermic reaction decreases. Calculate the value of K for the two temperatures and compare. For this reaction, $\Delta n = 0$ and $K_p = K_c$. We can ignore volume and use number of particles as a measure of moles and molarity. $K_c = [A][AB]/[A_2][B]$

(1) 300 K, 3A, 5AB, $1A_2$, 1B; $K_c = (3)(5)/(1)(1) = 15$

(2) 500 K, 1A, 3AB, $3A_2$, 3B; $K_c = (1)(3)/(3)(3) = 0.33$

K_c decreases as T increases, so the reaction is exothermic.

Equilibrium; the Equilibrium Constant

15.10 (a) $K_c = \frac{[C][D]}{[A][B]}$; if K_c is large, the numerator of the K_c expression is much greater than the denominator and products will predominate at equilibrium.

(b) $K_c = k_f/k_r$; if K_c is large, k_f is larger than k_r and the forward reaction has the greater rate constant.

15.12 (a) Yes. The algebraic form of the law of mass action depends only on the coefficients of a chemical equation, not on the reaction mechanism.

(b) $N_2(g) + 3H_2(g) \rightleftharpoons 2NH_3(g)$. The *Haber process* is the primary industrial method of nitrogen fixation, that is, of converting $N_2(g)$ into usable forms. The major use of $NH_3(g)$ from the Haber process is for fertilizer.

(c) $K_c = \frac{[NH_3]^2}{[N_2][H_2]^3}$

15.14 (a) $K_c = \frac{[NO]^2}{[N_2][O_2]}$ (b) $K_c = \frac{1}{[Cl_2]^2}$

(c) $K_c = \frac{[C_2H_6]^2[O_2]}{[C_2H_4]^2[H_2O]^2}$ (d) $K_c = \frac{[H_2O]}{[H_2]}$

(e) $K_c = \frac{[Cl_2]^2}{[HCl]^4[O_2]}$

homogeneous: (a), (c); heterogeneous: (b), (d), (e)

15.16 (a) equilibrium lies to right, favoring products ($K_c >> 1$)

(b) equilibrium lies to left, favoring reactants ($K_c << 1$)

15.18 $SO_2(g) + Cl_2(g) \rightleftharpoons SO_2Cl_2(g)$, $K_p = 34.5$. $\Delta n = 1 - 2 = -1$

$K_p = K_c(RT)^{\Delta n}$; $34.5 = K_c(RT)^{-1} = K_c/RT$;

$K_c = 34.5\ RT = 34.5(0.08206)(303) = 857.81 = 858$

15.20 (a) $K_p(\text{forward}) = \frac{P^2_{NO_2}}{P^2_{NO} \times P_{O_2}} = 1.48 \times 10^4$

$K_p(\text{reverse}) = \frac{P^2_{NO} \times P_{O_2}}{P^2_{NO_2}} = \frac{1}{1.48 \times 10^4} = 6.76 \times 10^{-5}$

(b) $K_p > 1$ when NO_2 is the product, and $K_p < 1$ when NO_2 is the reactant, so the equilibrium favors NO_2 at this temperature.

15.22 $K_p = \frac{P^4_{HCl} \times P_{O_2}}{P^2_{Cl_2} \times P^2_{H_2O}} = 0.0752$

(a) $K_p = \frac{P^2_{Cl_2} \times P^2_{H_2O}}{P^4_{HCl} \times P_{O_2}} = \frac{1}{0.0752} = 13.298 = 13.3$

(b) $K_p = \frac{P^2_{HCl} \times P^{1/2}_{O_2}}{P_{Cl_2} \times P_{H_2O}} = (0.0752)^{1/2} = 0.2742 = 0.274$

(c) $K_p = K_c(RT)^{\Delta n}$; $\Delta n = 2.5 - 2 = 0.5$; $T = 480°C + 273 = 753\ K$

$K_p = K_c(RT)^{1/2}$, $K_c = K_p/(RT)^{1/2} = 0.2742/[0.08206 \times 753]^{1/2} = 0.03488 = 0.0349$

15.24

$2NO(g) + Br_2(g) \rightleftharpoons 2NOBr(g)$ $K_1 = 2.0$

$N_2(g) + O_2(g) \rightleftharpoons 2NO(g)$ $K_2 = \frac{1}{2.1 \times 10^{30}}$

$2NO(g) + Br_2(g) + N_2(g) + O_2(g) \rightleftharpoons 2NOBr(g) + 2NO(g)$

$N_2(g) + O_2(g) + Br_2(g) \rightleftharpoons 2NOBr(g)$

$K_c = K_1 \times K_2 = 2.0 \times \frac{1}{2.1 \times 10^{30}} = 9.524 \times 10^{-31} = 9.5 \times 10^{-31}$

15.26 (a) $K_p = 1/\ P_{SO_2}$

(b) Na_2O is a pure solid. The molar concentration, the ratio of moles of a substance to volume occupied by the substance, is a constant for pure solids and liquids.

Calculating Equilibrium Constants

15.28 $$[CH_3OH] = \frac{0.0406 \text{ mol}}{2.00 \text{ L}} = 0.0203 \text{ M}$$

$$[CO] = \frac{0.170 \text{ mol CO}}{2.00 \text{ L}} = 0.0850\ M$$

$$[H_2] = \frac{0.302 \text{ mol } H_2}{2.00 \text{ L}} = 0.151\ M$$

$$K_c = \frac{[CH_3OH]}{[CO][H_2]^2} = \frac{0.0203}{(0.0850)(0.151)^2} = 10.4743 = 10.5$$

15.30 (a) $$K_p = \frac{P_{PCl_5}}{P_{PCl_3} \times P_{Cl_2}} = \frac{1.30 \text{ atm}}{0.124 \text{ atm} \times 0.157 \text{ atm}} = 66.8$$

(b) Since $K_p > 1$, products (the numerator of the K_p expression) are favored over reactants (the denominator of the K_p expression).

15.32 (a) Calculate the concentrations of $H_2(g)$ and $Br_2(g)$ and the equilibrium concentration of $H_2(g)$. M = mol/L.

$$[H_2]_{init} = 1.374 \text{ g } H_2 \times \frac{1 \text{ mol } H_2}{2.0159 \text{ g } H_2} \times \frac{1}{2.00 \text{ L}} = 0.34079 = 0.341\ M$$

$$[Br_2] = 70.31 \text{ g } Br_2 \times \frac{1 \text{ mol } Br_2}{159.81 \text{ g } Br_2} \times \frac{1}{2.00 \text{ L}} = 0.21998 = 0.220\ M$$

$$[H_2]_{equil} = 0.566 \text{ g } H_2 \times \frac{1 \text{ mol } H_2}{2.0159 \text{ g } H_2} \times \frac{1}{2.00 \text{ L}} = 0.14038 = 0.140\ M$$

	$H_2(g)$	+ $Br_2(g)$	$\rightleftharpoons$ $2HBr(g)$
initial	0.34079 M	0.21998 M	0
change	-0.20041 M	-0.20041 M	+2(0.20041) M
equil.	0.14038 M	0.01957 M	0.40082 M

The change in H_2 is (0.34079 – 0.14038 = 0.20041 = 0.200). The changes in $[Br_2]$ and [HBr] are set by stoichiometry, resulting in the equilibrium concentrations shown in the table.

(b) $$K_c = \frac{[HBr]^2}{[H_2][Br_2]} = \frac{(0.40082)^2}{(0.14038)(0.01957)} = \frac{(0.401)^2}{(0.140)(0.020)} = 58.48 = 58$$

The equilibrium concentration of Br_2 has 3 decimal places and 2 sig figs, so the value of K_c has 2 sig figs.

15.34 (a)

	$N_2O_4(g)$	$\rightleftharpoons$	$2NO_2(g)$
initial	1.500 atm		1.000 atm
change	+0.244 atm		–0.488 atm
equil	1.744 atm		0.512 atm

The change in P_{NO_2} is (1.000 - 0.512) = -0.488 atm, so the change in $P_{N_2O_4}$ is +(0.488/2) = +0.244 atm.

(b) $$K_p = \frac{P^2_{NO_2}}{P_{N_2O_4}} = \frac{(0.512)^2}{(1.744)} = 0.1503 = 0.150$$

Applications of Equilibrium Constants

15.36 (a) If the value of Q_c equals the value of K_c, the system is at equilibrium.

(b) In the direction of less products (more reactants), to the left.

(c) $Q_c = 0$ if the concentration of any product is zero.

15.38 Calculate the reaction quotient in each case, compare with

$$K_p = \frac{P^2_{NH_3}}{P_{N_2} \times P^3_{H_2}} = 4.51 \times 10^{-5}$$

(a) $$Q = \frac{(105)^2}{(35)\,(495)^3} = 2.6 \times 10^{-6}$$

Since $Q < K_p$, the reaction will shift to the right to attain equilibrium.

(b) $$Q = \frac{(35)^2}{(0)\,(595)^3} = \infty$$

Since $Q > K_p$, reaction must shift to the left to attain equilibrium. There must be **some** N_2 present to attain equilibrium. In this example, the only source of N_2 is the decomposition of NH_3.

(c) $$Q = \frac{(26)^2}{(42)^3\,(202)} = 4.52 \times 10^{-5};\; Q = K_p$$ Reaction is at equilibrium.

15.40 $$K_p = \frac{P^2_{SO_3}}{P^2_{SO_2} \times P_{O_2}};\; P_{SO_3} = \left(K_p \times P^2_{SO_2} \times P_{O_2}\right)^{1/2} = [(0.345)(0.165)^2(0.755)]^{1/2} = 0.0842 \text{ atm}$$

15.42 (a) $$K_c = \frac{[I]^2}{[I_2]} = 3.1 \times 10^{-5}$$

$$[I] = \frac{2.67 \times 10^{-2} \text{ g I}}{10.0 \text{ L}} \times \frac{1 \text{ mol I}}{126.9 \text{ g I}} = 2.1040 \times 10^{-5} = 2.10 \times 10^{-5}\,M$$

$$[I_2] = \frac{[I]^2}{K_c} = \frac{(2.104 \times 10^{-5})^2}{3.1 \times 10^{-5}} = 1.428 \times 10^{-5} = 1.43 \times 10^{-5}\,M$$

$$\frac{1.428\times10^{-5}\text{ mol } I_2}{L}\times 10.0\text{ L}\times\frac{253.8\text{ g } I_2}{\text{mol } I_2}=M=0.0362\text{ g } I_2$$

Check. $K_c=\dfrac{(2.104\times10^{-5})^2}{1.428\times10^{-5}}=3.1\times10^{-5}$

(b) $PV=nRT;\ P=\dfrac{gRT}{MM\ V}$

$$P_{SO_3}=\frac{1.57\text{ g } SO_3}{80.06\text{ g/mol}}\times\frac{0.08206\text{ L}\bullet\text{atm}}{\text{K}\bullet\text{mol}}\times\frac{700\text{ K}}{2.00\text{ L}}=0.5632=0.563\text{ atm}$$

$$P_{O_2}=\frac{0.125\text{ g } O_2}{32.00\text{ g/mol}}\times\frac{0.08206\text{ L}\bullet\text{atm}}{\text{K}\bullet\text{mol}}\times\frac{700\text{ K}}{2.00\text{ L}}=0.1122=0.112\text{ atm}$$

$$K_p=3.0\times10^4=\frac{P^2_{SO_3}}{P^2_{SO_2}\times P_{O_2}};\ P_{SO_2}=\left[P^2_{SO_3}/(K_p)(P_{O_2})\right]^{1/2}$$

$$P_{SO_2}=[(0.5632)^2/(3.0\times10^4)(0.1122)]^{1/2}=9.708\times10^{-3}=9.7\times10^{-3}\text{ atm}$$

$$g\ SO_2=\frac{MM\ PV}{RT}=\frac{64.06\text{ g } SO_2}{\text{mol } SO_2}\times\frac{\text{K}\bullet\text{mol}}{0.08206\text{ L}\bullet\text{atm}}\times\frac{9.708\times10^{-3}\text{ atm}\times2.00\text{ L}}{700\text{ K}}$$

$$=0.02165=0.022\text{ g } SO_2$$

Check. $K_p=[(9.708\times10^{-3})^2/(0.563)^2(0.1122)]=3.0\times10^4$

15.44 $[Br_2]=[Cl_2]=0.30\text{ mol}/1.0\text{ L}=0.30\ M$

	$Br_2(g)$ +	$Cl_2(g)$ ⇌	$2BrCl(g)$	$K_c=\dfrac{[BrCl]^2}{[Br_2][Cl_2]}=7.0$
initial	0.30 *M*	0.30 *M*	0	
change	−x	−x	+2x	
equil.	(0.30 − x)	(0.30 − x)	+2x	

$7.0=\dfrac{(2x)^2}{(0.30-x)^2}$ (We can solve this exactly by taking the square root of both sides.)

$(7.0)^{1/2}=\dfrac{2x}{0.30-x}$, $2.646(0.30-x)=2x$, $0.7937=4.646x$, $x=0.1709=0.17\ M$

$[BrCl]=2x=0.3417=0.34\ M$; $[Br_2]=[Cl_2]=0.30-x=0.1291=0.13\ M$

Check. $K_c=(0.3417)^2/(0.1291)^2=7.0$

15.46 $K_c=[NH_3][H_2S]=1.2\times10^{-4}$. Because of the stoichiometry, equilibrium concentrations of H_2S and NH_3 will be equal; call this quantity y. Then, $y^2=1.2\times10^{-4}$, $y=0.010954=0.011\ M$.

15.48 (a) *Analyze/Plan.* If only $PH_3BCl_3(s)$ is present initially, the equation requires that the equilibrium concentrations of $PH_3(g)$ and $BCl_3(g)$ are equal. Write the K_c expression and solve for $x=[PH_3]=[BCl_3]$. *Solve.*

$K_c=[PH_3][BCl_3]$; $1.87\times10^{-3}=x^2$; $x=0.043243=0.0432\ M$ PH_3 and BCl_3

(b) Since the mole ratios are 1:1:1, mol $PH_3BCl_3(s)$ required = mol PH_3 or BCl_3 produced.

$$\frac{0.043243 \text{ mol } PH_3}{L} \times 0.500 \text{ L} = 0.02162 = 0.0216 \text{ mol } PH_3 = 0.0216 \text{ mol } PH_3BCl_3$$

$$0.02162 \text{ mol } PH_3BCl_3 \times \frac{151.2 \text{ g } PH_3BCl_3}{1 \text{ mol } PH_3BCl_3} = 3.269 = 3.27 \text{ g } PH_3BCl_3$$

In fact, some $PH_3BCl_3(s)$ must remain for the system to be in equilibrium, so a bit more than 3.27 g PH_3BCl_3 is needed.

15.50 $CaCrO_4(s) \rightleftharpoons Ca^{2+}(aq) + CrO_4^{2-}(aq)$ $K_c = [Ca^{2+}][CrO_4^{2-}] = 7.1 \times 10^{-4}$

At equilibrium, $[Ca^{2+}] = [CrO_4^{2-}] = x$

$K_c = 7.1 \times 10^{-4} = x^2$, $x = 0.0266 = 0.027$ *M* Ca^{2+} and CrO_4^{2-}

LeChâtelier's Principle

15.52 (a) increase (b) increase (c) decrease

(d) no effect (e) no effect (f) no effect

15.54 (a) The reaction must be endothermic ($+\Delta H$) if heating increases the fraction of products.

(b) There must be more moles of gas in the products if increasing the volume of the vessel increases the fraction of products.

15.56 (a) $\Delta H° = \Delta H_f^\circ\ CH_3OH(g) - \Delta H_f^\circ\ CO(g) - 2\Delta H_f^\circ\ H_2(g)$

$= -201.2 \text{ kJ} - (-110.5 \text{ kJ}) - 0 \text{ kJ}$

$= -90.7 \text{ kJ}$

(b) The reaction is exothermic; an increase in temperature would decrease the value of K and decrease the yield. A low temperature is needed to maximize yield.

(c) Increasing total pressure would increase the partial pressure of each gas, shifting the equilibrium toward products. The extent of conversion to CH_3OH increases as the total pressure increases.

Additional Exercises

15.58 $CH_4(g) + H_2O(g) \rightarrow CO(g) + 3H_2(g)$

$$K_p = \frac{P_{CO} \times P_{H_2}^3}{P_{CH_4} \times P_{H_2O}}; P = \frac{g\,RT}{MM\,V}; T = 1000 \text{ K}$$

$$P_{CO} = \frac{8.62 \text{ g}}{28.01 \text{ g/mol}} \times \frac{0.08206 \text{ L} \bullet \text{atm}}{\text{mol} \bullet \text{K}} \bullet \frac{1000 \text{ K}}{5.00 \text{ L}} = 5.0507 = 5.05 \text{ atm}$$

$$P_{H_2} = \frac{2.60 \text{ g}}{2.016 \text{ g/mol}} \times \frac{0.08206 \text{ L} \bullet \text{atm}}{\text{mol} \bullet \text{K}} \bullet \frac{1000 \text{ K}}{5.00 \text{ L}} = 21.1663 = 21.2 \text{ atm}$$

$$P_{CH_4} = \frac{43.0\ g}{16.04\ g/mol} \times \frac{0.08206\ L \bullet atm}{mol \bullet K} \bullet \frac{1000\ K}{5.00\ L} = 43.9973 = 44.0\ atm$$

$$P_{H_2O} = \frac{48.4\ g}{18.02\ g/mol} \times \frac{0.08206\ L \bullet atm}{mol \bullet K} \bullet \frac{1000\ K}{5.00\ L} = 44.0811 = 44.1\ atm$$

$$K_p = \frac{(5.0507)\,(21.1663)^3}{(43.9973)\,(44.0811)} = 24.6949 = 24.7$$

$K_p = K_c(RT)^{\Delta n}$, $K_c = K_p/(RT)^{\Delta n}$; $\Delta n = 4 - 2 = 2$

$K_c = (24.6949)/[(0.08206)(1000)]^2 = 3.6673 \times 10^{-3} = 3.67 \times 10^{-3}$

15.60 (a) $H_2(g) + S(s) \rightleftharpoons H_2S(g)$ $K_c = [H_2S]/[H_2]$

(b) Calculate the molarities of H_2S and H_2.

$$[H_2S] = \frac{0.46\ g}{34.1\ g/mol} \times \frac{1}{1.0\ L} = 0.01349 = 0.013\ M$$

$$[H_2] = \frac{0.40\ g}{2.02\ g/mol} \times \frac{1}{1.0\ L} = 0.1980 = 0.20\ M$$

$K_c = 0.01349/0.1980 = 0.06812 = 0.068$

(c) Since S is a pure solid, its concentration doesn't change during the reaction, so [S] does not appear in the equilibrium expression.

15.61 (a) $K_p = \dfrac{P_{Br_2} \times P^2_{NO}}{P^2_{NOBr}}$; $P = \dfrac{gRT}{PV}$; $T = 100°C + 273 = 373$

$$P_{Br_2} = \frac{4.19\ g}{159.8\ g/mol} \times \frac{0.08206\ L \bullet atm}{K \bullet mol} \times \frac{373}{5.0\ L} = 0.16051 = 0.161\ atm$$

$$P_{NO} = \frac{3.08\ g}{30.01\ g/mol} \times \frac{0.08206\ L \bullet atm}{K \bullet mol} \times \frac{373}{5.0\ L} = 0.62828 = 0.628\ atm$$

$$P_{NOBr} = \frac{3.22\ g\ NOBr}{109.9\ g/mol} \times \frac{0.08206\ L \bullet atm}{K \bullet mol} \times \frac{373}{5.0\ L} = 0.17936 = 0.179\ atm$$

$$K_p = \frac{(0.16051)\,(0.62828)^2}{(0.17936)^2} = 1.9695 = 1.97 \qquad K_p = K_c(RT)^{\Delta n},\ \Delta n = 3 - 2 = 1$$

$K_c = K_p/RT = 1.9695/(0.08206)(373) = 0.064345 = 0.0643$

(b) $P_t = P_{Br_2} + P_{NO} + P_{NOBr} = 0.16051 + 0.62828 + 0.17936 = 0.96815 = 0.968\ atm$

15.63 (a) $K_p = \dfrac{P^2_{NH_3}}{P_{N_2} \times P^3_{H_2}} = 4.34 \times 10^{-3}$; $T = 300°C + 273 = 573\ K$

$$P_{NH_3} = \frac{gRT}{MM \times V} = \frac{1.05\ g}{17.03\ g/mol} \times \frac{0.08206\ L \bullet atm}{K \bullet mol} \times \frac{573\ K}{1.00\ L} = 2.899 = 2.90\ atm$$

	$N_2(g)$ +	$3H_2(g)$	⇌	$2NH_3(g)$
initial	0 atm	0 atm		?
change	x	3x		–2x
equil.	x atm	3x atm		2.899 atm

(Remember, only the change line reflects the stoichiometry of the reaction.)

$$K_p = \frac{(2.899)^2}{(x)(3x)^3} = 4.34\times10^{-3};\ 27\,x^4 = \frac{(2.899)^2}{4.34\times10^{-3}};\ x^4 = 71.725$$

$$x = 2.910 = 2.91\ \text{atm} = P_{N_2};\ P_{H_2} = 3x = 8.730 = 8.73\,\text{atm}$$

$$g_{N_2} = \frac{MM \times PV}{RT} = \frac{28.02\,g\ N_2}{mol\ N_2} \times \frac{K\bullet mol}{0.08206\ L\bullet atm} \times \frac{2.910\ atm \times 1.00\ L}{573\ K} = 1.73\ g\ N_2$$

$$g_{H_2} = \frac{2.016\ g\ H_2}{mol\ H_2} \times \frac{K\bullet mol}{0.08206\ L\bullet atm} \times \frac{8.730\ atm \times 1.00\ L}{573\ K} = 0.374\ g\ H_2$$

(b) The initial $P_{NH_3} = 2.899\,\text{atm} + 2(2.910\,\text{atm}) = 8.719 = 8.72\ \text{atm}$

$$g_{NH_3} = \frac{17.03\ g\ NH_3}{mol\ NH_3} \times \frac{K\bullet mol}{0.08206\ L\bullet atm} \times \frac{8.719\ atm \times 1.00\ L}{573\ K} = 3.16\ g\ NH_3$$

(c) $P_t = P_{N_2} + P_{H_2} + P_{NH_3} = 2.910\ \text{atm} + 8.730\ \text{atm} + 2.899\ \text{atm} = 14.54\ \text{atm}$

15.64

	2IBr	⇌	I_2 +	Br_2
initial	0.025 atm		0	0
change	–2x		x	x
equil.	(0.025 – 2x) atm		x	x

$$K_p = 8.5\times10^{-3} = \frac{P_{I_2}\times P_{Br_2}}{P^2_{IBr}} = \frac{x^2}{(0.025-2x)^2};\ \text{Taking the square root of both sides}$$

$$\frac{x}{0.025-2x} = (8.5 \times 10^{-3})^{1/2} = 0.0922;\ x = 0.0922(0.025-2x)$$

$x + 0.184x = 0.002305;\ 1.184x = 0.002305;\ x = 0.001947 = 1.9 \times 10^{-3}$

P_{IBr} at equilibrium = $0.025 - 2(1.947 \times 10^{-3}) = 0.02111 = 0.021$ atm

15.66 $K_p = P_{NH_3} \times P_{H_2S};\ P_t = 0.614$ atm

If the equilibrium amounts of NH_3 and H_2S are due solely to the decomposition of $NH_4HS(s)$, the equilibrium pressures of the two gases are equal, and each is 1/2 of the total pressure.

$P_{NH_3} = P_{H_2S} = 0.614\ \text{atm}/2 = 0.307$ atm

$K_p = (0.307)^2 = 0.0943$

15.67 Initial $P_{SO_3} = \frac{gRT}{MM\ V} = \frac{0.831\ g}{80.07\ g/mol} \times \frac{0.08206\ L \bullet atm}{K \bullet mol} \times \frac{1100\ K}{1.00\ L} = 0.9368 = 0.937$ atm

	$2SO_3$	$\rightleftharpoons$	$2SO_2$	+	O_2
initial	0.9368 atm		0		0
change	-2x		+2x		+x
equil.	0.9368–2x		2x		x
[equil.]	0.2104 atm		0.7264 atm		0.3632 atm

$P_t = (0.9368–2x) + 2x + x$; $0.9368 + x = 1.300$ atm; $x = 1.300 – 0.9368 = 0.3632 = 0.363$ atm

$$K_p = \frac{P^2_{SO_2} \times P_{O_2}}{P^2_{SO_3}} = \frac{(0.7264)^2\ (0.3632)}{(0.2104)^2} = 4.3292 = 4.33$$

$K_p = K_p = K_c(RT)^{\Delta n}$; $\Delta n = 3 - 2 = 1$; $K_p = K_c(RT)$

$K_c = K_p/RT = 4.3292/[(0.08206)(1100)] = 0.04796 = 0.0480$

15.69 $K_c = [CO_2] = 0.0108; [CO_2] = \frac{g\ CO_2}{44.01\ g/mol} \times \frac{1}{10.0\ L}$

In each case, calculate $[CO_2]$ and determine the position of the equilibrium.

(a) $[CO_2] = \frac{4.25\ g}{44.01\ g/mol} \times \frac{1}{10.0\ L} = 9.657 \times 10^{-3} = 9.66 \times 10^{-3}\ M$

$Q = 9.66 \times 10^{-3} > K_c$. The reaction proceeds to the right to achieve equilibrium and the amount of $CaCO_3(s)$ decreases.

(b) $[CO_2] = \frac{5.66\ g\ CO_2}{44.01\ g/mol} \times \frac{1}{10.0\ L} = 0.0129\ M$

$Q = 0.0129 > K_c$. The reaction proceeds to the left to achieve equilibrium and the amount of $CaCO_3(s)$ increases.

(c) 6.48 g CO_2 means $[CO_2] > 0.0129\ M$; $Q > 0.0129 > K_c$, the amount of $CaCO_3$ increases.

15.70 (a) $K_p = \frac{P_{Ni(CO)_4}}{P^4_{CO}}$

(b) Increasing the temperature to 200°C favors the reverse process (decomposition of $Ni(CO)_4(g)$ and thus the value of K_p is smaller at the higher temperature. This is the behavior expected from an exothermic reaction (heat is a product).

(c) At the temperature of the exhaust pipe, the $Ni(CO)_4$ product is a gas and is carried into the atmosphere with other exhaust gases. Thus, equilibrium is never established (we do not have a closed system) and the reaction proceeds to the right as $Ni(CO)_4$ product is removed.

15.72 (a)

	$CCl_4(g)$ ⇌	$C(s) + 2Cl_2(g)$
initial	2.00 atm	0 atm
change	–x atm	+2x atm
equil.	(2.00–x) atm	2x atm

$$K_P = 0.76 = \frac{P^2_{Cl_2}}{P_{CCl_4}} = \frac{(2x)^2}{(2.00-x)}$$

$1.52 - 0.76x = 4x^2$; $4x^2 + 0.76x - 1.52 = 0$

Using the quadratic formula, a = 4, b = 0.76, c = –1.52

$$x = \frac{-0.76 \pm \sqrt{(0.76)^2 - 4(4)(-1.52)}}{2(4)} = \frac{-0.76 + 4.99}{8} = 0.5287 = 0.53 \text{ atm}$$

$$\text{Fraction } CCl_4 \text{ reacted} = \frac{x \text{ atm}}{2.00 \text{ atm}} = \frac{0.53}{2.00} = 0.264 = 26\%$$

(b) $P_{Cl_2} = 2x = 2(0.5287) = 1.06 \text{ atm}$

$P_{CCl_4} = 2.00 - x = 2.00 - 0.5287 = 1.47 \text{ atm}$

15.73 (a) $$Q = \frac{P_{PCl_5}}{P_{PCl_3} \times P_{Cl_2}} = \frac{(0.20)}{(0.50)(0.50)} = 0.80$$

0.80 (Q) > 0.0870 (K), the reaction proceeds to the left.

(b)

	$PCl_3(g)$ +	$Cl_2(g)$ ⇌	$PCl_5(g)$
initial	0.50 atm	0.50 atm	0.20 atm
change	+x atm	+x atm	–x atm
equil.	(0.50 + x) atm	(0.50 + x) atm	(0.20 – x) atm

(Since the reaction proceeds to the left, P_{PCl_5} must decrease and P_{PCl_3} and P_{Cl_2} must increase.)

$$K_P = 0.0870 = \frac{(0.20-x)}{(0.50+x)(0.50+x)}; \quad 0.0870 = \frac{(0.20-x)}{(0.250 + 1.00x + x^2)}$$

$0.0870(0.250 + 1.00x + x^2) = 0.20 - x$; $-0.17825 + 1.0870x + 0.0870x^2 = 0$

$$x = \frac{-1.0870 \pm \sqrt{(1.0870)^2 - 4(0.0870)(-0.17825)}}{2(0.0870)} = \frac{-1.0870 + 1.1152}{0.174} = 0.162$$

$P_{PCl_3} = (0.50 + 0.162) \text{ atm} = 0.662$ $P_{Cl_2} = (0.50 + 0.162) \text{ atm} = 0.662 \text{ atm}$

$P_{PCl_5} = (0.20 - 0.162) \text{ atm} = 0.038 \text{ atm}$

To two decimal places, the pressures are 0.66, 0.66 and 0.04 atm, respectively. When substituting into the K_p expression, pressures to three decimal places yield a result much closer to 0.0870.

(c) Increasing the volume of the container favors the process where more moles of gas are produced, so the reverse reaction is favored and the equilibrium shifts to the left; the mole fraction of Cl_2 increases.

(d) For an exothermic reaction, increasing the temperature decreases the value of K; more reactants and fewer products are present at equilibrium and the mole fraction of Cl_2 increases.

15.75 (a) Since the volume of the vessel = 1.00 L, mol = *M*. The reaction will proceed to the left to establish equilibrium.

	A(g) +	2B(g) ⇌	2C(g)
initial	0 *M*	0 *M*	1.00 *M*
change	+x *M*	+2x *M*	-2x *M*
equil.	x *M*	2x *M*	(1.00 – 2x) *M*

At equilibrium, [C] = (1.00 – 2x) *M*, [B] = 2x *M*.

(b) x must be less than 0.50 *M* (so that [C], 1.00 -2x, is not less than zero).

(c) $K_c = \dfrac{[C]^2}{[A][B]^2}; \dfrac{(1.00-2x)^2}{(x)(2x)^2} = 0.25$

$1.00 - 4x + 4x^2 = 0.25(4x)^3; x^3 - 4x^2 + 4x - 1 = 0$

(d)

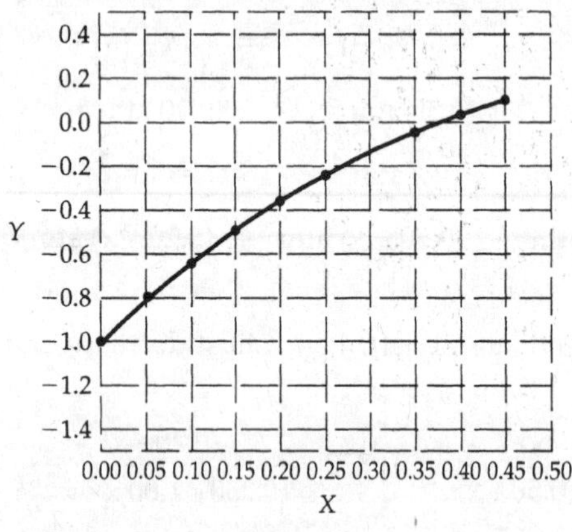

X	Y
0.0	−1.000
0.05	−0.810
0.10	−0.639
0.15	−0.487
0.20	−0.352
0.25	−0.234
0.35	−0.047
0.40	+0.024
0.45	+0.081
~0.383	0.00

(e) From the plot, x ≈ 0.383 *M*

[A] = x = 0.383 *M*; [B] = 2x = 0.766 *M*

[C] = 1.00 – 2x = 0.234 *M*

Using the K_c expression as a check:

$K_c = 0.25; \quad \dfrac{(0.234)^2}{(0.383)(0.766)^2} = 0.24;$ the estimated values are reasonable.

15.76　$K_P = \frac{P_{O_2} \times P_{CO}^2}{P_{CO_2}^2} \approx 1 \times 10^{-13}$; $P_{O_2} = (0.03)\,(1\text{ atm}) = 0.03\text{ atm}$

$P_{CO} = (0.002)\,(1\text{ atm}) = 0.002\text{ atm}$;　$P_{CO_2} = (0.12)\,(1\text{ atm}) = 0.12\text{ atm}$

$$Q = \frac{(0.03)(0.002)^2}{(0.12)^2} = 8.3 \times 10^{-6} = 8 \times 10^{-6}$$

Since $Q > K_p$, the system will shift to the left to attain equilibrium. Thus a catalyst that promoted the attainment of equilibrium would result in a lower CO content in the exhaust.

Integrative Exercises

15.79　(a)　$AgCl(s) \rightleftharpoons Ag^+(aq) + Cl^-(aq)$

(b)　$K_c = [Ag^+][Cl^-]$

(c)　Using thermodynamic data from Appendix C, calculate ΔH for the reaction in part (a).

$$\Delta H° = \Delta H_f^\circ\ Ag^+(aq) + \Delta H_f^\circ\ Cl^-(aq) - \Delta H_f^\circ\ AgCl(s)$$

$$\Delta H° = 105.90\text{ kJ} - 167.2\text{ kJ} - (-127.0\text{ kJ}) = 65.7\text{ kJ}$$

The reaction is endothermic (heat is a reactant), so the solubility of AgCl(s) in $H_2O(l)$ will increase with increasing temperature.

15.81　Consider the energy profile for an exothermic reaction.

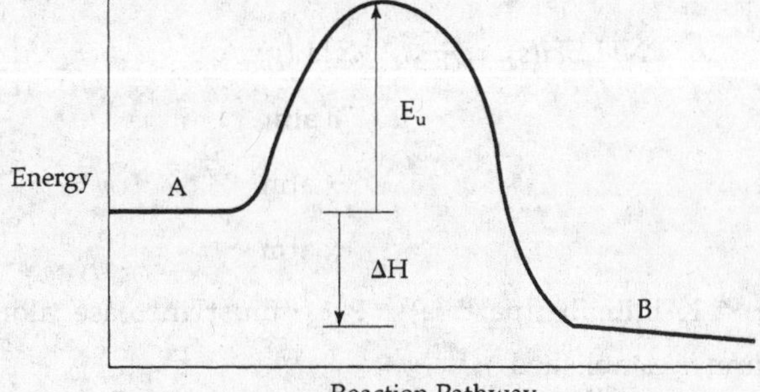

The activation energy in the forward direction, E_{af}, equals E_u, and the activation energy in the reverse reaction, E_{ar}, equals E_u - ΔH. (The same is true for an endothermic reaction because the sign of ΔH is the positive and $E_{ar} < E_{af}$). For the reaction in question,

$$K = \frac{k_f}{k_r} = \frac{A_f e^{-E_{af}/RT}}{A_r e^{-E_{ar}/RT}}$$

Since the ln form of the Arrhenius equation is easier to manipulate, we will consider ln K.

$$\ln K = \ln\left(\frac{k_f}{k_r}\right) = \ln k_f - \ln k_r = \frac{-E_{af}}{RT} + \ln A_f - \left[\frac{-E_{ar}}{RT} + \ln A_r\right]$$

Substituting E_u for E_{af} and $(E_u - \Delta H)$ for E_{ar}

$$\ln K = \frac{-E_u}{RT} + \ln A_f - \left[\frac{-(E_u - \Delta H)}{RT} + \ln A_r\right];\ \ln K = \frac{-E_u + (E_u - \Delta H)}{RT} + \ln A_f - \ln A_r$$

$$\ln K = \frac{-\Delta H}{RT} + \ln\frac{A_f}{A_r}$$

For the catalyzed reaction, $E_{cat} < E_u$ and $E_{af} = E_{cat}$, $E_{ar} = E_{cat} - \Delta H$. The catalyst does not change the value of ΔH.

$$\ln K_{cat} = \frac{-E_{cat} + (E_{cat} - \Delta H)}{RT} + \ln A_f - \ln A_r$$

$$\ln K_{cat} = \frac{-\Delta H}{RT} + \frac{\ln A_f}{A_r}$$

Thus, assuming A_f and A_r are not changed by the catalyst, $\ln K = \ln K_{cat}$ and $K = K_{cat}$.

15.82 (a) $P = \frac{gRT}{MM\ V} = \frac{0.300\text{ g } H_2S}{34.08\text{ g/mol } H_2S} \times \frac{298\text{ K}}{5.00\text{ L}} \times \frac{0.08206\text{ L}\bullet\text{atm}}{\text{mol}\bullet\text{K}} = 0.043053$

$= 0.0431$ atm

(b) $K_p = P_{NH_3} \times P_{H_2S}$. Before solid is added, $Q = P_{NH_3} \times P_{H_2S} = 0 \times 0.0431 = 0$.

$Q < K$ and the reaction will proceed to the right. However, no $NH_4SH(s)$ is present to produce $NH_3(g)$, so the reaction cannot proceed.

(c)

	$NH_4SH(s)$	$\rightleftharpoons$	$NH_3(g)$	+	$H_2S(g)$
initial			0 atm		0.043053 atm
change			+x atm		+x atm
equil.			+x atm		(0.043053+x) atm

Since $Q < K$ initially [part (a)], P_{H_2S} must increase along with P_{NH_3} until equilibrium is established.

$K_p = P_{NH_3} \times P_{H_2S}$; $0.120 = (x)(0.043053 + x)$; $0 = x^2 + 0.043053\,x - 0.120$

Solve for x using the quadratic formula.

$$x = \frac{-0.043053 \pm \sqrt{(0.043053)^2 - 4(1)(-0.120)}}{2(1)};\quad x = 0.3256 = 0.326\text{ atm}$$

$P_{NH_3} = 0.326$ atm; $P_{H_2S} = (0.043053 + 0.3256)$ atm $= 0.3686 = 0.369$ atm

(d) $\chi_{H_2S} = \frac{P_{H_2S}}{P_t} = \frac{0.3686\text{ atm}}{(0.3256 + 0.3686)\text{ atm}} = 0.531$

(e) The minimum amount of $NH_4HS(s)$ required is slightly greater than the number of moles NH_3 present at equilibrium. We can calculate the mol NH_3 present at equilibrium using the ideal-gas equation.

$$n_{NH_3} = \frac{P_{NH_3} V}{RT} = 0.3256 \text{ atm} \times \frac{K \bullet mol}{0.08206 \text{ L} \bullet atm} \times \frac{5.00 \text{ L}}{298 \text{ K}} = 0.06657$$

$$= 0.0666 \text{ mol } NH_3$$

$$0.06657 \text{ mol } H_2S \times \frac{1 \text{ mol } NH_4SH}{1 \text{ mol } H_2S} \times \frac{51.12 \text{ g } NH_4SH}{1 \text{ mol } NH_4SH} = 3.40 \text{ g } NH_4SH$$

The minimum amount is slightly greater than 3.40 g NH_4HS.

15.84 (a) $H_2O(l) \rightleftharpoons H_2O(g)$; $K_p = P_{H_2O}$

(b) At 30°C, the vapor pressure of $H_2O(l)$ is 31.82 torr. $K_P = P_{H_2O} = 31.82$ torr

$K_p = 31.82 \text{ torr} \times 1 \text{ atm}/760 \text{ torr} = 0.041868 = 0.04187$ atm

(c) From part (b), the value of K_p is the vapor pressure of the liquid at that temperature. By definition, vapor pressure = atmospheric pressure = 1 atm at the normal boiling point. $K_p = 1$ atm

15.85 (a)

```
     H   H
     |   |          H          H
Cl—C—C—Cl            \C=C/
     |   |          H/         \Cl
     H   H
```

C-C B.O. = 1   C=C B.O. = 2

(b) ΔH = D (bond breaking) – D (bond making)

E1: $\Delta H = D(C{=}Cl) - D(C{-}Cl) - D(C{-}C) - 2D(C{-}Cl)$

$\Delta H = 614 + 242 - 348 - 2(328) = -148$ kJ

E2: $\Delta H = D(C{-}C) + D(C{-}H) + D(C{-}Cl) - D(C{=}C) - D(H{-}Cl)$

$= 348 + 413 + 328 - 614 - 431 = 44$ kJ

(c) E1 is exothermic with $\Delta n = -1$. The yield of $C_2H_4Cl_2(g)$ would decrease with increasing temperature and with increasing container volume.

(d) E2 is endothermic with $\Delta n = 1$. The yield of C_2H_3Cl would increase with increasing temperature and with increasing container volume.

(e) The boiling points of the reactants and products are: C_2H_4, -103.7°C; Cl_2, -34.6°C, $C_2H_4Cl_2$, +83.5°C; C_2H_3Cl, -13.4°C; HCl, -84.9°C.

Because the products of E1 and E2 are optimized by different conditions, carry out the two equilibria in separate reactors. Since E1 is exothermic and $\Delta n = -1$, the reactor on the left should be as small and cold as possible to maximize yield of $C_2H_4Cl_2$. At temperatures below 83.5°C, $C_2H_4Cl_2$ will condense to the liquid and it can be easily transferred to the second (right) reactor.

Since E_2 is endothermic, the reactor on the right should be as large and hot as possible to optimize production of C_2H_3Cl. The outlet stream will be a mixture of $C_2H_3Cl(g)$, $C_2H_4Cl_2(g)$, and HCl(g). This mixture could be run through a heat exchanger to condense and subsequently recycle $C_2H_4Cl_2(l)$. Since HCl has a lower boiling point than C_2H_3Cl, it cannot be removed by condensation. The $C_2H_3Cl(g)$ / HCl(g) mixture could be bubbled through a basic aqueous solution such as $NaHCO_3(aq)$ or NaOH(aq) to remove HCl(g), leaving pure $C_2H_3Cl(g)$.

Other details of reactor design such as the use of catalysts to speed up these reactions, the exact costs and benefits of heat exchange, recycling unreacted components, and separation and recovery of products are issues best resolved by chemical engineers.

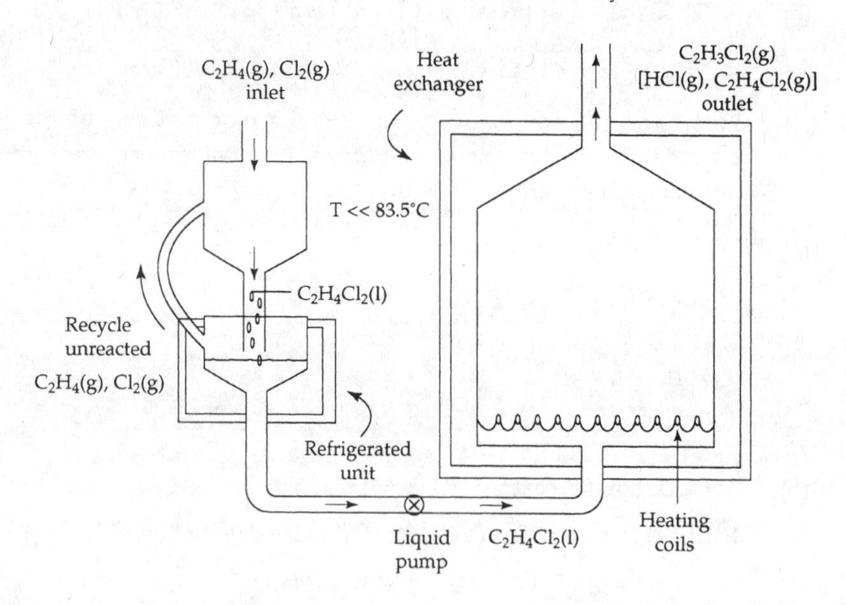

16 Acid-Base Equilibria

Visualizing Concepts

16.2 *Plan.* The stronger the acid, the greater the extent of ionization. The stronger the acid, the weaker its conjugate base. In an acid-base reaction, equilibrium will favor the side with the weaker acid and base. *Solve.*

(a) HY is stronger than HX. Starting with six HY molecules, four are dissociated; of six HX molecules, only two are dissociated. Because it is dissociated to a greater extent, HY is the stronger acid.

(b) If HY is the stronger acid, Y^- is the weaker base and X^- is the stronger base.

(c) HX and Y^-, the reactants, are the weaker acid and base. Equilibrium lies to the left, and $K_c < 1$.

16.3 *Plan.* Strong acids are completely ionized. The acid that is least ionized is weakest, and has the smallest K_a value. At equal concentrations, the weakest acid has the smallest $[H^+]$ and highest pH. *Solve.*

(a) HY is a strong acid. There are no neutral HY molecules in solution, only H^+ cations and Y^- anions.

(b) HX has the smallest K_a value. It has most neutral acid molecules and fewest ions.

(c) HX has the fewest H^+ and highest pH.

16.4 *Plan.* The definition of % ionization is $\frac{[H^+]}{[HA]_{Initial}} \times 100$. *Solve.*

(a) Curve C shows the effect of initial concentration on % ionization of a weak acid.

(b) The % ionization is inversely related to initial acid concentration; only curve C shows a decrease in % ionization as acid concentration increases.

16.6 *Plan.* Write the molecular formula so we can count the correct number of valence electrons. Use the atom connectivity shown to draw the Lewis structure. *Solve.*

(a) The molecular formula is $(CH_3)_2NH$, or C_2H_7N. The number of valence electrons is $2(4) + 7 + 5 = 20\ e^-$, $10\ e^-$ pr.

```
    H       H
    |   ..  |
H—C—N—C—H
    |   |   |
    H   H   H
```

(b) The compound is an amine. It is an ammonia molecule where two H atoms have been replaced by CH_3 groups.

16.7 *Plan.* Evaluate the interactions of Na^+ and X^- with H_2O.

Solve. Na^+ does not affect the $[H^+]$ or $[OH^-]$ of an aqueous solution. It is a "negligible" acid in water (which can be thought of as the conjugate acid of the strong base NaOH).

X^- is the conjugate base of HX. It is not a negligible base in water, because we see from the diagram that one X^- has gained an H^+ to form HX. In this solution, H_2O acts as the Brønsted acid, according to the hydrolysis equilibrium:

$$X^-(aq) + H_2O(l) \rightleftharpoons HX(aq) + OH^-(aq).$$

The missing ion is $OH^-(aq)$. According to the equilibrium reaction, the number of HX molecules and OH^- ions are equal. Since there is 1 HX molecule in the diagram, 1 OH^- should be shown.

16.9 The carboxyl group in the H-atom group at the "top" of the molecule.

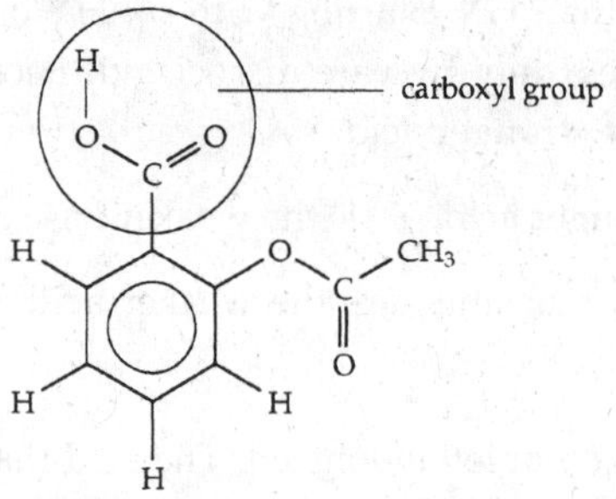

carboxyl group

The group on the right of the molecule is <u>not</u> a carboxyl group because it contains no ionizable H.

16.10 (a) *Plan.* Count valence electrons and draw the correct Lewis structures. Consider the definition of Lewis acids and bases. *Solve.*

PCl_4^- + Cl^- → PCl_5

32 e^-, 16 e^- pr; 8 e^-, 4 e^- pr; 40 e^-, 20 e^- pr

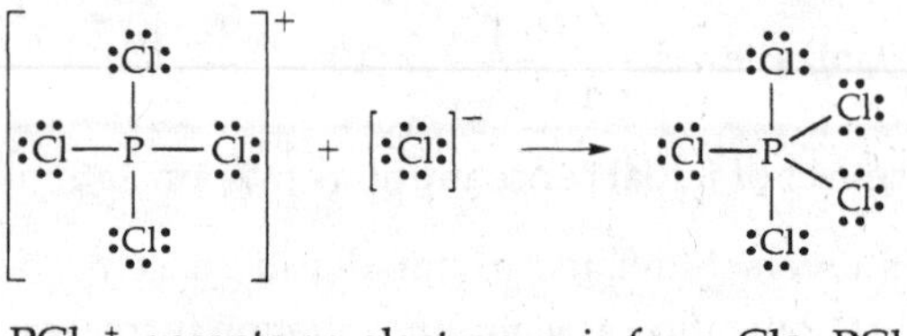

PCl_4^+ accepts an electron pair from Cl^-; PCl_4^+ is the Lewis acid and Cl^- is the Lewis base.

(b) The hydrated cation is an oxyacid: the ionizable H is attached to O, which is bound to the central cation. As the charge on the cation increases, it attracts more electron density from the O-H bond, which becomes weaker and more polar. The degree of ionization increases and the equilibrium constant (K_a) increases.

Arrhenius and Brønsted-Lowry Acids and Bases

16.12 When NaOH dissolves in water, it completely dissociates to form $Na^+(aq)$ and $OH^-(aq)$. CaO is the oxide of a metal; it dissolves in water according to the following process: $CaO(s) + H_2O(l) \rightarrow Ca^{2+}(aq) + 2OH^-(aq)$. Thus, the properties of both solutions are dominated by the presence of $OH^-(aq)$. Both solutions taste bitter, turn litmus paper blue (are basic), neutralize solutions of acids, and conduct electricity.

16.14 (a) According to the Arrhenius definition, a *base* when dissolved in water increases $[OH^-]$. According to the Brønsted-Lowry theory, a *base* is an H^+ acceptor regardless of physical state. A Brønsted-Lowry base is not limited to aqueous solution and need not contain OH^- or produce it in aqueous solution.

(b) $NH_3(g) + H_2O(l) \rightleftharpoons NH_4^+(aq) + OH^-(aq)$ When NH_3 dissolves in water, it accepts H^+ from H_2O (B-L definition). In doing so, OH^- is produced (Arrhenius definition).

Note that the OH^- produced was originally part of the H_2O molecule, not part of the NH_3 molecule.

16.16 A conjugate acid has one more H^+ than its conjugate base.

(a) HCN (b) OH (c) $H_2PO_4^-$ (d) $C_2H_5NH_3^+$

16.18

	B-L acid	+	B-L base	$\rightleftharpoons$	Conjugate acid	+	Conjugate base
(a)	$HBrO(aq)$		$H_2O(l)$		$H_3O^+(aq)$		$BrO^-(aq)$
(b)	$HSO_4^-(aq)$		$HCO_3^-(aq)$		$H_2CO_3(aq)$		$SO_4^{2-}(aq)$
(c)	$H_3O^+(aq)$		$HSO_3^-(aq)$		$H_2SO_3(aq)$		$H_2O(l)$

16.20 (a) $H_2C_6H_7O_5^-(aq) + H_2O(l) \rightleftharpoons H_3C_6H_7O_5(aq) + OH^-(aq)$

(b) $H_2C_6H_7O_5^-(aq) + H_2O(l) \rightleftharpoons HC_6H_7O_5^{2-}(aq) + H_3O^+(aq)$

(c) $H_3C_6H_7O_5$ is the conjugate acid of $H_2C_6H_7O_5^-$

$HC_6H_7O_5^{2-}$ is the conjugate base of $H_2C_6H_7O_5^-$

16.22 (a) $C_2H_3O_2^-$, weak base; $HC_2H_3O_2$, weak acid

(b) HCO_3^-, weak base; H_2CO_3, weak acid

(c) O_2^-, strong base; OH^-, negligible acid

(d) Cl^-, negligible base; HCl, strong acid

(e) NH_3, weak base; NH_4^+, weak acid

16.24 (a) HNO_3. It is one of the seven strong acids (Section 16.5). Also, in a series of oxyacids with the same central atom (N), the acid with more O atoms is stronger (Section 16.10).

(b) NH_3. When NH_3 and H_2O are combined, as in $NH_3(aq)$, NH_3 acts as the B-L base, accepting H^+ from H_2O. NH_3 has the greater tendency to accept H^+. For binary hydrides, base strength increases going to the left across a row of the periodic chart (Section 16.10).

16.26

	Base	+	Acid	$\rightleftharpoons$	Conjugate acid	+	Conjugate base
(a)	$OH^-(aq)$	+	$NH_4^+(aq)$	$\rightleftharpoons$	$H_2O(l)$	+	$NH_3(aq)$

OH^- is a stronger base than NH_3 (Figure 16.4), so the equilibrium lies to the right.

(b) $C_2H_3O_2^-(aq) + H_3O^+(aq) \rightleftharpoons HC_2H_3O_2(aq) + H_2O(l)$

H_3O^+ is a stronger acid than $HC_2H_3O_2$ (Figure 16.4), so the equilibrium lies to the right.

(c) $F^-(aq) + HCO_3^-(aq) \rightleftharpoons HC_2H_3O_2(aq) + H_2O(l)$

CO_3^{2-} is a stronger base than F^-, so the equilibrium lies to the left.

Autoionization of Water

16.28 (a) $H_2O(l) \rightleftharpoons H^+(aq) + OH^-(aq)$

(b) $K_w = [H^+][OH^-]$. The $[H_2O(l)]$ is omitted because water is a pure liquid. The molarity (mol/L) of pure solids or liquids does not change as equilibrium is established, so it is usually omitted from equilibrium expressions.

(c) If a solution is basic, it contains more OH^- than H^+ ($[OH^-] > [H^+]$).

16.30 In pure water at 25°C, $[H^+] = [OH^-] = 1 \times 10^{-7}$ *M*. If $[OH^-] > 1 \times 10^{-7}$ *M*, the solution is basic; if $[OH^-] < 1 \times 10^{-7}$ *M*, the solution is acidic.

(a) $[OH^-] = \dfrac{K_W}{[H^+]} = \dfrac{1.0 \times 10^{-14}}{7.5 \times 10^{-3}\, M} = 1.3 \times 10^{-12}\, M < 1 \times 10^{-7}\, M$; acidic

(b) $[OH^-] = \dfrac{K_W}{[H^+]} = \dfrac{1.0 \times 10^{-14}}{6.5 \times 10^{-10}\, M} = 1.5 \times 10^{-5}\, M > 1 \times 10^{-7}\, M$; basic

(c) $[H^+] = 10[OH^-]$; $K_w = 10[OH^-][OH^-] = 10[OH^-]^2$

$[OH^-] = (K_w/10)^{1/2} = 3.2 \times 10^{-8}\, M < 1 \times 10^{-7}\, M$; acidic

16.32 $K_w = [D^+][OD^-]$; for pure D_2O, $[D^+] = [OD^-]$; $8.9 \times 10^{-16} = [D^+]^2$;

$[D^+] = [OD^-] = 3.0 \times 10^{-8}\, M$

The pH Scale

16.34 $[H^+]_A = 500\,[H^+]_B$ From Solution 16.33, $\Delta pH = -\log \dfrac{[H^+]_B}{[H^+]_A}$

$$\Delta pH = -\log \frac{[H^+]_B}{500\,[H^+]_B} = -\log\left(\frac{1}{500}\right) = 2.70$$

The pH of solution A is 2.70 pH units lower than the pH of solution B, because $[H^+]_A$ is 500 times greater than $[H^+]_B$. The greater $[H^+]$, the lower the pH of the solution.

16.36 (a) $K_w = [H^+][OH^-]$. If HNO_3 is added to water, it ionizes to form $H^+(aq)$ and $NO_3^-(aq)$. This increases $[H^+]$ and necessarily decreases $[OH^-]$. When $[H^+]$ increases, pH decreases.

(b) On Figure 16.5, 1.4×10^{-2} M OH^- is between pH = 12 (1×10^{-2} M OH^-) and pH 13 (1×10^{-1} M OH^-), slightly higher than pH = 12, so we estimate pH = 12.1. If pH > 7, the solution is basic.

(c) pH = 6.6 is midway between pH 6 and pH 7 on Figure 16.5.

At pH = 7, $[H^+] = 1 \times 10^{-7}$; at pH = 6, $[H^+] = 1 \times 10^{-6} = 10 \times 10^{-7}$.

A reasonable estimate is 5×10^{-7} M H^+. By calculation:

pH = 6.6, $[H^+] = 10^{-pH} = 10^{-6.6} = 3 \times 10^{-7}$

At pH = 6, $[OH^-] = 1 \times 10^{-9}$; at pH = 7, $[OH^-] = 1 \times 10^{-7} = 10 \times 10^{-8}$.

A reasonable estimate is 5×10^{-8} M OH^-. By calculation:

pOH = 14.0 – 6.6 = 7.4; $[OH^-] = 10^{-pOH} = 10^{-7.4} = 4 \times 10^{-8}$ M OH^-.

16.38

pH	pOH	$[H^+]$	$[OH^-]$	acidic or basic
4.75	9.25	1.8×10^{-5} M	5.6×10^{-10} M	acidic
2.11	11.89	7.8×10^{-3} M	1.3×10^{-12} M	acidic
2.19	11.81	6.5×10^{-3} M	1.5×10^{-12} M	acidic
7.93	6.07	1.2×10^{-8} M	8.6×10^{-7} M	basic

16.40 The pH ranges from 5.2–5.6; pOH ranges from (14.0–5.2 =) 8.8 to (14.0–5.6 =) 8.4.

$[H^+] = 10^{-pH}$, $[OH^-] = 10^{-pOH}$

$[H^+] = 10^{-5.2} = 6.31 \times 10^{-6} = 6 \times 10^{-6}$ M; $[H^+] = 10^{-5.6} = 2.51 \times 10^{-6} = 3 \times 10^{-6}$ M

The range of $[H^+]$ is 6×10^{-6} M to 3×10^{-6} M.

$[OH^-] = 10^{-8.8} = 1.58 \times 10^{-9} = 2 \times 10^{-9}$ M; $[OH^-] = 10^{-8.4} = 3.98 \times 10^{-9} = 4 \times 10^{-9}$ M.

The range of $[OH^-]$ is 2×10^{-9} M to 4×10^{-9} M.

(The pH has one decimal place, so concentrations are reported to one sig fig.)

Strong Acids and Bases

16.42 (a) A strong base is completely dissociated in aqueous solution; a strong base is a strong electrolyte.

(b) $Sr(OH)_2$ is a soluble strong base.

$Sr(OH)_2(aq) \rightarrow Sr^{2+}(aq) + 2OH^-(aq)$

0.045 M $Sr(OH)_2(aq)$ = 0.090 M OH^-

(c) Base strength should not be confused with solubility. Base strength describes the tendency of a dissolved molecule (formula unit for ionic compounds such as $Mg(OH)_2$) to dissociate into cations and hydroxide ions. $Mg(OH)_2$ is a strong base because each $Mg(OH)_2$ unit that dissolves also dissociates into $Mg^{2+}(aq)$ and $OH^-(aq)$. $Mg(OH)_2$ is not very soluble, so relatively few $Mg(OH)_2$ units dissolve when the solid compound is added to water.

16.44 For a strong acid, which is completely ionized, $[H^+]$ = the initial acid concentration.

(a) $0.00835\ M\ HNO_3 = 0.00835\ M\ H^+$; $pH = -\log(0.00835) = 2.08$

(b) $$\frac{0.525 \text{ g } HClO_4}{2.00 \text{ L soln}} \times \frac{1 \text{ mol } HClO_4}{100.5 \text{ g } HClO_4} = 2.612 \times 10^{-3} = 2.61 \times 10^{-3}\ M\ HClO_4$$

$[H^+] = 2.61 \times 10^{-3}\ M$; $pH = -\log(2.612 \times 10^{-3}) = 2.583$

(c) $M_c \times V_c = M_d \times V_d$; $0.500 \text{ L} = 500 \text{ mL}$

$1.00\ M\ HCl \times 5.00 \text{ mL } HCl = M_d\ HCl \times 500 \text{ mL } HCl$

$$M_d\ HCl = \frac{1.00\ M \times 5.00 \text{ mL}}{500 \text{ mL}} = 1.00 \times 10^{-2}\ M\ HCl = 1.00 \times 10^{-2}\ M\ H^+$$

$pH = -\log(1.00 \times 10^{-2}) = 2.000$

(d) $$[H^+]_{total} = \frac{\text{mol } H^+ \text{ from } HCl + \text{mol } H^+ \text{ from } HI}{\text{total L solution}}; \ \text{mol} = M \times L$$

$$[H^+]_{total} = \frac{(0.020\ M\ HCl \times 0.0500 \text{ L}) + (0.010\ M\ HI \times 0.150 \text{ L})}{0.200 \text{ L}}$$

$$[H^+]_{total} = \frac{1.0 \times 10^{-3} \text{ mol } H^+ + 1.50 \times 10^{-3} \text{ mol } H^+}{0.200 \text{ L}} = 0.0125 = 0.013\ M$$

$pH = -\log(0.0125) = 1.90$

16.46 For a strong base, which is completely dissociated, $[OH^-]$ = the initial base concentration. Then, $pOH = -\log[OH^-]$ and $pH = 14 - pOH$.

(a) $0.012\ M\ KOH = 0.012\ M\ OH^-$; $pOH = -\log(0.012) = 1.92$; $pH = 14 - 1.92 = 12.08$

(b) $$\frac{1.565 \text{ g } KOH}{0.5000 \text{ L}} \times \frac{1 \text{ mol } KOH}{56.106 \text{ g } KOH} = 0.055787 = 0.05579\ M = [OH^-]$$

$pOH = -\log(0.055787) = 1.2535$; $pH = 14 - pOH = 12.7465$

(c) $M_c \times V_c = M_d \times V_d$

$0.250\ M\ Ca(OH)_2 \times 10.0 \text{ mL} = M_d\ Ca(OH)_2 \times 500 \text{ mL}$

$$M_d\ Ca(OH)_2 = \frac{0.0105\ M\ Ca(OH)_2 \times 10.0 \text{ mL}}{500.0 \text{ mL}} = 2.10 \times 10^{-4}\ M\ Ca(OH)_2$$

$Ca(OH)_2(aq) \rightarrow Ca^{2+}(aq) + 2OH^-(aq)$

$[OH^-] = 2[Ca(OH)_2] = 2(2.10 \times 10^{-4}\ M) = 4.20 \times 10^{-4}\ M$

$pOH = -\log(4.20 \times 10^{-4}) = 3.377$; $pH = 14 - pOH = 10.623$

(d) $$[OH^-]_{total} = \frac{\text{mol } OH^- \text{ from } NaOH + \text{mol } OH^- \text{ from } Ba(OH)_2}{\text{total L solution}}$$

$$\frac{(7.5 \times 10^{-3}\ M \times 0.0400 \text{ L}) + 2(0.015\ M \times 0.0100 \text{ L})}{0.0500 \text{ L}}$$

$$[OH^-]_{total} = \frac{3.00 \times 10^{-4} \text{ mol } OH^- + 3.0 \times 10^{-4} \text{ mol } OH^-}{0.0500 \text{ L}} = 0.01200 = 0.012\ M\ OH^-$$

$pOH = -\log(0.0120) = 1.92$; $pH = 14 - pOH = 12.08$

16.48 $pOH = 14 - pH = 14.00 - 12.05 = 1.95$

$pOH = 1.95 = -\log[OH^-]$; $[OH^-] = 10^{-1.95} = 0.01122 = 1.1 \times 10^{-2}\ M$

$[OH^-] = 2[Ca(OH)_2]$; $[Ca(OH)_2] = [OH^-] / 2 = 0.01122/2 = 5.6 \times 10^{-3}\ M$

16.50 Upon dissolving, Li_2O dissociates to form Li^+ and O^{2-}. According to Equation 16.22, O^{2-} is completely protonated in aqueous solution.

$$Li_2O(s) + H_2O(l) \rightarrow 2Li^+(aq) + 2OH^-(aq)$$

Thus, initial $[Li_2O] = [O_2^-]$; $[OH^-] = 2[O^{2-}] = 2[Li_2O]$

$$[Li_2O] = \frac{\text{mol } Li_2O}{\text{L solution}} = 2.50 \text{ g } Li_2O \times \frac{1 \text{ mol } Li_2O}{29.88 \text{ g } Li_2O} \times \frac{1}{1.500 \text{ L}} = 0.0558 = 0.0558\ M$$

$[OH^-] = 0.11156 = 0.112\ M$; $pOH = 0.9525 = 0.958$ $pH = 14.00 - pOH = 13.0475 = 13.048$

Weak Acids

16.52 (a) $$HC_6H_5O(aq) \rightleftharpoons H^+(aq) + C_6H_5O^-(aq);\quad K_a = \frac{[H^+][C_6H_5O^-]}{[HC_6H_5O]}$$

$$HC_6H_5O(aq) + H_2O(l) \rightleftharpoons H_3O^+(aq) + C_6H_5O^-(aq);\quad K_a = \frac{[H_3O^+][C_6H_5O^-]}{[HC_6H_5O]}$$

(b) $$HCO_3^-(aq) \rightleftharpoons H^+(aq) + CO_3^{2-}(aq);\quad K_a = \frac{[H^+][CO_3^{2-}]}{[HCO_3^-]}$$

$$HCO_3^-(aq) + H_2O(aq) \rightleftharpoons H_3O^+(aq) + CO_3^{2-}(aq);\quad K_a = \frac{[H_3O^+][CO_3^{2-}]}{[HCO_3^-]}$$

16.54 $$HC_8H_7O_2(aq) \rightleftharpoons H^+(aq) + C_8H_7O_2^-(aq);\quad K_a = \frac{[H^+][C_8H_7O_2^-]}{[HC_8H_7O_2]}$$

$[H^+] = [C_8H_7O_2^-] = 10^{-2.68} = 2.09 \times 10^{-3} = 2.1 \times 10^{-3}\ M$

$[HC_8H_7O_2] = 0.085 - 2.09 \times 10^{-3} = 0.0829 = 0.083\ M$

$$K_a = \frac{(2.09 \times 10^{-3})^2}{0.0829} = 5.3 \times 10^{-5}$$

16.56 $[H^+] = 0.132 \times [BrCH_2COOH]_{initial} = 0.0132\ M$

	$BrCH_2COOH(aq)$	$\rightleftharpoons$	$H^+(aq)$	+	$BrCH_2COO^-(aq)$
initial	0.100 M		0		0
equil.	0.087		0.0132 M		0.0132 M

$$K_a = \frac{[H^+][BrCH_2COO^-]}{[BrCH_2COOH]} = \frac{(0.0132)^2}{0.087} = 2.0 \times 10^{-3}$$

16.58 $[H^+] = 10^{-pH} = 10^{-3.25} = 5.623 \times 10^{-4} = 5.62 \times 10^{-4}$ *M*

$$K_a = 6.8 \times 10^{-4} = \frac{[H^+][F^-]}{[HF]} = \frac{(5.623 \times 10^{-4})^2}{x - 5.623 \times 10^{-4}}$$

$6.8 \times 10^{-4}(x - 5.623 \times 10^{-4}) = (5.623 \times 10^{-4})^2$;

$6.8 \times 10^{-4}\,x = 3.824 \times 10^{-7} + 3.162 \times 10^{-7} = 6.986 \times 10^{-7}$

$x = 1.027 \times 10^{-3} = 1.0 \times 10^{-3}$ *M* HF

mol = *M* × L = 1.027×10^{-3} *M* × 0.200 L = $2.055 \times 10^{-4} = 2.1 \times 10^{-4}$ mol HF

16.60

	$HClO(aq)$	$\rightleftharpoons$	$H^+(aq)$	+	$ClO^-(aq)$
initial	0.0075 *M*		0		0
equil.	(0.0075 – x) *M*		x *M*		x *M*

$$K_a = \frac{[H^+][ClO^-]}{[HClO]} = \frac{x^2}{(0.0075 - x)} \approx \frac{x^2}{0.0075} = 3.0 \times 10^{-5}$$

$x^2 = 0.0075\,(3.0 \times 10^{-8})$; $x = 1.5 \times 10^{-5}$ *M* $= [H^+] = [H_3O^+] = [ClO^-]$

$[HClO] = 7.5 \times 10^{-3} - 1.5 \times 10^{-6} = 7.485 \times 10^{-3} = 7.5 \times 10^{-3}$ *M*

Check. $\frac{4.7 \times 10^{-5}\ MH^+}{0.0075\ M\ HClO} \times 100 = 0.20\%$ ionization; the assumption is valid

16.62 (a)

	$HOCl(aq)$	$\rightleftharpoons$	$H^+(aq)$	+	$OCl^-(aq)$
initial	0.125 *M*		0		0
equil	(0.125 – x) *M*		x *M*		x *M*

$$K_a = \frac{[H^+][OCl^-]}{[HOCl]} = \frac{x^2}{(0.125 - x)} \approx \frac{x^2}{0.125} = 3.0 \times 10^{-8}$$

$x^2 = 0.125\,(3.0 \times 10^{-8})$; $x = [H^+] = 6.1 \times 10^{-5}$ *M*, pH = 4.21

Check. $\frac{6.1 \times 10^{-5}\ M\ H^+}{0.125\ M\ HOCl} \times 100 = 0.049\%$ ionization; the assumption is valid

(b) $$K_a = \frac{[H^+][C_6H_5O^-]}{[C_6H_5OH]} = \frac{x^2}{(0.0085 - x)} \approx \frac{x^2}{0.0085} = 1.3 \times 10^{-10}$$

$x^2 = 0.0085\,(1.3 \times 10^{-10})$; $x = [H^+] = 1.1 \times 10^{-6}$ *M*, pH = 5.98

Check. Clearly 1.1×10^{-6} *M* H^+ is small compared to 8.5×10^{-3} *M* C_6H_5OH, and the assumption is valid.

(c)

	$HONH_2(aq) + H_2O(l)$	$\rightleftharpoons$	$HONH_3^+(aq)$	+	$OH^-(aq)$
initial	0.095 *M*		0		0
equil	(0.095 – x) *M*		x *M*		x *M*

$$K_b = \frac{[HONH_3^+][OH^-]}{[HONH_2]} = \frac{x^2}{(0.095 - x)} \approx \frac{x^2}{0.095} = 1.1 \times 10^{-8}$$

$x^2 = 0.095\ (1.1 \times 10^{-8})$; $x = [OH^-] = 3.2 \times 10^{-5}\ M$, pH = 9.51

Check. $\frac{3.2 \times 10^{-5}\ M\ OH^-}{0.095\ M\ HONH_2} \times 100 = 0.034\%$ ionization; the assumption is valid

16.64 Calculate the initial concentration of $HC_9H_7O_4$.

$$2 \text{ tablets} \times \frac{500 \text{ mg}}{\text{tablet}} \times \frac{1 \text{ g}}{1000 \text{ mg}} \times \frac{1 \text{ mol } HC_9H_7O_4}{180.2 \text{ g } HC_9H_7O_4} = 0.005549 = 0.00555 \text{ mol } HC_9H_7O_4$$

$$\frac{0.005549 \text{ mol } HC_9H_7O_4}{0.250 \text{ L}} = 0.02220 = 0.0222\ M\ HC_9H_7O_4$$

	$HC_9H_7O_4$ (aq)	⇌	$C_9H_7O_4^-$	+	H^+ (aq)
Initial	0.0222 *M*		0 *M*		0 *M*
equil	(0.0222 – x)		x *M*		x *M*

$$K_a = 3.3 \times 10^{-4} = \frac{[H^+][C_7H_9O_4^-]}{[HC_7H_9O_4]} = \frac{x^2}{(0.0222 - x)}$$

Assuming x is small compared to 0.0222,

$x^2 = 0.0222\ (3.3 \times 10^{-4})$; $x = [H^+] = 2.7 \times 10^{-3}\ M$

$\frac{2.7 \times 10^{-3}\ M\ H^+}{0.0222\ M\ HC_9H_7O_4} \times 100 = 12\%$ ionization; the assumption is not valid

Using the quadratic formula, $x^2 + 3.3 \times 10^{-4} x - 7.325 \times 10^{-6} = 0$

$$x = \frac{-3.3 \times 10^{-4} \pm \sqrt{(3.3 \times 10^{-4})^2 - 4(1)(-7.325 \times 10^{-6})}}{2(1)} = \frac{-3.3 \times 10^{-4} \pm \sqrt{2.941 \times 10^{-5}}}{2}$$

$x = 2.547 \times 10^{-3} = 2.5 \times 10^{-3}\ M\ H^+$; pH = $-\log(2.547 \times 10^{-3}) = 2.594 = 2.59$

16.66 (a) $HC_3H_5O_2(aq) \rightleftharpoons H^+(aq) + C_3H_5O_2^-(aq)$

$$K_a = 1.3 \times 10^{-5} = \frac{[H^+][C_3H_5O_2^-]}{[HC_3H_5O_2]} = \frac{x^2}{0.250 - x}$$

$x^2 = 0.250\ (1.3 \times 10^{-5})$; $x = 1.803 \times 10^{-3} = 1.8 \times 10^{-3}\ M\ H^+$

$$\% \text{ ionization} = \frac{1.803 \times 10^{-3}\ M H^+}{0.250\ M\ HC_3H_5O_2} \times 100 = 0.721\%$$

(b) $\frac{x^2}{0.0800} \simeq 1.3 \times 10^{-5}$; $x = 1.020 \times 10^{-3} = 1.0 \times 10^{-3}\ M\ H^+$

$$\% \text{ ionization} = \frac{1.020 \times 10^{-3}\ M H^+}{0.0800\ M HC_3H_5O_2} \times 100 = 1.27\%$$

(c) $\dfrac{x^2}{0.0200} \approx 1.3 \times 10^{-5}$; $x = 5.099 \times 10^{-4} = 5.1 \times 10^{-4}$ *M* H^+

$$\%\text{ ionization} = \frac{5.099 \times 10^{-4}\ M\ H^+}{0.0200\ M\ HC_3H_5O_2} \times 100 = 2.55\%$$

16.68 $HX(aq) \rightleftharpoons H^+(aq) + X^-(aq);\ K_a = \dfrac{[H^+][X^-]}{[HX]}$

$[H^+] = [X^-]$; assume the % ionization is small; $K_a = \dfrac{[H^+]^2}{[HX]}$; $[H^+] = K_a^{1/2}\,[HX]^{1/2}$

$pH = -\log K_a^{1/2}\,[HX]^{1/2} = -\log K_a^{1/2} - \log [HX]^{1/2}$; $pH = -1/2 \log K_a - 1/2 \log [HX]$

This is the equation of a straight line, where the intercept is $-1/2 \log K_a$, the slope is $-1/2$, and the independent variable is $\log [HX]$.

16.70 $H_2C_4H_4O_6(aq) \rightleftharpoons H^+(aq) + HC_4H_4O_6^-(aq)$ $K_{a1} = 1.0 \times 10^{-3}$

$HC_4H_4O_6^-(aq \rightleftharpoons H^+(aq) + C_4H_4O_6^{2-}(aq)$ $K_{a2} = 4.6 \times 10^{-5}$

Begin by calculating the $[H^+]$ from the first ionization. The equilibrium concentrations are $[H^+] = [HC_4H_4O_6^-] = x$, $[H_2C_4H_4O_6] = 0.25 - x$.

$$K_{a1} = \frac{[H^+][HC_4H_4O_6^-]}{[H_2C_4H_4O_8]} = \frac{x^2}{0.25 - x};\ x^2 + 1.0 \times 10^{-3}\,x - 2.5 \times 10^{-4} = 0$$

Using the quadratic formula, $x = 1.532 \times 10^{-2} = 0.015$ *M* H^+ from the first ionization. Next calculate the H^+ contribution from the second ionization.

	$HC_4H_4O_6^-(aq)$	$\rightleftharpoons$	$H^+(aq)$	+	$C_4H_4O_6^{2-}(aq)$
initial	0.015		0.015		0
equil.	(0.015 – y)		(0.015 + y)		y

$$K_{a2} = \frac{(0.015 + y)(y)}{(0.015 - y)} = 4.6 \times 10^{-5};$$ assuming y is small compared to 0.015,

$y = 4.6 \times 10^{-5}$ *M* $HC_4H_4O_6^{2-}(aq)$

This assumption is reasonable, since 4.6×10^{-5} is only 0.3% of 0.015. $[H^+] = 0.015$ *M* (first ionization) + 4.6×10^{-5} (second ionization). Since 4.6×10^{-5} is 0.3% of 0.015 *M*, it can be safely ignored when calculating total $[H^+]$. Thus, pH = –log(0.01532) = 1.18148 = 1.181.

Assumptions:

1) The ionization can be treated as a series of steps (valid by Hess' law).

2) The extent of ionization in the second step (y) is small relative to that from the first step (valid for this acid and initial concentration). This assumption was used twice, to calculate the value of y from K_{a2} and to calculate total $[H^+]$ and pH.

Weak Bases

16.72 Organic amines (neutral molecules with nonbonded pairs on N atoms) and anions that are the conjugate bases of weak acids function as weak bases.

16.74 (a) $C_3H_7NH_2(aq) + H_2O(l) \rightleftharpoons C_3H_7NH_3^+(aq) + OH^-(aq); K_b = \frac{[C_3H_7NH_3^+][OH^-]}{[C_3H_7NH_2]}$

(b) $HPO_4^{2-}(aq) + H_2O(l) \rightleftharpoons H_2PO_4^-(aq) + OH^-(aq); K_b = \frac{[H_2PO_4][OH^-]}{[HPO_4^{2-}]}$

(c) $C_6H_5CO_2^-(aq) + H_2O(l) \rightleftharpoons C_6H_5CO_2H(aq) + OH^-(aq); K_b = \frac{[C_6H_5CO_2H][OH^-]}{[C_6H_5CO_2^-]}$

16.76

	$BrO^-(aq) + H_2O(l)$	$\rightleftharpoons$	$HOBr(aq)$	+	$OH^-(aq)$
initial	1.15 *M*		0		0
equil.	(1.15 – x) *M*		x *M*		x *M*

$$K_b = \frac{[HOBr][OH^-]}{[BrO^-]} = \frac{x^2}{1.15-x} \approx \frac{x^2}{1.15} = 4.0 \times 10^{-6}$$

$x^2 = 1.15\,(4.0 \times 10^{-6}); x = [OH^-] = 2.14 \times 10^{-3} = 2.1 \times 10^{-3}\ M$; pH = 11.33

Check. $\frac{2.1 \times 10^{-3}\ M\ OH^-}{1.15\ M\ C_3H_5O_2^-} \times 100 = 0.19\%$ hydrolysis; the assumption is valid

16.78 (a) $pOH = 14.00 - 9.95 = 4.05; [OH^-] = 10^{-4.05} = 8.91 \times 10^{-6} = 8.9 \times 10^{-5}\ M$

	$C_{18}H_{21}NO_3(aq) + H_2O(1)$	$\rightleftharpoons$	$C_{18}H_{21}NO_3H^+(aq)$	+	$OH^-(aq)$
initial	0.0050 *M*		0		0
equil.	$(0.0050 - 8.9 \times 10^{-5})$		$8.9 \times 10^{-5}\ M$		$8.9 \times 10^{-5}\ M$

$$K_b = \frac{[C_{18}H_{21}NO_3H^+][OH^-]}{[C_{18}H_{21}NO_3]} = \frac{(8.91 \times 10^{-5})^2}{(0.0050 - 8.91 \times 10^{-5})} = 1.62 \times 10^{-5} = 1.6 \times 10^{-6}$$

(b) $pK_b = -\log(K_b) = -\log(1.62 \times 10^{-5}) = 5.79$

The K_a - K_b Relationship; Acid-Base Properties of Salts

16.80 (a) We need K_a for the conjugate acid of CO_3^{2-}, K_a for HCO_3^-. K_a for HCO_3^- is K_{a2}.

(b) $K_b = K_w/K_a = 1.0 \times 10^{-14}/5.6 \times 10^{-11} = 1.8 \times 10^{-4}$

(c) K_b for CO_3^{2-} $(1.8 \times 10^{-4}) > K_b$ for NH_3 (1.8×10^{-5}).

CO_3^{2-} is a stronger base than NH_3.

16.82 (a) Ammonia is the stronger base because it has the larger K_b value.

(b) Hydroxylammonium is the stronger acid because the weaker base, hydroxylamine, has the stronger conjugate acid.

(c) K_a for $NH_4^+ = K_w/K_b$ for $NH_3 = 1.0 \times 10^{-14}/1.8 \times 10^{-5} = 5.6 \times 10^{-10}$

K_a for $HONH_3^+ = K_w/K_b$ for $HONH_2 = 1.0 \times 10^{-14}/1.1 \times 10^{-8} = 9.1 \times 10^{-7}$

Note that K_a for $HONH_3^+$ is larger than K_a for NH_4^+.

16.84 (a) Proceeding as in Solution 16.83(a):

	$F^-(aq) + H_2O(l)$	$\rightleftharpoons$	$HF(aq)$	+	$OH^-(aq)$
initial	0.085 *M*		0 *M*		0 *M*
equil	(0.085 – x) *M*		x *M*		x *M*

$$K_b \text{ for } F^- = \frac{[HF][OH^-]}{[F^-]} = \frac{K_w}{K_a \text{ for HF}} = \frac{1.0 \times 10^{-14}}{6.8 \times 10^{-4}} = 1.47 \times 10^{-11} = 1.5 \times 10^{-11}$$

$1.5 \times 10^{-11} = \dfrac{(x)(x)}{(0.085 - x)}$; assume the amount of F^- that hydrolyzes is small

$x^2 = 0.085(1.47 \times 10^{-11})$; $x = [OH^-] = 1.118 \times 10^{-6} = 1.1 \times 10^{-6}$ *M*

pOH = 5.95; pH = 14 – 5.95 = 8.05

(b) $Na_2S(aq) \rightarrow S^{2-}(aq) + 2Na^+(aq)$

$S^{2-}(aq) + H_2O(l) \rightleftharpoons HS^-(aq) + OH^-(aq)$

As in part (a) above, $[OH^-] = [HS^-] = x$; $[S^{2-}] = 0.055$ *M*

$$K_b = \frac{[HS^-][OH^-]}{[S^{2-}]} = \frac{K_w}{K_a \text{ for } HS^-} = \frac{1.0 \times 10^{-14}}{1 \times 10^{-19}} = 1 \times 10^5$$

Since $K_b >> 1$, the equilibrium above lies far to the right and $[OH^-] = [S^{2-}] = 0.055$ *M*. K_b for $HS^- = 1.05 \times 10^{-7}$; $[OH^-]$ produced by further hydrolysis of HS^- amounts to 7.6×10^{-5} *M*. The second hydrolysis step does not make a significant contribution to the total $[OH^-]$ and pH.

$[OH^-] = 0.055$ *M*; pOH = 1.26, pH = 12.74

(c) As in Solution 16.83(c), calculate total $[C_2H_3O_2^-]$.

$[C_2H_3O_2^-]_t = [C_2H_3O_2^-]$ from $NaC_2H_3O_2$ + $[C_2H_3O_2^-]$ from $Ba(C_2H_3O_2)_2$

$[C_2H_3O_2^-]_t = 0.045\ M + 2(0.055\ M) = 0.155\ M$

The hydrolysis equilibrium is $C_2H_3O_2^-(aq) + H_2O(l) \rightleftharpoons HC_2H_3O_2(aq) + OH^-(aq)$

$$K_b = \frac{[HC_2H_3O_2][OH^-]}{[C_2H_3O_2^-]} = \frac{K_w}{K_a \text{ for } HC_2H_3O_2} = \frac{1.0 \times 10^{-14}}{1.8 \times 10^{-5}} = 5.56 \times 10^{-10}$$

$$= 5.6 \times 10^{-10}$$

$[OH^-] = [HC_2H_3O_2] = x$, $[C_2H_3O_2^-] = 0.155 - x$

$K_b = 5.56 \times 10^{-10} = \dfrac{x^2}{(0.155 - x)}$; assume x is small compared to 0.155 *M*

$x^2 = 0.155\,(5.56 \times 10^{-10})$; $x = [OH^-] = 9.280 \times 10^{-6} = 9.3 \times 10^{-6}$

$pH = 14 + \log(9.280 \times 10^{-6}) = 8.97$

16.86 (a) acidic; Cr^{3+} is a highly charged metal cation and a Lewis acid; Br^- is negligible.

(b) neutral; both Li^+ and I^- are negligible.

(c) basic; PO_4^{3-} is the conjugate base of HPO_4^{2-}; K^+ is negligible.

(d) acidic; $CH_3NH_3^+$ is the conjugate acid of CH_3NH_2; Cl^- is negligible.

(e) acidic; HSO_4^- is a negligible base, but a fairly strong acid ($K_a = 1.2 \times 10^{-2}$). K^+ is negligible.

16.88 *Plan.* Estimate pH of salt solution by evaluating the ions in the salts. Calculate to confirm if necessary. *Solve.*

KBr: salt of strong acid and strong base, neutral solution. The unknown is probably KBr. Check the others to be sure.

NH_4Cl: salt of a weak base and a strong acid, acidic solution

KCN: salt of a strong base and a weak acid, basic solution

K_2CO_3: salt of a strong base and a weak acid (HCO_3^-), basic solution

Only KBr fits the acid-base properties of the unknown.

16.90 The solution is basic because of the hydrolysis of PO_4^{3-}. The molarity of PO_4^{3-} is

$$\frac{45.0 \text{ g } Na_3PO_4}{1.00 \text{ L soln}} \times \frac{1 \text{ mol } Na_3PO_4}{163.9 \text{ g } Na_3PO_4} = 0.2746 = 0.275\ M\ PO_4^{3-}$$

$$PO_4^{3-}(aq) + H_2O(l) \rightleftharpoons HPO_4^{2-}(aq) + OH^-(aq)$$

$$K_b = \frac{[HPO_4^{2-}][OH^-]}{[PO_4^{3-}]} = \frac{K_w}{K_a \text{ for } HPO_4^{2-}} = \frac{1.0 \times 10^{-14}}{4.2 \times 10^{-13}} = 0.0238 = 2.4 \times 10^{-2}$$

Ignoring the further hydrolysis of HPO_4^{2-},

$[OH^-] = [HPO_4^{2-}] = x$, $[PO_4^{3-}] = 0.275 - x$

$$2.4 \times 10^{-2} = \frac{x^2}{(0.275 - x)};\ x^2 + 2.4 \times 10^{-2}\ x - 0.0065 = 0$$

Since K_b is relatively large, we will not assume x is small compared to 0.275.

$$x = \frac{-0.024 \pm \sqrt{(0.024)^2 - 4(1)(-0.0065)}}{2(1)} = \frac{-0.024 \pm \sqrt{0.0267}}{2(1)}$$

$x = 0.070\ M\ OH^-$; pH = 14 + log (0.070) = 12.84

Acid-Base Character and Chemical Structure

16.92 (a) Acid strength increases as the polarity of the X–H bond increases and decreases as the strength of the X–H bond increases.

(b) Assuming the element, X, is more electronegative than H, as the electronegativity of X increases, the X–H bond becomes more polar and the strength of the acid increases. This trend holds true as electronegativity increases across a row of the periodic chart. However, as electronegativity decreases going down a family, acid strength increases because the strength of the H–X bond decreases, even though the H–X bond becomes less polar.

16.94 (a) For binary hydrides, acid strength increases going across a row, so HCl is a stronger acid than H_2S.

(b) For oxyacids, the more electronegative the central atom, the stronger the acid, so H_3PO_4 is a stronger acid than H_3AsO_4.

(c) $HBrO_3$ has one more nonprotonated oxygen and a higher oxidation number on Br, so it is a stronger acid than $HBrO_2$.

(d) The first dissociation of a polyprotic acid is always stronger because H^+ is more tightly held by an anion, so $H_2C_2O_4$ is a stronger acid than $HC_2O_4^-$.

(e) The conjugate base of benzoic acid, $C_7H_5O_2^-$, is stabilized by resonance, while the conjugate base of phenol, $C_6H_5O^-$, is not. $HC_7H_5O_2$ has greater tendency to form its conjugate base and is the stronger acid.

16.96 (a) NO_2^- (HNO_3 is the stronger acid because it has more nonprotonated O atoms, so NO_2^- is the stronger base.)

(b) PO_4^{3-} (K_a for $HAsO_4^{2-}$ is greater than K_a for HPO_4^{2-}, so K_b for PO_4^{3-} is greater and PO_4^{3-} is the stronger base. Note that P is more electronegative than As and H_3PO_4 is a stronger acid than H_3AsO_4, which could lead to the conclusion that AsO_4^{3-} is the stronger base. As in all cases, the measurement of base strength, K_b, supercedes the prediction. Chemistry is an experimental science.

(c) CO_3^{2-} (The more negative the anion, the stronger the attraction for H^+.)

16.98 (a) True.

(b) False. For oxyacids with the same structure but different central atom, the acid strength **increases** as the electronegativity of the central atom increases.

(c) False. HF is a weak acid, weaker than the other hydrogen halides, primarily because the H–F bond energy is exceptionally high.

Lewis Acids and Bases

16.100 No. If a substance is a Lewis acid, it is not necessarily a Brønsted or an Arrhenius acid. The Lewis definition of an acid, an electron pair acceptor, is most general. A Lewis acid does not necessarily fit the more narrow description of a Brønsted or Arrhenius acid. An electron pair acceptor isn't necessarily an H^+ donor, nor must it produce H^+ in aqueous solution. An example is Al^{3+}, which is a Lewis acid, but has no ionizable hydrogen.

16.102

	Lewis Acid	Lewis Base
(a)	HNO_2 (or H^+)	OH^-
(b)	$FeBr_3$ (Fe^{3+})	Br^-
(c)	Zn^{2+}	NH_3
(d)	SO_2	H_2O

16.104 (a) $ZnBr_2$, smaller cation radius, same charge

(b) $Cu(NO_3)_2$, higher cation charge

(c) $NiBr_2$, smaller cation radius, same charge

Additional Exercises

16.106 (a) Correct.

(b) Incorrect. A Brønsted acid must have ionizable hydrogen. Lewis acids are electron pair acceptors, but need not have ionizable hydrogen.

(c) Correct.

(d) Incorrect. K^+ is a negligible Lewis acid because it is the conjugate of strong base KOH. Its relatively large ionic radius and low positive charge render it a poor attractor of electron pairs.

(e) Correct.

16.108 Assume T = 25°C. For acid or base solute concentrations less than 1×10^{-6} *M*, we must consider the autoionization of water as a source of $[OH^-]$ and $[H^+]$.

	$H_2O(l)$	$\rightleftharpoons$	$[H^+]$	+	$[OH^-]$
initial	C		0		2.5×10^{-9} *M*
equil	C		x		$(x + 2.5 \times 10^{-9})$ *M*

$K_w = 1.0 \times 10^{-14} = [H^+][OH^-] = (x)(x + 2.5 \times 10^{-9})$; $x^2 + 2.5 \times 10^{-9}\,x - 1.0 \times 10^{-14} = 0$

From the quadratic formula, $x = \dfrac{-2.5 \times 10^{-9} \pm \sqrt{(2.5 \times 10^{-9})^2 - 4(-1 \times 10^{-14})}}{2}$

$= 9.876 \times 10^{-6} = 9.9 \times 10^{-6}\ M\ H^+$

$[H^+] = 9.9 \times 10^{-8}$ *M*; $[OH^-] = (9.9 \times 10^{-8} + 2.5 \times 10^{-9}) = 1.013 \times 10^{-7} = 1.0 \times 10^{-7}$ *M*

pH = 7.0054 = 7.01

Check: $[9.876 \times 10^{-8}][1.013 \times 10^{-7}] = 1.0 \times 10^{-14}$. Our answer makes sense. The very small concentration of OH^- from the solute raises the solution pH to slightly more than 7.

16.109 The solution with the higher pH has the lower $[H^+]$.

(a) For solutions with equal concentrations, the weaker acid will have a lower $[H^+]$ and higher pH.

(b) The acid with $K_a = 8 \times 10^{-5}$ is the weaker acid, so it has the higher pH.

(c) The base with $pK_b = 4.5$ is the stronger base, has greater $[OH^-]$ and smaller $[H^+]$, so higher pH.

16.111 (a) $H_2X \rightarrow H^+ + HX^-$

Assuming HX^- does not ionize, $[H^+] = 0.050$ *M*, pH = 1.30

(b) $H_2X \rightarrow 2H^+ + X^-$; 0.050 *M* H_2X = 0.10 *M* H^+; pH = 1.00

(c) The observed pH of a 0.050 *M* solution of H_2X is only slightly less than 1.30, the pH assuming no ionization of HX^-. HX^- is not completely ionized; H_2X, which is completely ionized, is a stronger acid than HX^-.

(d) Since H_2X is a strong acid, HX^- has no tendency to act like a base. HX^- does act like a weak acid, so a solution of NaHX would be acidic.

16.112 *Analyze/Plan.* Evaluate the acid-base properties of the cation and anion to determine whether a solution of the salt will be acidic, basic, or neutral. *Solve.*

(i) NH_4NO_3: NH_4^+, weak conjugate acid of NH_3; NO_3^-, negligible conjugate base of HNO_3; acidic solution.

(ii) $NaNO_3$: Na^+, negligible conjugate acid of NaOH; NO_3^-, negligible conjugate base of HNO_3; neutral solution.

(iii) $NH_4C_2H_3O_2$: NH_4^+, weak conjugate acid of NH_3, $K_a = K_w/1.8 \times 10^{-5} = 5.6 \times 10^{-10}$; $C_2H_3O_2^-$, weak conjugate base of $HC_2H_3O_2$, $K_b = K_w/1.8 \times 10^{-5} = 5.6 \times 10^{-10}$; neutral solution ($K_a$ for the cation and K_b for the anion are accidentally equal, producing a neutral solution).

(iv) NaF: Na^+, negligible conjugate acid of NaOH; F^-, weak conjugate base of HF, $K_b = K_w/6.8 \times 10^{-4} = 1.5 \times 10^{-11}$; basic solution.

(v) $NaC_2H_3O_2$: Na^+, negligible; $C_2H_3O_2^-$, weak base, $K_b = 5.6 \times 10^{-10}$, basic solution.

In order of increasing acidity and decreasing pH: 0.1 M $NaC_2H_3O_2$ > 0.1 M NaF > 0.1 M $NH_4C_2H_3O_2$ = 0.1 M $NaNO_3$ > 0.1 M NH_4Cl; (v) > (iv) > (iii) ~ (ii) > (i)

(iv) and (v) are both bases, and (v) has the greater K_b value and higher pH. (ii) and (iii) are both neutral and (i) is acidic.

16.114 Call each compound in the neutral form Q.

Then, $Q(aq) + H_2O(l) \rightleftharpoons QH^+(aq) + OH^-$. $K_b = [QH^+][OH^-]/[Q]$

The ratio in question is $[QH^+]/[Q]$, which equals $K_b/[OH^-]$ for each compound.
At pH = 2.5, pOH = 11.5, $[OH^-]$ = antilog (–11.5) = $3.16 \times 10^{-12} = 3 \times 10^{-12}$ *M*. Now calculate $K_b/[OH^-]$ for each compound:

Nicotine $$\frac{[QH^+]}{[Q]} = 7 \times 10^{-7}/3.16 \times 10^{-12} = 2 \times 10^5$$

Caffeine $$\frac{[QH^+]}{[Q]} = 4 \times 10^{-14}/3.16 \times 10^{-12} = 1 \times 10^{-2}$$

Strychnine $$\frac{[QH^+]}{[Q]} = 1 \times 10^{-6}/3.16 \times 10^{-12} = 3 \times 10^5$$

Quinine $$\frac{[QH^+]}{[Q]} = 1.1 \times 10^{-6}/3.16 \times 10^{-12} = 3.5 \times 10^5$$

For all the compounds except caffeine the protonated form has a much higher concentration than the neutral form. However, for caffeine, a very weak base, the neutral form dominates.

16.115 (a) Consider the formation of the zwitterion as a series of steps (Hess' law).

$$NH_2{-}CH_2{-}COOH + H_2O \rightleftharpoons NH_2{-}CH_2{-}COO^- + H_3O^+ \qquad K_a$$

$$NH_2{-}CH_2{-}COOH + H_2O \rightleftharpoons {}^+NH_3{-}CH_2{-}COOH + OH^- \qquad K_b$$

$$H_3O^+ + OH^- \rightleftharpoons 2H_2O \qquad 1/K$$

$$NH_2{-}CH_2{-}COOH \rightleftharpoons {}^+NH_3{-}CH_2{-}COO^- \qquad \frac{K_a \times K_b}{K_w}$$

$$K = \frac{K_a \times K_b}{K_w} = \frac{(4.3 \times 10^{-3})(6.0 \times 10^{-5})}{1.0 \times 10^{-14}} = 2.6 \times 10^7$$

The large value of K indicates that formation of the zwitterion is favorable. The assumption is that the same NH_2-CH_2-COOH molecule is acting like an acid (-COOH → H^+ + $-COO^-$) and a base ($-NH_2$ + H^+ → NH_3^+), simultaneously. Glycine is both a stronger acid and a stronger base than water, so the H^+ transfer should be intramolecular, as long as there are no other acids or bases in the solution.

(b) Since glycine exists as the zwitterion in aqueous solution, the pH is determined by the equilibrium below.

$${}^+NH_3{-}CH_2{-}COO^- + H_2O \rightleftharpoons NH_2{-}CH_2{-}COO^- + H_3O^+$$

$$K_a = \frac{[NH_2{-}CH_2{-}COO^-][H_3O^+]}{[{}^+NH_3{-}CH_2{-}COO^-]} = \frac{K_w}{K_b} = \frac{1.0 \times 10^{-14}}{6.0 \times 10^{-5}} = 1.67 \times 10^{-10} = 1.7 \times 10^{-10}$$

$$x = [H_3O^+] = [NH_2{-}CH_2{-}COO^-];\ K_a = 1.67 \times 10^{-10} = \frac{(x)(x)}{(0.050 - x)} \approx \frac{x^2}{0.050}$$

$x = [H_3O^+] = 2.89 \times 10^{-6} = 2.9 \times 10^{-6}$ *M*; pH = 5.54

(c) In a strongly acidic solution the $-CO_2^-$ function would be protonated, so glycine would exist as ${}^+H_3NCH_2COOH$. In strongly basic solution the $-NH_3^+$ group would be deprotonated, so glycine would be in the form $H_2NCH_2CO_2^-$.

16.116 The general Lewis structures for these acids and their conjugate bases are shown below.

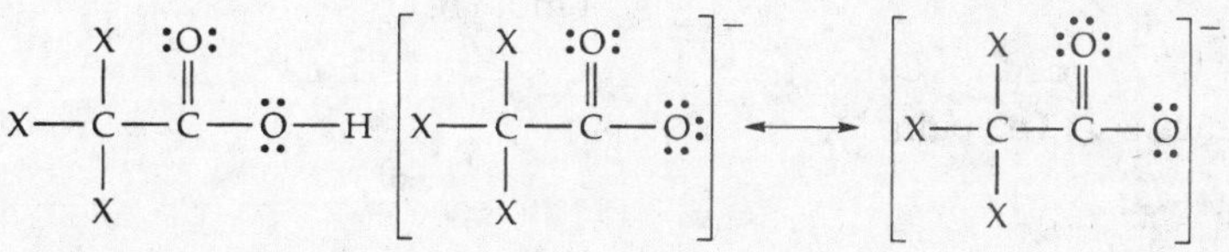

X :O:
| ||
X—C—C—Ö—H
|
X

[X—C(X)(X)—C(=O:)—Ö:]⁻ ⟷ [X—C(X)(X)—C(=Ö:)—Ö]⁻

where X = H or Cl.

Replacement of H on the acid by the more electronegative chlorine atoms causes the central carbon to become more positively charged, thus withdrawing more electrons from the attached COOH group, in turn causing the O–H bond to be more polar, so that H^+ is more readily ionized.

For the conjugate base (two resonance structures), the electronegative X atoms delocalize negative charge and stabilize these forms relative to the unsubstituted anions. This favors products in the ionization equilibrium and increases the value of K_a.

To calculate pH proceed as usual, except that the full quadratic formula must be used for all but acetic acid.

Acid	pH
acetic	3.37
chloroacetic	2.51
dichloroacetic	2.09
trichloroacetic	2.0

Integrative Exercises

16.118 *Analyze.* Given mass % and density of concentrated HCl, calculate volume of concentrated solution required to produce 10.0 L of HCl with pH = 2.05. *Plan.* Calculate molarity of concentrated solution from density and mass %. Calculate molarity of dilute solution from pH. Use the dilution formula to calculate volume (mL) concentrated solution required. *Solve.*

$$\frac{1.18 \text{ g conc. soln.}}{\text{mL conc. soln.}} \times \frac{36.0 \text{ g HCl}}{100 \text{ g conc. soln.}} \times \frac{1000 \text{ mL}}{1 \text{ L}} \times \frac{1 \text{ mol HCl}}{36.46 \text{ g HCl}} = 11.651 \text{ mol HCl/L}$$

$$= 11.7\ M \text{ HCl/L}$$

For the dilute HCl solution, $[H^+] = 10^{-pH} = 10^{-2.05} = 8.913 \times 10^{-3} = 8.9 \times 10^{-3}\ M$ HCl

$M_C \times L_C = M_D \times M_D;\ 11.651 \times L_C = 8.913 \times 10^{-3}\ M \times 10.0$ L;

$$L_C = 7.650 \times 10^{-3}; 7.650 \times 10^{-3} \text{ L} \times \frac{1000 \text{ mL}}{1 \text{ L}} = 7.65 = 7.7 \text{ mL conc. HCl}$$

16.119 $[H^+] = 10^{-pH} = 10^{-2} = 1 \times 10^{-2}\ M\ H^+;\ 1 \times 10^{-2}\ M \times 0.400 \text{ L} = 4.0 \times 10^{-3} = 4 \times 10^{-3} \text{ mol } H^+$

$HCl(aq) + HCO_3^-(aq) \rightarrow Cl^-(aq) + H_2O(l) + CO_2(g)$

$$4 \times 10^{-3} \text{ mol } H^+ = 4 \times 10^{-3} \text{ mol } HCO_3^- \times \frac{84.01 \text{ g } NaHCO_3}{1 \text{ mol } HCO_3^-} = 0.336 = 0.3 \text{ g } NaHCO_3$$

16.121 (a) 24 valence e^-, 12 e^- pairs

:Cl̈—Al—C̈l:
|
:C̤l:

The formal charges on all atoms are zero. Structures with multiple bonds lead to nonzero formal charges. There are three electron domains about Al. The electron-domain geometry and molecular structure are trigonal planar.

(b) The Al atom in $AlCl_3$ has an incomplete octet and is electron deficient. It "needs" to accept another electron pair, to act like a Lewis acid.

(c) $AlCl_3$ (:Cl: ligands on Al) + $:NH_3$ ⟶ $Cl_3Al—NH_3$

Both the Al and N atoms in the product have tetrahedral geometry.

(d) The Lewis theory is most appropriate. H^+ and $AlCl_3$ are both electron pair acceptors, Lewis acids.

16.122 *Plan.* Use acid ionization equilibrium to calculate the total moles of particles in solution. Use density to calculate kg solvent. From the molality (m) of the solution, calculate ΔT_b and T_b. *Solve.*

	$HSO_4^-(aq)$	$\rightleftharpoons$	$H^+(aq)$	+	$SO_4^{2-}(aq)$
Initial	0.10 M		0		0
equil.	0.10 – x M		x M		x M
	0.071 M		0.029 M		0.029 M

$$K_a = 1.2 \times 10^{-2} = \frac{[H^+][SO_4^{2-}]}{[HSO_4^-]} = \frac{x^2}{0.10 - x}$$; K_a is relatively large, so use the quadratic.

$$x^2 + 0.012\,x - 0.0012 = 0;\ x = \frac{-0.012 \pm \sqrt{(0.012)^2 - 4(1)(-0.0012)}}{2};\ x = 0.029\ M\ H^+, SO_4^{2-}$$

Total ion concentration = 0.10 M Na^+ + 0.071 M HSO_4^- + 0.029 M H^+ + 0.029 M SO_4^{2-} = 0.229 = 0.23 M.

Assume 100.0 mL of solution. 1.002 g/mL × 100.0 mL = 100.2 g solution.

$$0.10\ M\ NaHSO_4 \times 0.1000\ L = 0.010\ mol\ NaHSO_4 \times \frac{120.1\ g\ NaHSO_4}{mol\ NaHSO_4} = 1.201 = 12\ g\ NaHSO_4$$

100.2 g soln – 1.201 g $NaHSO_4$ = 99.0 g = 0.099 kg H_2O

$$m = \frac{mol\ ions}{kg\ H_2O} = \frac{0.229\ M \times 0.1000\ L}{0.0990\ kg} = 0.231 = 0.23\ m\ ions$$

$\Delta T_b = K_b(m) = 0.52°C/m \times (0.23\ m) = +0.12°C$: $T_b = 100.0 + 0.12 = 100.1°C$

16.124 (a) rate = $k[IO_3^-][SO_3^{2-}][H^+]$

(b) $\Delta pH = pH_2 - pH_1 = 3.50 - 5.00 = -1.50$

$\Delta pH = -\log[H^+]_2 - (-\log[H^+]_1)$; $-\Delta pH = \log[H^+]_2 - \log[H^+]_1$

$-\Delta pH = \log[H^+]_2 / [H^+]_1$; $[H^+]_2 / [H^+]_1 = 10^{-\Delta pH}$

$[H^+]_2/[H^+]_1 = 10^{1.50} = 31.6 = 32$. The rate will increase by a factor of 32 if $[H^+]$ increases by a factor of 32. The reaction goes faster at lower pH.

(c) Since H^+ does not appear in the overall reaction, it is either a catalyst or an intermediate. An intermediate is produced and then consumed during a reaction, so its contribution to the rate law can usually be written in terms of concentrations of other reactants (Sample Exercise 14.15). A catalyst is present at the beginning and end of a reaction and can appear in the rate law if it participates in the rate-determining step (Solution 14.72). This reaction is pH dependent because H^+ is a homogeneous catalyst that participates in the rate-determining step.

16.126 (a) The structures of the two acids are similar, but lactic acid has an -OH group on the C atom adjacent to the -COOH group. This electronegative substituent withdraws electron density from the -COOH group, and stabilizes its conjugate base, increasing the strength of lactic acid relative to propionic acid. The stronger the acid, the larger the K_a value and the smaller the pK_a.

(b) $pK_a = 3.85$, $K_a = 10^{-3.85} = 1.4 \times 10^{-4}$; $[H^+] = [C_3H_5O_3^-] = x$; $[HC_3H_5O_3] = 0.050 - x$

$$K_a = 1.4 \times 10^{-4} = \frac{[H^+][C_3H_5O_3^-]}{[HC_3H_5O_3]} = \frac{x^2}{0.050 - x} \approx \frac{x^2}{0.050}$$

$x^2 = 0.050(1.4 \times 10^{-4}) = 2.646 \times 10^{-3} = 2.6 \times 10^{-3}\ M\ C_3H_5O_3^-$

(This represents 5.3% dissociation; solution by the quadratic yields essentially the same result.)

(c) Strategy: Assume a 100 g sample. Calculate mol Cu in sample. Use mole ratios from formula to calculate mass of O and H not due to H_2O. Subtract masses of Cu, C, H, and O from 100 g to get mass of H_2O. Calculate mol H_2O and X.

Assume a 100 g sample, 22.9 g Cu.

$$22.9 \text{ g Cu} \times \frac{1 \text{ mol Cu}}{63.55 \text{ g Cu}} = 0.3603 = 0.360 \text{ mol Cu}$$

mole ratios of Cu, O, and H (not due to H_2O): 1 Cu:6 O:10 H

$$\text{g O} = 0.3603 \text{ mol Cu} \times \frac{6 \text{ mol O}}{1 \text{ mol Cu}} \times \frac{16.00 \text{ g O}}{1 \text{ mol O}} = 34.59 = 34.6 \text{ g O}$$

$$\text{g H} = 0.3603 \text{ mol Cu} \times \frac{10 \text{ mol H}}{1 \text{ mol Cu}} \times \frac{1.008 \text{ g H}}{1 \text{ mol H}} = 3.632 = 3.63 \text{ g H}$$

$$\text{g } H_2O = 100 \text{ g sample} - [22.9 \text{ g Cu} + 26.0 \text{ g C} + 34.6 \text{ g O} + 3.63 \text{ g H}]$$
$$= 12.87 = 12.9 \text{ g } H_2O$$

$$12.87 \text{ g } H_2O \times \frac{1 \text{ mol } H_2O}{18.02 \text{ g } H_2O} = 0.7142 \text{ mol } H_2O/0.3604 = 1.98 \approx 2$$

X = 2 mol H_2O in the hydrate.

(d) Compare K_a for $Cu^{2+}(aq)$ and K_b for $C_3H_5O_3^-(aq)$.

The ion with the larger K value will undergo hydrolysis to the greater extend and will determine the pH of the solution.

pK_b for $C_3H_5O_3^-$ = 14 - pK_a for $HC_3H_5O_3$ = 14.00 - 3.85 = 10.15

$K_b = 10^{-10.15} = 7.1 \times 10^{-11}$.

Since K_a for Cu^{2+} (1.0×10^{-8}) is greater than K_b for $C_3H_5O_3^-$ (7.1×10^{-11}) the solution will be slightly acidic.

17 Additional Aspects of Aqueous Equilibria

Visualizing Concepts

17.2 (a) According to Figure 16.7, methyl orange is yellow above pH 4.5 and red (really pink) below pH 3.5. The beaker on the left has a pH greater than 4.5, and the one on the right has pH less than 3.5. (By calculation, pH of left beaker = 4.7, pH of right beaker = 2.9.) The right beaker, with lower pH and greater $[H^+]$, is pure acetic acid. The left beaker contains equal amounts of the weak acid and its conjugate base, acetic acid and acetate ion. Adding the "common-ion" acetate (in the form of sodium acetate) shifts the acid ionization equilibrium to the left, decreases $[H^+]$, and raises pH.

(b) When small amounts of NaOH are added, the left beaker is better able to maintain its pH. For solutions of the same weak acid, pH depends on the **ratio** of conjugate base to conjugate acid. Small additions of base (or acid) have the least effect when this ratio is close to one. The left beaker is a buffer because it contains a weak conjugate acid-conjugate base pair and resists rapid pH change upon addition of small amounts of strong base or acid.

17.4 *Analyze/Plan.* Consider the reaction $HA + OH^- \rightarrow A^- + H_2O$. What are the major species present in solution at the listed stages of the titration? Which diagram represents these species? *Solve.*

(a) *Before addition of NaOH,* the solution is mostly HA. The only A^- is produced by the ionization equilibrium of HA and is too small to appear in the diagram. This situation is shown in diagram (iii), which contains only HA.

(b) *After addition of NaOH but before the equivalence point,* some, but not all, HA has been converted to A^-. The solution contains a mixture of HA and A^-; this is shown in diagram (i).

(c) *At the equivalence point,* all HA has been converted to A^-, with no excess HA or OH^- present. This is shown in diagram (iv).

(d) *After the equivalence point,* the same amount of A^- is present as at the equivalence point, plus some excess OH^-. This is diagram (ii).

17.5 *Analyze/Plan.* In each case, the first substance is in the buret, and the second is in the flask. If acid is in the flask, the initial pH is low; with base in the flask, the pH starts high. Strong acids have lower pH than weak acids; strong bases have higher pH than weak bases. Polyprotic acids and bases have more than one "jump" in pH.

(a) Strong base in flask, pH starts high, ends low as acid is added. Only diagram (ii) fits this description.

(b) Weak acid in flask, pH starts low, but not extremely low. Diagrams (i), (iii), and (iv) all start at low pH and get higher. Diagram (i) has very low initial pH, and likely has strong acid in the flask. Diagram (iv) has two pH jumps, so it has a polyprotic acid in the flask. Diagram (iii) best fits the profile of adding a strong base to a weak acid.

(c) Strong acid in the flask, pH starts very low, diagram (i).

(d) Polyprotic acid, more than one pH jump, diagram (iv).

17.7 *Analyze/Plan*. Common anions or cations decrease the solubility of salts. Ions that participate in acid-base or complex ion equilibria increase solubility. *Solve*.

(a) CO_2^{3-} from $BaCO_3$ reacts with H^+ from HNO_3, causing solubility of $BaCO_3$ to increase with increasing HNO_3 concentration. This behavior matches the right diagram.

(b) Extra CO_2^{3-} from Na_2CO_3 decreases the solubility of $BaCO_3$. Solubility of $BaCO_3$ decreases as $[Na_2CO_3]$ increases. This behavior matches the left diagram.

(c) $NaNO_3$ has no common ions, nor does it enter into acid-base or complex ion equilibria with Ba^{2+} or CO_3^{2-}; it does not affect the solubility of $BaCO_3$. This behavior is shown in the center diagram.

17.8 A metal hydroxide that is soluble at very low and very high pH's, that is, in strong acid or strong base, is called amphoteric.

Common-Ion Effect

17.10 (a) For a generic weak base B, $K_b = \dfrac{[BH^+][OH^-]}{[B]}$. If an external source of BH^+ such as BH^+Cl^- is added to a solution of B(aq), $[BH^+]$ increases, decreasing $[OH^-]$ and increasing [B], effectively suppressing the ionization (hydrolysis) of B.

(b) NH_4Cl

17.12 In general, when an acid is added to a solution pH decreases; when a base is added to a solution, pH increases.

(a) pH increases; $C_7H_5O_2^-$ decreases ionization of $HC_7H_5O_2$ and decreases $[H^+]$.

(b) pH decreases; $C_5H_5NH^+$ decreases ionization (hydrolysis) of C_5H_5N and decreases $[OH^-]$.

(c) pH increases; NH_3 reacts with HCl, decreasing $[H^+]$.

(d) pH increases; HCO_3^- decreases ionization of H_2CO_3 and decreases $[H^+]$.

(e) no change; ClO_4^- is a negligible base and Na^+ is a negligible acid.

17.14 *Analyze/Plan.* Follow the logic in Sample Exercise 17.1. *Solve.*

(a) $HCHO_2$ is a weak acid, and $NaCHO_2$ contains the common ion CHO_2^-, the conjugate base of $HCHO_2$. Solve the common-ion equilibrium problem.

	$HCHO_2(aq)$	$\rightleftharpoons$	$H^+(aq)$	+	$CHO_2^-(aq)$
i	0.260 *M*				0.160 *M*
c	–x		+x		+x
e	(0.260 – x) *M*		+x *M*		(0.160 + x) *M*

$$K_a = 1.8 \times 10^{-4} = \frac{[H^+][CHO_2^-]}{[HCHO_2]} = \frac{(x)(0.160 + x)}{(0.260 - x)} \approx \frac{0.160\,x}{0.260}$$

$x = 2.93 \times 10^{-4} = 2.9 \times 10^{-4}$ *M* = $[H^+]$, pH = 3.53

Check. Since the extent of ionization of a weak acid or base is suppressed by the presence of a conjugate salt, the 5% rule usually holds true in buffer solutions.

(b) C_5H_5N is a weak base, and C_5H_5NHCl contains the common ion $C_5H_5NH^+$, which is the conjugate acid of C_5H_5N. Solve the common ion equilibrium problem.

	$C_5H_5N(aq)$ + $H_2O(l$	$\rightleftharpoons$	$C_5H_5NH^+(aq)$	+	$OH^-(aq)$
i	0.210 *M*		0.350 *M*		
c	–x		+x		+x
e	(0.210 – x) *M*		(0.350 + x) *M*		+x *M*

$$K_b = 1.7 \times 10^{-9} = \frac{[C_5H_5NH^+][OH^-]}{[C_5H_5N]} = \frac{(0.350 + x)(x)}{(0.210 - x)} \approx \frac{0.350\,x}{0.210}$$

$x = 1.02 \times 10^{-9} = 1.0 \times 10^{-9}$ *M* = $[OH^-]$, pOH = 8.991, pH = 14.00 – 8.991 = 5.01

Check. In a buffer, if [conj. acid] > [conj. base], pH < pK_a of the conj. acid. If [conj. acid] < [conj. base], pH > pK_a of the conj. acid. In this buffer, pK_a of $(CH_3)_3NH^+$ is 9.81. $[(CH_3)_3NH^+] > [(CH_3)_3N]$ and pH = 9.61, less than 9.81.

(c) mol = *M* × L; mol HF = 0.050 *M* × 0.125 L = $6.25 \times 10^{-3} = 6.3 \times 10^{-3}$ mol;
mol F^- = 0.10 *M* × 0.0500 L = 0.0050 mol

	HF(aq)	$\rightleftharpoons$	$H^+(aq)$	+	$F^-(aq)$
i	6.25×10^{-3} mol		0		0.0050 mol
c	–x		+x		+x
e	$(6.25 \times 10^{-3} - x)$ mol		x		(0.0050 + x) mol

$[HF] = (6.25 \times 10^{-3} + x)/0.175$ L; $[F^-] = (0.0050 + x)$ 0.175 L

Note that the volumes will cancel when substituted into the K_a expression.

$$K_a = 6.8 \times 10^{-4} = \frac{[H^+][F^-]}{[HF]} = \frac{x(0.0050 + x)/0.175}{(6.25\,x - x)/0.175} \approx \frac{x(0.0050)}{0.00625}$$

$x = 8.50 \times 10^{-4} = 8.5 \times 10^{-4}\ M\ H^+$; pH = 3.07

Check. pK_a for HF = 3.17. [HF] > [F^-], pH of buffer = 3.07, less than 3.17.

17.16

$$HLac(aq) \rightleftharpoons H^+(aq) + Lac^-(aq) \qquad K_a = \frac{[H^+][Lac^-]}{[HLac]} = 1.4 \times 10^{-4}$$

	HLac(aq)	H^+(aq)	Lac^-(aq)
equil (a)	0.085 – x *M*	x *M*	x *M*
equil (b)	0.085 – x *M*	x *M*	0.050 + x *M*

(a) $K_a = 1.4 \times 10^{-4} = \frac{x^2}{0.085 - x} \approx \frac{x^2}{0.085}$; $x = [H^+] = 3.45 \times 10^{-3}\ M = 3.5 \times 10^{-3}\ M H^+$

$$\%\ \text{ionization} = \frac{3.5 \times 10^{-3}\ MH^+}{0.085\ M\ \text{Lac}} \times 100 = 4.1\%\ \text{ionization}$$

(b) $K_a = 1.4 \times 10^{-4} = \frac{(x)(0.050 + x)}{0.085 - x} \approx \frac{0.050\,x}{0.085}$; $x = 2.4 \times 10^{-4}\ M H^+$

$$\%\ \text{ionization} = \frac{2.4 \times 10^{-4}\ MH^+}{0.085\ M\ \text{Lac}} \times 100 = 0.28\%\ \text{ionization}$$

Buffers

17.18 NaOH is a strong base and will react with $HC_2H_3O_2$ to form $NaC_2H_3O_2$. As long as $HC_2H_3O_2$ is present in excess, the resulting solution will contain both the conjugate acid $HC_2H_3O_2$(aq) and the conjugate base $C_2H_3O_2^-$(aq), the requirements for a buffer.

mmol = *M* × mL; mmol $HC_2H_3O_2$ = 1.00 *M* × 100 mL = 10.0 mmol

mmol NaOH = 0.100 *M* × 50 mL = 5.0 mmol

	$HC_2H_3O_2$(aq) +	NaOH(aq) →	$NaC_2H_3O_2$(aq) + H_2O(l)
initial	10.0 mmol	5.0 mmol	
after rx	5.0 mmol	0	5.0 mmol

Mixing these two solutions has created a buffer by partial neutralization of the weak acid $HC_2H_3O_2$.

17.20 Assume that % ionization is small in these buffers (Solutions 17.15 and 17.16).

(a) The conjugate acid in this buffer is HCO_3^-, so use K_{a2} for H_2CO_3, 5.6×10^{-11}

$$K_a = \frac{[H^+][CO_3^{2-}]}{[HCO_3^-]};\ [H^+] = \frac{K_a[HCO_3^-]}{[CO_3^{2-}]} = \frac{5.6 \times 10^{-11}\ (0.120)}{(0.105)}$$

$[H^+] = 6.40 \times 10^{-11} = 6.4 \times 10^{-11}\ M$; pH = 10.19

(b) mol = $M \times$ L; total volume = 140 mL = 0.140 L

$$[H^+] = \frac{K_a\ (0.20\ M \times 0.065\ L)/0.120\ L}{(0.15\ M \times 0.075\ L)/0.120\ L} = \frac{5.6 \times 10^{-11}\ (0.20 \times 0.065)}{(0.15 \times 0.075)}$$

$[H^+] = 6.47 \times 10^{-11} = 6.5 \times 10^{-11}\ M$; pH = 10.19

17.22 NH_4^+/NH_3 is a basic buffer. Either the hydrolysis of NH_3 or the dissociation of NH_4^+ can be used to determine the pH of the buffer. Using the dissociation of NH_4^+ leads directly to $[H^+]$ and facilitates use of the Henderson-Hasselbach relationship.

(a) $NH_4^+(aq) \rightleftharpoons H^+(aq) + NH_3(aq)$

$$K_a = \frac{K_w}{K_b} = \frac{1.0 \times 10^{-14}}{1.8 \times 10^{-5}} = 5.56 \times 10^{-10} = 5.6 \times 10^{-10}$$

$$[NH_3] = \frac{5.0\ g\ NH_3}{2.50\ L\ soln} \times \frac{1\ mol\ NH_3}{17.0\ g\ NH_3} = 0.118 = 0.12\ M\ NH_3$$

$$[NH_4^+] = \frac{20.0\ g\ NH_4Cl}{2.50\ L} \times \frac{1\ mol\ NH_4Cl}{53.50\ g\ NH_4Cl} = 0.1495 = 0.15\ M\ NH_4^+$$

$$K_a = \frac{[H^+][NH_3]}{[NH_4^+]};\ [H^+] = \frac{K_a[NH_4^+]}{[NH_3]} = \frac{5.56 \times 10^{-10}\ (0.1495 - x)}{(0.118 + x)} \approx \frac{5.56 \times 10^{-10}\ (0.1495)}{(0.118)}$$

$[H^+] = 7.044 \times 10^{-10} = 7.0 \times 10^{-10}\ M$, pH = 9.15

(b) $NH_3(aq) + H^+(aq) + NO_3^-(aq) \rightarrow NH_4^+(aq) + NO_3^-(aq)$

(c) $NH_4^+(aq) + Cl^-(aq) + K^+(aq) + OH^-(aq) \rightarrow NH_3(aq) + H_2O(l) + Cl^-(aq) + K^+(aq)$

17.24 $HC_3H_5O_3(aq) \rightleftharpoons H^+(aq) + C_3H_5O_3^-(aq)$

$$[H^+] = \frac{K_a[HC_3H_5O_3]}{[C_3H_5O_3^-]};\ [H^+] = 10^{-4.00} = 1.0 \times 10^{-4}$$

$[HC_3H_5O_3] = 0.150\ M$; calculate $[C_3H_5O_3^-]$

$$[C_3H_5O_3^-] = \frac{K_a[HC_3H_5O_3]}{[H^+]} = \frac{1.4 \times 10^{-4}\ (0.150)}{1.0 \times 10^{-4}} = 0.2100 = 0.21\ M$$

$$\frac{0.210\ mol\ NaC_3H_5O_3}{1.00\ L} \times \frac{112.1\ g\ NaC_3H_5O_3}{1\ mol\ NaC_3H_5O_3} = 23.54 = 24\ g\ NaC_3H_5O_3$$

17.26 (a) $K_a = \frac{[H^+][C_3H_5O_2^-]}{[HC_3H_5O_2]};\ [H^+] = \frac{K_a[HC_3H_5O_2]}{[C_3H_5O_2^-]}$

Since this expression contains a ratio of concentrations, we can ignore total volume and work directly with moles.

$$[H^+] = \frac{1.3 \times 10^{-5}\ (0.12 - x)}{(0.10 + x)} \approx \frac{1.3 \times 10^{-5}\ (0.12)}{0.10} = 1.56 \times 10^{-5} = 1.6 \times 10^{-5}\ M,\ pH = 4.81$$

(b)	$HC_3H_5O_2(aq)$	+	$OH^-(aq)$	$\rightarrow$	$C_2H_3O_2^-(aq) + H_2O(l)$
	0.12 mol		0.01 mol		0.10 mol
	−0.01 mol		−0.01 mol		+0.01 mol
	0.11 mol		0 mol		0.11 mol

$$[H^+] \approx \frac{1.3 \times 10^{-5}\,(0.11)}{(0.11)} = 1.3 \times 10^{-5}\,M;\ pH = 4.89$$

(c)	$C_3H_5O_2^-(aq)$	+	$HI(aq)$	$\rightarrow$	$HC_3H_5O_2(aq) + I^-(aq)$
	0.10 mol		0.01 mol		0.12 mol
	−0.01 mol		−0.01 mol		+0.01 mol
	0.09 mol		0 mol		0.13 mol

$$[H^+] \approx \frac{1.3 \times 10^{-5}\,(0.13)}{(0.09)} = 1.88 \times 10^{-3} = 2 \times 10^{-3}\,M;\ pH = 4.73 = 4.7$$

17.28 $$\frac{6.5\text{ g }NaH_2PO_4}{0.355\text{ L soln}} \times \frac{1\text{ mol }NaH_2PO_4}{120\text{ g }NaH_2PO_4} = 0.153 = 0.15\,M$$

$$\frac{8.0\text{ g }Na_2HPO_4}{0.355\text{ L soln}} \times \frac{1\text{ mol }Na_2HPO_4}{142\text{ g }Na_2HPO_4} = 0.159 = 0.16\,M$$

Use Equation [17.9] to find the pH of the buffer. K_a for $H_2PO_4^-$ is K_{a2} for H_3PO_4, 6.2×10^{-8}

$$pH = -\log(6.2 \times 10^{-8}) + \log\frac{0.159}{0.153} = 7.2076 + 0.0167 = 7.22$$

17.30 The solutes listed contain three possible conjugate acid/conjugate base (CA/CB) pairs. These are:

$HCHO_2/NaCHO_2$, $pK_a = 3.74$

$HC_3H_5O_2/NaC_3H_5O_2$, $pK_a = 4.80$

H_3PO_4/NaH_2PO_4, $pK_a = 2.12$

For maximum buffer capacity, pK_a should be within 1 pH unit of the buffer. The propionic acid/propionate pair are most appropriate for a buffer with pH 4.80.

$$pH = pK_a + \log\frac{[CB]}{CA};\ 4.80 = 4.886 + \log\frac{[NaC_3H_5O_2]}{[HC_3H_5O_2]}$$

$$\log\frac{[NaC_3H_5O_2]}{[HC_3H_5O_2]} = -0.0861;\ \frac{[NaC_3H_5O_2]}{[HC_3H_5O_2]} = 0.8202 = 0.82$$

Since we are making a total of 1 L of buffer, let y = vol $NaC_3H_5O_2$ and (1 − y) = vol $HC_3H_5O_2$.

$$0.8202 = \frac{[NaC_3H_5O_2]}{[HC_3H_5O_2]} = \frac{(0.1\,M \times y)/1.0\text{ L}}{[0.10\,M \times (1-y)]/1.0\text{ L}} = \frac{0.10\,y}{0.10 - 0.10\,y}$$

0.8202(0.10 − 0.10 y) = 0.10 y; 0.08202 = 0.18202 y; y = 0.4506 = 0.45 L

450 mL of 0.10 *M* $NaC_3H_5O_2$, 550 mL $HC_3H_5O_2$

Check. pH (buffer) < pK_a (CA) and the calculated amount of CA in the buffer is greater than the amount of CB.

Acid-Base Titrations

17.32 (a) The quantity of base required to reach the equivalence point is the same in the two titrations.

(b) The pH is higher initially in the titration of a weak acid.

(c) The pH is higher at the equivalence point in the titration of a weak acid.

(d) The pH in excess base is essentially the same for the two cases.

(e) In titrating a weak acid, one needs an indicator that changes at a higher pH than for the strong acid titration. The choice is more critical because the change in pH close to the equivalence point is smaller for the weak acid titration.

17.34 (a) $HCHO_2(aq) + NaOH(aq) \rightarrow NaCHO_2(aq) + H_2O(l)$

At the equivalence point, the major species are Na^+ and CHO_2^-. Na^+ is negligible and CHO_2^- is the CB of $HCHO_2$. The solution is basic, above pH 7.

(b) $Ca(OH)_2(aq) + 2HClO_4(aq) \rightarrow Ca(ClO_4)_2(aq) + 2H_2O(l)$

At the equivalence point, the major species are Ca^{2+} and ClO_4^-; both are negligible. The solution is at pH 7.

(c) $C_5H_5N(aq) + HNO_3(aq) \rightarrow C_5H_5NH^+NO_3^-(aq)$

At the equivalence point, the major species are $C_5H_5NH^+$ and NO_3^-. NO_3^- is negligible and $C_5H_5NH^+$ is the CA of C_5H_5N. The solution is acidic, below pH 7.

17.36 (a) At the equivalence point, moles HX added = moles B initially present = $0.10\ M \times 0.0300\ L = 0.0030$ moles HX added.

(b) $BH^+(aq)$

(c) Both K_a for BH^+ and concentration BH^+ determine pH at the equivalence point.

(d) Because the pH at the equivalence point will be less than 7, methyl red would be more appropriate.

17.38 (a) $$55.0\text{ mL NaOH} \times \frac{0.0950\text{ mol NaOH}}{1000\text{ mL soln}} \times \frac{1\text{ mol HCl}}{1\text{ mol NaOH}} \times \frac{1000\text{ mL soln}}{0.105\text{ mol HCl}} = 49.8\text{ mL HCl soln}$$

(b) $$22.5\text{ mL NH}_3 \times \frac{0.118\text{ mol NH}_3}{1000\text{ mL soln}} \times \frac{1\text{ mol HCl}}{1\text{ mol NH}_3} \times \frac{1000\text{ mL soln}}{0.105\text{ mol HCl}} = 25.3\text{ mL HCl soln}$$

(c) $$125.0\text{ mL} \times \frac{1.35\text{ g NaOH}}{1000\text{ mL}} \times \frac{1\text{ mol NaOH}}{40.00\text{ g NaOH}} \times \frac{1\text{ mol HCl}}{1\text{ mol NaOH}} \times \frac{1000\text{ mL soln}}{0.105\text{ mol HCl}} = 40.2\text{ mL HCl soln}$$

17.40 moles OH^- = $M_{KOH} \times L_{KOH} = 0.150\ M \times 0.0300\ L = 4.50 \times 10^{-3}$ mol

moles $H^+ = M_{HClO_4} \times L_{HClO_4} = 0.125\ M \times L_{HClO_4}$

	mL_{KOH}	mL_{HClO_4}	Total Volume	Moles OH^-	Moles H^+	Molarity Excess Ion	pH
(a)	30.0	30.0	60.0	4.50×10^{-3}	3.75×10^{-3}	$0.0125(OH^-)$	12.10
(b)	30.0	35.0	65.0	4.50×10^{-3}	4.38×10^{-3}	$1.9 \times 10^{-3}(OH^-)$	11.28
(c)	30.0	36.0	66.0	4.50×10^{-3}	4.50×10^{-3}	$1.0 \times 10^{-7}(OH^-)$	7.00
(d)	30.0	37.0	67.0	4.50×10^{-3}	4.63×10^{-3}	$1.9 \times 10^{-3}(H^+)$	2.73
(e)	30.0	40.0	70.0	4.50×10^{-3}	5.00×10^{-3}	$7.1 \times 10^{-3}(H^+)$	2.15

$$\text{molarity of excess ion} = \frac{\text{moles ion}}{\text{total vol in L}}$$

(a) $$\frac{4.50 \times 10^{-3}\ \text{mol OH}^- - 3.75 \times 10^{-3}\ \text{mol H}^+}{0.0600\ \text{L}} = 0.0125 = 0.013\ M\ \text{OH}^-$$

(b) $$\frac{4.50 \times 10^{-3}\ \text{mol OH}^- - 4.38 \times 10^{-3}\ \text{mol H}^+}{0.0650\ \text{L}} = 1.92 \times 10^{-3} = 1.9 \times 10^{-3}\ M\ \text{OH}^-$$

(c) equivalence point, mol H^+ = mol OH^- M
$KClO_4$ does not hydrolyze, so $[H^+] = [OH^-] = 1 \times 10^{-7}$

(d) $$\frac{4.50 \times 10^{-3}\ \text{mol OH}^- - 4.63 \times 10^{-3}\ \text{mol H}^+}{0.0670\ \text{L}} = 1.87 \times 10^{-3} = 1.9 \times 10^{-3}\ M\ \text{H}^+$$

(e) $$\frac{4.50 \times 10^{-3}\ \text{mol OH}^- - 5.00 \times 10^{-3}\ \text{mol H}^+}{0.0700\ \text{L}} = 7.14 \times 10^{-3} = 7.1 \times 10^{-3}\ M\ \text{H}^+$$

17.42 (a) Weak base problem: $K_b = 1.8 \times 10^{-5} = \dfrac{[NH_4^+][OH^-]}{[NH_3]}$

At equilibrium, $[OH^-] = x$, $[NH_3] = (0.030 - x)$; $[NH_4^+] = x$

$$1.8 \times 10^{-5} = \frac{x^2}{(0.030 - x)} \approx \frac{x^2}{0.030};\ x = [OH^-] = 7.348 \times 10^{-4} = 7.3 \times 10^{-4}\ M$$

pH = 14.00 – 3.13 = 10.87

(b–f) Calculate mol NH_3 and mol NH_4^+ after the acid-base reaction takes place. 0.030 M $NH_3 \times 0.0300$ L = 9.0×10^{-4} mol NH_3 present initially.

		$NH_3(aq)$	+ $HCl(aq)$	→ $NH_4^+(aq) + Cl^-(aq)$
			$(0.025\ M \times 0.0100\ L) =$	
(b)	before rx	9.0×10^{-4} mol	2.5×10^{-4} mol	0 mol
	after rx	**6.5×10^{-4} mol**	**0 mol**	**2.5×10^{-4} mol**

		$NH_3(aq)$	+ $HCl(aq)$	$\rightarrow$ $NH_4^+(aq) + Cl^-(aq)$
			$(0.025\ M \times 0.0200\ L) =$	
(c)	before rx	9.0×10^{-4} mol	5.0×10^{-4} mol	0 mol
	after rx	**4.0×10^{-4} mol**	**0 mol**	**5.0×10^{-4} mol**
			$(0.025\ M \times 0.0350\ L) =$	
(d)	before rx	9.0×10^{-4} mol	8.75×10^{-4} mol	0 mol
	after rx	**0.25×10^{-4} mol**	**0 mol**	**8.75×10^{-4} mol**
			$(0.025\ M \times 0.0360\ L) =$	
(e)	before rx	9.0×10^{-4} mol	9.0×10^{-4} mol	0 mol
	after rx	**0 mol**	**0 mol**	**9.0×10^{-4} mol**
			$(0.025\ M \times 0.0370\ L) =$	
(f)	before rx	9.0×10^{-4} mol	9.25×10^{-4} mol	0 mol
	after rx	**0 mol**	**0.25×10^{-4} mol**	**9.0×10^{-4} mol**

(b) Using the acid dissociation equilibrium for NH_4^+ (so that we calculate $[H^+]$ directly), $NH_4^+(aq) \rightleftharpoons H^+(aq) + NH_3(aq)$

$$K_a = \frac{[H^+][NH_3]}{[NH_4^+]} = \frac{K_w}{K_b \text{ for } NH_3} = \frac{1.0 \times 10^{-14}}{1.8 \times 10^{-5}} = 5.56 \times 10^{-10} = 5.6 \times 10^{-10}$$

$$[NH_3] = \frac{6.5 \times 10^{-4}\ \text{mol}}{0.0400\ \text{L}} = 0.01625\ M;\ [NH_4^+] = \frac{2.50 \times 10^{-4}\ \text{mol}}{0.0400\ \text{L}} = 6.25 \times 10^{-3}\ M$$

$$[H^+] = \frac{5.56 \times 10^{-10}\ [NH_4^+]}{[NH_3]} \approx \frac{5.56 \times 10^{-10}\ (6.25 \times 10^{-3})}{(0.01625)} = 2.14 \times 10^{-10};\ \text{pH} = 9.67$$

(We will assume $[H^+]$ is small compared to $[NH_3]$ and $[NH_4^+]$.)

(c) $$[NH_3] = \frac{4.0 \times 10^{-4}\ \text{mol}}{0.0500\ \text{L}} = 0.0080\ M;\ [NH_4^+] = \frac{5.0 \times 10^{-4}\ \text{mol}}{0.0500\ \text{L}} = 0.010 M$$

$$[H^+] = \frac{5.56 \times 10^{-10}\ (0.010)}{(0.0080)} = 6.94 \times 10^{-10} = 6.9 \times 10^{-10}\ M;\ \text{pH} = 9.16$$

(d) $$[NH_3] = \frac{0.25 \times 10^{-4}\ \text{mol}}{0.0650\ \text{L}} = 3.846 \times 10^{-4} = 4 \times 10^{-4}\ M;\ [NH_4^+] = \frac{8.75 \times 10^{-4}\ \text{mol}}{0.0650\ \text{L}}$$

$$= 0.01346 = 0.013\ M$$

$$[H^+] = \frac{5.56 \times 10^{-10}\ (0.01346)}{3.846 \times 10^{-4}} = 1.946 \times 10^{-8} = 2 \times 10^{-8}\ M;\ \text{pH} = 7.7$$

(e) At the equivalence point, $[H^+] = [NH_3] = x$

$$[NH_4^+] = \frac{9.0 \times 10^{-4}\ M}{0.0660\ L} = 0.01364 = 0.014\ M$$

$$5.56 \times 10^{-10} = \frac{x^2}{0.01364};\ x = [H^+] = 2.754 \times 10^{-6} = 2.8 \times 10^{-6};\ pH = 5.56$$

(f) Past the equivalence point, $[H^+]$ from the excess HCl determines the pH.

$$[H^+] = \frac{0.25 \times 10^{-4}\ mol}{0.0670\ L} = 3.731 \times 10^{-4} = 4 \times 10^{-4}\ M;\ pH = 3.4$$

17.44 The volume of NaOH solution required in all cases is

$$V_{base} = \frac{V_{acid} \times M_{acid}}{M_{base}} = \frac{(0.100)\ V_{acid}}{(0.080)} = 1.25\ V_{acid}$$

The total volume at the equivalence point is $V_{base} + V_{acid} = 2.25\ V_{acid}$.

The concentration of the salt at the equivalence point is $\frac{M_{acid}\ V_{acid}}{2.25\ V_{acid}} = \frac{0.100}{2.25} = 0.0444\ M$

(a) $0.0444\ M$ NaBr, pH = 7.00

(b) $0.0444\ M$ Na^+ $C_3H_5O_3^-$; $C_3H_5O_3^-(aq) + H_2O(l) \rightleftharpoons HC_3H_5O_3(aq) + OH^-(aq)$

$$K_b = \frac{[HC_3H_5O_3][OH^-]}{[C_3H_5O_3^-]} = \frac{K_w}{K_a} = \frac{1.0 \times 10^{-14}}{1.4 \times 10^{-4}} = 7.14 \times 10^{-11} = 7.1 \times 10^{-11}$$

$[HC_3H_5O_3] = [OH^-]$; $[C_3H_5O_3^-] \approx 0.0444$

$[OH^-]^2 \approx 0.0444(7.14 \times 10^{-11})$; $[OH^-] = 1.78 \times 10^{-6} = 1.8 \times 10^{-6}\ M$, pOH = 5.75;

pH = 8.25

(c) $CrO_4^{2-}(aq) + H_2O(l) \rightleftharpoons HCrO_4^-(aq) + OH^-(aq)$

$$K_b = \frac{[HCrO_4^-][OH^-]}{[CrO_4^{2-}]} = \frac{K_w}{K_a} = \frac{1.0 \times 10^{-14}}{3.0 \times 10^{-7}} = 3.33 \times 10^{-8} = 3.3 \times 10^{-8}$$

$[OH^-]^2 \approx 0.0444(3.33 \times 10^{-8})$; $[OH^-] = 3.849 \times 10^{-5} = 3.8 \times 10^{-5}$, pH = 9.59

Solubility Equilibria and Factors Affecting Solubility

17.46 (a) Solubility is the amount (grams, moles) of solute that will dissolve in a certain volume of solution. Solubility-product constant is an equilibrium constant, the product of the molar concentrations of all the dissolved ions in solution.

(b) $K_{sp} = [Mn^{2+}][CO_3^{2-}]$; $K_{sp} = [Hg^{2+}][OH^-]^2$; $K_{sp} = [Cu^{2+}]^3[PO_4^{3-}]^2$

17.48 (a) $PbBr_2(s) \rightleftharpoons Pb^{2+}(aq) + 2Br^-(aq)$

$K_{sp} = [Pb^{2+}][Br^-]^2$; $[Pb^{2+}] = 1.0 \times 10^{-2}\ M$, $[Br^-] = 2.0 \times 10^{-2}\ M$

$K_{sp} = (1.0 \times 10^{-2}\ M)(2.0 \times 10^{-2}\ M)^2 = 4.0 \times 10^{-6}$

(b) $AgIO_3(s) \rightleftharpoons Ag^+(aq) + IO_3^-(aq)$; $K_{sp} = [Ag^+][IO_3^-]$

$$[Ag^+] = [IO_3^-] = \frac{0.0490 \text{ g } AgIO_3}{1.00 \text{ L soln}} \times \frac{1 \text{ mol } AgIO_3}{282.8 \text{ g } AgIO_3} = 1.733 \times 10^{-4} = 1.73 \times 10^{-4} \, M$$

$K_{sp} = (1.733 \times 10^{-4} \, M)(1.733 \times 10^{-4} \, M) = 3.00 \times 10^{-8}$

(c) $Cu(OH)_2(s) \rightleftharpoons Cu^{2+}(aq) + 2OH^-(aq)$; $K_{sp} = [Cu^{2+}][OH^-]^2$

$[Cu^{2+}] = x$, $[OH^-] = 2x$; $K_{sp} = 4.8 \times 10^{-20} = (x)(2x)^2$

$4.8 \times 10^{-20} = 4x^3$; $x = [Cu^{2+}] = 2.290 \times 10^{-7} = 2.3 \times 10^{-7} \, M$

$$\frac{2.290 \times 10^{-7} \text{ mol } Cu(OH)_2}{1 \text{ L}} \times \frac{97.56 \text{ g } Cu(OH)_2}{1 \text{ mol } Cu(OH)_2} = 2.2 \times 10^{-5} \text{ g } Cu(OH)_2$$

However, $[OH^-]$ from $Cu(OH)_2 = 4.58 \times 10^{-7}$ M; this is similar to $[OH^-]$ from the autoionization of water.

$K_w = [H^+][OH^-]$; $[H^+] = y$, $[OH^-] = (4.58 \times 10^{-7} + y)$

$1.0 \times 10^{-14} = y(4.58 \times 10^{-7} + y)$; $y^2 + 4.58 \times 10^{-7} y - 1.0 \times 10^{-14}$

$$y = \frac{-4.58 \times 10^{-7} \pm \sqrt{(4.58 \times 10^{-7})^2 - 4(1)(-1.0 \times 10^{-14})}}{2}; \; y = 2.09 \times 10^{-8}$$

$[OH^-]_{total} = 4.58 \times 10^{-7} \, M + 0.209 \times 10^{-7} \, M = 4.79 \times 10^{-7} \, M$

Recalculating $[Cu^{2+}]$ and thus molar solubility of $Cu(OH)_2(s)$:

$4.8 \times 10^{-20} = x(4.79 \times 10^{-7})^2$; $x = 2.09 \times 10^{-7} \, M \; Cu^{2+}$

$$\frac{2.09 \times 10^{-7} \text{ mol } Cu(OH)_2(s)}{1 \text{ L}} \times \frac{97.56 \text{ g } Cu(OH)_2}{1 \text{ mol } Cu(OH)_2} = 2.0 \times 10^{-5} \text{ g } Cu(OH)_2$$

Note that the presence of OH^- as a common ion decreases the water solubility of $Cu(OH)_2$.

17.50 $PbI_2(s) \rightleftharpoons Pb^{2+}(aq) + 2I^-(aq)$; $K_{sp} = [Pb^{2+}][I^-]^2$

$$[Pb^{2+}] = \frac{0.54 \text{ g } PbI_2}{1.00 \text{ L soln}} \times \frac{1 \text{ mol } PbI_2}{461.0 \text{ g } PbI_2} = 1.17 \times 10^{-3} = 1.2 \times 10^{-3} \, M$$

$[I^-] = 2[Pb^{2+}]$; $K_{sp} = [Pb^{2+}](2[Pb^{2+}])^2 = 4[Pb^{2+}]^3 = 4(1.17 \times 10^{-3})^3 = 6.4 \times 10^{-9}$

17.52 $LaF_3(s) \rightleftharpoons La^{3+}(aq) + 3F^-(aq)$; $K_{sp} = [La^{3+}][F^-]^3$

(a) molar solubility $= x = [La^{3+}]$; $[F^-] = 3x$

$K_{sp} = 2 \times 10^{-19} \, (x)(3x)3$; $2 \times 10^{-19} = 27 \, x^4$; $x = (7.41 \times 10^{-21})^{1/4}$, $x = 9.28 \times 10^{-6}$

$= 9 \times 10^{-6}$ M La^{3+}

$$\frac{9.28 \times 10^{-6} \text{ mol } LaF_3}{1 \text{ L}} \times \frac{195.9 \text{ g } LaF_3}{1 \text{ mol}} = 1.82 \times 10^{-3} = 2 \times 10^{-3} \text{ g } LaF_3/\text{L}$$

(b) molar solubility = x = $[La^{3+}]$

There are two sources of F^-: KF(0.010 *M*) and LaF_3 (3x *M*)

$K_{sp} = (x)(0.010 + 3x)^3$; assume x is small compared to 0.010 M.

$2 \times 10^{-19} = (0.010)^3$ x; $x = 2 \times 10^{-19}/1.0 \times 10^{-6} = 2 \times 10^{-13}$ *M* La^{3+}

$$\frac{2 \times 10^{-13} \text{ mol } LaF_3}{1 \text{ L}} \times \frac{195.9 \text{ g } LaF_3}{1 \text{ mol}} = 3.92 \times 10^{-11} = 4 \times 10^{-11} \text{ g } LaF_3/\text{L}$$

(c) molar solubility = x, $[F^-] = 3x$, $[La^{3+}] = 0.050\ M + x$

$K_{sp} = (0.050 + x)(3x)^3$; assume x is small compared to 0.050 *M*.

$2 \times 10^{-19} = (0.050)(27\ x^3) = 1.35\ x^3$; $x = (1.48 \times 10^{-19})^{1/3} = 5.29 \times 10^{-7} = 5 \times 10^{-7}\ M$

$$\frac{5.29 \times 10^{-7} \text{ mol } LaF_3}{1 \text{ L}} \times \frac{195.9 \text{ g } LaF_3}{1 \text{ mol}} = 1.04 \times 10^{-4} = 1 \times 10^{-4} \text{ g } LaF_3/\text{L}$$

17.54 $Fe(OH)_2(s) \rightleftharpoons Fe^{2+}(aq) + 2OH^-(aq)$; $K_{sp} = 8.0 \times 10^{-16}$

Since the $[OH^-]$ is set by the pH of the solution, the solubility of $Fe(OH)_2$ is just $[Fe^{2+}]$.

(a) pH = 7.0, pOH = 14 – pH = 7.0, $[OH^-] = 10^{-pOH} = 1.0 \times 10^{-7} = 1 \times 10^{-7}\ M$

$$K_{sp} = 7.9 \times 10^{-16} = [Fe^{2+}](1.0 \times 10^{-7})^2; [Fe^{2+}] = \frac{7.9 \times 10^{-16}}{1.0 \times 10^{-14}} = 7.9 \times 10^{-2} = 8 \times 10^{-2}\ M$$

Check. In pure water, $[OH^-]$ from $Fe(OH)_2$ is similar to (OH^-) from the autoionization of water, resulting in a cubic equation for $[Fe^{2+}]$. The solubility of $Fe(OH)_2$ at a buffered pH = 7.0 is actually greater than the solubility in pure water.

(b) pH = 10.0, pOH = 4.0, $[OH^-] = 1.0 \times 10^{-4} = 1 \times 10^{-4}\ M$

$$K_{sp} = 7.9 \times 10^{-16} = [Fe^{2+}][1.0 \times 10^{-4}]^2; [Fe^{2+}] = \frac{7.9 \times 10^{-16}}{1.0 \times 8^{-10}} = 7.9 \times 10^{-8} = 8 \times 10^{-8}\ M$$

(c) pH = 12.0, pOH = 2.0, $[OH^-] = 1.0 \times 10^{-2} = 1 \times 10^{-2}\ M$

$$K_{sp} = 7.9 \times 10^{-16} = [Fe^{2+}][1.0 \times 10^{-2}]^2; [Fe^{2+}] = \frac{7.9 \times 10^{-16}}{1.0 \times 10^{-4}} = 7.9 \times 10^{-12} = 8 \times 10^{-12}\ M$$

17.56 If the anion in the slightly soluble salt is the conjugate base of a strong acid, there will be no reaction.

(a) $MnS(s) + 2H^+(aq) \rightarrow H_2S(aq) + Mn^{2+}(aq)$

(b) $PbF_2(s) + 2H^+(aq) \rightarrow 2HF(aq) + Pb^{2+}(aq)$

(c) $AuCl_3(s) + H^+(aq) \rightarrow$ no reaction

(d) $Hg_2C_2O_4(s) + 2H^+(aq) \rightarrow H_2C_2O_4(aq) + Hg_2^{2+}(aq)$

(e) $CuBr(s) + H^+(aq) \rightarrow$ no reaction

17.58 $NiC_2O_4(s) \rightleftharpoons Ni^{2+}(aq) + C_2O_4^{2-}(aq)$; $K_{sp} = [Ni^{2+}][C_2O_4^{2-}] = 4 \times 10^{-10}$

When the salt has just dissolved, $[C_2O_4^{2-}]$ will be 0.020 *M*. Thus $[Ni^{2+}]$ must be less than $4 \times 10^{-10} / 0.020 = 2 \times 10^{-8}$ *M*. To achieve this low $[Ni^{2+}]$ we must complex the Ni^{2+} ion with NH_3: $Ni^{2+}(aq) + 6NH_3(aq)$ *f* $Ni(NH_3)_6^{2+}(aq)$. Essentially all Ni(II) is in the form of the complex, so $[Ni(NH_3)_6^{2+}] = 0.020$. Find K_f for $Ni(NH_3)_6^{2+}$ in Table 17.1.

$$K_f = \frac{[Ni(NH_3)_6^{2+}]}{[Ni^{2+}][NH_3]^6} = \frac{(0.020)}{(2 \times 10^{-8})[NH_3]^6} = 1.2 \times 10^9; [NH_3]^6 = 8.33 \times 10^{-4};$$

$$[NH_3] = 0.307 = 0.3\,M$$

17.60

$$Ag_2S(s) \rightleftharpoons 2Ag^+(aq) + S^{2-}(aq) \qquad K_{sp}$$

$$S^{2-}(aq) + 2H^+(aq) \rightleftharpoons H_2S(aq) \qquad 1/(K_{a1} \times K_{a2})$$

$$2[Ag]^+(aq) + 2Cl^-(aq) \rightleftharpoons AgCl_2^-(aq) \qquad K_f^2$$

$$Ag_2S(s) + 2H^+(aq) + 4Cl^-(aq) \rightleftharpoons 2AgCl_2^-(aq) + H_2S(aq)$$

$$K = \frac{K_{sp} \times K_f^2}{K_{a1} \times K_{a2}} = \frac{(6 \times 10^{-51})(1.1 \times 10^5)^2}{(9.5 \times 10^{-8})(1 \times 10^{-19})} = 7.64 \times 10^{-15} = 8 \times 10^{-15}$$

Precipitation; Qualitative Analysis

17.62 (a) $Co(OH)_2(s) \rightleftharpoons Co^{2+}(aq) + 2OH^-(aq)$; $K_{sp} = [Co^{2+}][OH^-]^2 = 1.3 \times 10^{-15}$

pH = 8.5; pOH = 14 – 8.5 = 5.5; $[OH^-] = 10^{-5.5} = 3.16 \times 10^{-6} = 3 \times 10^{-6}$ *M*

$Q = (0.020)(3.16 \times 10^{-6})^2 = 2 \times 10^{-13}$; $Q > K_{sp}$, $Co(OH)_2$ will precipitate

(b) $AgIO_3(s) \rightleftharpoons Ag+(aq) + IO_3^-(aq)$; $K_{sp} = [Ag+][IO_3^-] = 3.1 \times 10^{-8}$

$$[Ag^+] = \frac{0.010\,M\,Ag^+ \times 0.100\,L}{0.110\,L} = 9.09 \times 10^{-3} = 9.1 \times 10^{-3}\,M$$

$$[IO_3^-] = \frac{0.015\,M\,IO_3^- \times 0.010\,L}{0.110\,L} = 1.36 \times 10^{-3} = 1.4 \times 10^{-3}\,M$$

$Q = (9.09 \times 10^{-3})(1.36 \times 10^{-3}) = 1.2 \times 10^{-5}$; $Q > K_{sp}$, $AgIO_3$ will precipitate

17.64 $AgCl(s) \rightleftharpoons Ag^+(aq) + Cl^-(aq)$; $K_{sp} = [Ag^+][Cl^-] = 1.8 \times 10^{-10}$

$$[Ag^+] = \frac{0.10\,M \times 0.2\,mL}{10\,mL} = 2 \times 10^{-3}\,M; \quad [Cl^-] = \frac{1.8 \times 10^{-10}}{2 \times 10^{-3}\,M} = 9 \times 10^{-8}\,M$$

$$\frac{9 \times 10^{-8}\,mol\,Cl^-}{1\,L} \times \frac{35.45\,g\,Cl^-}{1\,mol\,Cl^-} \times 0.010\,L = 3.19 \times 10^{-8}\,g\,Cl^- = 3 \times 10^{-8}\,g\,Cl^-$$

17.66 (a) Precipitation will begin when $Q = K_{sp}$.

$BaSO_4$: $K_{sp} = [Ba^{2+}][SO_4^{2-}] = 1.1 \times 10^{-10}$

$1.1 \times 10^{-10} = (0.010)[SO_4^{2-}]$; $[SO_4^{2-}] = 1.1 \times 10^{-8}$ *M*

$SrSO_4$: $K_{sp} = [Sr^{2+}][SO_4^{2-}] = 3.2 \times 10^{-7}$

$3.2 \times 10^{-7} = (0.010)[SO_4^{2-}]$; $[SO_4^{2-}] = 3.2 \times 10^{-5}$ *M*

The $[SO_4^{2-}]$ necessary to begin precipitation is the smaller of the two values, 1.1×10^{-8} *M* SO_4^{2-}.

(b) Ba^{2+} precipitates first, because it requires the smaller $[SO_4^{2-}]$.

(c) Sr^{2+} will begin to precipitate when $[SO_4^{2-}]$ in solution (not bound in $BaSO_4$) reaches 3.2×10^{-5} *M*.

17.68 Initial solubility in water rules out CdS and HgO. Formation of a precipitate on addition of HCl indicates the presence of $Pb(NO_3)_2$ (formation of $PbCl_2$). Formation of a precipitate on addition of H_2S at pH 1 probably indicates $Cd(NO_3)_2$ (formation of CdS). (This test can be misleading because enough Pb^{2+} can remain in solution after filtering $PbCl_2$ to lead to visible precipitation of PbS.) Absence of a precipitate on addition of H_2S at pH 8 indicates that $ZnSO_4$ is not present. The yellow flame test indicates presence of Na^+. In summary, $Pb(NO_3)_2$ and Na_2SO_4 are definitely present, $Cd(NO_3)_2$ is probably present, and CdS, HgO and $ZnSO_4$ are definitely absent.

17.70 (a) Make the solution slightly basic and saturate with H_2S; CdS will precipitate, Na^+ remains in solution.

(b) Make the solution acidic, saturate with H_2S; CuS will precipitate, Mg^{2+} remains in solution.

(c) Add HCl, $PbCl_2$ precipitates. (It is best to carry out the reaction in an ice-water bath to reduce the solubility of $PbCl_2$.)

(d) Add dilute HCl; AgCl precipitates, Hg^{2+} remains in solution.

17.72 The addition of $(NH_4)_2HPO_4$ could result in precipitation of salts from metal ions of the other groups. The $(NH_4)_2HPO_4$ will render the solution basic, so metal hydroxides could form as well as insoluble phosphates. It is essential to separate the metal ions of a group from other metal ions before carrying out the specific tests for that group.

Additional Exercises

17.74 $K_a = \dfrac{[H^+][In^-]}{[HIn]}$; at pH = 4.68, $[HIn] = [In^-]$; $[H^+] = K_a$; $pH = pK_a = 4.68$

17.75 (a) $HA(aq) + B(aq) \rightleftharpoons HB^+(aq) + A^-(aq)$ $\quad K_{eq} = \dfrac{[HB^+][A^-]}{[HA][B]}$

(b) Note that the solution is slightly basic because B is a stronger base than HA is an acid. (Or, equivalently, that A^- is a stronger base than HB^+ is an acid.) Thus, a little of the A^- is used up in reaction: $A^-(aq) + H_2O(l) \rightleftharpoons HA(aq) + OH^-(aq)$. Since pH is not very far from neutral, it is reasonable to assume that the reaction in part (a) has gone far to the right, and that $[A^-] \approx [HB^+]$ and $[HA] \approx [B]$. Then

$$K_a = \frac{[A^-][H^+]}{[HA]} = 8.0 \times 10^{-5}; \text{ when pH} = 9.2, [H^+] = 6.31 \times 10^{-10} = 6 \times 10^{-10}\ M$$

$$\frac{[A^-]}{[HA]} = 8.0 \times 10^{-5} / 6.31 \times 10^{-10} = 1.268 \times 10^5 = 1 \times 10^5$$

From the assumptions above, $\frac{[A^-]}{[HA]} = \frac{[HB^+]}{[B]}$, so $K_{eq} \approx \frac{[A^-]^2}{[HA]^2} = 1.608 \times 10^{10} = 2 \times 10^{10}$

(c) K_b for the reaction $B(aq) + H_2O(l) \rightleftharpoons BH^+(aq) + OH^-(aq)$ can be calculated by noting that the equilibrium constant for the reaction in part (a) can be written as $K_{eq} = K_a\ (HA) \times K_b\ (B) / K_w$. (You should prove this to yourself.) Then,

$$K_b\ (B) = \frac{K_{eq} \times K_w}{K_a\ (HA)} = \frac{(1.608 \times 10^{10})(1.0 \times 10^{-14})}{8.0 \times 10^{-5}} = 2.010 = 2$$

K_b (B) is larger than K_a (HA), as it must be if the solution is basic.

17.77 $\frac{0.20 \text{ mol } HC_2H_3O_2}{1 \text{ L soln}} \times 0.750 \text{ L} = 0.150 = 0.15 \text{ mol } HC_2H_3O_2$

$$0.15 \text{ mol } HC_2H_3O_2 \times \frac{60.05 \text{ g } HC_2H_3O_2}{1 \text{ mol } HC_2H_3O_2} \times \frac{1 \text{ g gl acetic acid}}{0.99 \text{ g } HC_2H_3O_2} \times \frac{1.00 \text{ mL gl acetic acid}}{1.05 \text{ g gl acetic acid}}$$
$$= 8.7 \text{ mL glacial acetic acid}$$

At pH 4.50, $[H^+] = 10^{-4.50} = 3.16 \times 10^{-5} = 3.2 \times 10^{-5}\ M$; this is small compared to $0.20\ M\ HC_2H_3O_2$.

$$K_a = \frac{(3.16 \times 10^{-5})[C_2H_3O_2^-]}{0.20} = 1.8 \times 10^{-5}; [C_2H_3O_2^-] = 0.114 = 0.11\ M$$

$$\frac{0.114 \text{ mol } NaC_2H_3O_2}{1 \text{ L soln}} \times 0.750 \text{ L} \times \frac{82.03 \text{ g } NaC_2H_3O_2}{1 \text{ mol } NaC_2H_3O_2} = 7.004 = 7.0 \text{ g } NaC_2H_3O_2$$

17.79 At the equivalence point of a titration, moles strong base added equals moles weak acid initially present. $M_B \times V_B$ = mol base added = mol acid initial

At the half-way point, the volume of base is one-half of the volume required to reach the equivalence point, and the moles base delivered equals one-half of the mol acid initially present. This means that one-half of the weak acid HA is converted to the conjugate base A^-. If exactly half of the acid reacts, mol HA = mol A^- and [HA] = $[A^-]$ at the half-way point.

From Equation [17.9], $pH = pK_a + \log \frac{[\text{conj. base}]}{[\text{conj. acid}]} = pK_a + \log \frac{[A^-]}{[HA^-]}$.

If $[A^-]/[HA] = 1$, $\log(1) = 0$ and pH = pK_a of the weak acid being titrated.

17.80 (a) $\frac{0.4885 \text{ g KHP}}{0.100 \text{ L}} \times \frac{1 \text{ mol KHP}}{204.2 \text{ g KHP}} = 0.02392 = 0.0239\ M\ P^{2-}$ at the equivalence point

The pH at the equivalence point is determined by the hydrolysis of P^{2-}.

$P^{2-}(aq) + H_2O(l) \rightleftharpoons HP^-(aq) + OH^-(aq)$

$$K_b = \frac{[HP^-][OH^-]}{[P^{2-}]} = \frac{K_w}{K_a \text{ for } HP^-} = \frac{1.0 \times 10^{-14}}{3.1 \times 10^{-6}} = 3.23 \times 10^{-9} = 3.2 \times 10^{-9}$$

$$3.23 \times 10^{-9} = \frac{x^2}{(0.02392 - x)} \approx \frac{x^2}{0.02392}; X = [OH^-] = 8.8 \times 10^{-6}\ M$$

pH = 14 – 5.06 = 8.94. From Figure 16.7, either phenolphthalein (pH 8.2 – 10.0) or thymol blue (pH 8.0 - 9.6) could be used to detect the equivalence point. Phenolphthalein is usually the indicator of choice because the colorless to pink change is easier to see.

(b) $$0.4885 \text{ g KHP} \times \frac{1 \text{ mol KHP}}{204.2 \text{ g KHP}} \times \frac{1 \text{ mol NaOH}}{1 \text{ mol KHP}} \times \frac{1}{0.03855 \text{ L NaOH}} = 0.06206\ M \text{ NaOH}$$

17.82 The reaction involved is $HA(aq) + OH^-(aq) \rightleftharpoons A^-(aq) + H_2O(l)$. We thus have 0.080 mol A^- and 0.12 mol HA in a total volume of 1.0 L, so the "initial" molarities of A^- and HA are 0.080 *M* and 0.12 *M*, respectively. The weak acid equilibrium of interest is

$$HA(aq) \rightleftharpoons H^+(aq) + A^-(aq)$$

(a) $$K_a = \frac{[H^+][A^-]}{[HA]}; [H^+] = 10^{-4.80} = 1.58 \times 10^{-5} = 1.6 \times 10^{-5}\ M$$

Assuming $[H^+]$ is small compared to [HA] and $[A^-]$,

$$K_a \approx \frac{(1.58 \times 10^{-5})(0.080)}{(0.12)} = 1.06 \times 10^{-5} = 1.1 \times 10^{-5}, pK_a = 4.98$$

(b) At pH = 5.00, $[H^+] = 1.0 \times 10^{-5}\ M$. Let b = extra moles NaOH.

$[HA] = 0.12 - b$, $[A^-] = 0.080 + b$

$$1.06 \times 10^{-5} \approx \frac{(1.0 \times 10^{-5})(0.080 + b)}{(0.12 - b)}; 2.06 \times 10^{-5}\, b = 4.72 \times 10^{-7};$$

b = 0.023 mol NaOH

17.83 Assume that H_3PO_4 will react with NaOH in a stepwise fashion: (This is not unreasonable, since the three K_a values for H_3PO_4 are significantly different.)

	$H_3PO_4(aq)$	+ NaOH(aq)	→ $H_2PO_4^-(aq) + Na^+(aq) + H_2O(l)$
before	0.20 mol	0.30 mol	0 mol
after	0 mol	0.10 mol	0.20 mol

	$H_2PO_4^-(aq)$	+ NaOH(aq)	→ $HPO_4^-(aq) + Na^+(aq) + H_2O(l)$
before	0.20 mol	0.10 mol	0.25 mol
after	0.10 mol	0	0.35 mol

Thus, after all NaOH has reacted, the resulting 1.00 L solution is a buffer containing 0.10 mol $H_2PO_4^-$ and 0.35 mol HPO_4^{2-}. $H_2PO_4^-(aq) \rightleftharpoons H^+(aq) + HPO_4^{2-}(aq)$

$$K_a = 6.2 \times 10^{-8} = \frac{[HPO_4^{2-}][H^+]}{[H_2PO_4^-]}; [H^+] = \frac{6.2 \times 10^{-8} (0.10\ M)}{0.35\ M} = 1.77 \times 10^{-8} = 1.8 \times 10^{-8}\ M;$$

$$pH = 7.75$$

17.85　$C_3H_5O_3^-$ will be formed by reaction of $HC_3H_5O_3$ with NaOH.

$0.1000\ M \times 0.02500\ L = 2.500 \times 10^{-3}$ mol $HC_3H_5O_3$; b = mol NaOH needed

	$HC_3H_5O_3$	+	NaOH	→	$C_3H_5O_3^-$ + H_2O + Na^+
initial	2.500×10^{-3} mol		b mol		
rx	-b mol		-b mol		+b mol
after rx	$(2.500 \times 10^3 - b)$ mol		0		b mol

$$K_a = \frac{[H^+][C_3H_5O_3^-]}{[HC_3H_5O_3]}; K_a = 1.4 \times 10^{-4}; [H^+] = 10^{-pH} = 10^{-3.75} = 1.778 \times 10^{-4} = 1.8 \times 10^{-4}\ M$$

Since solution volume is the same for $HC_3H_5O_3$ and $C_3H_5O_3^-$, we can use moles in the equation for $[H^+]$.

$$K_a = 1.4 \times 10^{-4} = \frac{1.778 \times 10^{-4}\ (b)}{(2.500 \times 10^{-3} - b)}; 0.7874\ (2.500 \times 10^{-3} - b) = b, 1.969 \times 10^{-3} = 1.7874\ b,$$

$b = 1.10 \times 10^{-3} = 1.1 \times 10^{-3}$ mol OH^-

(The precision of K_a dictates that the result has 2 sig figs.)

Substituting this result into the K_a expression gives $[H^+] = 1.8 \times 10^{-4}$. This checks and confirms our result. Calculate volume NaOH required from M = mol/L.

$$1.10 \times 10^{-3}\ \text{mol OH}^- \times \frac{1\ L}{1.000\ \text{mol}} \times \frac{1\ \mu L}{1 \times 10^{-6}\ L} = 1.1 \times 10^3\ \mu L\ (1.1\ mL)$$

17.86　(a)　$H^+(aq) + HCO_3^-(aq) \rightleftharpoons H_2CO_3(aq) \rightleftharpoons H_2O(l) + CO_2(g)$

A person breathing normally exhales $CO_2(g)$. Rapid breathing causes excess $CO_2(g)$ to be removed from the blood. By LeChatelier's principle, this causes both equilibria above to shift right, reducing $[H^+]$ in the blood and raising blood pH.

(b)　Breathing in a paper bag traps the exhaled CO_2; the gas in the bag contains more CO_2 than ambient air. When a person inhales gas from the bag, a greater amount (partial pressure) of $CO_2(g)$ in the lungs shifts the equilibria left, increasing $[H^+]$ and lowering blood pH.

17.88　pH = 10.38; pOH = 14.00 - 10.38 = 3.62; $[OH^-] = 10^{-3.62}$

$[OH^-] = 2.40 \times 10^{-4} = 2.4 \times 10^{-4}\ M$; $[Mg^{2+}] = 0.5[OH^-] = 1.20 \times 10^{-4} = 1.2 \times 10^{-4}\ M$

$K_{sp} = [Mg^{2+}][OH^-]^2 \approx (1.20 \times 10^{-4})(2.40 \times 10^{-4})^2 \approx 6.9 \times 10^{-12}$

17.89　$Ca(OH)_2(aq) + 2HCl(aq) \rightarrow CaCl_2(aq) + 2H_2O$

mmol HCl = $M \times$ mL = 0.0983 $M \times$ 11.23 mL = 1.1039 = 1.10 mmol HCl

mmol $Ca(OH)_2$ = mmol HCl/2 = 1.1039/2 = 0.55195 = 0.552 mmol $Ca(OH)_2$

$$[Ca^{2+}] = \frac{0.55195 \text{ mmol}}{50.00 \text{ mL}} = 0.01104 = 0.0110\ M$$

$[OH^-] = 2[Ca^{2+}] = 0.02208 = 0.0221\ M$

$K_{sp} = [Ca^{2+}][OH^-]^2 = (0.01104)(0.02208)2 = 5.38 \times 10^{-6}$

The value in Appendix D is 6.5×10^{-6}, a difference of 17%. Since a change in temperature does change the value of an equilibrium constant, the solution may not have been kept at 25°C. It is also possible that experimental errors led to the difference in K_{sp} values.

17.91　$[Ca^{2+}][CO_3^{2-}] = 4.5 \times 10^{-9}$; $[Fe^{2+}][CO_3^{2-}] = 2.1 \times 10^{-11}$

Since $[CO_3^{2-}]$ is the same for both equilibria:

$$[CO_3^{2-}] = \frac{4.5 \times 10^{-9}}{[Ca^{2+}]} = \frac{2.1 \times 10^{-11}}{[Fe^{2+}]}; \text{ rearranging } \frac{[Ca^{2+}]}{[Fe^{2+}]} = \frac{4.5 \times 10^{-9}}{2.1 \times 10^{-11}} = 214 = 2.1 \times 10^2$$

17.92　$PbSO_4(s) \rightleftharpoons Pb^{2+}(aq) + SO_4^{2-}(aq)$;　$K_{sp} = 6.3 \times 10^{-7} = [Pb^{2+}]\ [SO_4^{2-}]$

$SrSO_4(s) \rightleftharpoons Sr^{2+}(aq) + SO_4^{2-}(aq)$;　$K_{sp} = 3.2 \times 10^{-7} = [Sr^{2+}]\ [SO_4^{2-}]$

Let $x = [Pb^{2+}]$, $y = [Sr^{2+}]$, $x + y = [SO_4^{2-}]$

$$\frac{x(x+y)}{y(x+y)} = \frac{6.3 \times 10^{-7}}{3.2 \times 10^{-7}}; \frac{x}{y} = 1.9688 = 2.0; x = 1.969\ y = 2.0\ y$$

$y(1.969\ y + y) = 3.2 \times 10^{-7}$; $2.969\ y^2 = 3.2 \times 10^{-7}$; $y = 3.283 \times 10^{-4} = 3.3 \times 10^{-4}$

$x = 1.969\ y$; $x = 1.969(3.283 \times 10^{-4}) = 6.464 \times 10^{-4} = 6.5 \times 10^{-4}$

$[Pb^{2+}] = 6.5 \times 10^{-4}\ M$, $[Sr^{2+}] = 3.3 \times 10^{-4}\ M$, $[SO_4^{2-}] = (3.283 + 6.464) \times 10^{-4} = 9.7 \times 10^{-4}\ M$

17.94　The student failed to account for the hydrolysis of the AsO_4^{3-} ion. If there were no hydrolysis, $[Mg^{2+}]$ would indeed be 1.5 times that of $[AsO_4^{3-}]$. However, as the reaction $AsO_4^{3-}(aq) + H_2O(l) \rightleftharpoons HAsO_4^{2-}(aq) + OH^-(aq)$ proceeds, the ion product $[Mg^{2+}]^3[AsO_4^{3-}]^2$ falls below the value for K_{sp}. More $Mg_3(AsO_4)_2$ dissolves, more hydrolysis occurs, and so on, until an equilibrium is reached. At this point $[Mg^{2+}]$ in solution is much greater than 1.5 times free $[AsO_4^{3-}]$. However, it is exactly 1.5 times the **total** concentration of all arsenic-containing species. That is,

$[Mg^{2+}] = 1.5\ ([AsO_4^{3-}] + [HAsO_4^{2-}] + [H_2AsO_4^-] + [H_3AsO_4])$

17.95

$$Zn(OH)_2(s) \rightleftharpoons Zn^{2+}(aq) + 2OH^-(aq) \qquad K_{sp} = 3.0 \times 10^{-16}$$

$$Zn^{2+}(aq) + 4OH^-(aq) \rightleftharpoons Zn(OH)_4{}^{2-}(aq) \qquad K_f = 4.6 \times 10^{17}$$

$$Zn(OH)_2(s) + 2OH^-(aq) \rightleftharpoons Zn(OH)_4{}^{2-}(aq) \qquad K = K_{sp} \times K_f = 138 = 1.4 \times 10^2$$

$$K = 138 = 1.4 \times 10^2 = \frac{[Zn(OH)_4^{2-}]}{[OH^-]^2}$$

If 0.015 mol $Zn(OH)_2$ dissolves, 0.015 mol $Zn(OH)_4{}^{2-}$ should be present at equilibrium.

$$[OH^-]^2 = \frac{(0.015)}{138}; [OH^-] = 1.043 \times 10^{-2}\ M\ \ [OH^-] \geq 1.0 \times 10^{-2}\ M \text{ or pH} \geq 12.02$$

Integrative Exercises

17.97 (a) For a monoprotic acid (one H^+ per mole of acid), at the equivalence point moles OH^- added = moles H^+ originally present

$M_B \times V_B$ = g acid/molar mass

$$MM = \frac{\text{g acid}}{M_B \times V_B} = \frac{0.1044\,g}{0.0500\,M \times 0.02210\,L} = 94.48 = 94.5\text{ g/mol}$$

(b) 11.05 mL is exactly half-way to the equivalence point (22.10 mL). When half of the unknown acid is neutralized, [HA] = $[A^-]$, $[H^+] = K_a$ and pH = pK_a.

$K_a = 10^{-4.89} = 1.3 \times 10^{-5}$

(c) From Appendix D, Table D.1, acids with K_a values close to 1.3×10^{-5} are

name	K_a	formula	molar mass
propionic	1.3×10^{-5}	$HC_3H_5O_2$	74.1
butanoic	1.5×10^{-5}	$HC_4H_7O_2$	88.1
acetic	1.8×10^{-5}	$HC_2H_3O_2$	60.1
hydroazoic	1.9×10^{-5}	HN_3	43.0

Of these, butanoic has the closest match for K_a and molar mass, but the agreement is not good.

17.99 Calculate the initial M of aspirin in the stomach and solve the equilibrium problem to find equilibrium concentrations of $C_8H_7O_2COOH$ and $C_8H_7O_2COO^-$. At pH = 2, $[H^+] = 1 \times 10^{-2}$.

$$\frac{325\text{ mg}}{\text{tablet}} \times 2\text{ tablets} \times \frac{1\text{ g}}{1000\text{ mg}} \times \frac{1\text{ mol } C_8H_7O_2COOH}{180.2\text{ g } C_8H_7O_2COOH} \times \frac{1}{1\text{ L}} = 3.61 \times 10^{-3} = 4 \times 10^{-3}\ M$$

	$C_8H_7O_2COOH(aq)$	$\rightleftharpoons$ $C_8H_7O_2COO^-$	+ $H^+(aq)$
initial	$3.61 \times 10^{-3}\ M$	0	$1 \times 10^{-2}\ M$
equil	$(3.61 \times 10^{-3} - x)\ M$	$x\ M$	$(1 \times 10^{-2} + x)\ M$

$$K_a = 3 \times 10^{-5} = \frac{[H^+][C_8H_7O_2COO^-]}{[C_8H_7O_2COOH]} = \frac{(0.01+x)(x)}{(3.61 \times 10^{-3} - x)} \approx \frac{0.01x}{3.61 \times 10^{-3}}$$

$x = [C_8H_7O_2COO^-] = 1.08 \times 10^{-5} = 1 \times 10^{-5}\ M$

$$\%\ \text{ionization} = \frac{1.08 \times 10^{-5}\ M\ C_8H_7O_2COO^-}{3.61 \times 10^{-3}\ M\ C_8H_7O_2COOH} \times 100 = 0.3\%$$

(% ionization is small, so the assumption was valid.)

% aspirin molecules = 100.0% – 0.3% = 99.7% molecules

17.100 According to Equation [13.4], $S_g = kP_g$

$$S_{CO_2} = 3.1 \times 10^2 \frac{\text{mol}}{\text{L} \bullet \text{atm}} \times 1.10\ \text{atm} = 0.0341 = \frac{0.034\ \text{mol}}{\text{L}} = 0.034\ M\ CO_2$$

$CO_2(g) + H_2O(l) \rightarrow H_2CO_3(aq)$; $0.0341\ M\ CO_2 = 0.0341\ M\ H_2CO_3$

Consider the stepwise dissociation of $H_2CO_3(aq)$.

	$H_2CO_3(aq)$	$\rightleftharpoons$ $H^+(aq)$	+ $HCO_3^-(aq)$
initial	$0.0341\ M$	0	0
equil.	$(0.0341\text{-}x)\ M$	x	x

$$K_{a1} = \frac{[H^+][HCO_3^-]}{[H_2CO_3]} = \frac{x^2}{(0.0341 - x)} \approx \frac{x^2}{0.0341} \approx 4.3 \times 10^{-7}$$

$x^2 = 1.47 \times 10^{-8}$; $x = 1.2 \times 10^{-4}\ M\ H^+$; pH = 3.92

$K_{a2} = 5.6 \times 10^{-11}$; assume the second ionization does not contribute significantly to $[H^+]$.

17.102 For very dilute aqueous solutions, assume the solution density is 1 g/mL.

$$\text{ppb} = \frac{\text{g solute}}{10^9\ \text{g solution}} = \frac{1 \times 10^{-6}\ \text{g solute}}{1 \times 10^3\ \text{g solution}} = \frac{\mu\text{g solute}}{\text{L solution}}$$

(a) $K_{sp} = [Ag^+][Cl^-] = 1.8 \times 10^{-10}$; $[Ag^+] = (1.8 \times 10^{-10})^{1/2} = 1.34 \times 10^{-5} = 1.3 \times 10^{-5}\ M$

$$\frac{1.34 \times 10^{-5}\ \text{mol}\ Ag^+}{\text{L}} \times \frac{107.9\ \text{g}\ Ag^+}{1\ \text{mol}\ Ag^+} \times \frac{1\ \mu\text{g}}{1 \times 10^{-6}\ \text{g}} = \frac{1.4 \times 10^3\ \mu\text{g}\ Ag^+}{\text{L}}$$

$$= 1.4 \times 10^3\ \text{ppb} = 1.4\ \text{ppm}$$

(b) $K_{sp} = [Ag^+][Br^-] = 5.0 \times 10^{-13}$; $[Ag^+] = (5.0 \times 10^{-13})^{1/2} = 7.07 \times 10^{-7} = 7.1 \times 10^{-7}$ *M*

$$\frac{7.07 \times 10^{-7} \text{ mol Ag}^+}{\text{L}} \times \frac{107.9 \text{ g Ag}^+}{1 \text{ mol Ag}^+} \times \frac{1 \ \mu\text{g}}{1 \times 10^{-6} \text{ g}} = 76 \text{ ppb}$$

(c) $K_{sp} = [Ag^+][I^-] = 8.3 \times 10^{-17}$; $[Ag^+] = (8.3 \times 10^{-17})^{1/2} = 9.11 \times 10^{-9} = 9.1 \times 10^{-9}$ *M*

$$\frac{9.11 \times 10^{-9} \text{ mol Ag}^+}{\text{L}} \times \frac{107.9 \text{ g Ag}^+}{1 \text{ mol Ag}^+} \times \frac{1 \ \mu\text{g}}{1 \times 10^{-6} \text{ g}} = 0.98 \text{ ppb}$$

$AgBr(s)$ would maintain $[Ag^+]$ in the correct range.

17.103 To determine precipitation conditions, we must know K_{sp} for $CaF_2(s)$ and calculate Q under the specified conditions. $K_{sp} = 3.9 \times 10^{-11} = [Ca^{2+}][F^-]^2$

$[Ca^{2+}]$ and $[F^-]$: The term 1 ppb means 1 part per billion or 1 g solute per billion g solution. Assuming that the density of this very dilute solution is the density of water:

$$1 \text{ ppb} = \frac{1 \text{ g solute}}{1 \times 10^9 \text{ g solution}} \times \frac{1 \text{ g solution}}{1 \text{ mL solution}} \times \frac{1 \times 10^3 \text{ mL}}{1 \text{ L}} = \frac{1 \times 10^{-6} \text{ g solute}}{1 \text{ L solution}}$$

$$\frac{1 \times 10^{-6} \text{ g solute}}{1 \text{ L solution}} \times \frac{1 \ \mu\text{g}}{1 \times 10^{-6} \text{ g}} = 1 \ \mu\text{g}/1 \text{ L}$$

$$8 \text{ ppb Ca}^{2+} \times \frac{1 \ \mu\text{g}}{1 \text{ L}} = \frac{8 \ \mu\text{g Ca}^{2+}}{1 \text{ L}} = \frac{8 \times 10^{-6} \text{ g Ca}^{2+}}{1 \text{ L}} \times \frac{1 \text{ mol Ca}^{2+}}{40 \text{ g}} = 2 \times 10^{-7} \ M \text{ Ca}^{2+}$$

$$1 \text{ ppb F}^- \times \frac{1 \ \mu\text{g}}{1 \text{ L}} = \frac{1 \ \mu\text{g F}^-}{1 \text{ L}} = \frac{1 \times 10^{-6} \text{ g F}^-}{1 \text{ L}} \times \frac{1 \text{ mol F}^-}{19.0 \text{ g}} = 5 \times 10^{-8} \ M \text{ F}^-$$

$Q = [Ca^{2+}][F^-]^2 = (2 \times 10^{-7})(5 \times 10^{-8})^2 = 5 \times 10^{-22}$

$5 \times 10^{-22} < 3.9 \times 10^{-11}$, $Q < K_{sp}$, no CaF_2 will precipitate

18 Chemistry of the Environment

Visualizing Concepts

18.2 Molecules in the upper atmosphere tend to have multiple bonds because they have sufficiently high bond dissociation enthalpies (Table 8.4) to survive the incoming high energy radiation from the sun. According to Table 8.4, for the same two bonded atoms, multiple bonds have higher bond dissociation enthalpies than single bonds. Molecules with single bonds are likely to undergo photodissociation in the presence of the high energy, short wavelength solar radiation present in the upper atmosphere.

18.4 *Analyze.* Given granite, marble, bronze, and other solid materials, what observations and measurements indicate whether the material is appropriate for an outdoor sculpture? If the material changes (erodes) over time, what chemical processes are responsible?

Plan. An appropriate material resists chemical and physical changes when exposed to environmental conditions. An inappropriate material undergoes chemical reactions with substances in the troposphere, degrading the structural strength of the material and the sculpture. *Solve.*

(a) The appearance and mass of the material upon environmental exposure are both indicators of chemical and physical changes. If the appearance and mass of the material are unchanged after a period of time, the material is well-suited for the sculpture because it is inert to chemical and physical changes. Changes in the color or texture of the material's surface indicate that a chemical reaction has occurred, because a different substance with different properties has formed. A decrease in mass indicates that some of the material has been lost, either by chemical reaction or physical change. An increase in mass indicates corrosion. If the mass of the material is unchanged, it is probably inert to chemical and physical environmental changes and suitable for sculpture.

(b) The two main chemical processes that lead to erosion are reaction with acid rain and corrosion or air oxidation, which is encouraged by acid conditions (see Section 20.8).

Acid rain is primarily H_2SO_3 and/or H_2SO_4, which reacts directly with carbonate minerals such as marble and limestone. Acidic conditions created by acid rain encourage corrosion of metals such as iron, steel, and bronze. Corrosion produces metal oxides which may or may not cling to the surface of the material. If the oxides are washed away, the material will lose mass after corrosion. Physical erosion due to the effects of wind and rain on soft materials such as sandstone also causes mass to decrease.

18.6 *Analyze/Plan.* Explain how an ion-exchange column "softens" water. See the Closer Look box on "Water Softening" in Section 18.6.

Solve. The plastic beads in an ion-exchange column contain covalently bound anionic groups such as $R–COO^-$ and $R–SO_3^-$. These groups have Na^+ cations associated with them for charge balance. When "hard" water containing Ca^{2+} and other divalent cations passes over the beads, the 2+ cations are attracted to the anionic groups and Na^+ is displaced. The higher charge on the divalent cations leads to greater electrostatic attractions, which promote the cation exchange. The "soft" water that comes out of the column contains two Na^+ ions in place of each divalent cation, mostly Ca^{2+} and Mg^{2+}, that remains in the column associated with the plastic beads.

18.8 Some of the missing CO_2 is absorbed by "land plants" (vegetation other than trees) and incorporated into the soil. Soil is the largest land-based carbon reservoir. The amount of carbon-storing capacity of soil is affected by erosion, soil fertility, and other complex factors. For more details, search the internet for "carbon budget."

Earth's Atmosphere

18.10 (a) Boundaries between regions of the atmosphere are at maxima and minima (peaks and valleys) in the atmospheric temperature profile. For example, in the troposphere, temperature decreases with altitude, while in the stratosphere, it increases with altitude. The temperature minimum is the tropopause boundary.

(b) Atmospheric pressure in the troposphere ranges from 1.0 atm to 0.4 atm, while pressure in the stratosphere ranges from 0.4 atm to 0.001 atm. Gas density (g/L) is directly proportional to pressure. The much lower density of the stratosphere means it has the smaller mass, despite having a larger volume than the troposphere.

18.12 $P_{Ar} = \chi_{Ar} \bullet P_{atm}$; $P_{Ar} = 0.00934\,(98.6\text{ kPa}) = 0.921\text{ kPa}$; $0.921\text{ kPa} \times \dfrac{760\text{ torr}}{101.325\text{ kPa}} = 6.91\text{ torr}$

$P_{CO_2} = \chi_{CO_2} \bullet P_{atm}$; $P_{CO_2} = 0.000375\,(98.6\text{ kPa}) = 0.0370\text{ kPa}$; $0.0350\text{ kPa} \times \dfrac{760\text{ torr}}{101.325\text{ kPa}} = 0.277\text{ torr}$

18.14 (a) ppm Ne = mol Ne/1×10^6 mol air; $\chi_{Ne} = 1.818 \times 10^{-5}$ mol Ne/mol air

$$\frac{1.818 \times 10^{-5}\text{ mol Ne}}{1\text{ mol air}} = \frac{x\text{ mol Ne}}{1 \times 10^6\text{ mol air}};\ x = 18.18\text{ ppm Ne}$$

(b) $P_{Ne} = \chi_{Ne} \bullet P_{atm} = 1.818 \times 10^{-5} \times 743\text{ torr} \times \dfrac{1\text{ atm}}{760\text{ torr}} = 1.7773 \times 10^{-5}$

$= 1.78 \times 10^{-5}$ atm

$T = 300°C + 273 = 573\text{ K}$

$$\frac{n_{Ne}}{V} = \frac{P_{Ne}}{RT} = \frac{1.7773 \times 10^{-5}\text{ atm}}{573\text{ K}} \times \frac{\text{K} \bullet \text{mol}}{0.08206\text{ L} \bullet \text{atm}} = 3.7799 \times 10^{-7} = 3.78 \times 10^{-7}\text{ mol/L}$$

$$\frac{3.7799 \times 10^{-7} \text{ mol Ne}}{\text{L}} \times \frac{6.022 \times 10^{23} \text{ atoms}}{\text{mol}} = 2.2763 \times 10^{17}$$

$$= 2.28 \times 10^{17} \text{ Ne atoms/L}$$

The Upper Atmosphere; Ozone

18.16 $$\frac{339 \times 10^{3} \text{ J}}{1 \text{ mol}} \times \frac{1 \text{ mol}}{6.022 \times 10^{23} \text{ molecules}} = 5.6294 \times 10^{-19} = 5.63 \times 10^{-19} \text{ J/molecule}$$

$$\lambda = \frac{hc}{E} = \frac{(6.626 \times 10^{-34} \text{ J} \bullet \text{sec})(3.00 \times 10^{8} \text{ m/sec})}{5.6294 \times 10^{-19} \text{ J}} = 3.53 \times 10^{-7} \text{ m} = 353 \text{ nm}$$

$$\frac{293 \times 10^{3} \text{ J}}{1 \text{ mol}} \times \frac{1 \text{ mol}}{6.022 \times 10^{23} \text{ molecules}} = 4.8655 \times 10^{-19} = 4.87 \times 10^{-19} \text{ J/molecule}$$

$$\lambda = \frac{(6.626 \times 10^{-34} \text{ J} \bullet \text{sec})(3.00 \times 10^{8} \text{ m/sec})}{4.8655 \times 10^{-19} \text{ J}} = 4.09 \times 10^{-7} \text{ m} = 409 \text{ nm}$$

Photons of wavelengths longer than 409 nm cannot cause rupture of the C–Cl bond in either CF_3Cl or CCl_4. Photons with wavelengths between 409 and 353 nm can cause C–Cl bond rupture in CCl_4, but not in CF_3Cl.

18.18 Photodissociation of N_2 is relatively unimportant compared to photodissociation of O_2 for two reasons. The bond dissociation energy of N_2, 941 kJ/mol, is much higher than that of O_2, 495 kJ/mol. Photons with a wavelength short enough to photodissociate N_2 are not as abundant as the ultraviolet photons that lead to photodissociation of O_2. Also, N_2 does not absorb these photons as readily as O_2 so even if a short-wavelength photon is available, it may not be absorbed by an N_2 molecule.

18.20 32 e^-, 16 e^- pr

```
       ..
      :F:
       |
  ..        ..
 :Cl — C — Cl:
  ..        ..
       |
      :Cl:
       ..
```

CFC–11, $CFCl_3$, contains C–Cl bonds that can be cleaved by UV light in the stratosphere to produce Cl atoms. It is chlorine in atomic form that catalyzes the destruction of stratospheric ozone. CFC–11 is chemically inert and resists decomposition in the troposphere, so that it eventually reaches the stratosphere in molecular form.

18.22 Yes. Assuming $CFBr_3$ reaches the stratosphere intact, it contains C–Br bonds that are even more susceptible to cleavage by UV light than C–Cl bonds. According to Table 8.4, the average C–Br bond dissociation energy is 273 kJ/mol, compared to 328 kJ/mol for C–Cl bonds. Once in atomic form, Br atoms catalyze the destruction of ozone by a mechanism similar to that of Cl atoms.

Chemistry of the Troposphere

18.24 Rainwater is naturally acidic due to the presence of $CO_2(g)$ in the atmosphere. All oxides of nonmetals produce acidic solutions when dissolved in water. Even in the absence of polluting gases such as SO_2, SO_3, NO, and NO_2, CO_2 causes rainwater to be acidic. The important equilibria are:

$$CO_2(g) + H_2O(l) \rightleftharpoons H_2CO_3(aq) \rightleftharpoons H^+(aq) + HCO_3^-(aq).$$

18.26 (a) $Fe(s) + 2H^+(aq) \rightarrow Fe^{2-}(aq) + H_2(g)$

(b) No. Silver is a "noble" metal. It is relatively resistant to oxidation, and much more resistant than iron. In Table 4.5, The Activity Series of Metals in Aqueous Solution, Ag is much, much lower than Fe and it is below hydrogen, while Fe is above hydrogen. This means that Fe is susceptible to oxidation by acid, while Ag is not.

18.28 (a) Visible (Figure 6.4)

(b) $$E_{photon} = hc/\lambda = \frac{6.626 \times 10^{-34}\ J\bullet s \times 3.00 \times 10^8\ m/s}{420 \times 10^{-9}\ m} = 4.733 \times 10^{-19}$$

$$= 4.73 \times 10^{-19}\ J/photon$$

$$\frac{4.733 \times 10^{-19}\ J}{1\ photon} \times \frac{6.022 \times 10^{23}\ photons}{1\ mol} \times \frac{1\ kJ}{1000\ J} = 285\ kJ/mol$$

(c) $\ddot{O}{=}N{-}\ddot{O}: + h\nu \longrightarrow \ddot{O}{=}\dot{N} + :\dot{O}\cdot$

18.30 (a) A *greenhouse gas* absorbs energy in the 10,000–30,000 nm or infrared region. It absorbs wavelengths of radiation emitted by earth and returns it as heat. A non-greenhouse gas is transparent to radiation in this wavelength range.

(b) Ar(g) is monatomic, while $CH_4(g)$ contains 4 C–H bonds. Infrared radiation has insufficient energy to cause electron transitions or bond cleavage; but it has an appropriate amount of energy to cause molecular deformations, bond stretching, and angle bending. Monatomic gases such as Ar cannot "use" infrared radiation and are transparent to it.

The World Ocean

18.32 If the phosphorous is present as phosphate, there is a 1:1 ratio between the molarity of phosphorus and molarity of phosphate. Thus, we can calculate the molarity based on the given mass of P.

$$\frac{0.07\ g\ P}{1 \times 10^6\ g\ H_2O} \times \frac{1\ mol\ P}{31\ g\ P} \times \frac{1\ mol\ PO_4^{3-}}{1\ mol\ P} \times \frac{1 \times 10^3\ g\ H_2O}{1\ L\ H_2O} = 2.26 \times 10^{-6} = 2 \times 10^{-6}\ M\ PO_4^{3-}$$

18.34 0.05 ppb Au = 0.05 g Au/1 × 10^9 g seawater

$$\$1{,}000{,}000 \times \frac{1\ oz\ Au}{\$400} \times \frac{1\ lb}{16\ oz} \times \frac{453.6\ g}{1\ lb} = 7.0875 \times 10^4\ g = 7.09 \times 10^4\ g\ Au\ needed$$

$$7.0875 \times 10^4 \text{ g Au} \times \frac{1 \times 10^9 \text{ g seawater}}{0.05 \text{ g Au}} \times \frac{1 \text{ mL seawater}}{1.03 \text{ g seawater}} \times \frac{1 \text{ L}}{1000 \text{ mL}} = 1.3762 \times 10^{12}$$

$$= 1 \times 10^{12} \text{ L seawater}$$

1×10^{12} L seawater are needed if the process is 100% efficient; since it is only 50% efficient, twice as much seawater is needed.

$1.3762 \times 10^{12} \times 2 = 2.7524 \times 10^{12} = 3 \times 10^{12}$ L seawater

Note that the 1 sig fig in 0.05 ppb Au limits the precision of the calculation.

18.36 Calculate the total ion concentration of sea water by summing the molarities given in Table 18.6. Then use $\pi = \Delta MRT$ to calculate pressure.

$$M_{total} = 0.55 + 0.47 + 0.028 + 0.054 + 0.010 + 0.010 + 2.3 \times 10^{-3} + 8.3 \times 10^{-4} + 4.3 \times 10^{-4} + 9.1 \times 10^{-5} + 7.0 \times 10^{-5} = 1.1257 = 1.13\ M$$

$$\pi = \frac{(1.1257 - 0.02) \text{ mol}}{\text{L}} \times \frac{0.08206 \text{ L} \times \text{atm}}{\text{mol} \bullet \text{K}} \times 305 \text{ K} = 27.674 = 27.7 \text{ atm}$$

Check. The largest numbers in the molarity sum have 2 decimal places, so M_{total} has 2 decimal places and 3 sig figs. ΔM also has 2 decimal places and 3 sig figs so the calculated pressure has 3 sig figs. Units are correct.

Fresh Water

18.38 (a) Decomposition of organic matter by aerobic bacteria depletes dissolved O_2. A low dissolved oxygen concentration indicates the presence of organic pollutants.

(b) According to Section 13.3, the solubility of $O_2(g)$ (or any gas) in water decreases with increasing temperature.

18.40 $$120{,}000 \text{ persons} \times \frac{59 \text{ g } O_2}{1 \text{ person}} \times \frac{1 \times 10^6 \text{ g } H_2O}{9 \text{ g } O_2} \times \frac{1 \text{ L } H_2O}{1 \times 10^3 \text{ g } H_2O} = 7.9 \times 10^8 = 8 \times 10^8 \text{ L } H_2O$$

18.42 (a) Ca^{2+}, Mg^{2+}, Fe^{2+}

(b) Divalent cations (ions with 2+ charges) contribute to water hardness. These ions react with soap to form scum on surfaces or leave undesirable deposits on surfaces, particularly inside pipes, upon heating.

18.44 $Ca(OH)_2$ is added to remove Ca^{2+} as $CaCO_3(s)$, and Na_2CO_3 removes the remaining Ca^{2+}.

$$Ca^{+2}(aq) + 2HCO_3^-(aq) + [Ca^{2+}(aq) + 2OH^-(aq)] \rightarrow 2CaCO_3(s) + 2H_2O(l).$$

One mole $Ca(OH)_2$ is needed for each 2 moles of $HCO_3^-(aq)$ present.

$$5.0 \times 10^7 \text{ L } H_2O \times \frac{1.7 \times 10^{-3} \text{ mol } HCO_3^-}{1 \text{ L } H_2O} \times \frac{1 \text{ mol } Ca(OH)_2}{2 \text{ mol } HCO_3^-} \times \frac{74 \text{ g } Ca(OH)_2}{1 \text{ mol } Ca(OH)_2}$$

$$= 3.1 \times 10^6 \text{ g } Ca(OH)_2$$

Half of the native HCO_3^- precipitates the added Ca^{2+} so this operation reduces the Ca^{2+} concentration from 5.7×10^{-3} *M* to $(5.7 \times 10^{-3} - 8.5 \times 10^{-4})$ $M = 4.85 \times 10^{-3} = 4.9 \times 10^{-3}$ *M*. Next we must add sufficient Na_2CO_3 to further reduce $[Ca^{2+}]$ to 1.1×10^{-3} *M* (20% of the original $[Ca^{2+}]$). We thus need to reduce $[Ca^{2+}]$ by $(4.85 \times 10^{-3} - 1.1 \times 10^{-3})$ $M = 3.75 \times 10^{-3} = 3.8 \times 10^{-3}$ *M*

$Ca^{2+}(aq) + CO_3^{-2}(aq) \rightarrow CaCO_3(s)$.

$$5.0 \times 10^7 \text{ L } H_2O \times \frac{3.75 \times 10^{-3} \text{ mol } Ca^{2+}}{1 \text{L} H_2O} \times \frac{1 \text{ mol } Na_2CO_3}{1 \text{ mol } Ca^{2+}} \times \frac{106 \text{ g } Na_2CO_3}{1 \text{ mol } Na_2CO_3}$$

$$= 2.0 \times 10^7 \text{ g } Na_2CO_3$$

18.46 $Al_2(SO_4)_3$ is a typical coagulant in municipal water purification. It reacts with OH^- in a slightly basic solution to form a gelatinous precipitate that occludes very small particles and bacteria. The precipitate settles slowly and is removed by sand filtration.

Properties of $Al_2(SO_4)_3$ and other useful coagulants are:

- They react with low concentrations of $OH^-(aq)$. That is, K_{sp} of the hydroxide precipitate is very small. The capacity to form a hydroxide precipitates means that no extra salts must be added to form the precipitate. Also, the $[OH^-]$ can be easily adjusted by $Ca(OH)_2$ and other reagents that are part of the purification process.

- The hydroxide precipitate is composed of very small, evenly dispersed particles that do not settle quickly. This is required to remove very small bacteria and viruses from all parts of the liquid, not just the sites of solid formation.

Green Chemistry

18.48 Catalysts increase the rate of a reaction by lowering activation energy, E_a. For an uncatalyzed reaction that requires extreme temperatures and pressures to generate product at a viable rate, finding a suitable catalyst reduces the required temperature and/or pressure, which reduces the amount of energy used to run the process. A catalyst can also increase rate of production, which would reduce the net time and thus energy required to generate a certain amount of product.

18.50
- In either solvent, the reaction is catalyzed, which usually leads to decreased processing temperatures and times, and greater energy efficiency.
- $scCO_2$ is the preferred solvent. It achieves maximum conversion much faster than CH_2Cl_2 solvent. $scCO_2$ reduces processing time, temperature, and energy requirements. It also means fewer unwanted by-products to be separated and processed. While use of $scCO_2$ increases the amount of a greenhouse gas released to the environment, it eliminates use of CH_2Cl_2, which is implicated in stratospheric ozone depletion. Use of $scCO_2$ rather than CH_2Cl_2 is a good green trade-off.

Additional Exercises

18.52 MM_{avg} at the surface = 83.8(0.17) + 16.0(0.38) + 32.0(0.45) = 34.73 = 35 g/mol.

Next, calculate the percentage composition at 200 km. The fractions can be "normalized" by saying that the 0.45 fraction of O_2 is converted into **two** 0.45 fractions of O atoms, then dividing by the total fractions, 0.17 + 0.38 + 0.45 + 0.45 = 1.45:

$$MM_{avg} = \frac{83.8(0.17) + 16.0(0.38) + 16.0(0.90)}{1.45} = 23.95 = 24 \text{ g/mol}$$

18.53 Stratospheric ozone is formed and destroyed in a cycle of chemical reactions. The decomposition of O_3 to O_2 and O produces oxygen atoms, an essential ingredient for the production of ozone. While single O_3 molecules exist for only a few seconds, new O_3 molecules are constantly reformed. This cyclic process ensures a finite concentration of O_3 in the stratosphere available to absorb ultraviolet radiation. (This explanation assumes that the cycle is not disrupted by outside agents such as CFCs.)

18.54

$$\begin{array}{lr} 2[Cl(g) + O_3(g) \rightarrow ClO(g) + O_2(g)] & [18.7] \\ 2Cl(g) + 2O_3(g) \rightarrow 2ClO(g) + 2O_2(g) & \\ 2ClO(g) \rightarrow O_2(g) + 2Cl(g) & [18.9] \\ \hline 2Cl(g) + 2O_3(g) + 2ClO(g) \rightarrow 2ClO(g) + 3O_2(g) + 2Cl(g) & \\ 2O_3(g) \xrightarrow{Cl} 3O_2(g) & [18.10] \end{array}$$

Note that Cl(g) fits the definition of a catalyst in this reaction.

18.56 (a) $\cdot\ddot{\underset{\cdot\cdot}{O}}—H$

(b) HNO_3 is a major component in acid rain.

(c) While it removes CO, the reaction produces NO_2. The photodissociation of NO_2 to form O atoms is the first step in the formation of tropospheric ozone and photochemical smog.

(d) Again, NO_2 is the initiator of photochemical smog. Also, methoxyl radical, OCH_3, is a reactive species capable of initiating other undesirable reactions.

18.58 Oxygen is present in the atmosphere to the extent of 209,000 parts per million. If CO binds 210 times more effectively than O_2, then the **effective** concentration of CO is 210 × 125 ppm = 26,250 = 26,300 ppm. The fraction of carboxyhemoglobin in the blood leaving the lungs is thus $\frac{26{,}250}{26{,}250 + 209{,}000} = 0.112$. Thus, 11.2 percent of the blood is in the form of carboxyhemoglobin, 88.8 percent as the O_2-bound oxyhemoglobin.

18.60 (a) According to Section 13.3, the solubility of gases in water decreases with increasing temperature. Thus, the solubility of $CO_2(g)$ in the ocean would decrease if the temperature of the ocean increased.

(b) If the solubility of $CO_2(g)$ in the ocean decreased because of global warming, more $CO_2(g)$ would be released into the atmosphere, perpetuating a cycle of increasing temperature and concomitant release of $CO_2(g)$ from the ocean.

18.62 Given 169 watts/m^2 at 10% efficiency, find the land area needed to produce 55 watts/m^2.

169 watts/m^2 (0.10) = 16.9 watts/m^2 solar energy possible with current technology.

$$\frac{55 \text{ watts}}{m^2} \times \frac{1 \text{ m}^2}{16.9 \text{ watts}} = 3.254 = 3.3$$

Solar energy must be harvested from 3.3 times the land area of New York City to supply its energy needs. According to *Wikipedia*, http://en.wikipedia.org/wiki/New_York_City, the land area of New York City is 831 km^2, which is 831×10^6 m^2 or 2.05×10^5 acres. The area needed for solar energy harvesting would then be 2.70×10^3 km^2, 2.70×10^9 m^2 or 6.68×10^5 acres.

18.64 (a) CO_3^{2-} is a relatively strong Brønsted base and produces OH^- in aqueous solution according to the hydrolysis reaction:

$$CO_3^{2-}(aq) + H_2O(l) \rightleftharpoons HCO_3^-(aq) + OH^-(aq), \quad K_b = 1.8 \times 10^{-4}$$

If $[OH^-(aq)]$ is sufficient to exceed K_{sp} for $Mg(OH)_2$, the solid will precipitate.

(b)
$$\frac{125 \text{ mg } Mg^{2+}}{1 \text{ kg soln}} \times \frac{1 \text{ g } Mg^{2+}}{1000 \text{ mg } Mg^{2+}} \times \frac{1.00 \text{ kg soln}}{1.00 \text{ L soln}} \times \frac{1 \text{ mol } Mg^{2+}}{24.305 \text{ g } Mg^{2+}} = 5.143 \times 10^{-3}$$

$$= 5.14 \times 10^{-3} \, M \, Mg^{2+}$$

$$\frac{4.0 \text{ g } Na_2CO_3}{1.0 \text{ L soln}} \times \frac{1 \text{ mol } CO_3^{2-}}{106.0 \text{ g } Na_2CO_3} = 0.03774 = 0.038 \, M \, CO_3^{2-}$$

$$K_b = 1.8 \times 10^{-4} = \frac{[HCO_3^-][OH^-]}{[CO_3^{2-}]} \approx \frac{x^2}{0.03774}; x = [OH^-] = 2.606 \times 10^{-3}$$

$$= 2.6 \times 10^{-3} \, M$$

(This represents 6.9% hydrolysis, but the result will not be significantly different using the quadratic formula.)

$Q = [Mg^{2+}][OH^-]^2 = (5.143 \times 10^{-3})(2.606 \times 10^{-3})^2 = 3.5 \times 10^{-8}$

K_{sp} for $Mg(OH)_2 = 1.6 \times 10^{-12}$; $Q > K_{sp}$, so $Mg(OH)_2$ will precipitate.

18.66 *Plan.* Calculate the volume of air above Los Angeles and the volume of pure O_3 that would be present at the 85 ppb level. For gases at the same temperature and pressure, volume fractions equal mole fractions. *Solve.*

$$V_{air} = 4000 \text{ mi}^2 \times \frac{(1.6093)^2 \text{ km}^2}{\text{mi}^2} \times \frac{(1000)^2 \text{ m}^2}{1 \text{ km}^2} \times 10 \text{ m} \times \frac{1 \text{ L}}{1 \times 10^{-3} \text{ m}^3} = 1.036 \times 10^{14}$$

$$= 1.0 \times 10^{14} \text{ L air}$$

$$85 \text{ ppb } O_3 = \frac{85 \text{ mol } O_3}{1 \times 10^9 \text{ mol air}} = 8.5 \times 10^{-8} = \chi_{O_3}$$

$V \text{ (pure } O_3) = 8.5 \times 10^{-8} (1.036 \times 10^{14} \text{ L air}) = 8.805 \times 10^6 = 8.8 \times 10^6 \text{ L } O_3$

Values for P and T are required to calculate mol O_3 from volume O_3, using the ideal-gas law. Since these are not specified in the exercise, we will make a reasonable assumption for a sunny April day in Los Angeles. The city is near sea level and temperatures are moderate throughout the year, so P = 1 atm and T = 25°C (78°F) are reasonable values.

PV = nRT, n = PV/RT

$$n = 1.000 \text{ atm} \times \frac{8.805 \times 10^6 \text{ L}}{298 \text{ K}} \times \frac{\text{K} \bullet \text{mol}}{0.08206 \text{ L} \bullet \text{atm}} = 3.601 \times 10^5 = 3.6 \times 10^5 \text{ mol } O_3$$

Check. Using known conditions to make reasonable estimates and assumptions is a valuable skill for problem solving. Knowing when assumptions are required is an important step in the learning process.

Integrative Exercises

18.68 (a) $$8{,}376{,}726 \text{ tons coal} \times \frac{83 \text{ ton C}}{100 \text{ ton coal}} \times \frac{44.01 \text{ ton } CO_2}{12.01 \text{ ton C}} = 2.5 \times 10^7 \text{ ton } CO_2$$

$$8{,}376{,}726 \text{ tons coal} \times \frac{2.5 \text{ ton S}}{100 \text{ ton coal}} \times \frac{64.07 \text{ ton } SO_2}{32.07 \text{ ton S}} = 4.2 \times 10^5 \text{ ton } SO_2$$

(b) $CaO(s) + SO_2(g) \rightarrow CaSO_3(s)$

$$4.18 \times 10^5 \text{ ton } SO_2 \times \frac{55 \text{ ton } SO_2 \text{ removed}}{100 \text{ ton } SO_2 \text{ produced}} \times \frac{120.15 \text{ ton } CaSO_3}{64.07 \text{ ton } SO_2}$$

$$= 4.3 \times 10^5 \text{ ton } CaSO_3$$

18.69 *Coarse sand* is removed by coarse sand filtration. *Finely divided particles* and some *bacteria* are removed by precipitation with aluminum hydroxide. Remaining *harmful bacteria* are removed by ozonation. *Trihalomethanes* are removed by either aeration or activated carbon filtration; use of activated carbon might be preferred because it does not involve release of TCMs into the atmosphere. *Dissolved organic substances* are oxidized (and rendered less harmful, but not removed) by both aeration and ozonation. Dissolved *nitrates* and *phosphates* are not removed by any of these processes, but are rendered less harmful by adequate aeration.

18.70 (a) $H—\ddot{\underset{..}{O}}—H \longrightarrow H\cdot + \cdot\ddot{\underset{..}{O}}—H$

(b) $\Delta H = 2D(O—H) - D(O—H) = D(O—H) = 463 \text{ kJ/mol}$

$$\frac{463 \text{ kJ}}{\text{mol } H_2O} \times \frac{1 \text{ mol } H_2O}{6.022 \times 10^{23} \text{ molecules}} \times \frac{1000 \text{ J}}{\text{kJ}} = 7.688 \times 10^{-19}$$

$$= 7.69 \times 10^{-19} \text{ J/}H_2O \text{ molecule}$$

$$\lambda = \frac{hc}{\Delta E} = \frac{6.626 \times 10^{-34} \text{ J} \bullet \text{sec} \times 2.998 \times 10^8 \text{ m/s}}{7.688 \times 10^{-19} \text{ J}} = 2.58 \times 10^{-7} \text{ m} = 258 \text{ nm}$$

This wavelength is in the UV region of the spectrum, close to the visible.

(c)

$$OH(g) + O_3(g) \rightarrow HO_2(g) + O_2(g)$$
$$HO_2(g) + O(g) \rightarrow OH(g) + O_2(g)$$
$$\overline{OH(g) + O_3(g) + HO_2(g) + O(g) \rightarrow HO_2(g) + 2O_2(g) + OH(g)}$$
$$O_3(g) + O(g) \rightarrow 2O_2(g)$$

$OH(g)$ is the catalyst is this overall reaction, another pathway for the destruction of ozone.

18.72 (i) $ClO(g) + O_3(g) \rightarrow ClO_2(g) + O_2(g)$

$$\Delta H_i = \Delta H_f^\circ\ ClO_2(g) + \Delta H_f^\circ\ O_2(g) - \Delta H_f^\circ\ ClO(g) - \Delta H_f^\circ\ O_3(g)$$

$$\Delta H_i = 102 + 0 - 101 - (142.3) = -141\ kJ$$

(ii) $ClO_2(g) + O(g) \rightarrow ClO(g) + O_2(g)$

$$\Delta H_{ii} = \Delta H_f^\circ\ ClO(g) + \Delta H_f^\circ\ O_2(g) - \Delta H_f^\circ\ ClO_2(g) + \Delta H_f^\circ\ O(g)$$

$$\Delta H_{ii} = 101 + 0 - 102 - (247.5) = -249\ kJ$$

(overall) $ClO(g) + O_3(g) + ClO_2(g) + O(g) \rightarrow ClO_2(g) + O_2(g) + ClO(g) + O_2(g)$

$$O_3(g) + O(g) \rightarrow 2O_2(g)$$

$$\Delta H = \Delta H_i + \Delta H_{ii} = -141\ kJ + (-249)\ kJ = -390\ kJ$$

Because the enthalpies of both (i) and (ii) are distinctly exothermic, it is possible that the ClO - ClO_2 pair could be a catalyst for the destruction of ozone.

18.74 (a) A rate constant of $M^{-1}s^{-1}$ is indicative of a reaction that is second order overall. For the reaction given, the rate law is probably rate = $k[O][O_3]$. (Although rate = $k[O]^2$ or $k[O_3]^2$ are possibilities, it is difficult to envision a mechanism consistent with either one that would result in two molecules of O_2 being produced.)

(b) Yes. Most atmospheric processes are initiated by collision. One could imagine an activated complex of four O atoms collapsing to form two O_2 molecules. Also, the rate constant is large, which is less likely for a multistep process. The reaction is analogous to the destruction of O_3 by Cl atoms (Equation 18.7), which is also second order with a large rate constant.

(c) According to the Arrhenius equation, $k = Ae^{-E_a/RT}$. Thus, the larger the value of k, the smaller the activation energy, E_a. The value of the rate constant for this reaction is large, so the activation energy is small.

(d) $\Delta H_f^\circ = 2\Delta H_f^\circ\ O_2(g) - \Delta H_f^\circ\ O(g) - \Delta H_f^\circ\ O_3(g)$

$$\Delta H_f^\circ = 0 - 247.5\ kJ - 142.3\ kJ = -389.8\ kJ$$

The reaction is exothermic, so energy is released; the reaction would raise the temperature of the stratosphere.

18.76 Calculate $[H_2SO_4]$ required to produce a solution with pH = 3.5. From the volume of rainfall, calculate the amount of H_2SO_4 present.

$[H^+] = 10^{-3.5} = 3.16 \times 10^{-4} = 3 \times 10^{-4}\ M$

Assume initially that both ionization steps are complete.

$$\underset{}{H_2SO_4(aq)} \rightleftharpoons \underset{x\ M}{H^+(aq)} + \underset{x\ M}{HSO_4^-(aq)}$$

$$HSO_4^-(aq) \rightleftharpoons \underset{+2x\ M}{H^+(aq)} + \underset{x\ M}{SO_4^{2-}(aq)}$$

Since $[HSO_4^-]$ at equilibrium is small but finite, let $[HSO_4^-] = y$.

	$HSO_4^-(aq)$	$\rightleftharpoons$	$H^+(aq)$	+	$SO_4^{2-}(aq)$	$K_a = 0.012$
equil	$y\ M$		$(2x - y)\ M$		$(x - y)\ M$	

$$K_a = 0.012 = \frac{[H^+][SO_4^{2-}]}{[HSO_4^-]} = \frac{(2x - y)(x - y)}{y}$$

But we know that $[H^+]$ at equilibrium $= 3.16 \times 10^{-4}\ M$.

$2x - y = 3.16 \times 10^{-4}$; $y = 2x - 3.16 \times 10^{-4}$; $(x - y) = [x - (2x - 3.16 \times 10^{-4})] = 3.16 \times 10^{-4} - x$

$$K_a = 0.012 = \frac{(3.16 \times 10^{-4})(3.16 \times 10^{-4} - x)}{2x - 3.16 \times 10^{-4}}$$

$(0.012)(2x - 3.16 \times 10^{-4}) = 1.00 \times 10^{-7} - 3.16 \times 10^{-4}x$;

$0.024x - 3.795 \times 10^{-6} = 1.00 \times 10^{-7} - 3.16 \times 10^{-4}x$;

$0.024316x = 3.895 \times 10^{-6}$; $x = 1.60 \times 10^{-4} = 2 \times 10^{-4}\ M\ H_2SO_4$

Check. This result is reasonable, since it is just slightly greater than $[H^+]/2$. The amount of HSO_4^- at equilibrium, $y = 4.1 \times 10^{-6}\ M$.

$$\frac{[H^+][SO_4^{2-}]}{[HSO_4^-]} = \frac{(3.16 \times 10^{-4})(1.60 \times 10^{-4} - 4.1 \times 10^{-6})}{4.1 \times 10^{-6}} = 0.012$$

The calculated results are reasonable and self-consistent.

Now proceed to find the volume of rainfall and corresponding mass of H_2SO_4 if $[H_2SO_4] = 1.60 \times 10^{-4} = 2 \times 10^{-4}\ M$.

$$V = 1.0\ \text{in} \times 1500\ \text{mi}^2 \times \frac{5280^2\ \text{ft}^2}{\text{mi}^2} \times \frac{12^2\ \text{in}^2}{\text{ft}^2} \times \frac{2.54^3\ \text{cm}^3}{\text{in}^3} \times \frac{1\ \text{L}}{1000\ \text{cm}^3} = 9.868 \times 10^{10}$$

$$= 9.9 \times 10^{10}$$

$$\frac{1.60 \times 10^{-4}\ \text{mol}\ H_2SO_4}{1\ \text{L rainfall}} \times 9.868 \times 10^{10}\ \text{L} \times \frac{98.1\ \text{g}\ H_2SO_4}{1\ \text{mol}\ H_2SO_4} \times \frac{1\ \text{kg}}{1000\ \text{g}} = 1.55 \times 10^6$$

$$= 2 \times 10^6\ \text{kg}\ H_2SO_4$$

18.78 (a) $Al(OH)_3(s) \rightleftharpoons Al^{3+}(aq) + 3OH^-(aq) \quad K_{sp} = 1.3 \times 10^{-33} = [Al^{3+}][OH^-]^3$

This is a precipitation conditions problem. At what $[OH^-]$ (we can get pH from $[OH^-]$) will $Q = 1.3 \times 10^{-33}$, the requirement for the onset of precipitation?

$Q = 1.3 \times 10^{-33} = [Al^{3+}][OH^-]^3$. Find the molar concentration of $Al_2(SO_4)_3$ and thus $[Al^{3+}]$.

$$\frac{5.0 \text{ lb } Al_2(SO_4)_3}{2000 \text{ gal } H_2O} \times \frac{453.6 \text{ g}}{1 \text{ lb}} \times \frac{1 \text{ mol } Al_2(SO_4)_3}{342.2 \text{ g } Al_2(SO_4)_3} \times \frac{1 \text{ gal}}{4 \text{ qt}} \times \frac{1 \text{ qt}}{0.946 \text{ L}}$$

$$= 8.758 \times 10^{-4} \, M \, Al_2(SO_4)_3 = 1.752 \times 10^{-3} = 1.8 \times 10^{-3} \, M \, Al^{3+}$$

$Q = 1.3 \times 10^{-33} = (1.752 \times 10^{-3})[OH^-]^3; \; [OH^-]^3 = 7.42 \times 10^{-31}$

$[OH^-] = 9.054 \times 10^{-11} = 9.1 \times 10^{-11} \, M$; pOH = 10.04; pH = 14 – 10.04 = 3.96

(b) $CaO(s) + H_2O(l) \rightarrow Ca^{2+}(aq) + 2OH^-(aq)$; $[OH^-] = 9.054 \times 10^{-11}$ mol/L

$$\text{mol } OH^- = \frac{9.054 \times 10^{-11} \text{ mol}}{1 \text{ L}} \times 2000 \text{ gal} \times \frac{4 \text{ qt}}{1 \text{ gal}} \times \frac{0.946 \text{ L}}{1 \text{ qt}} = 6.852 \times 10^{-7}$$

$$= 6.9 \times 10^{-7} \text{ mol } OH^-$$

$$6.852 \times 10^{-7} \text{ mol } OH^- \times \frac{1 \text{ mol CaO}}{2 \text{ mol } OH^-} \times \frac{56.1 \text{ g CaO}}{1 \text{ mol CaO}} \times \frac{1 \text{ lb}}{453.6 \text{ g}} = 4.2 \times 10^{-8} \text{ lb CaO}$$

This is a very small amount of CaO, about 20 μg.

19 Chemical Thermodynamics

Visualizing Concepts

19.2 (a) The process depicted is a change of state from a solid to a gas. ΔS increases because of the greater motional freedom of the particles. ΔH increases because both melting and boiling are endothermic processes. Since $\Delta G = \Delta H - T\Delta S$, and both ΔH and ΔS are positive, the sign of ΔG depends on temperature. This is true for all phase changes. If the temperature of the system is greater than the boiling point of the substance, the process is spontaneous and ΔG is negative. If the temperature is lower than the boiling point, the process is not spontaneous and ΔG is positive.

(b) If the process is spontaneous, the second law states that $\Delta S_{univ} \geq 0$. Since ΔS_{sys} increases, ΔS_{surr} must decrease. If the change occurs via a reversible pathway, $\Delta S_{univ} = 0$ and $\Delta S_{surr} = -\Delta S_{sys}$. If the pathway is irreversible, the magnitude of ΔS_{sys} is greater than the magnitude of ΔS_{surr}, but the sign of ΔS_{surr} is still negative.

19.3 In the depicted reaction, both reactants and products are in the gas phase (they are far apart and randomly placed). There are twice as many molecules (or moles) of gas in the products, so ΔS is positive for this reaction.

19.5 (a) *Analyze.* The boxes depict three different mixtures of reactants and products for the reaction $A_2 + B_2 \rightleftharpoons 2AB$.

Plan. Box 1 is an equilibrium mixture. By definition, $\Delta G = 0$ for box 1. Calculate K and $\Delta G°$ for the reaction from box 1. Boxes 2 and 3 are nonequilibrium mixtures. Calculate Q and ΔG for boxes 2 and 3.

Solve. $K = \frac{[AB]^2}{[A][B]}$. Use number of molecules as a measure of concentration.

Box 1: $K = \frac{(3)^2}{(3)(3)} = 1.$

$\Delta G = \Delta G° + RT \ln K; 0 = \Delta G° - RT \ln(1), 0 = \Delta G° - 0; \Delta G° = 0$

Box 2: $Q = \frac{(1)^2}{(4)(4)} = \frac{1}{16} = 0.0625 = 0.06$

$\Delta G = \Delta G° + RT \ln Q = 0 - RT \ln(0.0625) = 2.77\ RT = 3RT$

Box 3: $Q = \frac{(7)^2}{(1)(1)} = \frac{46}{1} = 49$

$\Delta G = \Delta G° + RT \ln Q = 0 - RT \ln(49) = -3.89\ RT = -4RT$

(b) The magnitudes of ΔG (ignoring sign) are: box 1, 0; box 2, 2.8 RT; box 3, 3.9 RT. The order of increasing *magnitude* of ΔG is: box 1 < box 2 < box 3.

The signs on ΔG indicate in which direction the reaction is spontaneous. The mixture in box 2 will react spontaneously in the forward direction, toward products. The mixture in box 3 will react spontaneously in the reverse direction, toward reactants. The driving force for the reverse reaction in box 3, the *magnitude* of ΔG, is greater than the driving force for the forward reaction in box 2.

19.6 (a) The minimum in the plot is the equilibrium position of the reaction, where $\Delta G = 0$.

(b) X is the difference in free energy between reactant and products in their standard states, $\Delta G°$.

Spontaneous Processes

19.8 (a) Spontaneous; a gas, in this case perfume vapor, expands to fill its container, the room.

(b) Nonspontaneous; a mixture cannot be separated without outside intervention.

(c) Nonspontaneous; an inflated balloon doesn't burst without external stress, such as a pin prick, a squeeze, or adding more gas.

(d) Spontaneous; see Figure 8.2.

(e) Spontaneous; the very polar HCl molecules readily dissolve in water to form concentrated HCl(aq).

19.10 Berthelot's suggestion is incorrect. Some examples of nonexothermic spontaneous processes are expansion of certain pressurized gases, dissolving of one liquid in another, and dissolving of many salts in water.

19.12 (a) Exothermic. If melting requires heat and is endothermic, freezing must be exothermic.

(b) At 1 atm (indicated by the term "normal" freezing point), the freezing of 1-propanol is spontaneous at temperatures below -127°C.

(c) At 1 atm, the freezing of 1-propanol is nonspontaneous at temperatures above -127°C.

(d) At 1 atm and -127°C, the normal freezing point of 1-propanol, the solid and liquid phases are in equilibrium. That is, at the freezing point, 1-propanol molecules escape to the liquid phase at the same rate as liquid 1-propanol solidifies, assuming no heat is exchanged between 1-propanol and the surroundings.

19.14 (a) A process is *irreversible* if the system cannot be returned to its original state by the same path that the forward process took place.

(b) Since the system returned to its initial state via a different path (different q_r and w_r than q_f and w_f), there is a net change in the surroundings.

(c) The condensation of a liquid will be irreversible if it occurs at any temperature other than the boiling point of the liquid, at a specified pressure.

19.16 (a) $\Delta E\,(1 \rightarrow 2) = -\Delta E\,(2 \rightarrow 1)$

(b) We can say nothing about the values of q and w because we have no information about the paths.

(c) If the changes of state are reversible, the two paths are the same and $w\,(1 \rightarrow 2) = -w\,(2 \rightarrow 1)$. This is the maximum realizable work from this system.

19.18 (a) The detonation of an explosive is definitely spontaneous, once it is initiated.

(b) The quantity q is related to ΔH. Since the detonation is highly exothermic, q is large and negative.

If only PV-work is done and P is constant, $\Delta H = q$. Although these conditions probably do not apply to a detonation, we can still predict the sign of q, based on ΔH, if not its exact magnitude.

(c) The sign (and magnitude) of w depend on the path of the process, the exact details of how the detonation is carried out. It seems clear, however, that work will be done by the system on the surroundings in almost all circumstances (buildings collapse, earth and air are moved), so the sign of w is probably negative.

(d) $\Delta E = q + w$. If q and w are both negative, then the sign of ΔE is negative, regardless of the magnitudes of q and w.

Entropy and the Second Law of Thermodynamics

19.20 (a) When a liquid freezes, the entropy of the system decreases.

(b) ΔS is negative.

(c) Entropy being a state function means that ΔS is independent of the path of the process. ΔS defined in terms of a reversible path must equal ΔS for any path.

19.22 (a) $Cs(l) \rightarrow Cs(s)$, ΔS is negative

(b) $$\Delta H = 15.0\text{ g Cs} \times \frac{1\text{ mol Cs}}{132.9\text{ g Cs}} \times \frac{2.09\text{ kJ}}{\text{mol Cs}} = 0.2359 = 0.236\text{ kJ}$$

$$\Delta S = \frac{\Delta H}{T} = 0.2359\text{ kJ} \times \frac{1000\text{ J}}{1\text{ kJ}} \times \frac{1}{(273.15 + 28.4)\text{K}} = 0.782\text{ J/K}$$

19.24 (a) For a spontaneous process, $\Delta S_{universe} > 0$. For a reversible process, $\Delta S_{universe} = 0$.

(b) $\Delta S_{surroundings}$ is positive and greater than the magnitude of the decrease in ΔS_{system}.

(c) $\Delta S_{system} = 78$ J/K.

19.26 According to Boyle's law, $P_1V_1 = P_2V_2$ at constant n and T.

$0.900 \text{ atm} \times V_1 = 3.00 \text{ atm} \times V_2; V_2/V_1 = 0.900 \text{ atm}/3.00 \text{ atm} = 0.300$

$\Delta S_{sys} = nR \ln (V_2/V_1) = 0.500 \text{ mol } (8.314 \text{ J/mol}\cdot\text{K})(\ln 0.300) = -5.0049 = -5.00 \text{ J/K}$

Check. An increase in pressure results in a decrease in volume at constant T, so we expect ΔS to be negative, and it is.

The Molecular Interpretation of Entropy

19.28 (a) ΔH_{vap} for H_2O at 25°C = 44.02 kJ/mol; at 100°C = 40.67 kJ/mol

$$\Delta S = \frac{q_{rev}}{T} = \frac{44.02 \text{ kJ}}{\text{mol}} \times \frac{1000 \text{ J}}{\text{kJ}} \times \frac{1}{298 \text{ K}} = 148 \text{ J/mol} \bullet \text{K}$$

$$\Delta S = \frac{q_{rev}}{T} = \frac{40.67 \text{ kJ}}{\text{mol}} \times \frac{1000 \text{ J}}{\text{kJ}} \times \frac{1}{373 \text{ K}} = 109 \text{ J/mol} \bullet \text{K}$$

(b) At both temperatures, the liquid → gas phase transition is accompanied by an increase in entropy, as expected. That the magnitude of the increase is greater at the lower temperature requires some explanation.

In the liquid state, there are significant hydrogen bonding interactions between H_2O molecules. This reduces the number of possible molecular positions and the number of microstates. Liquid water at 100° has sufficient kinetic energy to have broken many hydrogen bonds, so the number of microstates for $H_2O(l)$ at 100° is greater than the number of microstates for $H_2O(l)$ at 25°C. The difference in the number of microstates upon vaporization at 100°C is smaller, and the magnitude of ΔS is smaller.

19.30 (a) Solids are much more ordered than gases, so ΔS is negative.

(b) ΔS is positive for Exercise 19.7 (a), (b), and (e). [At room temperature and 1 atm pressure, H_2O is a liquid, so there are more moles of gas in the products in part (e) and $\Delta S > 0$.]

19.32 (a) When temperature increases, the range of accessible molecular speeds and kinetic energies increases. This produces more microstates and an increase in entropy.

(b) When the volume of a gas increases (even at constant T), there are more possible positions for the particles, more microstates, and greater entropy.

(c) When a solid dissolves in water, there are both more possible positions for the particles (ions or molecules) and more motional freedom. The number of microstates and entropy increases.

19.34 (a) Since CO_2 has more than one atom, the thermal energy can be distributed as translational, vibrational, or rotational motion.

z

O=C=O → O=C=O translational O=C=O

O=C=O rotational

O=C=O ↔ O=C=O ↔ O=C=O ↔ O=C=O

symmetric stretch — asymmetric stretch — bend

vibrational

(b) According to Boltzmann's Law, $S = k \ln W$. The number of microstates, W, is directly proportional to entropy, S. Thus, if the number of microstates for a system increases, the entropy of the system increases.

19.36 Melting = –126.5°C; boiling = 97.4°C.

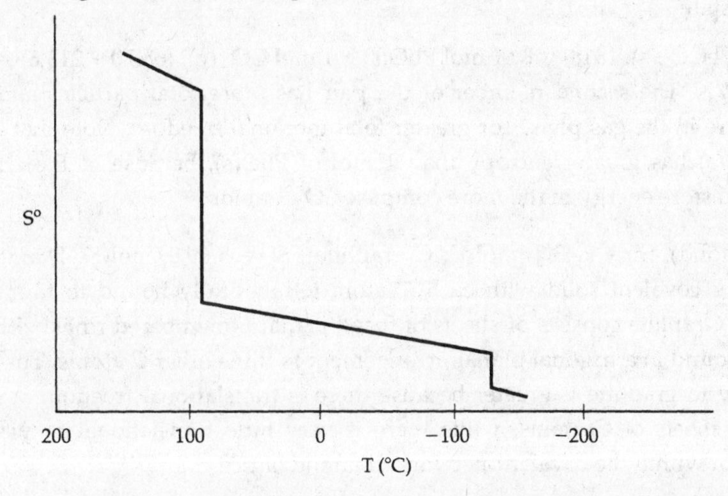

19.38 (a) 1 mol of $As_4(g)$ at 300°C, 0.01 atm (As_4 has more massive atoms in a comparable system at the same temperature.)

(b) 1 mol $H_2O(g)$ at 100°C, 1 atm (larger volume occupied by $H_2O(g)$)

(c) 0.5 mol $CH_4(g)$ at 298 K, 20-L volume (more complex molecule, more rotational and vibrational degrees of freedom)

(d) 100 g of $Na_2SO_4(aq)$ at 30°C (more motional freedom in aqueous solution)

19.40 (a) $Fe(l) \rightarrow Fe(s)$; ΔS is negative (less motional freedom)

(b) $2Li(s) + Cl_2(g) \rightarrow 2LiCl$; ΔS is negative (moles of gas decrease)

(c) $Zn(s) + 2HCl(aq) \rightarrow ZnCl_2(aq) + H_2(g)$; ΔS is positive (moles of gas increase)

(d) $AgNO_3(aq) + KBr(aq) \rightarrow AgBr(s) + KNO_3(aq)$; ΔS is negative (less motional freedom)

Entropy Changes in Chemical Reactions

19.42 Propylene will have a higher S° at 25°C. At this temperature, both are gases, so there are no lattice effects. Since they have the same molecular formula, only the details of their structures are different. In propylene, there is free rotation around the C—C single bond, while in cyclopropane the 3-membered ring severely limits rotation. The greater motional freedom of the propylene molecule leads to a higher absolute entropy.

19.44 (a) CuO(s), 42.59 J/mol•K; $Cu_2O(s)$, 92.36 J/mol•K. Molecules in the solid state have only vibrational motion available to them. The more complex Cu_2O molecule has more vibrational degrees of freedom and a larger standard entropy.

(b) 1 mol $N_2O_4(g)$, 304.3 J/K; 2 mol $NO_2(g)$, 2(240.45) = 480.90 J/K. More particles have a greater number of arrangements.

(c) $CH_3OH(g)$, 237.6 J/mol•K; $CH_3OH(l)$, 126.8 J/mol•K. Molecules in the gas phase occupy a larger volume and have more motional freedom than molecules in the liquid state.

(d) 1 mol $PbCO_3(s)$, 131.0 J/K; 1 mol PbO(s) + 1 mol $CO_2(g)$, (68.70 + 213.6) = 282.3 J/K. The second member of the pair has more total particles and half of them are in the gas phase for greater total motional freedom. Note that 1 mol of $PbCO_3(s)$ has greater entropy than 1 mol of PbO(s), because of the additional ways to store energy in the more complex CO_3^{2-} anion.

19.46 (a) C(diamond), S° = 2.43 J/mol•K; C(graphite), S° = 5.69 J/mol•K. Diamond is a network covalent solid with each C atom tetrahedrally bound to four other C atoms. Graphite consists of sheets of fused planar 6-membered rings with each C atom bound in a trigonal planar arrangement to three other C atoms. The internal entropy in graphite is greater because there is translational freedom among the planar sheets of C atoms, while there is very little translational or vibrational freedom within the covalent-network diamond lattice.

(b) S° for buckminsterfullerene will be ≥ 10 J/mol•K. S° for graphite is twice S° for diamond, and S° for the fullerene should be higher than that of graphite. The 60-atom "bucky" balls have more flexibility than graphite sheets. Also, the balls have translational freedom in three dimensions, while graphite sheets have it in only two directions. Because of the ball structure, there is more empty space in the fullerene lattice than in graphite or diamond; essentially, 60 C-atoms in fullerene occupy a larger volume than 60 C-atoms in graphite or diamond. Thus, the fullerene has additional "molecular" complexity, more degrees of translational freedom, and occupies a larger volume, all features that point to a higher absolute entropy.

19.48 (a) $\Delta S° = 2S°\ NH_3(g) - S°\ N_2H_4(g) - S°\ H_2(g)$

$= 2(192.5) - 238.5 - 130.58 = +15.9$ J/K

ΔS° is small because there are the same number of moles of gas in the products as in reactants. The slight increase is due to the relatively small S° value of $H_2(g)$, which has fewer degrees of freedom than molecules with more than two atoms.

(b) $\Delta S° = 2S°\ AlCl_3(s) - 2S°\ Al(s) - 3S°\ Cl_2(g)$

$= 2(109.3) - 2(28.32) - 3(222.96) = -506.9\ J/K$

$\Delta S°$ is negative because the products contain fewer (no) moles of gas.

(c) $\Delta S° = S°\ MgCl_2(s) + 2S° H_2O(l) - S°\ Mg(OH)_2(s) - 2S°\ HCl(g)$

$= 89.6 + 2(69.91) - 63.24 - 2(186.69) = -207.2\ J/K$

$\Delta S°$ is negative because the products contain fewer (no) moles of gas.

(d) $\Delta S° = S°\ C_2H_6(g) + S°\ H_2(g) - 2S°\ CH_4(g)$

$= 229.5 + 130.58 - 2(186.3) = -12.5\ J/K$

$\Delta S°$ is very small because there are the same number of moles of gas in the products and reactants. The slight decrease is related to the relatively small S° value for $H_2(g)$, which has fewer degrees of freedom than molecules with more than two atoms.

Gibbs Free Energy

19.50 (a) The *standard* free energy change, $\Delta G°$, represents the free energy change for the process when all reactants and products are in their standard states. When any or all reactants or products are not in their standard states, the free energy is represented simply as ΔG. The value for ΔG thus depends on the specific states of all reactants and products.

(b) When $\Delta G = 0$, the system is at equilibrium.

(c) The sign and magnitude of ΔG give no information about rate; we cannot predict whether the reaction will occur rapidly.

19.52 (a) $\Delta H°$ is negative; the reaction is exothermic.

(b) $\Delta S°$ is positive; the reaction leads to an increase in disorder.

(c) $\Delta G° = \Delta H° - T\Delta S° = -19.5\ kJ - 298\ K\ (0.0427\ kJ/K) = -32.225 = -32.2\ kJ$

(d) At 298 K, $\Delta G°$ is negative. If all reactants and products are present in their standard states, the reaction is spontaneous (in the forward direction) at this temperature.

19.54 (a) $\Delta H° = -305.3 - [0 + 0] = -305.3\ kJ$

$\Delta S° = 97.65 - [29.9 + 222.96] = -155.21 = -155.2\ J/K$

$\Delta G° = -259.0 - [0 + 0] = -259.0\ kJ$

$\Delta G° = -305.3\ kJ - 298(-0.15521)\ kJ = -259.047 = -259.0\ kJ$

(b) $\Delta H° = -635.5 + (-393.5) - (-1207.1) = 178.1\ kJ$

$\Delta S° = 39.75 + 213.6 - (92.88) = 160.47 = 160.5\ J/K$

$\Delta G° = -604.17 + (-394.4) - (-1128.76) = 130.19 = 130.2\ kJ$

$\Delta G° = 178.1\ kJ - 298(0.16047)\ kJ = 130.28 = 130.3\ kJ$

(c) $\Delta H° = 4(-1288.3) - [-2940.1 + 6(-285.83)] = -498.12 = -498.1$ kJ

$\Delta S° = 4(158.2) - [228.9 + 6(69.91)] = -15.56 = -15.6$ J/K

$\Delta G° = 4(-1142.6) - [-2675.2 + 6(-237.13)] = -472.42 = -472.4$ kJ

$\Delta G° = -498.12$ kJ $- 298(-0.01556)$ kJ $= -493.48 = -493.5$ kJ

(The discrepancy in $\Delta G°$ values is due to experimental uncertainties in the tabulated thermodynamic data.)

(d) $\Delta H° = 2(-393.5) + 4(-285.83) - [2(-238.6) + 3(0)] = -1453.1$ kJ

$\Delta S° = 2(213.6) + 4(69.91) - [2(126.8) + 3(205.0)] = -161.76 = -161.8$ J/K

$\Delta G° = 2(-394.4) + 4(-237.13) - [2(-166.23) + 3(0)] = -1404.86 = -1404.9$ kJ

$\Delta G° = -1453.2$ kJ $- 298(-0.16176)$ kJ $= -1404.996 = -1405.0$ kJ

19.56 (a) $\Delta G° = 2\Delta G°\ HCl(g) - [\Delta G°\ H_2(g) + \Delta G°\ Cl_2(g)]$

$= 2(-95.27$ kJ$) - 0 - 0 = -190.5$ kJ, spontaneous

(b) $\Delta G° = \Delta G°\ MgO(s) + 2\Delta G°\ HCl(g) - [\Delta G°\ MgCl_2(s) + \Delta G°\ H_2O(l)]$

$= -569.6 + 2(-95.27) - [-592.1 + (-237.13)] = +69.1$ kJ, nonspontaneous

(c) $\Delta G° = \Delta G°\ N_2H_4(g) + \Delta G°\ H_2(g) - 2\Delta G°\ NH_3(g)$

$= 159.4 + 0 - 2(-16.66) = +192.7$ kJ, nonspontaneous

(d) $\Delta G° = 2\Delta G°\ NO(g) + \Delta G°\ Cl_2(g) - 2\Delta G°\ NOCl(g)$

$= 2(86.71) + 0 - 2(66.3) = +40.8$ kJ, nonspontaneous

19.58 (a) $\Delta G°$ should be less negative than $\Delta H°$. Products contain fewer moles of gas, so $\Delta S°$ is negative. $\Delta G° = \Delta H° - T\Delta S°$; $-T\Delta S°$ is positive so $\Delta G°$ is less negative than $\Delta H°$.

(b) We can estimate $\Delta S°$ using a similar reaction and then use $\Delta G° = \Delta H° - T\Delta S°$ (estimate) to get a ballpark figure. There are no sulfite salts listed in Appendix C, so use a reaction such as $CO_2(g) + CaO(s) \rightarrow CaCO_3(s)$ or $CO_2(g) + BaO(s) \rightarrow BaCO_3(s)$. Or calculate both $\Delta S°$ values and use the average as your estimate.

19.60 $\Delta G° = \Delta H° - T\Delta S°$

(a) $\Delta G° = -844$ kJ $- 298$ K$(-0.165$ kJ/K$) = -795$ kJ, spontaneous

(b) $\Delta G° = +572$ kJ $- 298$ K$(0.179$ kJ/K$) = +519$ kJ, nonspontaneous

To be spontaneous, ΔG must be negative ($\Delta G < 0$).

Thus, $\Delta H° - T\Delta S° < 0$; $\Delta H° < T\Delta S°$; $T > \Delta H°/\Delta S°$; $T > \dfrac{572 \text{ kJ}}{0.179 \text{ kJ/K}} = 3.20 \times 10^3$ K

19.62 At $-25°C$ or 248 K, $\Delta G > 0$. $\Delta G = \Delta H - T\Delta S > 0$

$\Delta H - 248$ K $(95$ J/K$) > 0$; $\Delta H > +2.4 \times 10^4$ J; $\Delta H > +24$ kJ

19.64 ΔG is negative when $T\Delta S > \Delta H$ or $T > \Delta H/\Delta S$.

$\Delta H° = \Delta H°\ CH_4(g) + \Delta H°\ CO_2(g) - \Delta H°\ CH_3COOH(l)$

$= -74.8 + (-393.5) - (-487.0) = +18.7$ kJ

$\Delta S° = S°\ CH_4(g) + S°\ CO_2(g) - S°\ CH_3COOH(l) = +186.3 + 213.6 - 159.8 = +240.1\ J/K$

$$T > \frac{18.7\ kJ}{0.2401\ kJ/K} = 77.9\ K$$

The reaction is spontaneous above 77.9 K (–195°C).

19.66 (a) $\Delta H° = \Delta H_f° CH_3OH(g) - \Delta H_f° CH_4(g) - 1/2\ \Delta H_f° O_2(g)$

$= -201.2 - (-74.8) - 0 = -126.4\ kJ$

$\Delta S° = S° CH_3OH(g) - S° CH_4(g) - 1/2\ S° O_2(g)$

$= 237.6 - 186.3 - 1/2(205.0) = -51.2\ J/K = -0.0512\ kJ/K$

(b) $\Delta G° = \Delta H° - T\Delta S°$. $-T\Delta S°$ is positive, so $\Delta G°$ becomes more positive as temperature increases.

(c) $\Delta G° = \Delta H° - T\Delta S° = -126.4\ kJ - 298\ K(-0.0512\ kJ/K) = -111.1\ kJ$

The reaction is spontaneous at 298 K because $\Delta G°$ is negative at this temperature. In this case, $\Delta G°$ could have been calculated from $\Delta G_f°$ values in Appendix C, since these values are tabulated at 298 K.

(d) The reaction is at equilibrium when $\Delta G° = 0$.

$\Delta G° = \Delta H° - T\Delta S° = 0$. $\Delta H° = T\Delta S°$, $T = \Delta H°/\Delta S°$

$T = -126.4\ kJ/-0.0512\ kJ/K = 2469 = 2470\ K$.

This temperature is so high that the reactants and products are likely to decompose. At standard conditions, equilibrium is functionally unattainable for this reaction.

19.68 (a) As in Sample Exercise 19.9, $T_{sub} = \Delta H°_{sub}/\Delta S°_{sub}$

Use Data from Appendix C to calculate $\Delta H°_{sub}$ and $\Delta S°_{sub}$ for $I_2(s)$.

$I_2(s) \rightarrow I_2(l)$ melting
$I_2(l) \rightarrow I_2(g)$ boiling

$I_2(s) \rightarrow I_2(g)$ sublimation

$\Delta H°_{sub} = \Delta H_f° I_2(g) - \Delta H_f° I_2(s) = 62.25 - 0 = 62.25\ kJ$

$\Delta S°_{sub} = S° I_2(g) - S° I_2(s) = 260.57 - 116.73 = 143.84\ J/K = 0.14384\ kJ/K$

$$T_{sub} = \frac{\Delta H°_{sub}}{\Delta S°_{sub}} = \frac{62.25\ kJ}{0.14384\ kJ/K} = 432.8\ K = 159.6°C$$

(b) T_m for $I_2(s) = 386.85\ K = 113.7°C$; $T_b = 457.4\ K = 184.3°C$ (from WebElements™, 2005)

(c) The boiling point of I_2 is closer to the sublimation temperature. Both boiling and sublimation begin with molecules in a condensed phase (little space between molecules) and end in the gas phase (large intermolecular distances). Separation of the molecules is the main phenomenon that determines both ΔH and ΔS, so it is not surprising that the ratio of $\Delta H/\Delta S$ is similar for sublimation and boiling.

19.70 (a) $C_2H_4(g) + 3O_2(g) \rightarrow 2CO_2(g) + 2H_2O(l)$

$$\Delta H^\circ = 2\Delta H_f^\circ\ CO_2(g) + 2\Delta H_f^\circ\ H_2O(l) - \Delta H_f^\circ\ C_2H_4(g) - 3\Delta H_f^\circ\ O_2(g)$$

$$= 2(-393.5) + 2(-285.83) - 52.30 - 3(0) = -1410.96 = -1411.0\ \text{kJ/mol}\ C_2H_4\ \text{burned}$$

(b) $$w_{max} = \Delta G^\circ = 2\Delta G_f^\circ\ CO_2(g) + 2\Delta H_f^\circ\ H_2O(l) - \Delta G_f^\circ\ C_2H_4(g) - 3\Delta G_f^\circ\ O_2(g)$$

$$= 2(-394.4) + 2(-237.13) - 68.11 - 3(0) = -1331.2\ \text{kJ}$$

The system can accomplish at most 1331.2 kJ of work per mole of C_2H_4 on the surroundings.

Free Energy and Equilibrium

19.72 Consider the relationship $\Delta G = \Delta G^\circ + RT \ln Q$, where Q is the reaction quotient.

(a) $H_2(g)$ appears in the denominator of Q for this reaction. An increase in pressure of H_2 decreases Q and ΔG becomes smaller or more negative. Increasing the concentration of a reactant increases the tendency for a reaction to occur.

(b) $H_2(g)$ appears in the numerator of Q for this reaction. Increasing the pressure of H_2 increases Q and ΔG becomes more positive. Increasing the concentration of a product decreases the tendency for the reaction to occur.

(c) $H_2(g)$ appears in the denominator of Q for this reaction. An increase in pressure of H_2 decreases Q and ΔG becomes smaller or more negative.

19.74 (a) $\Delta G^\circ = 2\Delta G^\circ\ HF(g) - [\Delta G^\circ\ H_2(g) + \Delta G^\circ\ F_2(g)] = 2(-270.70) - [0 + 0] = -541.40\ \text{kJ}$

(b) $$\Delta G = \Delta G^\circ + RT \ln P_{HF}^2 / P_{H_2} \times P_{F_2}$$

$$= -541.40 + \frac{8.314 \times 10^{-3}\ \text{kJ}}{\text{K} \bullet \text{mol}} \times 298\ \text{K} \ln[(0.36)^2 / 8.0 \times 4.5] = -555.34 = -555\ \text{kJ}$$

19.76 $\Delta G^\circ = -RT \ln K$; $\ln K = -\Delta G^\circ / RT$; at 298 K, $RT = 2.4776 = 2.478$ kJ

(a) $\Delta G^\circ = \Delta G^\circ\ NaOH(s) + \Delta G^\circ\ CO_2(g) - \Delta G^\circ\ NaHCO_3(s)$

$$= -379.5 + (-394.4) - (-851.8) = +77.9\ \text{kJ}$$

$$\ln K = \frac{-\Delta G^\circ}{RT} = \frac{-77.9\ \text{kJ}}{2.478\ \text{kJ}} = -31.442 = -31.4;\quad K = 2 \times 10^{-14}$$

$$K = P_{CO_2} = 2 \times 10^{-14}$$

(b) $\Delta G^\circ = 2\Delta G^\circ\ HCl(g) + \Delta G^\circ\ Br_2(g) - 2\Delta G^\circ\ HBr(g) - \Delta G^\circ\ Cl_2(g)$

$$= 2(-95.27) + 3.14 - 2(-53.22) - 0 = -80.96\ \text{kJ}$$

$$\ln K = \frac{-(-80.96)}{2.4776} = +32.68;\ K = 1.6 \times 10^{14}$$

$$K = \frac{P_{HCl}^2 \times P_{Br_2}}{P_{HBr}^2 \times P_{Cl_2}} = 1.6 \times 10^{14}$$

(c) From Exercise 19.55(a), $\Delta G°$ at 298 K = -140.0 kJ.

$$\ln K = \frac{-\Delta G°}{RT} = \frac{-(-140.0)}{2.4776} = 56.51; \quad K = 3.5 \times 10^{24}$$

$$K = \frac{P^2_{SO_3}}{P^2_{SO_2} \times P_{O_2}} = 3.5 \times 10^{24}$$

19.78 $K = P_{CO_2}$. Calculate $\Delta G°$ at the two temperatures using $\Delta G° = \Delta H° - T\Delta S°$ and then calculate K and P_{CO_2}.

$\Delta H° = \Delta H°\ PbO(s) + \Delta H°\ CO_2(g) - \Delta H°\ PbCO_3(s)$

$= -217.3 - 393.5 + 699.1 = 88.3$ kJ

$\Delta S° = S°\ PbO(s) + S°\ CO_2(g) - S°\ PbCO_3(s)$

$= 68.70 + 213.6 - 131.0 = 151.3$ J/K or 0.1513 kJ/K

(a) $\Delta G° = \Delta H° - T\Delta S°$. At 393 K, $\Delta G° = 88.3$ kJ $- 393$ K(0.1513 kJ/K) = 28.84

= 28.8 kJ

$$\ln K = \frac{-\Delta G°}{RT} = \frac{-28.84 \times 10^3\ J}{8.314\ J/K \times 393\ K} = -8.82631 = -8.83$$

$K = P_{CO_2} = 1.5 \times 10^{-4}$ atm

(b) $\Delta G° = \Delta H° - T\Delta S°$. At 753 K, $\Delta G° = 88.3$ kJ $- 753$ K (0.1513 kJ) = -25.629

= -25.6 kJ

$$\ln K = \frac{-(-25.629 \times 10^3\ J)}{8.314\ J/K \times 753\ K} = 4.0938 = 4.09; \quad K = P_{CO_2} = 60\ atm$$

19.80 (a) $CH_3NH_2(aq) + H_20(l) \rightleftharpoons CH_3NH_3^+(aq) + OH^-(aq)$

(b) $\Delta G° = -RT \ln K_b = -(8.314 \times 10^{-3})(298) \ln (4.4 \times 10^{-4}) = 19.148 = 19.1$ kJ

(c) $\Delta G = 0$ at equilibrium

(d) $\Delta G = \Delta G° + RT \ln Q$; $[OH^-] = 1 \times 10^{-14}/1.5 \times 10^{-8} = 6.7 \times 10^{-7}$

$$= 19.148 + (8.314 \times 10^{-3})(298) \ln \frac{(5.5 \times 10^{-4})(6.67 \times 10^{-7})}{0.120} = -29.43 = -29\ kJ$$

Additional Exercises

19.81 (a) False. The essential question is whether the reaction proceeds far to the right before arriving at equilibrium. The position of equilibrium, which is the essential aspect, is not only dependent on ΔH but on the entropy change as well.

(b) True.

(c) True.

(d) False. Nonspontaneous processes in general require that work be done to force them to proceed. Spontaneous processes occur without application of work.

(e) False. Such a process **might** be spontaneous, but would not necessarily be so. Spontaneous processes are those that are exothermic and/or that lead to increased disorder in the system.

19.83 There is no inconsistency. The second law states that in any spontaneous process there is an increase in the entropy of the universe. While there may be a decrease in entropy of the system, as in the present case, this decrease is more than offset by an increase in entropy of the surroundings.

19.84 If $NH_4NO_3(s)$ dissolves spontaneously in water, $\Delta G = \Delta H - T\Delta S$. If ΔG is negative and ΔH is positive, the sign of ΔS must be positive. Furthermore, $T\Delta S > \Delta H$ at room temperature.

19.85 At the normal boiling point of a liquid, $\Delta G = 0$ and $\Delta H_{vap} = T\Delta S_{vap}$; $T = \Delta H_{vap}/\Delta S_{vap}$. By Trouton's rule, ΔS_{vap} = 88 J/mol•K. The process of vaporization is:

(a) $Br_2(l) \rightleftharpoons Br_2(g)$

$$\Delta H_{vap} = \Delta H^\circ_f\ Br_2(g) - \Delta H^\circ_f\ Br_2(l) = 30.71\ kJ - 0 = 30.71\ kJ$$

$$T_b = \frac{\Delta H_{vap}}{\Delta S_{vap}} = \frac{30.71\ kJ}{88\ J/mol \bullet K} \times \frac{1000\ J}{kJ} = 349 = 3.5 \times 10^2\ K$$

(b) According to WebElements™ 2005, the normal boiling pont of $Br_2(l)$ is 332 K. Trouton's rule provides a good"ballpark" estimate.

19.87 (a) (i) $2RbCl(s) + 3O_2(g) \rightarrow 2RbClO_3(s)$

$$\Delta H^\circ = 2\Delta H^\circ\ RbClO_3(s) - 3\Delta H^\circ\ O_2(g) - 2\Delta H^\circ\ RbCl(s)$$

$$= 2(-392.4) - 3(0) - 2(-430.5) = +76.2\ kJ$$

$$\Delta S^\circ = 2(152) - 3(205.0) - 2(92) = -495\ J/K = -0.495\ kJ/K$$

$$\Delta G^\circ = 2(-292.0) - 3(0) - 2(-412.0) = +240.0\ kJ$$

(ii) $C_2H_2(g) + 4Cl_2(g) \rightarrow 2CCl_4(l) + H_2(g)$

$$\Delta H^\circ = 2\Delta H^\circ\ CCl_4(l) + \Delta H^\circ\ H_2(g) - \Delta H^\circ\ C_2H_2(g) - 4\Delta H^\circ\ Cl_2(g)$$

$$= 2(-139.3) + 0 - (226.7) - 4(0) = -505.3\ kJ$$

$$\Delta S^\circ = 2(214.4) + 130.58 - (200.8) - 4(222.96) = -533.3\ J/K = -0.5333\ kJ/K$$

$$\Delta G^\circ = 2(-68.6) + 0 - (209.2) - 4(0) = -346.4\ kJ$$

(iii) $TiCl_4(l) + 2H_2O(l) \rightarrow TiO_2(s) + 4HCl(aq)$

$$\Delta H^\circ = \Delta H^\circ\ TiO_2(s) + 4\Delta H^\circ\ HCl(aq) - \Delta H^\circ\ TiCl_4(l) - 2\Delta H^\circ\ H_2O(l)$$

$$= -944.7 + 4(-167.2) - (-804.2) - 2(-285.83) = -237.6\ kJ$$

$$\Delta S^\circ = 50.29 + 4(56.5) - 221.9 - 2(69.91) = -85.43\ J/K = -0.0854\ kJ/K$$

$$\Delta G^\circ = -889.4 + 4(-131.2) - (-728.1) - 2(-237.13) = -211.8\ kJ$$

(b) (i) $\Delta G°$ is (+), nonspontaneous

(ii) $\Delta G°$ is (–), spontaneous

(iii) $\Delta G°$ is (-), spontaneous

(c) In each case the manner in which free energy change varies with temperature depends mainly on ΔS: $\Delta G = \Delta H - T\Delta S$. When ΔS is substantially positive, ΔG becomes more negative as temperature increases. When ΔS is substantially negative, ΔG becomes more positive as temperature increases.

(i) $\Delta S°$ is negative, $\Delta G°$ becomes more positive with increasing temperature.

(ii) $\Delta S°$ is negative, $\Delta G°$ becomes more positive with increasing temperature. (The reaction will become nonspontaneous at some temperature.)

(iii) $\Delta S°$ is negative, $\Delta G°$ becomes more positive with increasing temperature. (The reaction will become nonspontaneous at some temperature.)

19.88 $\Delta G = \Delta G° + RT \ln Q$

(a) $$Q = \frac{P^2_{NH_3}}{P_{N_2} \times P^3_{H_2}} = \frac{(1.2)^2}{(2.6)(5.9)^3} = 2.697 \times 10^{-3} = 2.7 \times 10^{-3}$$

$$\Delta G° = 2\Delta G°\ NH_3(g) - \Delta G°\ N_2(g) - 3\Delta G°\ H_2(g)$$

$$= 2(-16.66) - 0 - 3(0) = -33.32\ kJ$$

$$\Delta G = -33.32\ kJ + \frac{8.314 \times 10^{-3}\ kJ}{K \bullet mol} \times 298\ K \times \ln(2.69 \times 10^{-3})$$

$$\Delta G = -33.32 - 14.66 = -47.98 = -48.0\ kJ$$

(b) $$Q = \frac{P^3_{N_2} \times P^4_{H_2O}}{P^2_{N_2H_4} \times P^2_{NO_2}} = \frac{(0.5)^3(0.3)^4}{(5.0 \times 10^{-2})^2(5.0 \times 10^{-2})^2} = 162 = 2 \times 10^2$$

$$\Delta G° = 3\Delta G°\ N_2(g) + 4\Delta G°\ H_2O(g) - 2\Delta G°\ N_2H_4(g) - 2\Delta G°\ NO_2(g)$$

$$= 3(0) + 4(-228.57) - 2(159.4) - 2(51.84) = -1336.8\ kJ$$

$$\Delta G = -1336.8\ kJ + 2.478 \ln 162 = -1324.2 = -1.32 \times 10^3\ kJ$$

(c) $$Q = \frac{P_{N_2} \times P^2_{H_2}}{P_{N_2H_4}} = \frac{(1.5)(2.5)^2}{0.5} = 18.75 = 2 \times 10^1$$

$$\Delta G° = \Delta G°\ N_2(g) + 2\Delta G°\ H_2(g) - \Delta G°\ N_2H_4(g)$$

$$= 0 + 2(0) - 159.4 = -159.4\ kJ$$

$$\Delta G = -159.4\ kJ + 2.478 \ln 18.75 = -152.1 = -152\ kJ$$

19.89

Reaction	(a) Sign of ΔH°	(a) Sign of ΔS°	(b) $K > 1$?	(c) Variation in K as Temp. Increases
(i)	–	–	yes	decrease
(ii)	+	+	no	increase
(iii)	+	+	no	increase
(iv)	+	+	no	increase

(a) Note that at a particular temperature, positive ΔH° leads to a smaller value of K, while positive ΔS° increases the value of K.

19.91 (a) First calculate ΔG° for each reaction:

For $C_6H_{12}O_6(s) + 6O_2(g) \rightleftharpoons 6CO_2(g) + 6H_2O(l)$ (A)

$$\Delta G^\circ = 6(-237.13) + 6(-394.4) - (-910.4) + 6(0) = -2878.8 \text{ kJ}$$

For $C_6H_{12}O_6(s) \rightleftharpoons 2C_2H_5OH(l) + 2CO_2(g)$ (B)

$$\Delta G^\circ = 2(-394.4) + 2(-174.8) - (-910.4) = -228.0 \text{ kJ}$$

For (A), $\ln K = 2879 \times 10^3/(8.314)(298) = 1162$; $K = 5 \times 10^{504}$

For (B), $\ln K = 228 \times 10^3/(8.314)(298) = 92.026 = 92.0$; $K = 9 \times 10^{39}$

(b) Both these values for K are unimaginably large. However, K for reaction (A) is larger, because ΔG° is more negative. The magnitude of the work that can be accomplished by coupling a reaction to its surroundings is measured by ΔG. According to the calculations above, considerably more work can in principle be obtained from reaction (A), because ΔG° is more negative.

19.92 (a) $\Delta G^\circ = -RT \ln K$ (Equation [19.22]); $\ln K = -\Delta G^\circ/RT$

Use $\Delta G^\circ = \Delta H^\circ - T\Delta S^\circ$ to get ΔG° at the two temperatures. Calculate ΔH° and ΔS° using data in Appendix C.

$2CH_4(g) \rightarrow C_2H_6(g) + H_2(g)$

$$\Delta H^\circ = \Delta H^\circ\ C_2H_6(g) + \Delta H^\circ\ H_2(g) - 2\Delta H^\circ\ CH_4(g) = -84.68 + 0 - 2(-74.8) = 64.92 = 64.9 \text{ kJ}$$

$$\Delta S^\circ = S^\circ\ C_2H_6(g) + S^\circ\ H_2(g) - 2S^\circ\ CH_4(g) = 229.5 + 130.58 - 2(186.3) = -12.52 = -12.5 \text{ J/K}$$

at 298 K, $\Delta G = 64.92 \text{ kJ} - 298 \text{ K}(-12.52 \times 10^{-3} \text{ kJ/K}) = 68.65 = 68.7 \text{ kJ}$

$$\ln K = \frac{-68.65 \text{ kJ}}{(8.314 \times 10^{-3} \text{ kJ/K})(298 \text{ K})} = -27.709 = -27.7, K = 9.25 \times 10^{-13} = 9 \times 10^{-13}$$

at 773 K, $\Delta G = 64.9 \text{ kJ} - 773 \text{ K}(-12.52 \times 10^{-3} \text{ J/K}) = 74.598 = 74.6 \text{ kJ}$

$$\ln K = \frac{-74.598 \text{ kJ}}{(8.314 \times 10^{-3} \text{ kJ/K})(773 \text{ K})} = -11.607 = -11.6, \quad K = 9.1 \times 10^{-6}$$

Because the reaction is endothermic, the value of K increases with an increase in temperature.

$2CH_4(g) + 1/2\ O_2(g) \rightarrow C_2H_6(g) + H_2O(g)$

$\Delta H° = \Delta H°\ C_2H_6(g) + \Delta H°\ H_2O(g) - 2\Delta H°\ CH_4(g) - 1/2\ \Delta H°\ O_2(g)$

$= -84.68 + (-241.82) - 2(-74.8) - 1/2\ (0) = -176.9$ kJ

$\Delta S° = S°\ C_2H_6(g) + S°\ H_2O(g) - 2S°\ CH_4(g) - 1/2\ S°\ O_2(g)$

$= 229.5 + 188.83 - 2(186.3) - 1/2\ (205.0) = -56.77 = -56.8$ J/K

at 298 K, $\Delta G = -176.9$ kJ $- 298$ K$(-56.77 \times 10^{-3}$ kJ/K$) = -159.98 = -160.0$ kJ

$$\ln K = \frac{-(-159.98\ \text{kJ})}{(8.314 \times 10^{-3}\ \text{kJ/K})(298\ \text{K})} = 64.571 = 64.57;\ \ K = 1.1 \times 10^{28}$$

at 773 K, $\Delta G = -176.9$ kJ $- 773$ K $(-56.77 \times 10^{-3}$ kJ/K$) = -133.02 = -133.0$ kJ

$$\ln K = \frac{-(-133.02\ \text{kJ})}{(8.314 \times 10^{-3}\ \text{kJ/K})(773\ \text{K})} = 20.698 = 20.70;\ K = 9.750 \times 10^{8} = 9.8 \times 10^{8}$$

Because this reaction is exothermic, the value of K decreases with increasing temperature.

(b) The difference in $\Delta G°$ for the two reactions is primarily enthalpic; the first reaction is endothermic and the second exothermic. Both reactions have $-\Delta S°$, which inhibits spontaneity.

(c) This is an example of coupling a useful but nonspontaneous reaction with a spontaneous one to spontaneously produce a desired product.

$$2CH_4(g) \rightarrow C_2H_6(g) + H_2(g) \qquad \Delta G°_{298} = +68.7\ \text{kJ, nonspontaneous}$$
$$H_2(g) + 1/2\ O_2(g) \rightarrow H_2O(g) \qquad \Delta G°_{298} = -228.57\ \text{kJ, spontaneous}$$
$$2CH_4(g) + 1/2\ O_2(g) \rightarrow C_2H_6(g) + H_2O(g) \qquad \Delta G°_{298} = -159.9\ \text{kJ, spontaneous}$$

(d) $CH_4(g) + 2O_2(g) \rightarrow CO_2(g) + 2H_2O(g)$

19.93 $\Delta G°$ for the metabolism of glucose is:

$6\Delta G°\ CO_2(g) + 6\Delta G°\ H_2O(l) - \Delta G°\ C_6H_{12}O_6(s) - 6\Delta G°\ O_2(g)$

$\Delta G° = 6(-394.4) + 6(-237.13) - (-910.4) + 6(0) = -2878.8$ kJ

moles ATP $= -2878.8$ kJ $\times$ 1 mol ATP / $(-30.5$ kJ$) = 94.4$ mol ATP / mol glucose

Note that this calculation is done at standard conditions, not metabolic conditions. A more accurate answer would be obtained using ΔG values that reflect actual concentration, partial pressure, and pH in a cell.

19.95 (a) To obtain $\Delta H°$ from the equilibrium constant data, graph lnK at various temperatures vs 1/T, being sure to employ absolute temperature. The slope of the linear relationship that should result is $-\Delta H°/R$; thus, $\Delta H°$ is easily calculated.

(b) Use $\Delta G° = \Delta H° - T\Delta S°$ and $\Delta G° = -RT \ln K$. Substituting the second expression into the first, we obtain

$$-RT \ln K = \Delta H° - T\Delta S°;\quad \ln K = \frac{-\Delta H°}{RT} - \frac{-\Delta S°}{R} = \frac{-\Delta H°}{RT} + \frac{\Delta S°}{R}$$

Thus, the constant in the equation given in the exercise is $\Delta S°/R$.

19.96 $S = k \ln W$ (Equation [19.5]), $k = R/N$, $W \propto V^m$

$\Delta S = S_2 - S_1; S_1 = k \ln W_1, S_2 = k \ln W_2$

$\Delta S = k \ln W_2 - k \ln W_1; W_2 = cV_2^m; W_1 = cV_1^m$

(The number of particles, m, is the same in both states.)

$\Delta S = k \ln cV_2^m - k \ln cV_1^m; \ln a^b = b \ln a$

$\Delta S = k\, m \ln cV_2 - k\, m \ln cV_1; \ln a - \ln b = \ln (a/b)$

$$\Delta S = k\, m \ln\left(\frac{cV_2}{cV_1}\right) = k\, m \ln\left(\frac{V_2}{V_1}\right) = \frac{R}{N} m \ln\left(\frac{V_2}{V_1}\right)$$

$$\frac{m}{N} = \frac{\text{particles}}{6.022 \times 10^{23}} = n(\text{mol}); \quad \Delta S = nR \ln\left(\frac{V_2}{V_1}\right)$$

19.97 Absolute entropy is a fundamental property of matter at a specified set of conditions, that is, a state. In order to lower the entropy of the fuel, either the structure of the molecules or the conditions (temperature, pressure, amount) must be changed. Any of these changes would require energy, which would reduce the amount of energy available to drive the car, not increase it.

Integrative Exercises

19.99 (a) Polymerization is the process of joining many small molecules (monomers) into a few very large molecules (polymers). Polyethylene in particular can have extremely high molecular weights. In general, reducing the number of particles in a system reduces entropy, so ΔS_{poly} is expected to be negative.

(b) $\Delta G_{poly} = \Delta H_{poly} - T\Delta S_{poly}$. If the polymerization of ethylene is spontaneous, ΔG_{poly} is negative. If ΔS_{poly} is negative, $-T\Delta S_{poly}$ is positive, so ΔH_{poly} must be negative for ΔG_{poly} to be negative. The enthalpy of polymerization must be exothermic.

(c) According to Equation [12.1], polymerization of ethylene requires breaking one C=C and forming 2C—C per monomer (1C—C between the C-atoms of the monomer and 2 × 1/2 C—C to two other monomers).

$$\Delta H = D(C{=}C) - 2D(C{-}C) = 614 - 2(348) = -82 \text{ kJ/mol } C_2H_4$$

$$\frac{-82 \text{ kJ}}{\text{mol } C_2H_4} \times \frac{1 \text{ mol}}{6.022 \times 10^{23} \text{ molecules}} \times \frac{1000 \text{ J}}{1 \text{ kJ}} = 1.36 \times 10^{-19} \text{ J}/C_2H_4 \text{ monomer}$$

(d) The products of a condensation polymerization are the polymer and a small molecule, typically H_2O; there is usually one small molecule formed per monomer unit. Unlike addition polymerization, the total number of particles is not reduced. A condensation polymer does impose more order on the monomer or monomers than an addition polymer. If there is a single monomer, it has different functional groups at the two ends and only one end can react to join the polymer, so orientation is required. If there are two different monomers, as in nylon, the monomers alternate in the polymer, so only the correct monomer can react to join the polymer. In terms of structure, the condensation polymer imposes more order on the monomer(s) than an addition polymer. But, condensation polymerization does not lead to a reduction in the number of particles in the system, so ΔS_{poly} will be less negative than for addition polymerization.

19.100 The activated complex in Figure 14.13 is a single "particle" or entity that contains four atoms. It is formed from an atom A and a triatomic molecule, ABC, that must collide with exactly the correct energy and orientation to form the single entity. There are many fewer degrees of freedom for the activated complex than the separate reactant particules, so the *entropy of activation* is negative.

19.102 (a) 16 e^-, 8 e^- pairs. The C-S bond order is approximately 2.

$\ddot{S}=C=\ddot{S}$

(b) 2 e^- domains around C, linear e^- domain geometry, linear molecular structure

(c) $CS_2(l) + 3O_2(g) \rightarrow CO_2(g) + 2SO_2(g)$

(d) $\Delta H° = \Delta H°\ CO_2(g) + 2\Delta H°\ SO_2(g) - \Delta H°\ CS_2(l) - 3\ \Delta H\ \Delta H°\ O_2(g)$

$= -393.5 + 2(-296.9) - (89.7) - 3(0) = -1077.0\ kJ$

$\Delta G° = \Delta G°\ CO_2(g) + 2\Delta G°\ SO_2(g) - \Delta G°\ CS_2(l) - 3\ \Delta G°\ O_2(g)$

$= -394.4 + 2(-300.4) - (65.3) - 3(0) = -1060.5\ kJ$

The reaction is exothermic ($-\Delta H°$) and spontaneous ($-\Delta G°$) at 298 K.

(e) vaporization: $CS_2(l) \rightarrow CS_2(g)$

$\Delta G°_{vap} = \Delta H°_{vap} - T\Delta S°_{vap};\ \Delta S°_{vap} = (\Delta H°_{vap} - \Delta G°_{vap})/T$

$\Delta G°_{vap} = \Delta G°\ CS_2(g) - \Delta G°\ CS_2(l) = 67.2 - 65.3 = 1.9\ kJ$

$\Delta H°_{vap} = \Delta H°\ CS_2(g) - \Delta H°\ CS_2(l) = 117.4 - 89.7 = 27.7\ kJ$

$\Delta S°_{vap} = (27.7 - 1.9)\ kJ/298\ K = 0.086577 = 0.0866\ kJ/K = 86.6\ J/K$

ΔS_{vap} is always positive, because the gas phase occupies a greater volume, has more motional freedom and a larger absolute entropy than the liquid.

(f) At the boiling point, $\Delta G = 0$ and $\Delta H_{vap} = T_b \Delta S_{vap}$.

$T_b = \Delta H_{vap}/\Delta S_{vap} = 27.7\ kJ/0.086577\ kJ/K = 319.9 = 320\ K$

$T_b = 320\ K = 47°C$. CS_2 is a liquid at 298 K, 1 atm

19.103 (a) $Ag(s) + 1/2\ N_2(g) + 3/2\ O_2(g) \rightarrow AgNO_3(s)$; S decreases because there are fewer moles of gas in the product.

(b) $\Delta G°_f = \Delta H°_f - T\Delta S°_f;\ \Delta S°_f = (\Delta G°_f - \Delta H°_f)/(-T) = (\Delta H°_f - \Delta G°_f)/T$

$\Delta S°_f = -124.4\ kJ - (-33.4\ kJ)/298\ K = -0.305\ kJ/K = -305\ J/K$

$\Delta S°_f$ is relatively large and negative, as anticipated from part (a).

(c) Dissolving of $AgNO_3$ can be expressed as

$AgNO_3(s) \rightarrow AgNO_3$ (aq, 1 m)

$\Delta H° = \Delta H°\ AgNO_3(aq) - \Delta H°\ AgNO_3(s) = -101.7 - (-124.4) = +22.7$ kJ

$\Delta H° = \Delta H°\ MgSO_4(aq) - \Delta H°\ MgSO_4(s) = -1374.8 - (-1283.7) = -91.1$ kJ

Dissolving $AgNO_3(s)$ is endothermic ($+\Delta H°$), but dissolving $MgSO_4(s)$ is exothermic ($-\Delta H°$).

(d) $AgNO_3$: $\Delta G° = \Delta G_f°\ AgNO_3(aq) - \Delta G_f°\ AgNO_3(s) = -34.2 - (-33.4) = -0.8$ kJ

$\Delta S° = (\Delta H° - \Delta G°) / T = [22.7\ kJ - (-0.8\ kJ)] / 298\ K = 0.0789\ kJ/K = 78.9\ J/K$

$MgSO_4$: $\Delta G° = \Delta G_f°\ MgSO_4(aq) - \Delta G_f°\ MgSO_4(s) = -1198.4 - (-1169.6) = -28.8$ kJ

$\Delta S° = (\Delta H° - \Delta G°) / T = [-91.1\ kJ - (-28.8\ kJ)] / 298\ K = -0.209\ kJ/K = -209\ J/K$

(e) In general, we expect dissolving a crystalline solid to be accompanied by an increase in positional disorder and an increase in entropy; this is the case for $AgNO_3$ ($\Delta S° = +\ 78.9$ J/K). However, for dissolving $MgSO_4(s)$, there is a substantial decrease in entropy ($\Delta S = -209$ J/K). According to Section 13.5, ion-pairing is a significant phenomenon in electrolyte solutions, particularly in concentrated solutions where the charges of the ions are greater than 1. According to Table 13.5, a 0.1 m $MgSO_4$ solution has a van't Hoff factor of 1.21. That is, for each mole of $MgSO_4$ that dissolves, there are only 1.21 moles of "particles" in solution instead of 2 moles of particles. For a 1 m solution, the factor is even smaller. Also, the exothermic enthalpy of mixing indicates substantial interactions between solute and solvent. Substantial ion-pairing coupled with ion-dipole interactions with H_2O molecules lead to a decrease in entropy for $MgSO_4(aq)$ relative to $MgSO_4(s)$.

19.105 (a) $\Delta G° = 3\Delta G_f°\ S(s) + 2\Delta G_f°\ H_2O(g) - \Delta G_f°\ SO_2(g) - 2\Delta G_f°\ H_2S(g)$

$= 3(0) + 2(-228.57) - (-300.4) - 2(-33.01) = -90.72 = -90.7$ kJ

$$\ln K = \frac{-\Delta G°}{RT} = \frac{-(-90.72\ kJ)}{(8.314 \times 10^{-3}\ kJ/K)(298\ K)} = 36.6165 = 36.6;\ K = 7.99 \times 10^{15}$$

$$= 8 \times 10^{15}$$

(b) The reaction is highly spontaneous at 298 K and feasible in principle. However, use of $H_2S(g)$ produces a severe safety hazard for workers and the surrounding community.

(c) $$P_{H_2O} = \frac{25\ torr}{760\ torr/atm} = 0.033\ atm$$

$$K = \frac{P^2_{H_2O}}{P_{SO_2} \times P^2_{H_2S}};\ P_{SO_2} = P_{H_2S} = x\ atm$$

$$K = 7.99 \times 10^{15} = \frac{(0.033)^2}{x(x)^2};\ x^3 = \frac{(0.033)^2}{7.99 \times 10^{15}}$$

$x = 5 \times 10^{-7}$ atm

(d) $\Delta H^\circ = 3\Delta H_f^\circ\ S(s) + 2\Delta H_f^\circ\ H_2O(g) - \Delta H_f^\circ\ SO_2(g) - 2\Delta H_f^\circ\ H_2S(g)$

$= 3(0) + 2(-241.82) - (-296.9) - 2(-20.17) = -146.4$ kJ

$\Delta S^\circ = 3S^\circ\ S(s) + 2S^\circ\ H_2O(g) - S^\circ\ SO_2(g) - 2S^\circ\ H_2S(g)$

$= 3(31.88) + 2(188.83) - 248.5 - 2(205.6) = -186.4$ J/K

The reaction is exothermic ($-\Delta H$), so the value of K_{eq} will decrease with increasing temperature. The negative ΔS° value means that the reaction will become nonspontaneous at some higher temperature. The process will be less effective at elevated temperatures.

19.106 (a) When the rubber band is stretched, the molecules become more ordered, so the entropy of the system decreases, ΔS_{sys} is negative.

(b) $\Delta S_{sys} = q_{rev}/T$. Since ΔS_{sys} is negative, q_{rev} is negative and heat is evolved by the system.

20 Electrochemistry

Visualizing Concepts

20.2 (a) If a Zn(s) strip was placed in a $CdSO_4(aq)$ solution, Cd(s) would form on the strip. Although E°_{red} for $Zn^{2+}(aq)$, –0.763 V, and $Cd^{2+}(aq)$, -0.403 V, are both negative, the value for Cd^{2+} is larger (less negative), so it will be the reduced species in the redox reaction.

(b) If a Cu(s) strip was placed in a $AgNO_3(aq)$ solution, Ag(s) would form on the strip. Although E°_{red} for $Cu^{2+}(aq)$, 0.337 V, and $Ag^+(aq)$, 0.799 V, are both positive, the species with the larger E°_{red} value, Ag^+, will be reduced in the reaction.

20.4 The species with the largest E°_{red} is easiest to reduce, while the species with the smallest, most negative E°_{red} is easiest to oxidize.

(a) The species easiest to oxidize is at the bottom of Figure 20.14.

(b) The species easiest to reduce is at the top of Figure 20.14.

20.6

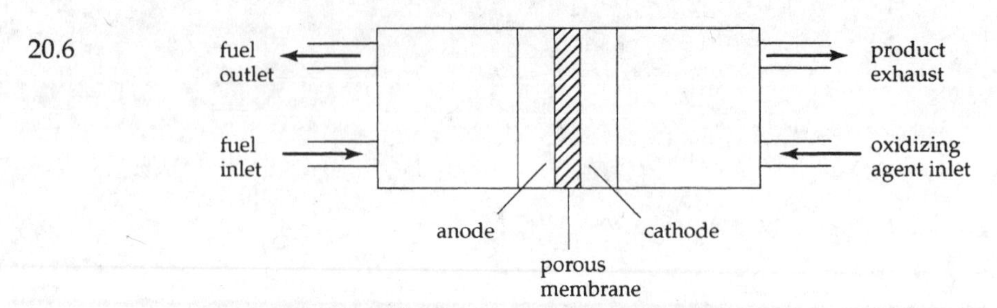

The main difference between a fuel cell and a battery is that a fuel cell is not self-contained. That is, there is a continuous supply of fuel (reductant) and oxidant to the cell, and continuous exhaust of products. The fuel cell produces electrical current as long as reactants are supplied. It never goes "dead."

20.8 Unintended oxidation reactions in the body lead to unwanted health effects, just as unwanted oxidation of metals leads to corrosion. Antioxidants probably have modes of action similar to anti-corrosion agents. They can preferentially react with oxidizing agents (cathodic protection), create conditions that are unfavorable to the oxidation-reduction reaction, or physically coat or surround the molecule being oxidized to prevent the oxidant from attacking it. The first of these modes of action is likely to be safest in biological systems. Adjusting reaction conditions in our body can be dangerous, and physical protection is unlikely to provide lasting protection against oxidation. Anti-oxidants are likely to be reductants that preferentially react with oxidizing agents.

Oxidation States

20.10 (a) *Reduction* is the gain of electrons.

(b) The electrons appear on the reactants side (left side) of a reduction half-reaction.

(c) The *reductant* is the reactant that is oxidized; it provides the electrons that are gained by the substance being reduced.

(d) A *reducing agent* is the substance that promotes reduction. It donates the electrons gained by the substance that is reduced. It is the same as the reductant.

20.12 (a) False. If something is reduced, it gains electrons.

(b) True.

(c) True. Oxidation can be thought of as a gain of oxygen atoms. Looking forward, this view will be useful for organic reactions, Chapter 25.

20.14 (a) No oxidation-reduction

(b) I is oxidized from –1 to +5; Cl is reduced from +1 to –1.

(c) S is oxidized from +4 to +6; N is reduced from +5 to +2.

(d) S is reduced from +6 to +4; Br is oxidized from –1 to 0.

Balancing Oxidation-Reduction Reactions

20.16 (a) $2N_2H_4(g) + N_2O_4(g) \rightarrow 3N_2(g) + 4H_2O(g)$

(b) $N_2H_4(g)$ is oxidized; $N_2O_4(g)$ is reduced.

(c) $N_2O_4(g)$ serves as the oxidizing agent; it is itself reduced. $N_2H_4(g)$ serves as the reducing agent; it is itself oxidized.

20.18 (a) $Mo^{3+}(aq) + 3e^- \rightarrow Mo(s)$, reduction

(b) $H_2SO_3(aq) + H_2O(l) \rightarrow SO_4^{2-}(aq) + 4H^+(aq) + 2e^-$, oxidation

(c) $NO_3^-(aq) + 4H^+(aq) + 3e^- \rightarrow NO(g) + 2H_2O(l)$, reduction

(d) $Mn^{2+}(aq) + 4OH^-(aq) \rightarrow MnO_2(s) + 2H_2O(l) + 2e^-$, oxidation

(e) $Cr(OH)_3(s) + 5OH^-(aq) \rightarrow CrO_4^{2-}(aq) + 4H_2O(l) + 3e^-$, oxidation

(f) $O_2(g) + 4H^+(aq) + 4e^- \rightarrow 2H_2O(l)$, reduction

(g) $O_2(g) + 2H_2O(l) + 4e^- \rightarrow 4OH^-(aq)$, reduction

20.20 (a)

$$3[NO_2^-(aq) + H_2O(l) \rightarrow NO_3^-(aq) + 2H^+(aq) + 2e^-]$$

$$Cr_2O_7^{2-}(aq) + 14H^+(aq) + 6e^- \rightarrow 2Cr^{3+}(aq) + 7H_2O(l)$$

Net: $3NO_2^-(aq) + Cr_2O_7^{2-}(aq) + 8H^+(aq) \rightarrow 3NO_3^-(aq) + 2Cr^{3+}(aq) + 4H_2O(l)$

oxidizing agent, $Cr_2O_7^{2-}$; reducing agent, NO_2^-

(b)

$$4[As(s) + 3H_2O(l) \rightarrow H_3AsO_3(aq) + 3H^+(aq) + 3e^-]$$
$$3[ClO_3^-(aq) + 5H^+(aq) + 4e^- \rightarrow HClO(aq) + 2H_2O(l)]$$
$$4As(s) + 3ClO_3^-(aq) + 6H_2O(l) + 3H^+(aq) \rightarrow 4H_3AsO_3(aq) + 3HClO(aq)$$

oxidizing agent, ClO_3^-; reducing agent, As

(c)

$$2[Cr_2O_7^{2-}(aq) + 14H^+(aq) + 6e^- \rightarrow 2Cr^{3+}(aq) + 7H_2O(l)]$$
$$3[CH_3OH(aq) + H_2O(l) \rightarrow HCO_2H(aq) + 4H^+(aq) + 4e^-]$$

Net: $2Cr_2O_7^{2-}(aq) + 3CH_3OH(aq) + 16H^+(aq) \rightarrow 4Cr^{3+}(aq) + 3HCO_2H(aq) + 11H_2O(l)$

oxidizing agent, $Cr_2O_7^{2-}$; reducing agent, CH_3OH

(d)

$$2[MnO_4^-(aq) + 8H^+(aq) + 5e^- \rightarrow Mn^{2+}(aq) + 4H_2O(l)]$$
$$5[2Cl^-(aq) \rightarrow Cl_2(aq) + 2e^-]$$

Net: $2MnO_4^-(aq) + 10Cl^-(aq) + 16H^+(aq) \rightarrow 2Mn^{2+}(aq) + 5Cl_2(g) + 8H_2O(l)$

oxidizing agent, MnO_4^-; reducing agent, Cl^-

(e) $H_2O_2(aq) + 2e^- \rightarrow O_2(g) + 2H^+(aq)$

Since the reaction is in base, the H^+ can be "neutralized" by adding $2OH^-$ to each side of the equation to give $H_2O_2(aq) + 2OH^-(aq) \rightarrow O_2(g) + 2H_2O(l) + 2e^-$. The other half reaction is $2[ClO_2(aq) + e^- \rightarrow ClO_2^-(aq)]$.

Net: $H_2O_2(aq) + 2ClO_2(aq) + 2OH^-(aq) \rightarrow O_2(g) + 2ClO_2^-(aq) + 2H_2O(l)$

oxidizing agent, ClO_2; reducing agent, H_2O_2

(f)

$$4[H_2O_2(aq) + 2OH^-(aq) \rightarrow O_2(g) + 2H_2O(l) + 2e^-]$$
$$Cl_2O_7(aq) + 3H_2O(l) + 8e^- \rightarrow 2ClO_2^-(aq) + 6OH^-(aq)$$

Net: $Cl_2O_7(aq) + 4H_2O_2(aq) + 2OH^-(aq) \rightarrow 2ClO_2^-(aq) + 4O_2(g) + 5H_2O(l)$

oxidizing agent, Cl_2O_7; reducing agent, H_2O_2

Voltaic Cells

20.22 (a) The porous glass dish in Figure 20.4 provides a mechanism by which ions not directly involved in the redox reaction can migrate into the anode and cathode compartments to maintain charge neutrality of the solutions. Ionic conduction within the cell, through the glass disk, completes the cell circuit.

(b) In the anode compartment of Figure 20.5, Zn atoms are oxidized to Zn^{2+} cations, increasing the number of positively charged particles in the compartment. NO_3^- anions migrate into the compartment to maintain charge balance as Zn^{2+} ions are produced.

20.24 (a) Al(s) is oxidized, $Ni^{2+}(aq)$ is reduced.

(b) $Al(s) \rightarrow Al^{3+}(aq) + 3e^-$; $Ni^{2+}(aq) + 2e^- \rightarrow Ni(s)$

(c) Al(s) is the anode; Ni(s) is the cathode.

(d) Al(s) is negative (–); Ni(s) is positive (+).

(e) Electrons flow from the Al(–) electrode toward the Ni(+) electrode.

(f) Cations migrate toward the Ni(s) cathode; anions migrate toward the Al(s) anode.

Cell EMF under Standard Conditions

20.26 (a) In a voltaic cell, the anode has the higher potential energy for electrons. To achieve a lower potential energy, electrons flow from the anode to the cathode.

(b) The units of electrical potential are volts. A potential of one volt imparts one joule of energy to one coulomb of charge.

(c) A *standard* cell potential describes the potential of an electrochemical cell where all components are present at standard conditions: elements in their standard states, gases at 1 atm pressure and 1 *M* aqueous solutions.

20.28 (a) $H_2(g) \rightarrow 2H^+(aq) + 2e^-$

(b) The platinum electrode serves as a reaction surface; the greater the surface area, the more H_2 or H^+ that can be adsorbed onto the surface to facilitate the flow of electrons.

(c)

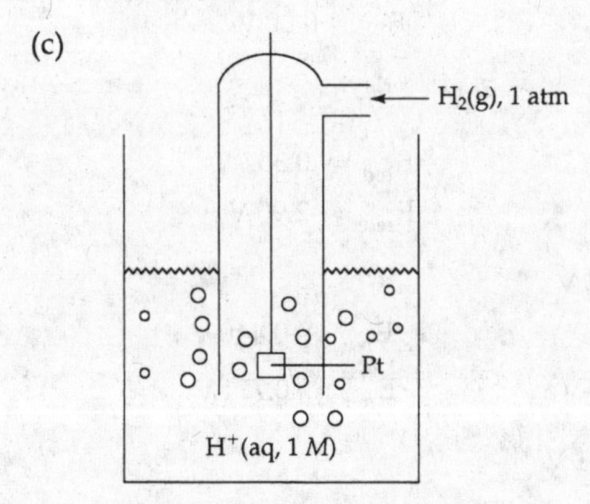

20.30 (a) It is not possible to measure the standard reduction potential of a single half-reaction because each voltaic cell consists of two half-reactions and only the potential of a complete cell can be measured.

(b) The standard reduction potential of a half-reaction is determined by combining it with a reference half-reaction of known potential and measuring the cell potential. Assuming the half-reaction of interest is the reduction half-reaction:

$E^\circ_{cell} = E^\circ_{red}(\text{cathode}) - E^\circ_{red}(\text{anode}) = E^\circ_{red}(\text{unknown}) - E^\circ_{red}(\text{reference})$;

$E^\circ_{red}(\text{unknown}) = E^\circ_{cell} + E^\circ_{red}(\text{reference})$.

(c) $Cd^{2+}(aq) + 2e^- \rightarrow Cd(s)$ $E° = -0.403$ V

$Ca^{2+}(aq) + 2e^- \rightarrow Ca(s)$ $E° = -2.87$ V

The reduction of $Ca^{2+}(aq)$ to Ca(s) is the more energetically unfavorable reduction because it has a more negative E° value.

20.32 (a) $PdCl_4^{2-}(aq) + 2e^- \rightarrow Pd(s) + 4Cl^-$ cathode $E^\circ_{red} = ?$

$Cd(s) \rightarrow Cd^{2+}(aq) + 2e^-$ anode $E^\circ_{red} = -0.403$ V

(b) $E^\circ_{cell} = E^\circ_{red}$ (cathode) $- E^\circ_{red}$ (anode); $1.03\text{ V} = E^\circ_{red} - (-0.403\text{ V})$;

$E^\circ_{red} = 1.03\text{ V} - 0.403 = 0.63\text{ V}$

(c)

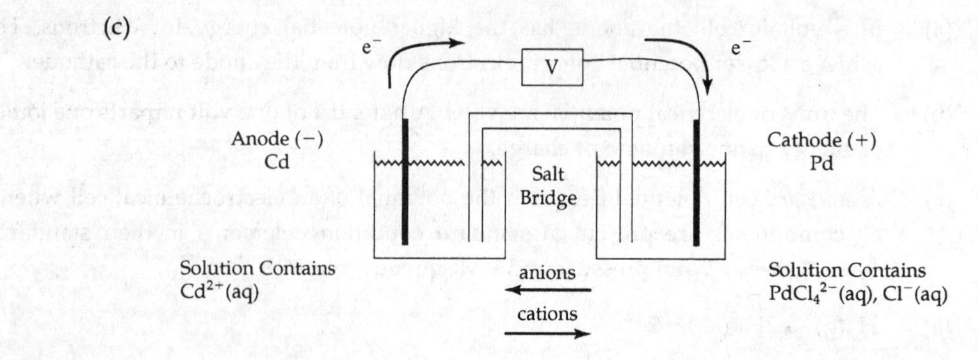

20.34 (a) $F_2(g) + 2e^- \rightarrow 2F^-(aq)$ $E^\circ_{red} = 2.87$ V

$H_2(g) \rightarrow 2H^+(aq) + 2e^-$ $E^\circ_{red} = 0.00$ V

E° = 2.87 V - 0.00 V = 2.87 V

(b) $Cu(s) \rightarrow Cu^{2+}(aq) + 2e^-$ $E^\circ_{red} = 0.337$ V

$Ba^{2+}(aq) + 2e^- \rightarrow Ba(s)$ $E^\circ_{red} = -2.90$ V

E° = –2.90 V - (0.337 V) = –3.24 V

(c) $Fe^{2+}(aq) + 2e^- \rightarrow Fe(s)$ $E^\circ_{red} = -0.440$ V

$2[Fe^{2+}(aq) \rightarrow Fe^{3+}(aq) + 1e^-]$ $E^\circ_{red} = 0.771$ V

E° = –0.440 V - 0.771 V = –1.211 V

(d) $Hg_2^{2+}(aq) + 2e^- \rightarrow 2Hg(l)$ $E^\circ_{red} = 0.789$ V

$2[Cu^+(aq) \rightarrow Cu^{2+}(aq) + 1e^-]$ $E^\circ_{red} = 0.153$ V

E° = 0.789 V - 0.153 V = 0.636 V

20.36 (a)

$$2[Au(s) + 4Br^-(aq) \rightarrow AuBr_4^-(aq) + 3e^-] \quad E^\circ_{red} = -0.858\text{ V}$$
$$3[2e^- + IO^-(aq) + H_2O(l) \rightarrow I^-(aq) + 2OH^-(aq)] \quad E^\circ_{red} = 0.49\text{ V}$$
$$2Au(s) + 8Br^-(aq) + 3IO^-(aq) + 3H_2O(l) \rightarrow 2AuBr_4^-(aq) + 3I^-(aq) + 6OH^-(aq)$$

E° = 0.49 - (–0.858) = 1.35 V

(b)

$$2[Eu^{2+}(aq) \rightarrow Eu^{3+}(aq) + 1e^-] \quad E^\circ_{red} = -0.43\text{ V}$$
$$Sn^{2+}(aq) + 2e^- \rightarrow Sn(s) \quad E^\circ_{red} = -0.14\text{ V}$$
$$2Eu^{2+}(aq) + Sn^{2+}(aq) \rightarrow 2Eu^{3+}(aq) + Sn(s) \quad E^\circ = -0.14 - (-0.43) = 0.29\text{ V}$$

20.38 (a) The two half-reactions are:

$$Pb^{2+}(aq) + 2e^- \rightarrow Pb(s) \qquad E^\circ = -0.126\ V$$
$$Cl_2(g) + 2e^- \rightarrow 2Cl^-(aq) \qquad E^\circ = 1.359\ V$$

Because E° for the reduction of Cl_2 is higher, the reduction of Cl_2 occurs at the Pt cathode. The Pb electrode is the anode.

(b) The Pb anode loses mass as $Pb^{2+}(aq)$ is produced.

(c) $Cl_2(g) + Pb(s) \rightarrow Pb^{2+}(aq) + 2Cl^-(aq)$

(d) E° = 1.359 V - (-0.126 V) = 1.485 V

Strengths of Oxidizing and Reducing Agents

20.40 The more readily a substance is oxidized, the stronger it is as a reducing agent. In each case choose the half-reaction with the more negative reduction potential and the given substance on the right.

(a) Mg(s) (-2.37 V vs. -0.440 V)

(b) Ca(s) (-2.87 V vs. -1.66 V)

(c) H_2(g, acidic) (0.000 V vs. 0.141 V)

(d) $H_2C_2O_4(aq)$ (-0.49 V vs. 0.17 V)

20.42 If the substance is on the left of a reduction half-reaction, it will be an oxidant; if it is on the right, it will be a reductant. The sign and magnitude of the E°_{red} determine whether it is strong or weak.

(a) Na(s): strong reductant (on the right, $E^\circ_{red} = -2.71\ V$)

(b) $O_3(g)$: strong oxidant (on the left, $E^\circ_{red} = 2.07\ V$)

(c) $Ce^{3+}(aq)$: very weak reductant (on the right, $E^\circ_{red} = 1.61\ V$)

(d) $Sn^{2+}(aq)$: reductant (on the right, $E^\circ_{red} = 0.154\ V$) **or** weak oxidant (on the left, -0.136 V)

20.44 (a) The strongest oxidizing agent is the species most readily reduced, as evidenced by a large, positive reduction potential. That species is H_2O_2. The weakest oxidizing agent is the species that least readily accepts an electron. We expect that it will be very difficult to reduce Zn(s); indeed, Zn(s) acts as a comparatively strong **reducing** agent. No potential is listed for reduction of Zn(s), but we can safely assume that it is less readily reduced than any of the other species present.

(b) The strongest reducing agent is the species most easily oxidized (the largest negative reduction potential). Zn, $E^\circ_{red} = -0.76\ V$, is the strongest reducing agent and F^-, $E^\circ_{red} = 2.87\ V$, is the weakest.

Free Energy and Redox Reactions

20.46 Any oxidized species from Table 20.1 or Appendix E with a reduction potential greater than 0.59 V will oxidize RuO_4^{2-} to RuO_4^-. From the list of possible oxidants in the exercise, $Cr_2O_7^{2-}$(aq) and ClO^-(aq) will oxidize RuO_4^{2-} to RuO_4^-.

20.48 (a)

$$\begin{array}{ll} 2I^-(aq) \rightarrow I_2(s) + 2e^- & E^\circ_{red} = 0.536\ V \\ Hg_2^{2+}(aq) + 2e^- \rightarrow 2Hg(l) & E^\circ_{red} = 0.789\ V \\ \hline 2I^-(aq) + Hg_2^{2+}(aq) \rightarrow I_2(s) + 2Hg(l) & E^\circ = 0.789 - 0.536 = 0.253\ V \end{array}$$

$$\Delta G^\circ = -nFE^\circ = -2\ mol\ e^- \times \frac{96.5\ kJ}{V \bullet mol\ e^-} \times 0.253\ V = -48.829 = -48.8\ kJ$$

$$\ln K = \frac{-(-4.8829 \times 10^4\ J)}{(8.314\ J/mol \bullet K)(298\ K)} = 19.708 = 19.7;\ K = e^{19.7} = 3.61 \times 10^8 = 3.6 \times 10^8$$

(b)

$$\begin{array}{ll} 3[Cu^+(aq) \rightarrow Cu^{2+}(aq) + 1e^-] & E^\circ_{red} = 0.153\ V \\ NO_3^-(aq) + 4H^+(aq) + 3e^- \rightarrow NO(g) + H_2O(l) & E^\circ_{red} = 0.96\ V \\ \hline 3Cu^+(aq) + NO_3^-(aq) + 4H^+(aq) \rightarrow 3Cu^{2+}(aq) + NO(g) + 2H_2O(l) & \end{array}$$

$E^\circ = 0.96 - 0.153 = 0.81$ V; $\Delta G^\circ = -3(96.5)(0.81) = -2.345 \times 10^2$ kJ $= -2.3 \times 10^5$ J

$$\ln K = \frac{-(-2.345 \times 10^5\ J)}{(8.314\ J/mol \bullet K)(298\ K)} = 94.65 = 95;\ K = e^{95} = 1.3 \times 10^{41} = 10^{41}$$

(c)

$$\begin{array}{ll} 2[Cr(OH)_3(s) + 5OH^-(aq) \rightarrow CrO_4^{2-}(aq) + 4H_2O(l) + 3e^-] & E^\circ_{red} = -0.13\ V \\ 3[ClO^-(aq) + H_2O(l) + 2e^- \rightarrow Cl^-(aq) + 2OH^-(aq)] & E^\circ_{red} = 0.89\ V \\ \hline 2Cr(OH)_3(s) + 3ClO^-(aq) + 4OH^-(aq) \rightarrow 2CrO_4^{2-}(aq) + 3Cl^-(aq) + 5H_2O(l) & \end{array}$$

$E^\circ = 0.89 - (-0.13) = 1.02$ V; $\Delta G^\circ = -6(96.5)(1.02) = -590.58$ kJ $= -5.91 \times 10^5$ J

$$\ln K = \frac{-(-5.9058 \times 10^5\ J)}{(8.314\ J/mol \bullet K)(298\ K)} = 238.37 = 238;\ K = 3.3 \times 10^{103} = 10^{103}$$

This is an unimaginably large number.

20.50 $K = 3.7 \times 10^6$; $\Delta G^\circ = -RT \ln K$; $E^\circ = -\Delta G^\circ/nF$; $n = 1$; $T = 298$ K

$\Delta G^\circ = -8.314\ J/mol \bullet K \times 298\ K \times \ln(3.7 \times 10^6) = -3.747 \times 10^4\ J = -37.5$ kJ

$$E^\circ = -\Delta G^\circ/nF = \frac{-(-37.47\ kJ)}{1e^- \times 96.5\ kJ/V \bullet mol\ e^-} = 0.388\ V$$

20.52 $E^\circ = \frac{0.0592\ V}{n} \log K$; $\log K = \frac{nE^\circ}{0.0592\ V.}$ See Solution 20.51 for a more complete explanation.

(a) $E^\circ = 1.00\ V - 0.799\ V = 0.201 = 0.20$ V; $n = 1$ ($VO_2^+ + 1e^- \rightarrow VO^{2+}$)

$$\log K = \frac{1(0.210\ V)}{0.0592\ V} = 3.3953 = 3.40;\ K = 2.48 \times 10^3 = 2.5 \times 10^3$$

Note: The correctly balanced chemical equation for this reaction is
$VO_2^+(aq) + 2H^+(aq) + Ag(s) \rightarrow 2\ VO^{2+}(aq) + H_2O(l) + Ag^+(aq)$, for which n = 1.

(b) $E° = 1.61\ V - 0.32\ V = 1.29\ V; n = 3\ (3Ce^{4+} + 3e^- \rightarrow 3Ce^{3+})$

$$\log K = \frac{3(1.29)}{0.0592} = 65.372 = 65.4; K = 2.35 \times 10^{65} = 2 \times 10^{65}$$

(c) $E° = 0.36\ V - (-0.23\ V) = 0.59\ V; n = 4\ (4Fe(CN)_6^{3-} + 4e^- \rightarrow 4Fe(CN)_6^{4-})$

$$\log K = \frac{4(0.59)}{0.0592} = 39.865 = 40;\ K = 7.3 \times 10^{39} = 10^{40}$$

20.54 $E^{\circ} = \frac{0.0592\ V}{n} \log K;\ n = \frac{0.0592\ V}{E^{\circ}} \log K$. See Solution 20.51 for a more complete development.

$$n = \frac{0.0592\ V}{0.17\ V} \log(5.5 \times 10^5); n = 2$$

Cell EMF under Nonstandard Conditions

20.56 (a) No. As the spontaneous chemical reaction of the voltaic cell proceeds, the concentrations of products increase and the concentrations of reactants decrease, so standard conditions are not maintained.

(b) Yes. The Nernst equation is applicable to cell EMF at nonstandard conditions, so it must be applicable at temperatures other than 298 K. There are two terms in the Nernst Equation. First, values of E° at temperatures other than 298 K are required. Then, in the form of Equation [20.14], there is a variable for T in the second term. In the short-hand form of Equation [20.16], the value 0.0592 assumes 298 K. A different coefficient would apply to cells at temperatures other than 298 K.

(c) If concentration of products increases, Q increases, and E decreases.

20.58 $Al(s) + 3Ag^+(aq) \rightarrow Al^{3+}(aq) + 3Ag(s);\ E = E^{\circ} - \frac{0.0592}{n} \log Q;\ Q = \frac{[Al^{3+}]}{[Ag^+]^3}$

Any change that causes the reaction to be less spontaneous (that causes Q to increase and ultimately shifts the equilibrium to the left) will result in a less positive value for E.

(a) Increases E by decreasing $[Al^{3+}]$ on the right side of the equation, which decreases Q.

(b) No effect; the "concentrations" of pure solids and liquids do not influence the value of K for a heterogeneous equilibrium.

(c) No effect; the concentration of Ag^+ and the value of Q are unchanged.

(d) Decreases E; forming AgCl(s) decreases the concentration of Ag^+, which increases Q.

20.60 (a)

$$\begin{array}{ll} 3[Ce^{4+}(aq) + 1e^- \rightarrow Ce^{3+}(aq)] & E^{\circ}_{red} = 1.61\ V \\ Cr(s) \rightarrow Cr^{3+}(aq) + 3e^- & E^{\circ}_{red} = -0.74\ V \\ \hline 3Ce^{4+}(aq) + Cr(s) \rightarrow 3Ce^{3+}(aq) + Cr^{3+}(aq) & E^{\circ} = 1.61 - (-0.74) = 2.35\ V \end{array}$$

(b) $$E = E^\circ - \frac{0.0592}{n} \log \frac{[Ce^{3+}]^3 [Cr^{3+}]}{[Ce^{4+}]^3}; \; n = 3$$

$$E = 2.35 - \frac{0.0592}{3} \log \frac{(0.010)^3 (0.010)}{(2.0)^3} = 2.35 - \frac{0.0592}{3} \log (1.250 \times 10^{-9})$$

$$E = 2.35 - \frac{0.0592(-8.903)}{3} = 2.35 + 0.176 = 2.53 \text{ V}$$

(c) $$E = 2.35 - \frac{0.0592}{3} \log \frac{(0.85)^3 (1.2)}{(0.35)^3} = 2.35 - 0.0244 = 2.33 \text{ V}$$

20.62 (a)

$$2[Fe^{3+}(aq) + 1e^- \rightarrow Fe^{2+}(aq)] \qquad E^\circ_{red} = 0.771 \text{ V}$$

$$H_2(g) \rightarrow 2H^+(aq) + 2e^- \qquad E^\circ_{red} = 0.000 \text{ V}$$

$$2Fe^{3+}(aq) + H_2(g) \rightarrow 2Fe^{2+}(aq) + 2H^+(aq) \qquad E^\circ = 0.771 - 0.000 = 0.771 \text{ V}$$

(b) $$E = E^\circ - \frac{0.0592}{n} \log \frac{[Fe^{2+}]^2 [H^+]^2}{[Fe^{3+}]^2 P_{H_2}}; \; [H^+] = 10^{-pH} = 1.0 \times 10^{-5}, \; n = 2$$

$$E = 0.771 - \frac{0.0592}{2} \log \frac{(0.0010)^2 (1.0 \times 10^{-5})^2}{(1.50)^2 (0.50)} = 0.771 - \frac{0.0592}{2} \log (8.9 \times 10^{-17})$$

$$E = 0.771 - \frac{0.0592(-16.05)}{2} = 0.771 + 0.4751 = 1.246 \text{ V}$$

20.64 (a) The compartment with 0.0150 *M* Cl^- (aq) is the cathode.

(b) $E^\circ = 0$ V

(c) $$E = E^\circ - \frac{0.0592}{n} \log Q; \; Q = [Cl^-, \text{dilute}]/[Cl^-, \text{conc.}]$$

$$E = 0 - \frac{0.0592}{1} \log \frac{(0.0150)}{(2.55)} = -0.13204 = -0.1320 \text{ V}$$

(d) In the anode compartment, $[Cl^-]$ will decrease from 2.55 *M*. In the cathode, $[Cl^-]$ will increase from 0.0150 *M*.

20.66 (a) $E^\circ = -0.136 \text{ V} - (-0.126 \text{ V}) = -0.010 \text{ V}; n = 2$

$$0.22 = -0.010 - \frac{0.0592}{2} \log \frac{[Pb^{2+}]}{[Sn^{2+}]} = -0.010 - \frac{0.0592}{2} \log \frac{[Pb^{2+}]}{1.00}$$

$$\log [Pb^{2+}] = \frac{-0.23(2)}{0.0592} = -7.770 = -7.8; \; [Pb^{2+}] = 1.7 \times 10^{-8} = 2 \times 10^{-8} \, M$$

(b) For $PbSO_4(s)$, $K_{sp} = [Pb^{2+}][SO_4^{2-}] = (1.0)(1.7 \times 10^{-8}) = 1.7 \times 10^{-8}$

Batteries and Fuel Cells

20.68 First, H_2O is a reactant in the cathodic half-reaction, so it must be present in some form. Additionally, liquid water enhances mobility of the hydroxide ion in the alkaline battery. OH^- is produced in the cathode compartment and consumed in the anode compartment. It must be available at all points where Zn(s) is being oxidized. If the

$Zn(s)$ near the separator is mostly reacted, OH^- must diffuse through the gel until it reaches fresh $Zn(s)$. A small amount of $H_2O(l)$ mobilizes OH^- so that redox can continue until reactants throughout the battery are depleted.

20.70 The overall cell reaction is:

$$2MnO_2(s) + Zn(s) + 2H_2O(l) \rightarrow 2MnO(OH)(s) + Zn(OH)_2(s)$$

$$12.6 \text{ g Zn} \times \frac{1 \text{ mol Zn}}{65.39 \text{ g Zn}} \times \frac{2 \text{ mol } MnO_2}{1 \text{ mol Zn}} \times \frac{86.94 \text{ g } MnO_2}{1 \text{ mol } MnO_2} = 33.5 \text{ g } MnO_2 \text{ reduced}$$

20.72 (a) $HgO(s) + Zn(s) \rightarrow Hg(l) + ZnO(s)$

(b) $E^\circ_{cell} = E^\circ_{red}(\text{cathode}) - E^\circ_{red}(\text{anode})$

$E^\circ_{red}(\text{anode}) = E^\circ_{red} - E^\circ_{cell} = 0.098 - 1.35 = -1.25 \text{ V}$

(c) E°_{red} is different from $Zn^{2+}(aq) + 2e^- \rightarrow Zn(s)$ (–0.76 V) because in the battery the process happens in the presence of base and Zn^{2+} is stabilized as $ZnO(s)$. Stabilization of a reactant in a half-reaction decreases the driving force, so E°_{red} is more negative.

20.74 (a) The alkali metal Li has much greater metallic character than Zn, Cd, Pb or Ni. The reduction potential for Li is thus more negative, leading to greater overall cell emf for the battery. Also, Li is less dense than the other metals, so greater total energy for a battery can be achieved for a given total mass of material. One disadvantage is that Li is very reactive and the cell reactions are difficult to control.

(b) Li has a much smaller molar mass (6.94 g/mol) than Ni (58.69 g/mol). A Li-ion battery can have many more charge-carrying particles than a Ni-based battery with the same mass. That is, Li-ion batteries have a greater *energy density* than Ni-based batteries.

20.76 No. The fuel in a fuel cell must be fluid, either gas or liquid. Because fuel must be continuously supplied to the fuel cell, it must be capable of flow; the fuel cannot be solid.

Corrosion

20.78 (a) Calculate E°_{cell} for the given reactants at standard conditions.

$$\begin{array}{ll} O_2(g) + 4H^+(aq) + 4e^- \rightarrow 2H_2O(l) & E^\circ_{red} = 1.23 \text{ V} \\ 2[Cu(s) \rightarrow Cu^{2+}(aq) + 2e^-] & E^\circ_{red} = 0.337 \text{ V} \\ \hline 2Cu(s) + O_2(g) + 4H^+(aq) \rightarrow 2Cu^{2+}(aq) + 2H_2O(l) & E^\circ = 1.23 - 0.337 = 0.89 \text{ V} \end{array}$$

At standard conditions with $O_2(g)$ and $H^+(aq)$ present, the oxidation of $Cu(s)$ has a positive E° value and is spontaneous. $Cu(s)$ will oxidize (corrode) in air in the presence of acid.

(b) Fe^{2+} has a more negative reduction potential (–0.440 V) than Cu^{2+} (+0.337 V), so Fe(s) is more readily oxidized than Cu(s). If the two metals are in contact, Fe(s) would act as a sacrificial anode and oxidize (corrode) in preference to Cu(s); this would weaken the iron support skeleton of the statue. The teflon spacers prevent contact between the two metals and insure that the iron skeleton doesn't corrode when the Cu(s) skin comes in contact with atmospheric O_2(g) and H^+(aq).

20.80 No. To afford cathodic protection, a metal must be more difficult to reduce (have a more negative reduction potential) than Fe^{2+}. E°_{red} Co^{2+} = –0.28 V, E°_{red} Fe^{2+} = –0.44 V.

20.82 The principal metallic component of steel is Fe. E°_{red} for Fe, –0.763 V, is more negative than that of Cu, 0.337 V. When the two are in contact, Fe acts as the sacrificial anode and corrodes (oxidizes) preferentially in the presence of O_2(g).

$2Fe(s) + O_2(g) + 4H^+(aq) \rightarrow 2Fe^{2+}(aq) + 2H_2O(l)$

$E^\circ = 1.23\ V - (-0.440\ V) = 1.67\ V$

$2Cu(s) + O_2(g) + 4H^+(aq) \rightarrow 2Cu^{2+}(aq) + 2H_2O(l)$

$E^\circ = 1.23\ V - (0.337\ V) = 0.893\ V$

Both reactions are spontaneous, but the corrosion of Fe has the larger E° value and happens preferentially.

Electrolysis; Electrical Work

20.84 (a) An *electrolytic cell* is the vessel in which electrolysis occurs. It consists of a power source and two electrodes in a molten salt or aqueous solution.

(b) It is the cathode. In an electrolysis cell, as in a voltaic cell, electrons are consumed (via reduction) at the cathode. Electrons flow from the negative terminal of the voltage source and then to the cathode.

(c) A small amount of H_2SO_4(aq) present during the electrolysis of water acts as a change carrier, or supporting electrolyte. This facilitates transfer of electrons through the solution and at the electrodes, speeding up the reaction. (Considering H^+(aq) as the substance reduced at the cathode changes the details of the half-reactions, but not the overall E° for the electrolysis. SO_4^{2-}(aq) cannot be oxidized.)

20.86 Coulombs = **amps** ·s; since this is a $2e^-$ reduction, each mole of Mg(s) requires 2 Faradays.

(a) $$5.25\ A \times 2.50\ d \times \frac{24\ hr}{1\ d} \times \frac{60\ min}{1\ hr} \times \frac{60\ s}{1\ min} \times \frac{1\ C}{1\ amp \bullet s} \times \frac{1\ F}{96{,}500\ C} \times \frac{1\ mol\ Mg}{2\ F} \times \frac{24.31\ g\ Mg}{1\ mol\ Mg} = 143\ g\ Mg$$

(b) $$10.00\ g\ Mg \times \frac{1\ mol\ Mg}{24.31\ g\ Mg} \times \frac{2\ F}{1\ mol\ Mg} \times \frac{96{,}500\ C}{F} \times \frac{1\ amp \bullet s}{C} \times \frac{1\ min}{60\ s} \times \frac{1}{3.50\ A} = 378\ min$$

20.88 For this cell at standard conditions, E° = 1.10 V.

$$w_{max} = \Delta G° = -nFE° = -2(96.5)(1.10) = -212.3 = -212 \text{ kJ/mol Cu}$$

$$50.0 \text{ g Cu} \times \frac{1 \text{ mol Cu}}{63.55 \text{ g Cu}} \times \frac{-212.3 \text{ kJ}}{\text{mol Cu}} = -167 \text{ kJ} = -1.67 \times 10^5 \text{ J}$$

20.90 (a) $6.5 \times 10^3 \text{ A} \times 48 \text{ hr} \times \frac{3600 \text{ s}}{1 \text{ hr}} \times \frac{1 \text{ C}}{1 \text{ amp} \bullet \text{s}} \times \frac{1 \text{ F}}{96{,}500 \text{ C}} \times \frac{1 \text{ mol Ca}}{2 \text{ F}}$

$$\times \frac{40.08 \text{ g Ca}}{1 \text{ mol Ca}} \times 0.68 = 1.586 \times 10^5 = 1.6 \times 10^5 \text{ g Ca}$$

(b) If the cell is 68% efficient, $\frac{96{,}500 \text{ C}}{\text{F}} \times \frac{2 \text{ F}}{0.68 \text{ mol Ca}} = 2.838 \times 10^5$

$= 2.8 \times 10^5$ C/mol Ca required

$$\text{Energy} = 5.00 \text{ V} \times \frac{2.838 \times 10^5 \text{ C}}{\text{mol Ca}} \times \frac{1 \text{ J}}{\text{C} \bullet \text{V}} \times \frac{1 \text{ kWh}}{3.6 \times 10^6 \text{ J}} = 0.3942 = 0.39 \text{ kWh}$$

Additional Exercises

20.92 (a)

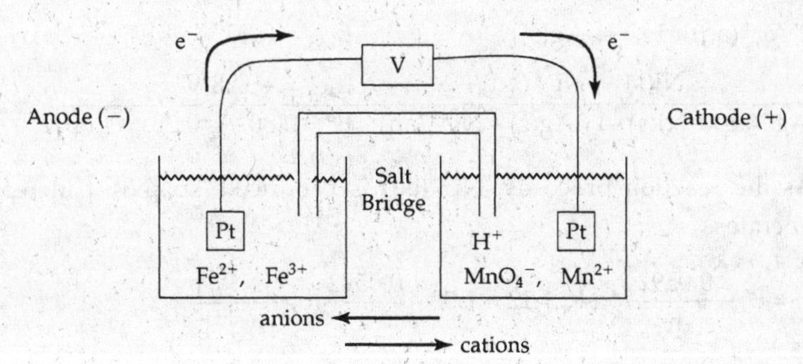

(b)

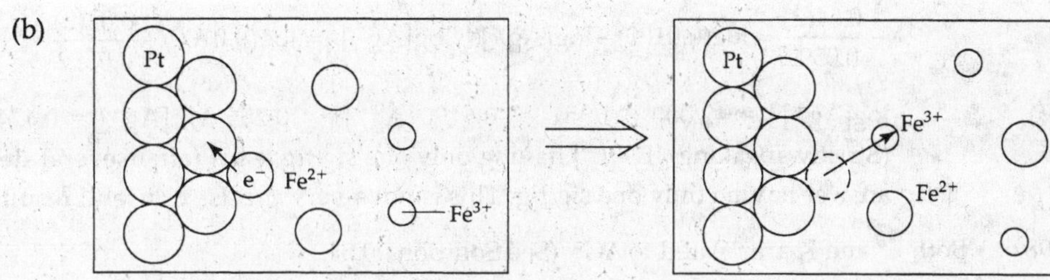

(c) $MnO_4^-(aq) + 8H^+(aq) + 5e^- \rightarrow Mn^{2+}(aq) + 4H_2O(l)$ $E°_{red} = 1.51$ V

$5[Fe^{2+}(aq) \rightarrow Fe^{3+}(aq) + 1e^-]$ $E°_{red} = 0.771$ V

E° = 1.51 V - 0.771 V = 0.74 V

(d) $E = E° - \frac{0.0592}{5} \log \frac{[Fe^{3+}]^5[Mn^{2+}]}{[Fe^{2+}]^5[MnO_4^-][H^+]^8}$; pH = 0.0, $[H^+] = 1.0$

$$E = 0.74 \text{ V} - \frac{0.0592}{5} \log \frac{(2.5 \times 10^{-4})^5 (0.010)}{(0.10)^5 (1.50)(1.0)^8}; Q = 6.510 \times 10^{-16} = 6.5 \times 10^{-16}$$

$$E = 0.74 \text{ V} - \frac{0.0592(-15.1864)}{5} = 0.74 \text{ V} + 0.18 \text{ V} = 0.92 \text{ V}$$

20.94 We need in each case to determine whether E° is positive (spontaneous) or negative (nonspontaneous).

(a)

$$\begin{array}{ll} I_2(s) + 2e^- \rightarrow 2I^-(aq) & E^\circ_{red} = 0.536\ V \\ Sn(s) \rightarrow Sn^{2+}(aq) + 2e^- & E^\circ_{red} = -0.136 \\ \hline Sn(s) + I_2(s) \rightarrow Sn^{2+}(aq) + 2I^-(aq) & E^\circ = 0.536 - (-0.136) = 0.672\ V\text{, spontaneous} \end{array}$$

(b)

$$\begin{array}{ll} Ni^{2+}(aq) + 2e^- \rightarrow Ni(s) & E^\circ_{red} = -0.28\ V \\ 2I^-(aq) \rightarrow I_2(s) + 2e^- & E^\circ_{red} = 0.536\ V \\ \hline Ni^{2+}(aq) + 2I^-(aq) \rightarrow Ni(s) + I_2(s) & E^\circ = -0.28 - 0.536 = -0.82\ V\text{, nonspontaneous} \end{array}$$

(c)

$$\begin{array}{ll} 2[Ce^{4+}(aq) + 1e^- \rightarrow Ce^{3+}(aq)] & E^\circ_{red} = 1.61\ V \\ H_2O_2(aq) \rightarrow O_2(g) + 2H^+(aq) + 2e^- & E^\circ_{red} = 0.68\ V \\ \hline 2Ce^{4+}(aq) + H_2O_2(aq) \rightarrow 2Ce^{3+}(aq) + O_2(g) + 2H^+(aq) & E^\circ = 1.61 - 0.68 = 0.93\ V\text{, spontaneous} \end{array}$$

(d)

$$\begin{array}{ll} Cu^{2+}(aq) + 2e^- \rightarrow Cu(s) & E^\circ_{red} = 0.337\ V \\ Sn^{2+}(aq) \rightarrow Sn^{4+}(aq) + 2e^- & E^\circ_{red} = 0.154\ V \\ \hline Cu^{2+}(aq) + Sn^{2+}(aq) \rightarrow Cu(s) + Sn^{4+}(aq) & E^\circ = 0.337 - 1.54 = 0.183\ V\text{, spontaneous} \end{array}$$

20.96 (a)

$$\begin{array}{ll} 2[Ag^+(aq) + 1e^- \rightarrow Ag(s)] & E^\circ_{red} = 0.80\ V \\ Ni(s) \rightarrow Ni^{2+}(aq) + 2e^- & E^\circ_{red} = -0.28\ V \\ \hline 2Ag^+(aq) + Ni(s) \rightarrow 2Ag(s) + Ni^{2+}(aq) & E^\circ = 0.80 - (-0.28) = 1.08\ V \end{array}$$

(b) As the reaction proceeds, $Ni^{2+}(aq)$ is produced, so $[Ni^{2+}]$ increases as the cell operates.

(c) $E = E^\circ - \frac{0.0592}{n} \log K;\ 1.12 = 1.08 - \frac{0.0592}{2} \log \frac{[Ni^{2+}]}{[Ag^+]^2}$

$$-\frac{0.04(2)}{0.0592} = \log(0.0100) - \log[Ag^+]^2;\ \log[Ag^+]^2 = \log(0.0100) + \frac{0.04(2)}{0.0592}$$

$\log[Ag^+]^2 = -2.000 + 1.351 = -0.649$; $[Ag^+]^2 = 0.255\ M$; $[Ag^+] = 0.474 = 0.5\ M$ (Strictly speaking, $[E - E^\circ]$ having only one sig fig leads (after several steps) to the answer having only one sig fig. This is not a very precise or useful result.)

20.98 Both E° and K are related to ΔG°. (See Solution 20.51.)

$\Delta G^\circ = -nFE^\circ$; $\Delta G^\circ = -RT \ln K$

$$-nFE^\circ = -RT \ln K,\ E^\circ = \frac{RT}{nF} \ln K$$

In terms of base 10 logs, $\ln K = 2.303 \log K$.

$$E^\circ = \frac{2.303\ RT}{nF} \log K$$

From the development of the Nernst equation,

$$\frac{2.303\ RT}{F} = 0.0592,\ E^\circ = \frac{0.0592}{n} \log K$$

20.99 Use the relationship developed in Solution 20.98 to calculate K from E°. Use data from Appendix E to calculate E° for the disproportionation.

$$Cu^+(aq) + 1e^- \rightarrow Cu(s) \qquad E^\circ_{red} = 0.521\ V$$
$$Cu^+(aq) \rightarrow Cu^{2+}(aq) + 1e^- \qquad E^\circ_{red} = 0.153\ V$$
$$2Cu^+(aq) \rightarrow Cu(s) + Cu^{2+}(aq) \qquad E^\circ = 0.521\ V - 0.153\ V = 0.368\ V$$

$$E^\circ = \frac{0.0592}{n}\log K, \log K = \frac{nE^\circ}{0.0592} = \frac{1 \times 0.368}{0.0592} = 6.216 = 6.22$$

$K = 10^{6.216} = 1.6 \times 10^6$

20.100 (a) False. For standard cell potentials derived from E°_{red} values listed in Appendix E, the maximum E°_{cell} value is 5.92 V. This reaction involves $F_2(g)$, the element with the largest positive E°_{red}, and Li, the element with the most negative E°_{red}. Other species stable in aqueous solution will have E°_{red} values between these two limiting values, so other possible E°_{cell} values would be less than 5.92 V. Aqueous E°_{cell} values are not likely to exceed 5.92 V.

(b) False. The effect of concentration of voltage is given by the Nernst equation; $E = E^\circ - \frac{0.0592}{n}\log Q$, where Q is the reaction quotient which incorporates concentrations. Clearly E is not directly proportional to either individual concentrations or Q.

(c) True. Oxidizing agents are themselves reduced. The strength of an oxidizing agent is measured by the magnitude of its reduction potential, E°_{red}.

20.102 The ship's hull should be made negative. By keeping an excess of electrons in the metal of the ship, the tendency for iron to undergo oxidation, with release of electrons, is diminished. The ship, as a negatively charged "electrode," becomes the site of reduction, rather than oxidation, in an electrolytic process.

20.103 It is well established that corrosion occurs most readily when the metal surface is in contact with water. Thus, moisture is a requirement for corrosion. Corrosion also occurs more readily in acid solution, because O_2 has a more positive reduction potential in the presence of $H^+(aq)$. SO_2 and its oxidation products dissolve in water to produce acidic solutions, which encourage corrosion. The anodic and cathodic reactions for the corrosion of Ni are:

$$Ni(s) \rightarrow Ni^{2+}(aq) + 2e^- \qquad E^\circ_{red} = -0.28\ V$$
$$O_2(g) + 4H^+(aq) + 4e^- \rightarrow 2H_2O(l) \qquad E^\circ_{red} = 1.23\ V$$

Nickel(II) oxide, NiO(s), can form by the dry air oxidation of Ni. This NiO coating serves to protect against further corrosion. However, NiO dissolves in acidic solutions such as those produced by SO_2 or SO_3, according to the reaction:
$NiO(s) + 2H^+(aq) \rightarrow Ni^{2+}(aq) + H_2O(l)$. This exposes Ni(s) to further wet corrosion.

20.104 A battery is a voltaic cell, so the cathode compartment contains the positive terminal of the battery. In the alkaline, Ni–Cd and NiMH batteries, $OH^-(aq)$ is produced at the cathode. The wire that turns the indicator pink is in contact with $OH^-(aq)$, so the rightmost wire is connected to the positive terminal of the battery. [The battery could not be a lead-acid battery because $OH^-(aq)$ is not present in either compartment, so neither wire would turn the indicator pink.]

20.106 (a) The work obtainable is given by the product of the voltage, which has units of J/C, times the number of Coulombs of electricity produced:

$$w_{max} = 300 \text{ amp} \bullet \text{hr} \times \frac{3600 \text{ s}}{1 \text{ hr}} \times \frac{1 \text{ C}}{1 \text{ amp} \bullet \text{s}} \times \frac{6 \text{ J}}{1 \text{ C}} \times \frac{1 \text{kWh}}{3.6 \times 10^6 \text{ J}} = 1.8 \text{ kWh} \approx 2 \text{ kWh}$$

(b) This maximum amount of work is never realized because some of the electrical energy is dissipated in overcoming the internal resistance of the battery, because the cell voltage does not remain constant as the reaction proceeds, and because the systems to which the electrical energy is delivered are not capable of completely converting electrical energy into work.

20.107 (a) $7 \times 10^8 \text{ mol } H_2 \times \frac{2 \text{ F}}{1 \text{ mol } H_2} \times \frac{96{,}500 \text{ C}}{1 \text{ F}} = 1.35 \times 10^{14} = 1 \times 10^{14} \text{ C}$

(b)

$$\begin{array}{ll} 2H_2O(l) \rightarrow O_2(g) + 4H^+(aq) + 4e^- & E^\circ_{red} = 1.23 \text{ V} \\ 2[2H^+(aq) + 2e^- \rightarrow H_2(g)] & E^\circ_{red} = 0 \text{ V} \\ \hline 2H_2O(l) \rightarrow O_2(g) + 2H_2(g) & E^\circ = 0.00 - 1.23 = -1.23 \text{ V} \end{array}$$

$P_t = 300 \text{ atm} = P_{O_2} + P_{H_2}$. Since $H_2(g)$ and $O_2(g)$ are generated in a 2:1 mole ratio, $P_{H_2} = 200 \text{ atm}$ and $P_{O_2} = 100 \text{ atm}$.

$$E = E^\circ - \frac{0.0592}{4} \log (P_{O_2} \times P_{H_2}^2) = -1.23 \text{ V} - \frac{0.0592}{4} \log [100 \times (200)^2]$$

$E = -1.23 \text{ V} - 0.100 \text{ V} = -1.33 \text{ V}; E_{min} = 1.33 \text{ V}$

(c) Energy $= nFE = 2(7 \times 10^8 \text{ mol})(1.33 \text{ V}) \frac{96{,}500 \text{ J}}{\text{V} \bullet \text{mol}} = 1.80 \times 10^{14} = 2 \times 10^{14} \text{ J}$

(d) $1.80 \times 10^{14} \text{ J} \times \frac{1 \text{ kWh}}{3.6 \times 10^6 \text{ J}} \times \frac{\$0.85}{\text{kWh}} = \$4.24 \times 10^7 = \$4 \times 10^7$

It would cost more than $40 million for the electricity alone.

Integrative Exercises

20.108

$$\begin{array}{ll} 2[NO_3^-(aq) + 4H^+(aq) + 3e^- \rightarrow NO(g) + 2H_2O(l)] & E^\circ_{red} = 0.96 \text{ V} \\ 3[Cu(s) \rightarrow Cu^{2+}(aq) + 2e^-] & E^\circ_{red} = 0.34 \text{ V} \\ \hline 3Cu(s) + 2NO_3^-(aq) + 8H^+(aq) \rightarrow 3Cu^{2+}(aq) + 2NO(g) + 4H_2O(l) & \\ & E^\circ = 0.96 - 0.34 = 0.62 \text{ V} \end{array}$$

$$\begin{array}{ll} 2H^+(aq) + 2e^- \rightarrow H_2(g) & E^\circ_{red} = 0 \text{ V} \\ Cu(s) \rightarrow Cu^{2+}(aq) + 2e^- & E^\circ_{red} = 0.34 \text{ V} \\ \hline Cu(s) + 2H^+(aq) \rightarrow Cu^{2+}(aq) + H_2(g) & E^\circ = 0 - 0.34 = -0.34 \text{ V} \end{array}$$

The overall cell potential for the oxidation of Cu(s) by HNO_3 is positive and the reaction is spontaneous. The cell potential for the oxidation of Cu(s) by HCl is negative and the reaction is nonspontaneous. Note that in the reaction with HNO_3, it is NO_3^- that is reduced, not H^+; Cl^- from HCl cannot be further reduced.

20.110 The redox reaction is: $2Ag^+(aq) + H_2(g) \rightarrow 2Ag(s) + 2H^+(aq)$. n = 2 for this reaction.

$$E^\circ_{cell} = E^\circ_{red}\ \text{cathode} - E^\circ_{red}\ \text{anode} = 0.799\ V - 0\ V = 0.799$$

$$E = E^\circ - \frac{0.0592}{n}\log\frac{[H^+]^2}{[Ag^+]^2\, P_{H_2}}$$

$[H^+]$ in the cell is held essentially constant by the benzoate buffer.

$$C_6H_5COOH(aq) \rightleftharpoons H^+(aq) + C_6H_5COO^-(aq) \quad K_a = ?$$

$$K_a = \frac{[H^+][C_6H_5COO^-]}{[C_6H_5COOH]};\ [H^+] = \frac{K_a[C_6H_5COOH]}{[C_6H_5COO^-]} = \frac{0.10\ M}{0.050\ M} \times K_a = 2K_a$$

Solve the Nernst expression for $[H^+]$ and calculate K_a and pK_a as shown above.

$$1.030\ V = 0.799\ V - \frac{0.0592}{n}\log\frac{[H^+]^2}{(1.00)^2(1.00)}$$

$$0.231 \times \frac{2}{0.0592} = -\log[H^+]^2 = -2\log[H^+]$$

$$\frac{0.231}{0.0592} = -\log[H^+] = pH;\ pH = 3.9020 = 3.90;\ [H^+] = 10^{-3.902} = 1.253 \times 10^{-4} = 1.3 \times 10^{-4};$$

$[H^+] = 2K_a$, $K_a = [H^+]/2 = 6.265 \times 10^{-5} = 6.3 \times 10^{-5}$; $pK_a = 4.20$

Check. According to Appendix D, K_a for benzoic acid is 6.3×10^{-5}.

20.111 (a) The oxidation potential of A is equal in magnitude but opposite in sign to the reduction potential of A^+.

(b) Li(s) has the highest oxidation potential, Au(s) the lowest.

(c) The relationship is reasonable because both oxidation potential and ionization energy describe removing electrons from a substance. Ionization energy is a property of gas phase atoms or ions, while oxidation potential is a property of the bulk material.

20.112 (a)

$$NO_3^-(aq) + 4H^+(aq) + 3e^- \rightarrow NO(g) + 2H_2O(l) \quad E^\circ_{red} = 0.96\ V$$
$$Au(s) \rightarrow Au^{3+}(aq) + 3e^- \quad E^\circ_{red} = 1.498\ V$$
$$Au(s) + NO_3^-(aq) + 4H^+(aq) \rightarrow Au^{3+}(aq) + NO(g) + 2H_2O(l)$$

$E^\circ = 0.96 - 1.498 = -0.54$ V; E° is negative, the reaction is not spontaneous.

(b)

$$3[2H^+(aq) + 2e^- \rightarrow H_2(g)] \quad E^\circ_{red} = 0.000\ V$$
$$2[Au(s) + 4Cl^-(aq) \rightarrow AuCl_4^-(aq) + 3e^-] \quad E^\circ_{red} = 1.002\ V$$
$$2Au(s) + 6H^+(aq) + 8Cl^-(aq) \rightarrow 2AuCl_4^-(aq) + 3H_2(g)$$

$E^\circ = 0.000 - 1.002 = -1.002$ V; E° is negative, the reaction is not spontaneous.

(c)

$$NO_3^-(aq) + 4H^+(aq) + 3e^- \rightarrow NO(g) + 2H_2O(l) \quad E^\circ_{red} = 0.96\ V$$
$$Au(s) + 4Cl^-(aq) \rightarrow AuCl_4^-(aq) + 3e^- \quad E^\circ_{red} = 1.002\ V$$
$$Au(s) + NO_3^-(aq) + 4Cl^-(aq) + 4H^+(aq) \rightarrow AuCl_4^-(aq) + NO(g) + 2H_2O(l)$$

$E^\circ = 0.96 - 1.002 = -0.04$; E° is small but negative, the process is not spontaneous.

(d) $E = E° - \frac{0.0592}{3} \log \frac{[AuCl_4^-] P_{NO}}{[NO_3^-][Cl^-]^4[H^+]^4}$

If $[H^+]$, $[Cl^-]$ and $[NO_3^-]$ are much greater than 1.0 *M*, the log term is negative and the correction to E° is positive. If the correction term is greater than 0.042 V, the value of E is positive and the reaction at nonstandard conditions is spontaneous.

20.114 (a) $\Delta H° = 2\Delta H°\ H_2O(l) - 2\Delta H°\ H_2(g) - \Delta H°\ O_2(g) = 2(-285.83) - 2(0) - 0 = -571.66$ kJ

$\Delta S° = 2S°\ H_2O(l) - 2S°\ H_2(g) - \Delta S°\ O_2(g)$

$= 2(69.91) - 2(130.58) - (205.0) = -326.34$ J

(b) Since $\Delta S°$ is negative, $-T\Delta S$ is positive and the value of ΔG will become more positive as T increases. The reaction will become nonspontaneous at a fairly low temperature, because the magnitude of $\Delta S°$ is large.

(c) $\Delta G = w_{max}$. The larger the negative value of ΔG, the more work the system is capable of doing on the surroundings. As the magnitude of ΔG decreases with increasing temperature, the usefulness of H_2 as a fuel decreases.

(d) The combustion method increases the temperature of the system, which quickly decreases the magnitude of the work that can be done by the system. Even if the effect of temperature on this reaction could be controlled, only about 40% of the energy from any combustion can be converted to electrical energy, so combustion is intrinsically less efficient than direct production of electrical energy via a fuel cell.

20.115 First balance the equation:

$4CyFe^{2+}(aq) + O_2(g) + 4H^+(aq) \rightarrow 4CyFe^{3+}(aq) + 2H_2O(l)$; E = +0.60 V; n = 4

(a) From Equation [20.11] we can calculate ΔG for the process under the conditions specified for the measured potential E:

$$\Delta G = -nFE = -(4 \text{ mol } e^-) \times \frac{96.5 \text{ kJ}}{1 \text{ V} \bullet \text{mol } e^-}(0.60 \text{ V}) = -231.6 = -232 \text{ kJ}$$

(b) The moles of ATP synthesized per mole of O_2 is given by:

$$\frac{231.6 \text{ kJ}}{O_2 \text{ molecule}} \times \frac{1 \text{ mol ATP formed}}{37.7 \text{ kJ}} = \text{approximately 6 mol ATP/mol } O_2$$

20.116 $AgSCN(s) + e^- \rightarrow Ag(s) + SCN^-(aq)$ $E°_{red} = 0.0895$ V

$Ag(s) \rightarrow Ag^+(aq) + e^-$ $E°_{red} = 0.799$ V

$AgSCN(s) \rightarrow Ag^+(aq) + SCN^-(aq)$ $E° = 0.0895 - 0.799 = -0.710$ V

$$E° = \frac{0.0592}{n} \log K_{sp};\ \log K_{sp} = \frac{(-0.710)(1)}{0.0592} = -11.993 = -12.0$$

$K_{sp} = 10^{-11.993} = 1.02 \times 10^{-12} = 1 \times 10^{-12}$

20.117 The reaction can be written as a sum of the steps:

$$\begin{array}{ll} Pb^{2+}(aq) + 2e^- \rightarrow Pb(s) & E^\circ_{red} = -0.126\ V \\ PbS(s) \rightarrow Pb^{2+}(aq) + S^{2-}(aq) & \text{"}E^\circ\text{"} = ? \\ \hline PbS(s) + 2e^- \rightarrow Pb(s) + S^{2-}(aq) & E^\circ_{red} = ? \end{array}$$

"E°" for the second step can be calculated from K_{sp}.

$$E^\circ = \frac{0.0592}{n}\log K_{sp} = \frac{0.0592}{2}\log(8.0 \times 10^{-28}) = \frac{0.0592}{2}(-27.10) = -0.802\ V$$

E° for the half-reaction = −0.126 V + (−0.802 V) = −0.928 V

Calculating an imaginary E° for a nonredox process like step 2 may be a disturbing idea. Alternatively, one could calculate K for step 1 (5.4×10^{-5}), K for the reaction in question ($K = K_1 \times K_{sp} = 4.4 \times 10^{-32}$) and then E° for the half-reaction. The result is the same.

20.119 The two half-reactions in the electrolysis of $H_2O(l)$ are:

$$\begin{array}{l} 2[2H_2O(l) + 2e^- \rightarrow H_2(g) + 2OH^-] \\ 2H_2O(l) \rightarrow O_2(g) + 4H^+ + 4e^- \\ \hline 2H_2O(l) \rightarrow 2H_2(g) + O_2(g) \end{array}$$

4 mol e^-/2 mol $H_2(g)$ or 2 mol e^-/mol $H_2(g)$

Using partial pressures and the ideal-gas law, calculate the mol $H_2(g)$ produced, and the current required to do so.

$P_t = P_{H_2} + P_{H_2O}$. From Appendix B, P_{H_2O} at 25.5°C is approximately 24.5 torr.

$$P_{H_2} = 768\ \text{torr} - 24.5\ \text{torr} = 743.5 = 744\ \text{torr}$$

$$n = PV/RT = \frac{(743.5/760)\ \text{atm} \times 0.0123\ \text{L}}{298.5\ \text{K} \times 0.08206\ \text{L} \bullet \text{atm/mol} \bullet \text{K}} = 4.912 \times 10^{-4} = 4.91 \times 10^{-4}\ \text{mol}\ H_2$$

$$4.912 \times 10^{-4}\ \text{mol}\ H_2 \times \frac{2\ \text{mol}\ e^-}{\text{mol}\ H_2} \times \frac{96{,}500\ \text{C}}{1\ \text{mol}\ e^-} \times \frac{1\ \text{amp} \bullet \text{s}}{1\ \text{C}} \times \frac{1\ \text{min}}{60\ \text{s}} \times \frac{1}{2.00\ \text{min}} = 0.790\ \text{amp}$$

21 Nuclear Chemistry

Nuclear Chemistry

21.2 *Analyze/Plan.* From the diagram, determine the atomic number (number of protons) and mass number (number of protons plus neutrons) of the two nuclides involved. Based on the relationship between the two nuclides, decide whether the reaction is α or β-decay, positron emission or electron capture. Complete the nuclear reaction, balancing atomic numbers and mass numbers.

Solve. The two nuclides in the diagram are $^{109}_{46}Pd$ and $^{109}_{47}Ag$ so the second product is a β-particle. The balanced reaction is:

$^{109}_{46}Pd \rightarrow ^{109}_{47}Ag + ^{0}_{-1}e$

Check. Atomic number and mass number balance.

21.3 *Analyze/Plan.* Determine the number of protons and neutrons present in the two heavy nuclides in the reaction. Draw a graph with appropriate limits and plot the two points. Draw an arrow from reactant to product. *Solve.*

Bi: 83 p, 128 n; Tl: 81 p, 126 n.

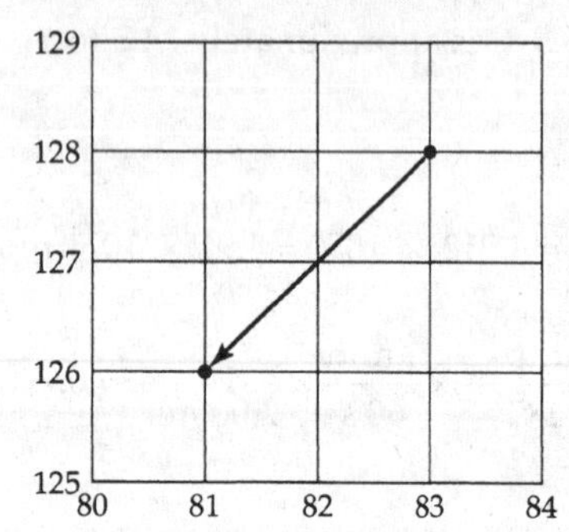

Check. An α-particle has 2 p and 2 n. The diagram shows a decrease in 2 p and 2 n for the reaction.

21.5 (a) The difference in mass between a nuclide and its component nucleons is the mass defect. It corresponds to the energy required to separate the nuclide into individual nucleons, according to the relationship $E = \Delta mc^2$.

(b) On Figure 21.13, ^{56}Fe has the largest binding energy per nucleon. ^{100}Ru is near this maximum and has more total nucleons than ^{56}Fe, so we expect the mass defect for ^{100}Ru to be significant.

(c) *Plan.* Calculate the total mass of the separate nucleons. Subtract the mass of the nuclide to get Δm, the mass defect. Convert Δm to binding energy, divide by 100 to get the binding energy per nucleon.

Δm = mass of individual protons and neutrons – mass of nuclide

$\Delta m = 44(1.0072765 \text{ amu}) + 56(1.0086649 \text{ amu}) - 99.90422 \text{ amu}$

$\Delta m = 0.9011804 = 0.90118 \text{ amu}$

$$\Delta E = (2.9979246 \times 10^8 \text{ m/s})^2 \times 0.9011804 \text{ amu} \times \frac{1 \text{ g}}{6.022 \times 10^{23} \text{ amu}} \times \frac{1 \text{ kg}}{1000 \text{ g}} = 1.34497 \times 10^{-10} \text{ J}$$

$$1.34497 \times 10^{-10} \frac{\text{J}}{\text{nuclide}} \times \frac{1 \text{ nuclide}}{100 \text{ nucleons}} = 1.34497 \times 10^{-12} \text{ J/nucleon}$$

21.6 *Analyze/Plan.* Express the particles in the diagram as a nuclear reaction. Determine the mass number and atomic number of the unknown particle by balancing these quantities in the nuclear reaction. *Solve.*

(a) $^{239}_{94}\text{Pu} + ^{1}_{0}\text{n} \rightarrow ^{95}_{40}\text{Zr} + ? + 2\,^{1}_{0}\text{n}$

The unknown particle has an atomic number of (94–40) = 54; it is Xe. The mass number of the nuclide is [(239 + 1) – (95 + 2)] = 143. The unknown particle is $^{143}_{54}\text{Xe}$.

(b) ^{95}Zr: 40 p, 55 n is stable. ^{143}Xe: 54 p, 89 n is above the belt of stability and is not stable; it will probably undergo β-decay.

Radioactivity

21.8 p = protons, n = neutrons, e = electrons; number of protons = atomic number; number of neutrons = mass number – atomic number

(a) $^{126}_{55}\text{Cs}$: 55p, 71n (b) ^{119}Sn: 50p, 69n (c) ^{141}Ba: 56p, 85n

21.10 (a) $^{1}_{0}\text{n}$ (b) $^{0}_{-1}\text{e}$ or $^{0}_{-1}\beta$ (c) $^{4}_{2}\text{He}$ or $^{4}_{2}\alpha$

21.12 (a) $^{141}_{60}\text{Nd} + ^{0}_{-1}\text{e}$ (orbital electron) $\rightarrow ^{141}_{59}\text{Pr}$ (b) $^{201}_{79}\text{Au} \rightarrow ^{201}_{80}\text{Hg} + ^{0}_{-1}\beta$

(c) $^{81}_{34}\text{Se} \rightarrow ^{81}_{35}\text{Br} + ^{0}_{-1}\beta$ (d) $^{83}_{38}\text{Sr} \rightarrow ^{83}_{37}\text{Rb} + ^{0}_{1}\text{e}$

21.14 (a) $^{24}_{11}\text{Na} \rightarrow ^{24}_{12}\text{Mg} + ^{0}_{-1}\text{e}$; a β particle is produced

(b) $^{188}_{80}\text{Hg} \rightarrow ^{188}_{79}\text{Au} + ^{0}_{1}\text{e}$; a positron is produced

(c) $^{122}_{53}\text{I} \rightarrow ^{122}_{54}\text{Xe} + ^{0}_{-1}\text{e}$; a β particle is produced

(d) $^{242}_{94}\text{Pu} \rightarrow ^{238}_{92}\text{U} + ^{4}_{2}\text{He}$; an α particle is produced

21.16 This decay series represents a change of (232–208 =) 24 mass units. Since only alpha emissions change the nuclear mass, and each changes the mass by four, there must be a total of 6 α emissions. Each alpha emission causes a decrease of two in atomic number. Therefore, the 6 alpha emissions, by themselves, would cause a decrease in atomic number of 12. The series as a whole involves a decrease of 8 in atomic number. Thus, there must be a total of 4 β emissions, each of which increases atomic number by one. Overall, there are 6 α emissions and 4 β emissions.

Nuclear Stability

21.18 (a) $^{66}_{32}Ge$ - low neutron/proton ratio, positron emission

(b) $^{105}_{45}Rh$ - high neutron/proton ratio, beta emission

(c) $^{137}_{53}I$ - high neutron/proton ratio, beta emission

(d) $^{133}_{58}Ce$ - low neutron/proton ratio, positron emission

21.20 Use criteria listed in Table 21.3.

(a) $^{112}_{48}Cd$ even, even more abundant

(b) $^{27}_{13}Al$ odd proton, even neutron more abundant

(c) $^{106}_{46}Pd$ even, even more abundant

(d) $^{128}_{54}Xe$ even proton, even neutron much more abundant than odd proton; odd neutron

21.22 $^{112}_{50}Sn$ has a magic number of protons and even numbers of protons and neutrons, good indications of nuclear stability. $^{112}_{49}In$ has no magic numbers and odd numbers of protons and neutrons, indicators of nuclear instability or radioactivity.

21.24 The criterion employed in judging whether the nucleus is likely to be radioactive is the position of the nucleus on the plot shown in Figure 21.2. If the neutron/proton ratio is too high or low, or if the atomic number exceeds 83, the nucleus will be radioactive.

Radioactive: $^{58}_{29}Cu$ – odd proton, odd neutron, low neutron/proton ratio

^{206}Po – high atomic number

Stable: $^{62}_{28}Ni$ – even proton, even neutron, stable neutron/proton ratio

$^{108}_{47}Ag$ – stable neutron/proton ratio, (one of 5 stable odd proton/odd neutron nuclides)

^{184}W – even proton, even neutron, stable neutron/proton ratio

Nuclear Transmutations

21.26 A major difference is that the charge on the nitrogen nucleus, +7, is much smaller than on the gold nucleus, +79. Thus, the alpha particle could more easily penetrate the coulomb barrier (that is, the repulsive energy barrier due to like charges) to make contact with the nitrogen nucleus than the gold nucleus. Rutherford used alpha particles that were being emitted from some radioactive source. He did not have access to machines that can accelerate particles to very high energy. It would be necessary to do just that to observe reaction of an alpha particle with a gold nucleus.

21.28 (a) $^{252}_{98}Cf + ^{10}_{5}B \rightarrow 3\ ^{1}_{0}n + ^{259}_{103}Lr$

(b) $^{2}_{1}H + ^{3}_{2}He \rightarrow ^{4}_{2}He + ^{1}_{1}H$

(c) $^{1}_{1}H + ^{11}_{5}B \rightarrow 3\ ^{4}_{2}He$

(d) $^{122}_{53}I \rightarrow ^{122}_{54}Xe + ^{0}_{-1}e$

(e) $^{59}_{26}Fe \rightarrow ^{0}_{-1}e + ^{59}_{27}Co$

21.30 (a) ${}^{238}_{92}U + {}^{4}_{2}He \rightarrow {}^{241}_{94}Pu + {}^{1}_{0}n$ (b) ${}^{14}_{7}N + {}^{4}_{2}He \rightarrow {}^{17}_{8}O + {}^{1}_{1}H$

(c) ${}^{56}_{26}Fe + {}^{4}_{2}He \rightarrow {}^{60}_{29}Cu + {}^{0}_{-1}e$

Rates of Radioactive Decay

21.32 The suggestion is not reasonable. The energies of nuclear states are very large relative to ordinary temperatures. Thus, merely changing the temperature by less than 100 K would not be expected to significantly affect the behavior of nuclei with regard to nuclear decay rates.

21.34 Calculate the decay constant, k, and then $t_{1/2}$.

$$k = \frac{-1}{t} \ln \frac{N_t}{N_o} = \frac{-1}{5.2 \text{ min}} \ln \frac{0.250 \text{ g}}{1.000 \text{ g}} = 0.2666 = 0.27 \text{ min}^{-1}$$

Using Equation [21.20], $t_{1/2} = 0.693/k = 0.693/0.02666 \text{ min}^{-1} = 2.599 = 2.6$ min

21.36 $k = 0.693 / t_{1/2} = 0.693/5.26 \text{ yr} = 0.1317 = 0.132 \text{ yr}^{-1}$; $N_t/N_o = 0.75$

$$t = \frac{-1}{k} \ln \frac{N_t}{N_o} = -(1/0.1317 \text{ yr}^{-1}) \ln (0.75) = 2.18 \text{ yr}$$

2.18 yr = 26.2 mo = 797 d. The source would have been replaced sometime in the fall of 2007, probably in October.

21.38 (a) Proceeding as in Solution 21.37, calculate k in s^{-1}.

$$5.26 \text{ yr} \times \frac{365 \text{ d}}{1 \text{ yr}} \times \frac{24 \text{ hr}}{1 \text{ d}} \times \frac{3600 \text{ sec}}{1 \text{ hr}} = 1.659 \times 10^8 = 1.66 \times 10^8 \text{ s}$$

$$k = \frac{0.693}{t_{1/2}} = \frac{0.693}{1.659 \times 10^8} = 4.178 \times 10^{-9} = 4.18 \times 10^{-9} \text{ s}^{-1}$$

$$\ln \frac{N_t}{N_o} = -kt = -(4.178 \times 10^{-9} \text{ s}^{-1})(45.5 \text{ s}) = -1.901 \times 10^{-7} = -1.90 \times 10^{-7}$$

$$\frac{N_t}{N_o} = e^{-1.90 \times 10^{-7}} = (1.000 - 1.90 \times 10^{-7}); N_t = 2.44 \times 10^{-3} \text{ g} (1.000 - 1.90 \times 10^{-7})$$

[$e^{-1.90 \times 10^{-7}}$ is a number very close to 1. In this calculation, it is conveneint to express the number as $(1 - 1.90 \times 10^{-7})$].

The amount that decays is $N_o - N_t$:

$$2.44 \times 10^{-3} \text{ g} - [2.44 \times 10^{-3} \text{ g} (1.000 - 1.90 \times 10^{-7})] = 2.44 \times 10^{-3} \text{ g} (1.90 \times 10^{-7})$$

$$= 4.638 \times 10^{-10} = \sim 5 \times 10^{-10} \text{ g Co}$$

(In terms of sig figs, $[N_o - N_t]$ is a very small number, found by subtracting two numbers known only to three sig figs and two decimal place. At best, we can express the result as an order of magnitude.)

$$N_o - N_t = 4.638 \times 10^{-10}\ g\ Co \times \frac{1\ mol\ Co}{60\ g\ Co} \times \frac{6.022 \times 10^{23}\ Co\ atoms}{1\ mol\ Co} \times \frac{1\beta}{1\ Co\ atom}$$

$= 4.655 \times 10^{12} = \sim 5 \times 10^{12}$

Between 10^{11} and 10^{12} β particles are emitted in 45.5 seconds

(b) $\frac{4.655 \times 10^{12}\ dis}{45.5\ s} \times \frac{1\ Bq}{1\ dis/s} = 1.02 \times 10^{11} = \sim 10^{11}\ Bq$

The activity of the sample is approximately 10^{11} Bq.

21.40 Calculate k in yr^{-1} and solve Equation 21.19 for t. N_o = 15.2/min/g, N_t = 8.9/min/g
$k = 0.693/t_{1/2} = 0.693/5715\ yr = 1.213 \times 10^{-4} = 1.21 \times 10^{-4}\ yr^{-1}$

$$t = \frac{-1}{k} \ln \frac{N_t}{N_o} = \frac{-1}{1.213 \times 10^{-4}\ yr^{-1}} \ln \frac{8.9}{15.2} = 4.414 \times 10^3 = 4.4 \times 10^3\ yr$$

21.42 $k = 0.693/1.27 \times 10^9\ yr = 5.457 \times 10^{-10} = 5.46 \times 10^{-10}\ yr^{-1}$

If the mass of ^{40}Ar is 3.6 times that of ^{40}K, then the original mass of ^{40}K must have been 3.6 + 1 = 4.6 times that now present.

$$t = \frac{-1}{5.457 \times 10^{-10}\ yr^{-1}} \times \ln \frac{1}{(4.6)} = 2.8 \times 10^9\ yr$$

Energy Changes

21.44 $\Delta E = c^2 \Delta m = (3.0 \times 10^8\ m/s)^2 \times 0.1\ mg \times \frac{1\ g}{1000\ mg} \times \frac{1\ kg}{1000\ g} \times \frac{1\ kJ}{1000\ J} = 9 \times 10^6\ kJ$

21.46 Δm = mass of individual protons and neutrons – mass of nucleus

Δm = 10(1.0072765 amu) + 11(1.0086649 amu) – 20.98846 amu = 0.1796189 = 0.17962 amu

$$\Delta E = (2.9979246 \times 10^8\ m/s)^2 \times 0.1796189\ amu \times \frac{1\ g}{6.0221421 \times 10^{23}\ amu} \times \frac{1\ kg}{1000\ g}$$

$= 2.680664 \times 10^{-11} = 2.6807 \times 10^{-11}\ J/^{21}Ne$ nucleus required

$$2.680664 \times 10^{-11} \frac{J}{nucleus} \times \frac{6.0221421 \times 10^{23}\ nuclei}{mol}$$

$= 1.6143 \times 10^{13}\ J/mol\ ^{21}Ne$ binding energy

21.48 In each case, calculate the mass defect, total nuclear binding energy and then binding energy per nucleon.

(a) Δm = 7(1.0072765) + 7(1.0086649) – 13.999234 = 0.1123558 = 0.112356 amu

$$\Delta E = 0.1123558\ amu \times \frac{1\ g}{6.0221421 \times 10^{23}\ amu} \times \frac{1\ kg}{1000\ g} \times \frac{8.987551 \times 10^{16}\ m^2}{s^2}$$

$= 1.676817 \times 10^{-11} = 1.67682 \times 10^{-11}\ J$

binding energy/nucleon = $1.676817 \times 10^{-11}\ J / 14 = 1.19773 \times 10^{-12}$ J/nucleon

(b) $\Delta m = 22(1.0072765) + 26(1.0086649) - 47.935878 = 0.4494924 = 0.449492$ amu

$$\Delta E = 0.4494924 \text{ amu} \times \frac{1 \text{ g}}{6.0221421 \times 10^{23} \text{ amu}} \times \frac{1 \text{ kg}}{1000 \text{ g}} \times \frac{8.987551 \times 10^{16} \text{ m}^2}{\text{s}^2}$$

$$= 6.708304 \times 10^{-11} = 6.70830 \times 10^{-11} \text{ J}$$

binding energy/nucleon = 6.708304×10^{-11} J / 48 = 1.39756×10^{-12} J/nucleon

(c) Calculate the nuclear mass by subtracting the electron mass from the atomic mass. $200.970277 - 80(5.485799 \times 10^{-4} \text{ amu}) = 200.926391$ amu

$\Delta m = 80(1.0072765) + 121(1.0086649) - 200.926391 = 1.7041819 = 1.704182$ amu

$$\Delta E = 1.7041819 \text{ amu} \times \frac{1 \text{ g}}{6.0221421 \times 10^{23} \text{ amu}} \times \frac{1 \text{ kg}}{1000 \text{ g}} \times \frac{8.987551 \times 10^{16} \text{ m}^2}{\text{s}^2}$$

$$= 2.543351 \times 10^{-10} \text{ J}$$

binding energy/nucleon = 2.543351×10^{-10} J/201 = 1.265348×10^{-12} J/nucleon

21.50 The calculated Δm is for one group of single nuclides involved in a reaction, labeled ΔM/'atomic reaction'. Multiplying by Avogadro's number changes the quantity to 'mol of reaction'. Since energy is released, the sign of ΔE is negative.

(a) $\Delta m = 4.00260 + 1.0086649 - 3.01605 - 2.01410 = -0.0188851 = -0.01889$ amu

$$\Delta E = \frac{-0.0188851 \text{ amu}}{\text{'atomic reaction'}} \times \frac{1 \text{ g}}{6.022 \times 10^{23} \text{ amu}} \times \frac{6.022 \times 10^{23} \text{ 'atomic reaction'}}{\text{mol of reaction}}$$

$$\times \frac{1 \text{ kg}}{10^3 \text{ g}} \times (2.99792458 \times 10^8 \text{ m/sec})^2 = -1.697 \times 10^{12} \text{ J/mol}$$

(b) $\Delta m = 3.01605 + 1.0086649 - 2(2.01410) = -3.4851 \times 10^{-3} = -3.49 \times 10^{-3}$ amu

$\Delta E = -3.13 \times 10^{11}$ J/mol

(c) $\Delta m = 4.00260 + 1.00782 - 3.01605 - 2.01410 = -1.973 \times 10^{-2}$ amu

$\Delta E = -1.773 \times 10^{12}$ J/mol

21.52 In a fission reactor absorption of neutrons causes a single heavy nucleus to undergo fission, producing two medium mass nuclei. These are seen in Figure 21.13 to have a larger total mass defect than the starting single heavier nucleus, so energy is released.

Effects and Uses of Radioisotopes

21.54 (a) Add ^{36}Cl to water as a chloride salt. Then dissolve ordinary CCl_3COOH. After a time, distill the volatile materials away from the salt; CCl_3COOH is volatile, and will distill with water. Count radioactivity in the volatile material. If chlorine exchange has occurred, there will be radioactivity.

(b) Prepare a saturated solution of $BaCl_2$ containing a small amount of solid $BaCl_2$. Add to this solution solid $BaCl_2$ containing ^{36}Cl. If the solid-solution equilibrium is dynamic, some of the ^{36}Cl in the solid will find itself in solution as chloride ion. After allowing some time for equilibrium to become established, filter the solution, measure radioactivity in the solution that is separated from the solid. If there were no dynamic equilibrium, the $^{36}Cl^-$ would remain in the added solid, since the solution is already saturated before the addition of more solid.

(c) Utilize ^{36}Cl in soils of various pH values; grow plants for a given period of time. Remove plants, and directly measure radioactivity in samples from stems, leaves, and so forth, or reduce the volume of plant sample by some form of digestion and evaporation of solution to give a dry residue that can be counted.

21.56 (a) In a chain reaction, one neutron initiates a nuclear transformation that produces more than one neutron. The product neutrons initiate more transformations, so that the reaction is self-sustaining.

(b) Critical mass is the mass of fissionable material required to sustain a chain reaction so that only one product neutron is effective at initiating a new transformation.

21.58 (a) $^{2}_{1}H + ^{2}_{1}H \rightarrow ^{3}_{2}He + 4\ ^{1}_{0}n$

(b) $^{233}_{92}U + ^{1}_{0}n \rightarrow ^{133}_{51}Sb + ^{98}_{41}Nb + 3\ ^{1}_{0}n$

21.60 (a) If the spent fuel rods are more radioactive than the original rods, the products of fission must lie outside the belt of stability and be radioactive themselves.

(b) The heavy (Z > 83) nucleus has a high neutron/proton ratio. The lighter radioactive fission products, (e.g., barium-142 and krypton-91) also have high neutron/proton ratios, since only 2 or 3 free neutrons are produced during fission. The preferred decay mode to reduce the neutron/proton ratio is β decay, which has the effect of converting a neutron into a proton. Both barium-142 (86 n, 56 p) and krypton-91 (55 n, 36 p) undergo β decay.

21.62 $[H{-}\dot{O}{\cdot}{-}H]^+ + H{-}\ddot{O}{:}{-}H \longrightarrow [H{-}\ddot{O}{-}H(H)]^+ + {\cdot}\ddot{O}{-}H$

H_2O^+ is a free radical because it contains seven valence electrons. Around the central O atom there are two bonding and one nonbonding electron pairs and a single unpaired electron (3(2) + 1 = 7 valence electrons).

21.64 (a) 1 Ci = 3.7×10^{10} dis/s; 1 Bq = 1 dis/s

$$21\text{ mCi} \times \frac{1\text{ Ci}}{1000\text{ mCi}} \times 3.7 \times 10^{10}\text{ dis/s} = 7.77 \times 10^8 = 7.8 \times 10^8\text{ dis/s} = 7.8 \times 10^8\text{ Bq}$$

(b) 1 Gy = 1 J/kg; 1 Gy = 100 rad

$$7.77 \times 10^8\text{ dis/s} \times 116\text{ s} \times 0.065 \times \frac{8.75 \times 10^{-14}\text{ J}}{\text{dis}} \times \frac{1}{65\text{ kg}} = 7.887 \times 10^{-6} = 7.9 \times 10^{-6}\text{ J/kg}$$

$$7.9 \times 10^{-6}\text{ J/kg} \times \frac{1\text{ Gy}}{1\text{ J/kg}} = 7.9 \times 10^{-6}\text{ Gy};\ 7.9 \times 10^{-6}\text{ Gy} \times \frac{100\text{ rad}}{1\text{Gy}} = 7.9 \times 10^{-4}\text{ rad}$$

(c) rem = rad (RBE); Sv = Gy (RBE)

$$7.9 \times 10^{-4} \text{ rad } (1.0) = 7.9 \times 10^{-4} \text{ rem} \times \frac{1000 \text{ mrem}}{1 \text{ rem}} = 0.79 \text{ mrem}$$

$$7.9 \times 10^{-6} \text{ Gy } (1.0) = 7.9 \times 10^{-6} \text{ Sv}$$

(d) From Figure 21.23, the average annual background radiation is 360 mrem, or about 1 mrem/day. This 0.79 mrem exposure is less than the average background radiation for a day.

Additional Exercises

21.66 (a) $^{1}_{0}\text{n} \rightarrow {}^{1}_{1}\text{p} + {}^{0}_{-1}\text{e}$ (or $^{0}_{-1}\beta$)

The other product of neutron decay is a β particle (with the mass and charge of an electron).

(b) Neutrons in atomic nuclei do not decay at this rate because they are stabilized by the strong forces among subatomic particles in a nucleus. Evidence for strong forces in the nucleus includes nuclear binding energies and the coexistence of like-charged protons in the very small volume of the nucleus.

21.68 (a) $^{36}_{17}\text{Cl} \rightarrow {}^{36}_{18}\text{Ar} + {}^{0}_{-1}\text{e}$

(b) According to Table 21.3, nuclei with even numbers of both protons and neutrons, or an even number of one kind of nucleon, are more stable. ^{35}Cl and ^{37}Cl both have an odd number of protons **but** an even number of neutrons. ^{36}Cl has an odd number of protons and neutrons (17 p, 19 n), so it is less stable than the other two isotopes. Also, ^{37}Cl has 20 neutrons, a nuclear closed shell.

21.69 (a) $^{6}_{3}\text{Li} \rightarrow {}^{56}_{28}\text{Ni} + {}^{62}_{31}\text{Ga}$

(b) $^{40}_{20}\text{Ca} + {}^{248}_{96}\text{Cm} \rightarrow {}^{147}_{62}\text{Sm} + {}^{141}_{54}\text{Xe}$

(c) $^{88}_{38}\text{Sr} + {}^{84}_{36}\text{Kr} \rightarrow {}^{116}_{46}\text{Pd} + {}^{56}_{28}\text{Ni}$

(d) $^{40}_{20}\text{Ca} + {}^{238}_{92}\text{U} \rightarrow {}^{70}_{30}\text{Zn} + 4\,{}^{1}_{0}\text{n} + 2\,{}^{102}_{41}\text{Nb}$

21.72 First calculate k in s^{-1}

$$k = \frac{0.693}{2.4 \times 10^{4} \text{ yr}} \times \frac{1 \text{ yr}}{365 \times 24 \times 3600 \text{ s}} = 9.16 \times 10^{-13} = 9.2 \times 10^{-13} \text{ s}^{-1}$$

Now calculate N:

$$N = 0.173 \text{ g Pu} \times \frac{1 \text{ mol Pu}}{239 \text{ g Pu}} \times \frac{6.022 \times 10^{23} \text{ Pu atoms}}{1 \text{ mol Pu}} = 4.36 \times 10^{20} = 4.4 \times 10^{20} \text{ Pu atoms}$$

rate = $(9.16 \times 10^{-13} \text{ s}^{-1})(4.36 \times 10^{20} \text{ Pu atoms}) = 3.99 \times 10^{8} = 4.0 \times 10^{8}$ dis/s

21.73 The C—OH bond of the acid and the O—H bond of the alcohol break in this reaction. Initially, ^{18}O is present in the C—^{18}OH group of the alcohol. In order for ^{18}O to end up in the ester, the ^{18}O—H bond of the alcohol must break. This requires that the C—OH bond in the acid also breaks. The unlabeled O from the acid ends up in the H_2O product.

21.75 First, calculate k in s^{-1}

$$k = \frac{0.693}{12.3 \text{ yr}} \times \frac{1 \text{ yr}}{365 \text{ d}} \times \frac{1 \text{ d}}{24 \text{ hr}} \times \frac{1 \text{ hr}}{3600 \text{ sec}} = 1.7866 \times 10^{-9} = 1.79 \times 10^{-9} \text{ s}^{-1}$$

From Equation [21.18], $1.50 \times 10^3 \text{ s}^{-1} = (1.7866 \times 10^{-9} \text{ s}^{-1})(N)$;

$N = 8.396 \times 10^{11} = 8.40 \times 10^{11}$. In 26.00 g of water, there are

$$26.00 \text{ g } H_2O \times \frac{1 \text{ mol } H_2O}{18.02 \text{ g } H_2O} \times \frac{6.022 \times 10^{23} \text{ } H_2O}{1 \text{ mol } H_2O} \times \frac{2 \text{ H}}{1 H_2O} = 1.738 \times 10^{24} \text{ H atoms}$$

The mole fraction of ${}^3_1H$ atoms in the sample is thus

$8.396 \times 10^{11} / 1.738 \times 10^{24} = 4.831 \times 10^{-13} = 4.83 \times 10^{-13}$

21.77 (a) $$\Delta m = \Delta E/c^2;\ \Delta m = \frac{3.9 \times 10^{26} \text{ J/s}}{(3.00 \times 10^8 \text{ m/s})^2} \times \frac{1 \text{ kg} \bullet \text{m}^2/\text{s}^2}{1 \text{ J}} = 4.3 \times 10^9 \text{ kg/s}$$

The rate of mass loss is 4.3×10^9 kg/s.

(b) The mass loss arises from fusion reactions that produce more stable nuclei from less stable ones, e.g., Equations [21.26-21.29].

21.78 $$1000 \text{ Mwatts} \times \frac{1 \times 10^6 \text{ watts}}{1 \text{ Mwatt}} \times \frac{1 \text{ J}}{1 \text{ watt} \bullet \text{s}} \times \frac{1 \text{ } ^{235}\text{U atom}}{3 \times 10^{-11} \text{ J}} \times \frac{1 \text{ mol U}}{6.02 \times 10^{23} \text{ atoms}}$$

$$\times \frac{235 \text{ g U}}{1 \text{ mol}} \times \frac{3600 \text{ s}}{1 \text{ hr}} \times \frac{24 \text{ hr}}{1 \text{ d}} \times \frac{365 \text{ d}}{1 \text{ yr}} \times \frac{100}{40} (\text{efficiency}) = 1.03 \times 10^6 = 1 \times 10^6 \text{ g U/yr}$$

21.79 $$2 \times 10^{-12} \text{ curies} \times \frac{3.7 \times 10^{10} \text{ dis/s}}{1 \text{ curie}} = 7.4 \times 10^{-2} = 7 \times 10^{-2} \text{ dis/s}$$

$$\frac{7.4 \times 10^{-2} \text{ dis/s}}{75 \text{ kg}} \times \frac{8 \times 10^{-13} \text{ J}}{\text{dis}} \times \frac{1 \text{ rad}}{1 \times 10^{-2} \text{ J/g}} \times \frac{3600 \text{ s}}{\text{hr}} \times \frac{24 \text{ hr}}{1 \text{ d}}$$

$$\times \frac{365 \text{ d}}{1 \text{ yr}} = 2.49 \times 10^{-6} = 2 \times 10^{-6} \text{ rad/yr}$$

Recall that there are 10 rem/rad for alpha particles.

$$\frac{2.49 \times 10^{-6} \text{ rad}}{1 \text{ yr}} \times \frac{10 \text{ rem}}{1 \text{ rad}} = 2.49 \times 10^{-5} = 2 \times 10^{-5} \text{ rem/yr}$$

Integrative Exercises

21.80 Calculate the molar mass of $NaClO_4$ that contains 25.6% ^{36}Cl. Atomic mass of the enhanced Cl is 0.256(36.0) + 0.744(35.453) = 35.593 = 35.6. The molar mass of $NaClO_4$ is then (22.99 + 35.593 + 64.00) = 122.58 = 122.6. Calculate N, the number of ^{36}Cl nuclei, the value of k in s^{-1}, and the activity in dis/s.

$$36.9\text{ mg NaClO}_4 \times \frac{1\text{g}}{1000\text{ mg}} \times \frac{1\text{ mol NaClO}_4}{122.58\text{ g NaClO}_4} \times \frac{1\text{ mol Cl}}{1\text{ mol NaClO}_4} \times \frac{6.022\times10^{23}\text{ Cl atoms}}{\text{mol Cl}}$$

$$\times \frac{25.6\ {}^{36}\text{Cl atoms}}{100\text{ Cl atoms}} = 4.641\times10^{19} = 4.64\times10^{19}\ {}^{36}\text{Cl atoms}$$

$$k = 0.693/t_{1/2} = \frac{0.693}{3.0\times10^{5}\text{ yr}} \times \frac{1\text{ yr}}{365\times24\times3600\text{ s}} = 7.32\times10^{-14} = 7.3\times10^{-14}\text{ s}^{-1}$$

$$\text{rate} = kN = (7.32\times10^{-14}\text{ s}^{-1})(4.641\times10^{19}\text{ nuclei}) = 3.40\times10^{6} = 3.4\times10^{6}\text{ dis/s}$$

21.82 (a) $0.18\text{ Ci} \times \frac{3.7\times10^{10}\text{ dis/s}}{\text{Ci}} \times \frac{3600\text{ s}}{\text{hr}} \times \frac{24\text{ hr}}{\text{d}} \times 235\text{ d} = 1.35\times10^{17} = 1.4\times10^{17}\ \alpha\text{ particles}$

(b) $P = nRT/V = 1.35\times10^{17}\text{ He atoms} \times \frac{1\text{ mol He}}{6.022\times10^{23}\text{ atoms}} \times \frac{295\text{ K}}{0.0150\text{ L}} \times \frac{0.08206\text{ L}\bullet\text{atm}}{\text{K}\bullet\text{mol}}$

$= 3.62\times10^{-4} = 3.6\times10^{-4}\text{ atm} = 0.28\text{ torr}$

21.83 Calculate N_t in dis/min/g C from 1.5×10^{-2} dis/0.788 g $CaCO_3$. N_o = 15.3 dis/min/g C. Calculate k from $t_{1/2}$, calculate t from $\ln(N_t/N_o) = -kt$.

$C(s) + O_2(g) \rightarrow CO_2(g) + Ca(OH_2)(aq) \rightarrow CaCO_3(s) + H_2O(l)$

1 C atom $\rightarrow$ 1 $CaCO_3$ molecule

$$\frac{1.5\times10^{-2}\text{ Bq}}{0.788\text{ g CaCO}_3} \times \frac{1\text{ dis/s}}{1\text{ Bq}} \times \frac{60\text{s}}{1\text{min}} \times \frac{100.1\text{ g CaCO}_3}{12.01\text{ g C}} = 9.52 = 9.5\text{ dis/min/g C}$$

$k = 0.693/t_{1/2} = 0.693/5.715\times10^{3}\text{ yr} = 1.213\times10^{-4} = 1.21\times10^{-4}\text{ yr}^{-1}$

$$t = -\frac{1}{k}\ln\frac{N_t}{N_o} = \frac{-1}{1.213\times10^{-4}\text{ yr}^{-1}}\ln\frac{9.52\text{ dis/min/g C}}{15.3\text{ dis/min/g C}} = 3.91\times10^{3}\text{ yr}$$

21.85 (a) $Ba(NO_3)_2(aq) + Na_2SO_4(aq) \rightarrow BaSO_4(s) + 2NaNO_3(aq)$

(b) 1.25 mmol Ba^{2+} + 1.25 mmol SO_4^{2-} $\rightarrow$ 1.25 mmol $BaSO_4$

Neither reactant is in excess, so the activity of the filtrate is due entirely to $[SO_4^{2-}]$ from dissociation of $BaSO_4(s)$. Calculate $[SO_4^{2-}]$ in the filtrate by comparing the activity of the filtrate to the activity of the reactant.

$$\frac{0.050\ M\ SO_4^{2-}}{1.22\times10^{6}\text{ Bq/mL}} = \frac{x\ M\text{ filtrate}}{250\text{ Bq/mL}}$$

$[SO_4^{2-}]$ in the filtrate = $1.0246\times10^{-5} = 1.0\times10^{-5}\ M$

$K_{sp} = [Ba^{2+}][SO_4^{2-}]$; $[SO_4^{2-}] = [Ba^{2+}]$

$K_{sp} = (1.0246\times10^{-5})^2 = 1.0498\times10^{-10} = 1.0\times10^{-10}$

22 Chemistry of the Nonmetals

Chemistry of Nonmetals

22.1 C_2H_4, the structure on the left, is the stable compound. Carbon, with a relatively small covalent radius owing to its location in the second row of the periodic chart, is able to closely approach other atoms. This close approach enables significant π overlap, so carbon can form strong multiple bonds to satisfy the octet rule. Silicon, in the third row of the periodic table, has a covalent radius too large for significant π overlap. Si does not form stable multiple bonds and Si_2H_4 is unstable.

22.3 *Analyze.* The structure is a trigonal bipyramid where one of the five positions about the central atom is occupied by a lone pair, often called a see-saw.

Plan A: Count the valence electrons in each molecule, draw a correct Lewis structure, and count the electron domains about the central atom.

Plan B: Molecules (a)–(d) each contain four F atoms bound to a central atom through a single bond (F is unlikely to form multiple bonds because of its high electronegativity). This represents 16 electron pairs; the fifth position is occupied by a lone pair, for a total of 17 e^- pairs. A valence e^- count for (a)–(d) will tell us which molecules are likely to have the designated structure. Molecule (e), $HClO_4$, is not exactly of the type AX_4, so a Lewis structure will be required. *Solve.*

(a) XeF_4 36 e^-, 16 e^- pairs. Plan B predicts that this molecule **will not** adopt the see-saw structure.

```
      :F:
       |
:F — Xe — F:
       |
      :F:
```

6 e^- domains about the Xe
octahedral domain geometry
square planar structure

(b) BrF_4^- 34 e^-, 17 e^- pairs; structure will be see-saw.

```
[     :F:   ]-
  :F   |
     > Br:
  :F   |
      :F:   ]
```

5 e^- domains about Br trigonal
bipyramidal domain geometry
see-saw structure

(c) SiF_4 32 e^-, 16 e^- pairs; structure **will not** be see-saw

```
      :F:
       |
:F — Si — F:
       |
      :F:
```

4 e^- domains about Si
tetrahedral domain geometry and structure

(d) $TeCl_4$ 34 e^-, 17 e^- pairs; structure will be see-saw

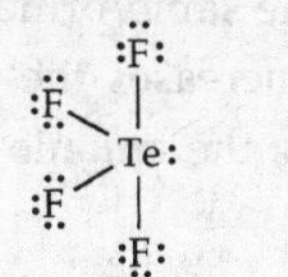

5 e^- domains about Te
trigonal bipyramidal domain geometry
see-saw structure

(e) $HClO_4$ 32 e^-, 16 e^- pairs; no prediction

:Ö:
|
:Ö—Cl—Ö—H
|
:Ö:

($HClO_4$ is an oxyacid, so H is bound to O, not Cl. Other Lewis structures that optimize formal charges are possible; structure predictions are the same.)

22.4 *Analyze.* Equation [22.33] is: $O_3 \rightarrow O_2 + O$ $\Delta H = 105$ kJ

Plan. Use $\Delta H°$ and the dissociation energy barrier to draw the profile.

Solve. $\Delta H°$ tells us that the products are 105 kJ higher in energy than the reactants. The dissociation energy barrier is the total energy difference between reactants and the activated complex (the top of the peak).

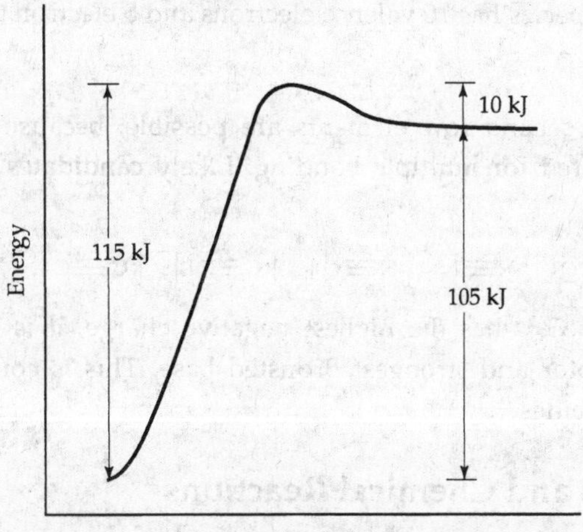

22.6 The graph is applicable only to (c) density. Density depends on both atomic mass and volume (radius). Both increase going down a family, but atomic mass increases to a greater extent. Density, the ratio of mass to volume, increases going down the family; this trend is consistent with the data in the figure.

According to periodic trends, (a) electronegativity and (b) first ionization energy both decrease rather than increase going down the family. According to Table 22.5 both (d) X—X single bond enthalpy and (e) electron affinity are somewhat erratic, with the trends decreasing from S to Po, and anomalous values for the properties of O, probably owing to its small covalent radius.

22.7 (a) Atomic radius increases moving downward in a group because the principal quantum number (n) of the valence electrons increases. As n increases, the average distance of an electron from the nucleus increases and so does atomic radius.

(b) Anionic radii are greater than atomic radii because of increased electrostatic repulsions among electrons. Additional electrons in the same principle quantum level lead to additional electrostatic repulsion. This increases the energy of the electrons, and their average distance from the nucleus; the anionic radii are thus greater than the atomic radii.

(c) The anion that is the strongest base in water is the conjugate base of the weakest conjugate acid. The conjugate acids are OH^-, SH^-, and SeH^-. According to trends in binary hydrides, the acid with the longest X—H bond will be the most readily ionized and the strongest acid. SeH^- is thus the strongest acid and OH^- the weakest. Therefore O^{2-} is the strongest base in water.

22.9 White phosphorus consists of tetrahedral P_4 molecules. This molecular geometry requires al P—P—P bond angles to be much smaller than the idealized 109°. This angular strain is relieved when P_4 reacts with another substance, which it does readily. The chains in red phosphorus require much less extreme bond angles, and there is less need to relieve steric repulsion by chemical reaction.

22.10 *Analyze/Plan.* The structure shown is a diatomic molecule or ion, depending on the value of n. Each species has 10 valence electrons and 5 electron pairs.

Solve.

(a) Only second row elements are possible, because of the small covalent radius required for multiple bonding. Likely candidates are CO, N_2, NO^+, CN^-, and C_2^{2-}.

$$:C\equiv O: \quad :N\equiv N: \quad [:N\equiv O:]^+ \quad [:C\equiv N:]^- \quad [:C\equiv C:]^{2-}$$

(b) Since C_2^{2-} has the highest negative charge, it is likely to be the strongest H^+ acceptor and strongest Brønsted base. This is confirmed in Section 22.9 under "Carbides."

Periodic Trends and Chemical Reactions

22.12 Metals: (a) Re, (d) Zr, (f) Ga nonmetals: (c) Ar metalloid: (b) As, (e) Te

22.14 (a) Cl (b) K

(c) K in the gas phase (lowest ionization energy), Li in aqueous solution (most positive E° value)

(d) Ne; Ne and Ar are difficult to compare to the other elements because they do not form compounds and their radii are not measured in the same way as other elements. However, Ne is several rows to the right of C and surely has a smaller atomic radius. The next smallest is C.

(e) C

22.16 (a) Nitrogen is a highly electronegative element. In HNO_3 it is in its highest oxidation state, +5, and thus is more readily reduced than phosphorus, which forms stable P—O bonds.

(b) The difference between the third row element and the second lies in the smaller size of C as compared with Si, and the fact that Si has 3d orbitals available to form an sp^3d^2 hybrid set that can accommodate more than an octet of electrons.

(c) Two of the carbon compounds, C_2H_4 and C_2H_2, contain C—C π bonds. Si does not readily form π bonds (to itself or other atoms), so Si_2H_4 and Si_2H_2 are not known as stable compounds.

22.18 (a) $NaOCH_3(s) + H_2O(l) \rightarrow NaOH(aq) + CH_3OH(aq)$

(b) $CuO(s) + 2HNO_3(aq) \rightarrow Cu(NO_3)_2(aq) + H_2O(l)$

(c) $WO_3(s) + 3H_2(g) \rightarrow W(s) + 3H_2O(g)$

(d) $4NH_2OH(l) + O_2(g) \rightarrow 6H_2O(l) + 2N_2(g)$

(e) $Al_4C_3(s) + 12H_2O(l) \rightarrow 4Al(OH)_3(s) + 3CH_4(g)$

Hydrogen, the Noble Gases, and the Halogens

22.20 Tritium is radioactive. ${}^{3}_{1}H \rightarrow {}^{3}_{2}He + {}^{0}_{-1}e$

22.22 Halogens are nonmetals with high electronegativity that need one additional valence electron to have a complete octet. They typically form negative ions with a 1- charge, homonuclear diatomic molecules, interhalogen compounds, or are the more electronegative atom in a polar covalent bond. Hydrogen also needs one additional valence electron to have a complete duet. In compounds (c) NaH and (d) H_2, hydrogen is acting like a halogen in two of the bonding situations described above.

(c) NaH is ionic, with H acting as hydride anion, H^-.

$NaH(s) + H_2O(l) \rightarrow H_2(g) + NaOH(aq)$

$NaF(s) + H_2O(l) \rightarrow HF(aq) + NaOH(aq)$

(d) Hydrogen forms the nonpolar homonuclear diatomic molecule H_2.

$2Li(s) + H_2(g) \rightarrow 2LiH(s)$

$2Li(s) + F_2(g) \rightarrow 2LiF(s)$

22.24 (a) Electrolysis of brine; reaction of carbon with steam; reaction of methane with steam; by-product in petroleum refining

(b) Synthesis of ammonia; synthesis of methanol; reducing agent; hydrogenation of unsaturated vegetable oils

22.26 (a) $2Al(s) + 6H^+(aq) \rightarrow 2Al^{3+}(aq) + 3H_2(g)$

(b) $Mg(s) + H_2O(g) \rightarrow MgO(s) + H_2(g)$

(c) $MnO_2(s) + H_2(g) \rightarrow MnO(s) + H_2O(g)$

(d) $CaH_2(s) + 2H_2O(l) \rightarrow Ca(OH)_2(aq) + 2H_2(g)$

22.28 (a) molecular (b) ionic (c) metallic

22.30 Xe(l), an atomic liquid, is capable only of London dispersion forces. All polar covalent molecules experience both dipole-dipole and dispersion forces, so dispersion forces are the most important intermolecular interactions between Xe(l) and polar covalent solutes. This is particularly true for CH_3I. The iodine atom has a large, very polarizable electron cloud, which makes dispersion forces the dominant type of intermolecular interaction for this molecule.

22.32 (a) $Ca(O\mathbf{Br})_2$, +1 (b) $H\mathbf{Br}O_3$, +5 (c) $\mathbf{Xe}O_3$; Xe, +6

(d) $\mathbf{Cl}O_4^-$, +7 (e) $H\mathbf{I}O_2$, +3 (f) $\mathbf{IF}_5$; I, +5; F, –1

22.34 (a) potassium chlorate (b) calcium iodate

(c) aluminum chloride (d) bromic acid

(e) paraperiodic acid (f) xenon tetrafluoride

22.36 (a) The more electronegative the central atom, the greater the extent to which it withdraws charge from oxygen, in turn making the O—H bond more polar, and enhancing ionization of H^+.

(b) HF reacts with the silica which is a major component of glass:

$6HF(aq) + SiO_2(s) \rightarrow SiF_6^{2-}(aq) + 2H_2O(l) + 2H^+(aq)$

(c) Iodide is oxidized by sulfuric acid, as shown in Figure 22.12.

(d) The major factor is size; there is not room about Br for the three chlorides plus the two unshared electron pairs that would occupy the bromine valence shell orbitals.

22.38 Cracking is a process where higher molecular weight hydrocarbons are broken down into smaller molecules (Section 25.3). One model reaction is

$$CH_4(g) \xrightarrow{\text{catalyst}} C(s) + 2H_2(g) \ \Delta H \sim 75 \text{ kJ}.$$

An analogous reaction can be written for "cracking" ethanol, where a third product is $O_2(g)$. The overall process is endothermic and requires energy. The question remains, what energy source drives the cracking? If energy to drive the cracking can be provided by some nonpetroleum fuel (such as grain-based ethanol) or power source (such as wind, geothermal, etc.), this could provide a satisfactory basis for a hydrogen economy.

Additionally, cracking would produce C(s) or soot and other C-based pollutants, which would have to be treated. This is probably easier at the power-plant level than the individual car level, but not insignificant.

22.40 Perchlorate salts are highly soluble, so the reaction is probably not a precipitation reaction. Perchloric acid, $HClO_4$, is a strong acid and ClO_4^- is a negligible base in water so the decomposition is probably not an acid-base reaction. Because oxyanions of the halogens are known to participate in oxidation-reduction reactions, the microorganisms probably destroy ClO_4^- via a redox reaction or series of reactions. While several lower oxidation states are accessible to Cl in ClO_4^-, intermediate species ClO_3^-, ClO_2^-, ClO^-, and Cl_2 are quite reactive, so the logical fate of the perchlorate anion is $Cl^-(aq)$ and $O_2(g)$ or $H_2O(l)$ with some oxidized organic compound.

Oxygen and the Group 6A Elements

22.42 $\ddot{O}$ resonance structures: $:\ddot{O}{=}\ddot{O}{-}\ddot{O}: \longleftrightarrow :\ddot{O}{-}\ddot{O}{=}\ddot{O}:$

Ozone has two resonance forms (Section 8.7); the molecular structure is bent, with an O—O—O bond angle of approximately 120°. The π bond in ozone is delocalized over the entire molecule; neither individual O—O bond is a full double bond, so the observed O—O distance of 1.28 Å is greater than the 1.21 Å distance in O_2, which has a full O—O double bond.

22.44
(a) $CaO(s) + H_2O(l) \rightarrow Ca^{2+}(aq) + 2OH^-(aq)$
(b) $Al_2O_3(s) + 6H^+(aq) \rightarrow 2Al^{3+}(aq) + 3H_2O(l)$
(c) $Na_2O_2(s) + 2H_2O(l) \rightarrow 2Na^+(aq) + 2OH^-(aq) + H_2O_2(aq)$
(d) $N_2O_3(g) + H_2O(l) \rightarrow 2HNO_2(aq)$
(e) $2KO_2(s) + 2H_2O(l) \rightarrow 2K^+(aq) + 2OH^-(aq) + O_2(g) + H_2O_2(aq)$
(f) $NO(g) + O_3(g) \rightarrow NO_2(g) + O_2(g)$

22.46
(a) Mn_2O_7 (higher oxidation state of Mn)
(b) SnO_2 (higher oxidation state of Sn)
(c) SO_3 (higher oxidation state of S)
(d) SO_2 (more nonmetallic character of S)
(e) Ga_2O_3 (more nonmetallic character of Ga)
(f) SO_2 (more nonmetallic character of S)

22.48
(a) SeO_2, +4 (b) $Na_2S_2O_3$, +2 (c) SF_6, +6
(d) H_2S, -2 (e) H_2SO_4, +6

Oxygen (a group 6A element) is in the -2 oxidation state in compounds (a), (b), and (e).

22.50 An aqueous solution of SO_2 contains H_2SO_3 and is acidic. Use H_2SO_3 as the reducing agent and balance assuming acid conditions.

(a)
$$2[MnO_4^-(aq) + 8H^+(aq) + 5e^- \rightarrow Mn^{2+}(aq) + 4H_2O(l)]$$
$$5[H_2SO_3(aq) + H_2O(l) \rightarrow SO_4^{2-}(aq) + 4H^+(aq) + 2e^-]$$
$$2MnO_4^-(aq) + 5H_2SO_3(aq) \rightarrow 2MnSO_4(aq) + 3SO_4^{2-}(aq) + 3H_2O(l) + 4H^+(aq)$$

(b)
$$Cr_2O_7^{2-}(aq) + 14H^+(aq) + 6e^- \rightarrow 2Cr^{3+}(aq) + 7H_2O(l)$$
$$3[H_2SO_3(aq) + H_2O(l) \rightarrow SO_4^{2-}(aq) + 4H^+(aq) + 2e^-]$$
$$Cr_2O_7^{2-}(aq) + 3H_2SO_3(aq) + 2H^+(aq) \rightarrow 2Cr^{3+}(aq) + 3SO_4^{2-}(aq) + 4H_2O(l)$$

(c)
$$Hg_2^{2+}(aq) + 2e^- \rightarrow 2Hg(l)$$
$$H_2SO_3(aq) + H_2O(l) \rightarrow SO_4^{2-}(aq) + 4H^+(aq) + 2e^-$$
$$Hg_2^{2+}(aq) + H_2SO_3(aq) + H_2O(l) \rightarrow 2Hg(l) + SO_4^{2-}(aq) + 4H^+(aq)$$

22.52 SF_4, 34 e^- SF_5^-, 42 e^-

(lone pairs on F atoms omitted for clarity)

trigonal bipyramidal electron pair geometry; see-saw molecular geometry

octahedral electron pair geometry; square pyramidal molecular geometry

22.54 (a) $Al_2Se_3(s) + 6H^+(aq) \rightarrow 2Al^{3+}(aq) + 3H_2Se(g)$

(b) $Cl_2(aq) + S_2O_3^{2-}(aq) + H_2O(l) \rightarrow 2Cl^-(aq) + S(s) + SO_4^{2-}(aq) + 2H^+(aq)$

Nitrogen and the Group 5A Elements

22.56 (a) HNO_3, +5 (b) N_2H_4, -2 (c) KCN, -3

(d) $NaNO_3$, +5 (e) NH_4Cl, -3 (f) Li_3N, -3

22.58 (a) $[NH_4]^+$

tetrahedral

(b) :Ö—N(=O:)—Ö—H ⟷ Ö=N(—Ö:)—Ö—H ⟷ :Ö—N(—Ö:)=O—H

The geometry around nitrogen is trigonal planar, but the hydrogen atom is not required to lie in this plane. The third resonance form makes a much smaller contribution to the structure than the first two.

(c) :N̈=N=Ö: ⟷ :N≡N—Ö: ⟷ :N̈—N≡O:

The molecule is linear. Again, the third resonance form makes less contribution to the structure because of the high formal charges involved.

(d) Ö=Ṅ—Ö: ⟷ :Ö—Ṅ=Ö:

The molecule is bent (nonlinear).

22.60 (a) $4Zn(s) + 2NO_3^-(aq) + 10H^+(aq) \rightarrow 4Zn^{2+}(aq) + N_2O(g) + 5H_2O(l)$

(b) $4NO_3^-(aq) + S(s) + 4H^+(aq) \rightarrow 4NO_2(g) + SO_2(g) + 2H_2O(l)$

(or $6NO_3^-(aq) + S(s) + 4H^+(aq) \rightarrow 6NO_2(g) + SO_4^{2-}(aq) + 2H_2O(l)$)

(c) $2NO_3^-(aq) + 3SO_2(g) + 2H_2O(l) \rightarrow 2NO(g) + 3SO_4^{2-}(aq) + 4H^+(aq)$

(d) $N_2H_4(g) + 5F_2(g) \rightarrow 2NF_3(g) + 4HF(g)$

(e) $4CrO_4^{2-}(aq) + 3N_2H_4(aq) + 4H_2O(l) \rightarrow 4Cr(OH)_4^-(aq) + 4OH^-(aq) + 3N_2(g)$

22.62 E°_{red} for both half-reactions can be read directly from Figure 22.28.

(a) $NO_3^-(aq) + 4H^+(aq) + 3e^- \rightarrow NO(g) + 2H_2O(l)$ $E^\circ_{red} = +0.96$ V

(b) $HNO_2(aq) \rightarrow NO_2(g) + H^+(aq) + 1e^-$ $E^\circ_{red} = 1.12$ V

22.64 (a) $H_3\mathbf{P}O_4$, +5 (b) $H_3\mathbf{As}O_3$, +3 (c) $\mathbf{Sb}_2S_3$, +3

(d) $Ca(H_2\mathbf{P}O_4)_2$, +5 (e) $K_3\mathbf{P}$, -3

22.66 (a) Only two of the hydrogens in H_3PO_3 are bound to oxygen. The third is attached directly to phosphorus, and not readily ionized, because the H—P bond is not very polar.

(b) The smaller, more electronegative nitrogen withdraws more electron density from the O—H bond, making it more polar and more likely to ionize.

(c) Phosphate rock consists of $Ca_3(PO_4)_2$, which is only slightly soluble in water. The phosphorus is unavailable for plant use.

(d) N_2 can form stable π bonds to complete the octet of both N atoms. Because phosphorus atoms are larger than nitrogen atoms, they do not form stable π bonds with themselves and must form σ bonds with several other phosphorus atoms (producing P_4 tetrahedral or sheet structures) to complete their octets.

(e) In solution Na_3PO_4 is completely dissociated into Na^+ and PO_4^{3-}. PO_4^{3-}, the conjugate base of the very weak acid HPO_4^{2-}, has a K_b of 2.4×10^{-2} and produces a considerable amount of OH^- by hydrolysis of H_2O.

22.68 (a) $PCl_5(l) + 4H_2O(l) \rightarrow H_3PO_4(aq) + 5HCl(aq)$

(b) $2H_3PO_4(aq) \xrightarrow{\Delta} H_4P_2O_7(aq) + H_2O(l)$

(c) $P_4O_{10}(s) + 6H_2O(l) \rightarrow 4H_3PO_4(aq)$

Carbon, the Other Group 4A Elements, and Boron

22.70 (a) H_2CO_3 (b) NaCN (c) $KHCO_3$ (d) C_2H_2

22.72 (a) $CO_2(g) + OH^-(aq) \rightarrow HCO_3^-(aq)$

(b) $NaHCO_3(s) + H^+(aq) \rightarrow Na^+(aq) + H_2O(l) + CO_2(g)$

(c) $2CaO(s) + 5C(s) \xrightarrow{\Delta} 2CaC_2(s) + CO_2(g)$

(d) $C(s) + H_2O(g) \xrightarrow{\Delta} H_2(g) + CO(g)$

(e) $CuO(s) + CO(g) \rightarrow Cu(s) + CO_2(g)$

22.74 (a) $2Mg(s) + CO_2(g) \rightarrow 2MgO(s) + C(s)$

(b) $6CO_2(g) + 6H_2O(l) \xrightarrow{h\nu} C_6H_{12}O_6(aq) + 6O_2(g)$

(c) $CO_3^{2-}(aq) + H_2O(l) \rightarrow HCO_3^-(aq) + OH^-(aq)$

22.76 (a) $\mathbf{Si}O_2$, +4 (b) $\mathbf{Ge}Cl_4$, +4 (c) $Na\mathbf{B}H_4$, +3

(d) $\mathbf{Sn}Cl_2$, +2 (e) $\mathbf{B}_2H_6$, +3

22.78 (a) carbon (b) lead (c) germanium

22.80 (a)

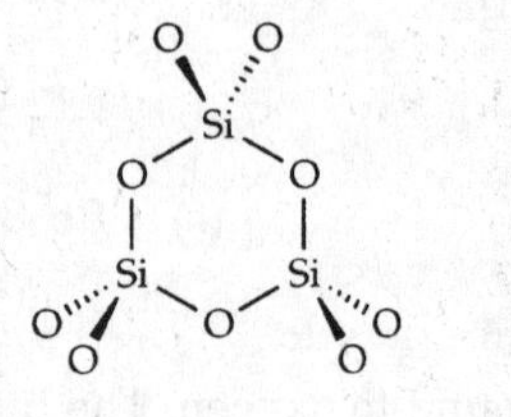

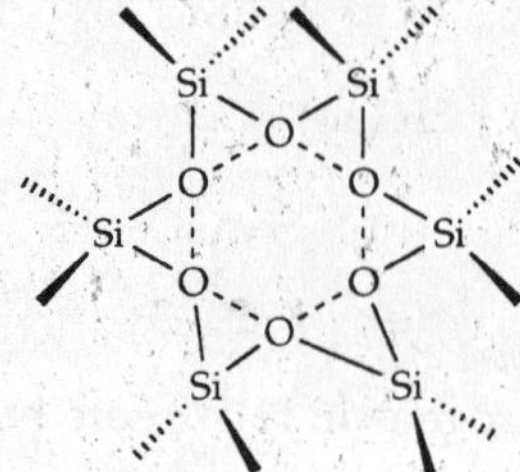

(b) $Si_3O_9^{6-}$ $Si_6O_{18}^{12-}$

22.82 (a) $B_2H_6(g) + 6H_2O(l) \rightarrow 2H_3BO_3(aq) + 6H_2(g)$

(b) $4H_3BO_3(s) \xrightarrow{\Delta} H_2B_4O_7(s) + 5H_2O(g)$

(c) $B_2O_3(s) + 3H_2O(l) \rightarrow 2H_3BO_3(aq)$

Additional Exercises

22.83 (a) A *reducing agent* agent is an electron-rich substance in a low oxidation state that loses electrons to the reactant being reduced in a redox reaction.

(b) *Allotropes* are different structural forms of the same element. They are composed of atoms of a single element bound into different structures. For example, graphite, diamond, and buckey balls are all allotropes of carbon.

(c) *Disproportionation* is an oxidation-reduction process where the same element is both oxidized and reduced.

(d) *Interhalogen* is a compound formed from atoms of two or more halogens.

(e) An *acidic anhydride* is a neutral molecule that is the oxide of a nonmetal. It reacts with water to produce an acid.

(f) A *condensation reaction* is the combination of two molecules to form a large molecule and a small one such as H_2O or HCl.

22.85 (a) React an ionic nitride with D_2O, e.g.,

$Mg_3N_2(s) + 6D_2O(l) \rightarrow 2ND_3(aq) + 3Mg(OD)_2(s)$

(b) React SO_3 with D_2O: $SO_3(g) + D_2O(l) \rightleftharpoons D_2SO_4(aq)$

(c) React Na_2O with D_2O: $Na_2O(s) + D_2O(l) \rightarrow 2NaOD(aq)$

(d) Dissolve $N_2O_5(g)$ in D_2O: $N_2O_5(g) + D_2O(l) \rightarrow 2DNO_3(aq)$

(e) React CaC_2 with D_2O: $CaC_2(s) + 2D_2O(l) \rightarrow Ca^{2+}(aq) + 2OD^-(aq) + C_2D_2(g)$

(f) Add NaCN to the D_2SO_4 solution prepared in (b):

$$NaCN(s) + D^+(aq) \xrightarrow{\Delta} DCN(aq) + Na^+(aq)$$

The DCN can be removed as gas from the reaction.

22.86 $BrO_3^-(aq) + XeF_2(aq) + H_2O(l) \rightarrow Xe(g) + 2HF(aq) + BrO_4^-(aq)$

22.88 (a) $SO_2(g) + H_2O(l) \rightleftharpoons H_2SO_3(aq)$

(b) $Cl_2O_7(g) + H_2O(l) \rightleftharpoons 2HClO_4(aq)$

(c) $Na_2O_2(s) + 2H_2O \rightarrow H_2O_2(aq) + 2NaOH(aq)$

(d) $BaC_2(s) + 2H_2O(l) \rightarrow Ba^{2+}(aq) + 2OH^-(aq) + C_2H_2(g)$

(e) $2RbO_2(s) + 2H_2O(l) \rightarrow 2Rb^+(aq) + 2OH^-(aq) + O_2(g) + H_2O_2(aq)$

(f) $Mg_3N_2(s) + 6H_2O(l) \rightarrow 3Mg(OH)_2(s) + 2NH_3(g)$

(g) $NaH(s) + H_2O \rightarrow NaOH(aq) + H_2(g)$

22.90 $S_8(s) + 8Fe(s) \rightarrow 8FeS(s)$

$S_8(s) + 16F_2(g) \rightarrow 8SF_4(g)$ or $S_8(s) + 24F_2(g) \rightarrow 8SF_6(g)$

$S_8(s) + 8O_2(g) \rightarrow 8SO_2(g)$

$S_8(s) + 8H_2(g) \rightarrow 8H_2S(g)$

Sulfur acts as an oxidizing agent in reactions with Fe or H_2 and as a reducing agent in reactions with O_2 or F_2. Incidentally, these reactions are often written using the symbol S rather than S_8 for sulfur.

22.91

$S(g) + O_2(g) \rightarrow SO_2(g)$	$\Delta H = -296.9$ kJ	(1)
$SO_2(g) + 1/2\, O_2(g) \rightarrow SO_3(g)$	$\Delta H = -98.3$ kJ	(2)
$SO_3(g) + H_2O(l) \rightarrow H_2SO_4(aq)$	$\Delta H = -130$ kJ	(3)
$S(g) + 3/2\, O_2(g) + H_2O(l) \rightarrow H_2SO_4(aq)$	$\Delta H = -525$ kJ	

$$1 \text{ ton } H_2SO_4 \times \frac{2000 \text{ lb}}{\text{ton}} \times \frac{453.6 \text{ g}}{1 \text{ lb}} \times \frac{1 \text{ mol } H_2SO_4}{98.09 \text{ g}} \times \frac{-525 \text{ kJ}}{\text{mol } H_2SO_4}$$

$= -4.86 \times 10^6$ kJ of heat/ton H_2SO_4

22.92 (a) PO_4^{3-}, + 5; NO_3^-, + 5

(b) The Lewis structure for NO_4^{3-} would be:

```
        ..      3-
   [   :O:        ]
        |
   [:O—N—O:       ]
        |
   [   :O:        ]
        ..
```

The formal charge on N is +1 and on each O atom is –1. The four electronegative oxygen atoms withdraw electron density, leaving the nitrogen deficient. Since N can form a maximum of four bonds, it cannot form a π bond with one or more of the O atoms to regain electron density, as the P atom in PO_4^{3-} does. Also, the short N—O distance would lead to a tight tetrahedron of O atoms subject to steric repulsion.

22.93 (a) Although P_4, P_4O_6 and P_4O_{10} all have four P atoms in a tetrahedral arrangement, the bonding **between** P atoms and **by** P atoms is not the same in the three molecules. In P_4, the 4 P atoms are bound only to each other by P—P single bonds and strained bond angles of approximately 60°. In the two oxides, the 4 P atoms are directly bound to oxygen atoms, not to each other. Bonding by P atoms in P_4O_6 and P_4O_{10} is very similar. Each contains the P_4O_6 cage, formed by four P_3O_3 rings which share a P—O—P edge. Phosphorus bonding to oxygen maintains the overall P_4 tetrahedron but allows the P atoms to move away from each other so that the angle strain is relieved relative to molecular P_4. The P—O—P and O—P—O angles in both oxides are near the ideal 109°. In P_4O_6, each P is bound to 3 O atoms and has a lone pair completing its octet. In P_4O_{10}, the lone pair is replaced by a terminal O atom and each P is bound to 3 bridging and 1 terminal O atom.

(b) [structures with charges 3− and 4−]

In both structures there are unshared pairs on all oxygens to give octets and the geometry around each P is approximately tetrahedral.

22.94 (a) Si Si Si

To complete their octets, the two terminal Si atoms each require three H atoms and the central Si requires two, for a total of 8 H atoms. The molecular formula is Si_3H_8.

(b) $Si_3H_8 + 5O_2 \rightarrow 3SiO_2 + 4H_2O$

22.96 (a)

$$2[5e^- + MnO_4^-(aq) + 8H^+(aq) \rightarrow Mn^{2+}(aq) + 4H_2O(l)]$$
$$5[H_2O_2(aq) \rightarrow O_2(g) + 2H^+(aq) + 2e^-]$$
$$2MnO_4^-(aq) + 5H_2O_2(aq) + 6H^+(aq) \rightarrow 2Mn^{2+}(aq) + 5O_2(g) + 8H_2O(l)$$

(b)

$$2[Fe^{2+}(aq) \rightarrow Fe^{3+}(aq) + e^-]$$
$$H_2O_2(aq) + 2H^+(aq) + 2e^- \rightarrow 2H_2O(l)$$
$$2Fe^{2+}(aq) + H_2O_2(aq) + 2H^+(aq) \rightarrow 2Fe^{3+}(aq) + 2H_2O(l)$$

(c)

$$2I^-(aq) \rightarrow I_2(s) + 2e^-$$
$$H_2O_2(aq) + 2H^+(aq) + 2e^- \rightarrow 2H_2O(l)$$
$$2\,I^-(aq) + H_2O_2(aq) + 2H^+(aq) \rightarrow I_2(s) + 2H_2O(l)$$

(d)

$$Cu(s) \rightarrow Cu^{2+}(aq) + 2e^-$$
$$H_2O_2(aq) + 2H^+(aq) + 2e^- \rightarrow 2H_2O(l)$$
$$Cu(s) + H_2O_2(aq) + 2H^+(aq) \rightarrow Cu^{2+}(aq) + 2H_2O(l)$$

(e)

$$2I^-(aq) \rightarrow I_2(s) + 2e^-$$
$$O_3(g) + H_2O(l) + 2e^- \rightarrow O_2(g) + 2OH^-(aq)$$
$$\overline{2I^-(aq) + O_3(g) + H_2O(l) \rightarrow O_2(g) + I_2(s) + 2OH^-(aq)}$$

22.98 (a) $Li_3N(s) + 3H_2O(l) \rightarrow 3Li^+(aq) + 3OH^-(aq) + NH_3(aq)$

(b) $NH_3(aq) + H_2O(l) \rightleftharpoons NH_4^+(aq) + OH^-(aq)$

(c) $3NO_2(g) + H_2O(l) \rightarrow NO(g) + 2H^+(aq) + 2NO_3^-(aq)$

(d) $2NO_2(g) \rightleftharpoons N_2O_4(g)$

(e) $4NH_3(g) + 5O_2(g) \xrightarrow{\text{catalyst}} 4NO(g) + 6H_2O(g)$

(f) $2CO(g) + O_2(g) \rightarrow 2CO_2(g)$

(g) $H_2CO_3(aq) \xrightarrow{\Delta} H_2O(g) + CO_2(g)$

(h) $Ni(s) + CO(g) \rightarrow NiO(s) + C(s)$

(i) $CS_2(g) + O_2(g) \rightarrow CO_2(g) + S_2(g)$

(j) $CaO(s) + SO_2(g) \rightarrow CaSO_3(s)$

(k) $2Na(s) + 2H_2O(l) \rightarrow 2NaOH(aq) + H_2(g)$

(l) $CH_4(g) + H_2O(g) \xrightarrow{\Delta} CO(g) + 3H_2(g)$

(m) $LiH(s) + H_2O(l) \rightarrow LiOH(aq) + H_2(g)$

(n) $Fe_2O_3(s) + 3H_2(g) \rightarrow 2Fe(s) + 3H_2O(g)$

Integrative Exercises

22.100 (a) $H_2(g) + 1/2\,O_2(g) \rightarrow H_2O(l);\ \Delta H = -285.83\ kJ$

$CH_4(g) + 2O_2(g) \rightarrow CO_2(g) + 2H_2O(l)$

$\Delta H = 2(-285.83) - 393.5 - (-74.8) = -890.4\ kJ$

(b) for H_2: $$\frac{-285.83\ kJ}{1\ mol\ H_2} \times \frac{1\ mol\ H_2}{2.0159\ g\ H_2} = -141.79\ kJ/g\ H_2$$

for CH_4: $$\frac{-890.4\ kJ}{1\ mol\ CH_4} \times \frac{1\ mol\ CH_4}{16.043\ g\ CH_4} = -55.50\ kJ/g\ CH_4$$

(c) Find the number of moles of gas that occupy 1 m^3 at STP:

$$n = \frac{1\ atm \times 1\ m^3}{273\ K} \times \frac{1\ K \bullet mol}{0.08206\ L \bullet atm} \times \left[\frac{100\ cm}{1\ m}\right]^3 \times \frac{1\ L}{10^3\ cm^3} = 44.64\ mol$$

$$\text{for } H_2: \frac{-285.83 \text{ kJ}}{1 \text{ mol } H_2} \times \frac{44.64 \text{ mol } H_2}{1 \text{ m}^3 \; H_2} = 1.276 \times 10^4 \text{ kJ/m}^3 \; H_2$$

$$\text{for } CH_4: \frac{-890.4 \text{ kJ}}{1 \text{ mol } CH_4} \times \frac{44.64 \text{ mol } CH_4}{1 \text{ m}^3 \; CH_4} = 3.975 \times 10^4 \text{ kJ/m}^3 \; CH_4$$

22.102 (a) $2NH_4ClO_4(s) \xrightarrow{\Delta} N_2(g) + 2HCl(g) + 3H_2O(g) + 5/2\ O_2(g)$

$NH_4ClO_4(s) \xrightarrow{\Delta} 1/2\ N_2(g) + HCl(g) + 3/2\ H_2O(g) + 5/4\ O_2(g)$

(b) $\Delta H° = \Sigma\, \Delta H_f^\circ \text{ prod} - \Sigma\, \Delta H \text{ react}$

$$\Delta H° = \Delta H_f^\circ\, HCl(g) + 3/2\, \Delta H_f^\circ\, H_2O(g) + 1/2\, \Delta H_f^\circ\, N_2(g) + 5/4\, \Delta H_f^\circ\, O_2(g) - \Delta H_f^\circ\, NH_4ClO_4 \Delta H°$$
$$= -92.30 \text{ kJ} + 3/2(-241.82 \text{ kJ}) + 1/2\,(0 \text{ kJ}) + 5/4\,(0 \text{ kJ}) - (-295.8 \text{ kJ})$$
$$= -159.2 \text{ kJ/mol } NH_4ClO_4$$

(c) The aluminum reacts exothermically with $O_2(g)$ and $HCl(g)$ produced in the decomposition, providing additional heat and thrust.

22.103 (a) $N_2H_4(g) + O_2(g) \rightarrow N_2(g) + 2H_2O(l)$

(b) $\Delta H° = \Delta H_f^\circ\, N_2(g) + 2\Delta H_f^\circ\, H_2O(l) - \Delta H_f^\circ\, N_2H_4(aq) - \Delta H_f^\circ\, O_2(g)$

$= 0 + 2(-285.83) - 95.40 - 0 = -667.06$ kJ

(c) $$\frac{9.1 \text{ g } O_2}{1 \times 10^6 \text{ g } H_2O} \times \frac{1.0 \text{ g } H_2O}{1 \text{ mL} H_2O} \times \frac{1000 \text{ mL}}{1\text{L}} \times 3.0 \times 10^4 \text{ L} = 273 = 2.7 \times 10^2 \text{ g } O_2$$

$$2.73 \times 10^2 \text{ g } O_2 \times \frac{1 \text{ mol } O_2}{32.00 \text{ g } O_2} \times \frac{1 \text{ mol } N_2H_4}{1 \text{ mol } O_2} \times \frac{32.05 \text{ g } N_2H_4}{1 \text{ mol } N_2H_4} = 2.7 \times 10^2 \text{ g } N_2H_4$$

22.105 *Plan.* vol air $\rightarrow$ kg air $\rightarrow$ g $H_2S \rightarrow$ g FeS. Use the ideal-gas equation to change volume of air to mass of air, (assuming 1.00 atm, 298 K and an average molar mass (MM) for air of 29.0 g/mol. Use (20 mg H_2S/kg) air to find the mass of H_2S in the given mass of air. *Solve.*

$$V_{air} = 2.7 \text{ m} \times 4.3 \text{ m} \times 4.3 \text{ m} \times \frac{(100)^3 \text{ cm}^3}{1 \text{ m}^3} \times \frac{1 \text{ L}}{1000 \text{ cm}^3} = 4.9923 \times 10^4 = 5.0 \times 10^4 \text{ L}$$

$$g_{air} = \frac{PV\, MM}{RT}; \text{ assume P} = 1.00 \text{ atm, T} = 298 \text{ K, } MM_{air} = 29.0 \text{ g/mol}$$

$$g_{air} = \frac{1.00 \text{ atm} \times 4.9923 \times 10^4 \text{ L} \times 29.0 \text{ g/mol}}{298 \text{ K}} \times \frac{\text{K} \bullet \text{mol}}{0.08206 \text{ L} \bullet \text{atm}} = 59{,}204 = 5.9 \times 10^4 \text{ g air}$$

$$5.9203 \times 10^4 \text{ g air} \times \frac{1 \text{ kg}}{1000 \text{ g}} \times \frac{20 \text{ mg } H_2S}{1 \text{ kg air}} \times \frac{1 \text{ g}}{1000 \text{ mg}} = 1.184 = 1.2 \text{ g } H_2S$$

$FeS(s) + 2HCl(aq) \rightarrow FeCl_2(aq) + H_2S(g)$

$$1.184 \text{ g } H_2 \times \frac{1 \text{ mol } H_2}{34.08 \text{ g } H_2S} \times \frac{1 \text{ mol FeS}}{1 \text{ mol } H_2S} \times \frac{87.91 \text{ g FeS}}{1 \text{ mol FeS}} = 3.054 = 3.1 \text{ g FeS}$$

22.107 (a) MnSi: more than one element, so not metallic; high melting, so not molecular; insoluble in water, so not ionic; therefore covalent network.

(b) $MnSi(s) + HF(aq) \rightarrow SiH_4(g) + MnF_4(s)$

Reduction of Mn(IV) to Mn(II) is unlikely, because F^- is an extremely weak reducing agent. E°_{red} for $F_2(g) + 2e^- \rightarrow 2F^-(aq) = 2.87$ V

22.108 $N_2H_5^+(aq) \rightarrow N_2(g) + 5H^+(aq) + 4e^-$ $E^\circ_{red} = -0.23$ V

Reduction of the metal should occur when E°_{red} of the metal ion is more positive than about –0.15 V. This is the case for (b) Sn^{2+} (marginal), (c) Cu^{2+} and (d) Ag^+.

22.110 First write the balanced equation to give the number of moles of gaseous products per mole of hydrazine.

(A) $(CH_3)_2NNH_2 + 2N_2O_4 \rightarrow 3N_2(g) + 4H_2O(g) + 2CO_2(g)$

(B) $(CH_3)HNNH_2 + 5/4\, N_2O_4 \rightarrow 9/4\, N_2(g) + 3H_2O(g) + CO_2(g)$

In case (A) there are nine moles gas per one mole $(CH_3)_2NNH_2$ plus two moles N_2O_4. The total mass of reactants is 60 + 2(92) = 244 g. Thus, there are

$$\frac{9 \text{ mol gas}}{244 \text{ g reactants}} = \frac{0.0369 \text{ mol gas}}{1 \text{ g reactants}}$$

In case (B) there are 6.25 moles of gaseous product per one mole $(CH_3)HNNH_2$ plus 1.25 moles N_2O_4. The total mass of this amount of reactants is 46.0 + 1.25(92.0) = 161 g.

$$\frac{6.25 \text{ mol gas}}{161 \text{ g reactants}} = \frac{0.0388 \text{ mol gas}}{1 \text{ g reactants}}$$

Thus the methylhydrazine (B) has marginally greater thrust.

22.111 (a) $HOOC{-}CH_2{-}COOH \xrightarrow{P_2O_5} C_3O_2 + 2H_2O$

(b) 24 valence e^-, 12 e^- pair $\ddot{O}{=}C{=}C{=}C{=}\ddot{O}$

(c) C=O, about 1.23 Å; C=C, 1.34 Å or less. Since consecutive C=C bonds require sp hybrid orbitals on C (as in allene, C_3H_4), we might expect the orbital overlap requirements of this bonding arrangement to require smaller than usual C=C distances.

(d) The product has the formula $C_3H_4O_2$.

28 valence e^-, 14 e^- pr

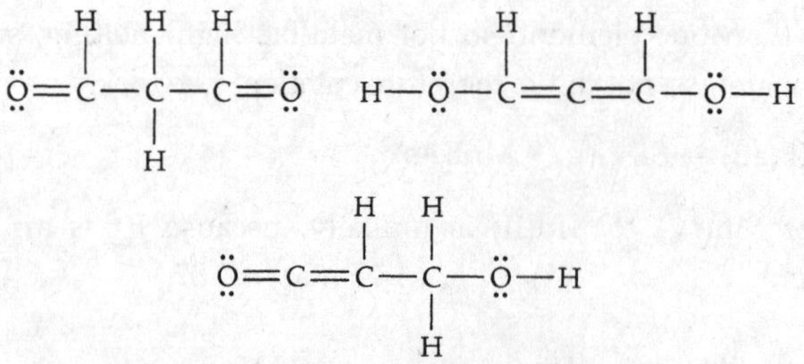

```
 H  H  H              H     H
 |  |  |              |     |
O=C—C—C=O      H—O—C=C=C—O—H
    |
    H

        H  H
        |  |
   O=C=C—C—O—H
           |
           H
```

Two possibilities are shown above. The O=C=C group in the structure on the right is uncommon and less likely than the two symmetrical structures.

22.112 BN has the same number of valence electrons per formula unit as carbon. (Three from B, five from N, for an average of four per atom.) To the extent that we can neglect the difference in nuclear charges between B and N, we can think of BN as carbon-like. Indeed, BN takes on the same structural forms as carbon. However, because the B—N bonds are somewhat polar, BN is in fact even harder than diamond.

23 Metals and Metallurgy

Visualizing Concepts

23.2 The diagram indicates that the roasting of ZnS is exothermic. The roasting reaction, once under way, will increase the temperature of the oven. The thermodynamic characteristics of the reaction do not affect its rate. Although heating will decrease the value of the equilibrium constant for an exothermic reaction, it is required so that the roasting reaction occurs at a practical rate.

23.3 (a)

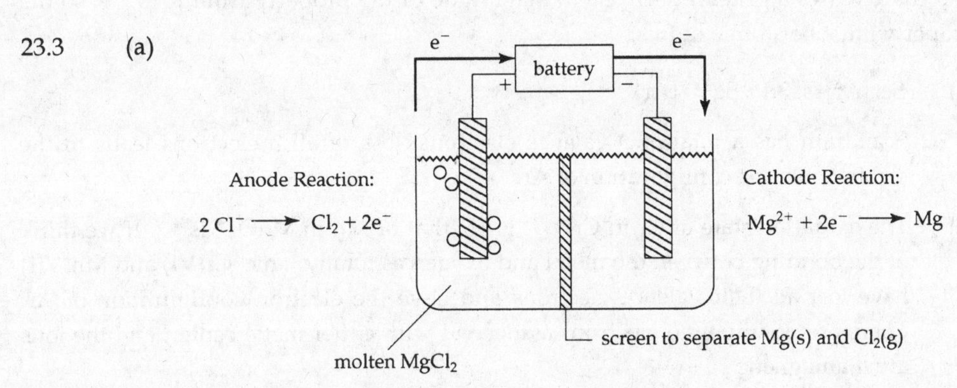

(b)

$MgCl_2(l) \rightarrow Mg(s) + Cl_2(g)$ overall

$2Cl^- \rightarrow Cl_2(g) + 2e^-$ anode (oxidation)

$Mg^{2+} + 2e^- \rightarrow Mg(s)$ cathode (reduction)

(c) Magnesium is an active metal. It must be separated from the $Cl_2(g)$ that is also formed by electrolysis (see screen in diagram) or $MgCl_2$ will form. The Mg(s) also should not come in contact with air (O_2) or moisture (H_2O).

23.4 (a) Cr: $[Ar]4s^13d^5$; Cd: $[Ar]4s^23d^{10}$

According to the molecular orbital model of metallic bonding, valence orbitals from each atom of a metallic solid combine to form a metal-metal bonding band and a metal-metal anti-bonding band. Chromium atoms have valence orbitals that are exactly half-filled. This means that the bonding band is full, and the anti-bonding band is empty; Cr has the strongest metal-metal bonding of the transition metals because it has maximum occupancy of the bonding band and minimum occupancy of the anti-bonding bond. Cadmium atoms have filled valence orbitals, resulting in filled bonding and anti-bonding bands. Metal-metal bonding in Cd is much weaker than that in Cr. The strength of metal-metal bonding in Cr causes it to be much harder than Cd.

(b) The chemical equation that represents formation of gaseous elements is $M(s) \rightarrow M(g)$, where M(s) represents the metal in its standard state and M(g) is the gaseous metal. In order for this change to occur, metal-metal bonds in the solid must be broken as the atoms move apart to enter the gas phase. The stronger the metal-metal bonding of the element, the greater the magnitude of $\Delta H_f°$. The metals shown in the diagram have valence orbitals that are less than half full and the number of valence electrons increases from K to V. The metal-metal bonding band contains progressively more electrons, the strength of metallic bonding increases, and $\Delta H_f°$ increases.

23.6 Periodic properties are explained in terms of effective nuclear charge, Z_{eff}. Moving from left to right in a period, Z_{eff} increases because the increase in Z is not offset by a significant increase in shielding. Increasing Z_{eff} leads to increasing ionization energy and electronegativity, but decreasing atomic radius.

The chart shows a general decrease in magnitude of the property from K to Ge, so the property must be atomic radius.

23.7 (a) Sc: $[Ar]4s^23d^1$; Sc^{3+}: [Ar]

Scandium has a total of 3 valence electrons; loss of all 3 electrons leads to the stable noble gas configuration of Ar.

(b) The oxidation state of Cr in CrO_4^{2-} is +6; that of Mn in MnO_4^- is +7. If we think of the bonding between the metal and oxygen as totally ionic, Cr(VI) and Mn(VII) have lost all their valence electrons and have the electron configuration of Ar. There are no unpaired electrons associated with either metal center, and the ions are diamagnetic.

Assuming perfectly covalent bonding between the metals and oxygen, we can draw the Lewis structures below.

CrO_4^{2-}, valence electrons: 6 from Cr + 24 from O + 2 from charge = 32 e^-, 16 e^- pairs

MnO_4^{2-}, valence electrons: 7 from Mn + 24 from O + 1 from charge = 32 e^-, 16 e^- pairs

```
[    :Ö:    ]2-    [    :Ö:     ]-
[     |     ]      [     |      ]
[:Ö—Cr—Ö:]         [:Ö—Mn—Ö:]
[     |     ]      [     |      ]
[    :Ö:    ]      [    :Ö:     ]
```

In both Lewis structures, all valence electrons are paired; both metals are in the diamagnetic state. Note that both extreme models predict a diamagnetic state for the metal centers.

(c) "Inorganic" ions are those not based on carbon. SO_4^{2-} and ClO_4^- have the same formula type and charge as CrO_4^{2-} and MnO_4^-. S has 6 valence electrons like Cr, and Cl has 7 like Mn. Both SO_4^- and ClO_4^- are tetrahedral, so we predict that CrO_4^{2-} and MnO_4^- are also tetrahedral. The Lewis structures in part (b) support this prediction.

23.8 (a) $Ni(s) + H_2SO_4(aq) \rightarrow NiSO_4(aq) + H_2(g)$

(b) $Ni(s) + Br_2(l) \rightarrow NiBr_2(s)$

Metallurgy

23.10 (a) +4

(b) $MnO_2(s) + 4H^+(aq) + 2e^- \rightarrow Mn^{2+}(aq) + 2H_2O(l)$ $E^\circ_{red} = 1.23$ V

$Mn^{2+}(aq) + 2e^- \rightarrow Mn(s)$ $E^\circ_{red} = -1.18$ V

Standard reduction potentials indicate that a very strong reducing agent is required to reduce the ore to Mn(s), at least if $Mn^{2+}(aq)$ is an intermediate product. According to Appendix E, only Group I and Group II metals (Li, Na, Mg, etc.) are strong enough to reduce Mn^{2+} to Mn. In practice, $MnO_2(s)$ is reduced by coke in blast furnaces, into which MnO_2 is added to incorporate Mn into steel.

23.12 (a) *Calcination* is heating an ore to decompose the mineral of interest into a simple solid and volatile compound. Calcination usually produces a metal oxide and a gas that is a nonmetal oxide.

(b) *Leaching* is dissolving the mineral of interest to remove it from an ore. The solvent is usually water or an aqueous solution of acid, base or salt.

(c) *Smelting* is heating an ore, often in a reducing atmosphere, to a very high temperature so that two immiscible liquid layers form. The layers are usually the molten metal or metals of interest and slag.

(d) *Slag* is the unwanted layer of the smelting process. It contains molten silicate, aluminate, phosphate or fluoride compounds.

23.14 (a) $2PbS(s) + 3O_2(s) \xrightarrow{\Delta} 2PbO(s) + 2SO_2(g)$

(b) $CoCO_3(s) \xrightarrow{\Delta} CoO(s) + CO_2(g)$

(c) $WO_3(s) + 3H_2(g) \xrightarrow{\Delta} W(s) + 3H_2O(g)$

(d) $VCl_3(g) + 3K(l) \rightarrow V(s) + 3KCl(s)$

(e) $3BaO(s) + P_2O_5(l) \rightarrow Ba_3(PO_4)_2(l)$

23.16 (a) $CoO(s) + CO_2(g)$

(b) $CO(g)$

(c) $CoCO_3(s) \xrightarrow{\Delta} CoO(s) + CO_2(g)$; $CoO(s) + CO(g) \rightarrow Co(s) + CO_2(g)$

23.18 $FeO(s) + H_2(g) \rightarrow Fe(s) + H_2O(g)$

$FeO(s) + CO(g) \rightarrow Fe(s) + CO_2(g)$

$Fe_2O_3(s) + 3H_2(g) \rightarrow 2Fe(s) + 3H_2O(g)$

$Fe_2O_3(s) + 3CO(g) \rightarrow 2Fe(s) + 3CO_2(g)$

23.20 (a) In the *converter*, oxidation of C, Si and metals by O_2 are exothermic reactions that raise the temperature.

(b) $2C(s) + O_2(g) \rightarrow 2CO(g)$; $S(s) + O_2(g) \rightarrow SO_2(g)$; $Si(s) + O_2(g) \rightarrow SiO_2(l)$

23.22 $SnO_2(s) + C(s) \xrightarrow{\Delta} Sn(l) + CO_2(g)$

$Sn(s) \rightarrow Sn^{2+}(aq) + 2e^-$ (anode)

$Sn^{2+}(aq) + 2e^- \rightarrow Sn(s)$ (cathode)

Metals and Alloys

23.24 Since silicon has the same crystal structure as diamond, it is a covalent-network solid with its bonding electrons localized between Si atoms. Since there is no significant delocalization in Si, it is not likely to have the metallic properties of malleability, ductility and high electrical and thermal conductivity. It is likely to be hard and high-melting like other covalent-network solids.

23.26 (a) Cr: $[Ar]4s^1 3d^5$, Z = 24; Se: $[Ar]4s^2 3d^{10} 4p^4$, Z = 34

Both elements have the [Ar] core configuration and both have six valence electrons. The orbital locations of the six valence electrons are different in the two elements, because Se has more total electrons.

(b) Different Z and Z_{eff} for the two elements, and the different orbital locations of the valence electrons, are the main factors that lead to the differences in properties. In Cr, the 4s and 3d electrons are the valence electrons. Its Z and Z_{eff} are smaller than those of Se and it is not likely to gain enough electrons to achieve a noble-gas configuration. Thus, Cr loses electrons when it forms ions, acting like a metal. Se is in the same row of the periodic table as Cr, but its 3d subshell is filled, so its valence electrons are in 4s and 4p. Because Se has a larger Z and Z_{eff}, it is more likely to hold its own valence electrons and gain other electrons when it forms ions. That Se needs only two additional electrons to achieve the noble-gas configuration of Kr is also a driving force for it gaining electrons when it forms ions, acting like a nonmetal.

23.28 Moving across the fifth period from Y to Mo, the melting points of the metals increase. The number of valence electrons also increases, from 3 for Y to 6 for Mo. More valence electrons (up to 6) mean increased occupancy of the bonding molecular orbital band, and increased strength of metallic bonding. Melting requires that atoms are moving relative to each other. Stronger metallic bonding requires more energy to break bonds and mobilize atoms, resulting in higher melting points from Y to Mo.

23.30 According to band theory, an *insulator* has a completely filled valence band and a large energy gap between the valence band and the nearest empty band; electrons are localized within the lattice. A *conductor* must have a partially filled energy band; a small excitation will promote electrons to previously empty levels within the band and allow them to move freely throughout the lattice, giving rise to the property of conduction. A *semiconductor* has a filled valence band, but the gap between the filled and empty bands is small enough to jump to the empty conduction band. The presence of an impurity may also place an electron in an otherwise empty band (producing an n-type semiconductor), or create a vacancy in an otherwise full band (producing a p-type semiconductor), providing a mechanism for conduction.

23.32 Electrical conductivity is related to the extent of valence electron delocalization in the material.

In the hexagonal close-packed structure of titanium, each Ti atom has twelve nearest neighbors. The four valence electrons of a Ti atom are delocalized over bonding interactions with twelve neighbors. In the diamond structure of silicon, each Si atom has four nearest neighbors and four valence electrons. These four valence electrons are essentially localized in four covalent (sigma) bonds to the four nearest neighbors. The much more extensive electron delocalization in Ti leads to its significantly greater electrical conductivity.

23.34 Substitutional and interstitial alloys are both solution alloys. In a *substitutional* alloy, the atoms of the "solute" take positions normally occupied by the "solvent." Substitutional alloys tend to form when solute and solvent atoms are of comparable size and have similar bonding characteristics. In an *interstitial* alloy, the atoms of the "solute" occupy the holes or interstitial positions between "solvent" atoms. Solute atoms are necessarily much smaller than solvent atoms.

Transition Metals

23.36 (b) NiCo alloy and (c) W will have metallic properties. The lattices of these substances are composed of neutral metal atoms. Delocalization of valence electrons via the MO model of metallic bonding produces metallic properties.

(d) Ge is a metalloid, not a metal. (a) $TiCl_4$ is an ionic compound and (e) Hg_2^{2+} is a metal ion. In ions and ionic compounds, electrons are localized on the individual ions, precluding metallic properties.

23.38 *Analyze/Plan.* Define lanthanide contraction (Section 23.7). Based on the definition, list properties related to atomic radius. *Solve.*

The *lanthanide contraction* is the name given to the decrease in atomic size due to the build-up in effective nuclear charge as we move through the lanthanides (elements 58–71) and beyond them. This effect offsets the expected increase in atomic size going from the second to the third transition series. The lanthanide contraction affects size-related properties such as ionization energy, electron affinity, and density.

23.40 (a) CdO (b) TiO (c) Nb_2O_5 (d) NiO_2

23.42 V: $[Ar]4s^23d^3$, (V^{2+}: $[Ar]3d^3$); Sc: $[Ar]4s^23d^1$; (Sc^{2+}: $[Ar]3d^1$)

V has a slightly larger Z (23) and Z_{eff} than Sc (Z > 21), so the 3d electrons in V are more tightly held than those in Sc. Also, losing the lone 3d electron in Sc^{2+} leads to the stable noble-gas configuration of [Ar] for Sc^{3+}.

23.44 (a) Ti^{2+}: $[Ar]3d^2$ (b) Co^{3+}: $[Ar]3d^6$ (c) Pd^{2+}: $[Kr]4d^8$

(d) Mo^{3+}: $[Kr]4d^3$ (e) Ru^{3+}: $[Kr]4d^5$ (f) Ni^{4+}: $[Ar]3d^6$

23.46 The stronger reducing agent is more easily oxidized; Cr^{2+} is more easily oxidized (see Solution 23.45), so it is the stronger reducing agent.

23.48 (a) Chromate ion, CrO_4^{2-}, is bright yellow. Dichromate, $Cr_2O_7^{2-}$, is orange.

(b) $Cr_2O_7^{2-}$ is more stable in acid solution than CrO_4^{2-}.

(c) Their interconversion in solution involves the acid-base equilibrium

$$2CrO_4^{2-}(aq) + 2H^+(aq) \rightleftharpoons Cr_2O_7^{2-}(aq) + H_2O(l).$$

23.50 (a) $MnO_2(s) + 4HCl(aq) \rightarrow MnCl_2(aq) + Cl_2(g) + 2H_2O(l)$

(b) Yes. MnO_2 is the oxidizing agent; HCl is the reducing agent.

23.52 (a) *Ferromagnetic* materials can form "permanent" magnets, whereas *paramagnetic* materials cannot.

(b) In order for a substance to be ferromagnetic, the magnetic moments on sites throughout the lattice must interact with one another. That is, the sites must be physically close and overlap of orbitals must enable the individual magnetic sites to couple forming a much larger magnetic moment throughout the solid. Because of these interactions, the sustained existence of the magnetic moment does not require application of an external magnetic field.

(c) No. Other metals (Co and Ni, for example), alloys and some oxides (CrO_2) are also ferromagnetic.

Additional Exercises

23.54 Al^{3+}, Mg^{2+}, and Na^+ all have large negative reduction potentials (Al, Mg, and Na are very active metals). A substance with a more negative reduction potential would have to be used to chemically reduce them. All such substances are more expensive and difficult to obtain than Al, Mg, and Na. Electrolysis is thus the most cost-efficient way to reduce Al^{3+}, Mg^{2+}, and Na^+ to their metallic states.

23.56 (a) $2VCl_3(s) + O_2(g) \rightarrow 2VOCl_3(s)$

(b) $Nb_2O_5(s) + 5H_2(g) \rightarrow 2Nb(s) + 5H_2O(l)$

(c) $2Fe^{3+}(aq) + Zn(s) \rightarrow 2Fe^{2+}(aq) + Zn^{2+}(aq)$

(d) $NbCl_5(s) + 3H_2O(l) \rightarrow HNbO_3(s) + 5HCl(aq)$

23.57 (a) $NiO(s) + 2H^+(aq) \rightarrow Ni^{2+}(aq) + H_2O(l)$

(b) The simple answer is that the solid is subjected to acid hydrolysis:

$$CuCo_2S_4(s) + 8H^+(aq) \rightarrow Cu^{2+}(aq) + 2Co^{3+}(aq) + 4H_2S(g)$$

However, in the absence of a strong complexing ligand, Co^{3+} is not stable in water. It oxidizes water according to the following reaction:

$$4Co^{3+}(aq) + 2H_2O(l) \rightarrow 4Co^{2+}(aq) + O_2(g) + 4H^+(aq)$$

(c) $TiO_2(s) + C(s) + 2Cl_2(g) \rightarrow TiCl_4(g) + CO_2(g)$

(d) In this reaction O_2 is reduced and sulfide is oxidized. Writing the sulfur product as S_8, the balanced equation is:

$$8ZnS(s) + 4O_2(g) + 16H^+(aq) \rightarrow 8Zn^{2+}(aq) + S_8(s) + 8H_2O(l)$$

23.59 All transition metals have the generic electron configuration $ns^2(n-1)d^x$. Regardless of the number of d electrons, each transition metal has 2 *ns* valence electrons that are the first electrons lost when metal ions are formed. Thus, almost every transition metal has a stable +2 oxidation state.

After the 2 *ns* electrons are lost, a varying number of $(n-1)$d electrons can be lost, depending on the identity of the transition metal. The availability of different numbers of d electrons leads to a wide variety of accessible oxidation states for the transition metals.

23.60 Assuming that SO_2 and N_2 are the nonmetallic products, two half-reactions can be written:

$$5[MoS_2(s) + 7H_2O(l) \rightarrow MoO_3(s) + 2SO_2(g) + 14H^+(aq) + 14e^-]$$
$$7[12H^+(aq) + 2NO_3^-(aq) + 10e^- \rightarrow N_2(g) + 6H_2O(l)]$$
$$5MoS_2(s) + 14H^+(aq) + 14NO_3^-(aq) \rightarrow 5MoO_3(s) + 10SO_2(g) + 7N_2(g) + 7H_2O(l)$$

$$MoO_3(s) + 2NH_3(aq) + H_2O(l) \rightarrow (NH_4)_2MoO_4(s)$$

$$(NH_4)_2MoO_4(s) \xrightarrow{\Delta} 2NH_3(g) + H_2O(g) + MoO_3(s)$$
$$MoO_3(s) + 3H_2(g) \xrightarrow{\Delta} Mo(s) + 3H_2O(g)$$

23.62 Antimony is a metalloid with 5 valence electrons ($5s^25p^3$). It is not a semiconductor like graphite, Si and Ge, group 4A elements with 4 valence electrons. As a poor electrical conductor (insulator), it must have a large band gap between the valence band (bonding band) and the conduction band (antibonding band). Valence electrons are localized in covalent Sb–Sb bonds. Niobium is a metal with 5 valence electrons ($5s^24d^3$). As a good electrical conductor, it has a very small or zero band gap. Its valence electrons are delocalized over several (more than 5) Nb–Nb close contacts. The extent of valence electron delocalization determines the electrical conductivity of a solid.

23.63 The metallic properties of malleability, ductility, and high electrical and thermal conductivity are results of the delocalization of valence electrons throughout the lattice. Delocalization occurs because metal atom valence orbitals of nearest-neighbor atoms interact to produce nearly continuous molecular orbital energy bands. When C atoms are introduced into the metal lattice, their valence orbitals do not have the same energies as metal orbitals, and their interaction is different. This causes a discontinuity in the band structure and limits delocalization of electrons. The properties of the carbon-infused metal begin to resemble those of a covalent-network lattice with localized electrons (Solution 23.24). The substance is harder and less conductive than the pure metal.

23.65 (a) Nb^{5+}: [Ar]; diamagnetic, no unpaired electrons

(b) Cr^{2+}: $[Ar]3d^4$ paramagnetic, unpaired electrons

(c) Cu^+: $[Ar]3d^{10}$; diamagnetic, no unpaired electrons

(d) Ru^{8+}: [Ar]; diamagnetic, no unpaired electrons

(e) Ni^{2+}: $[Ar]3d^8$; paramagnetic, unpaired electrons

23.66 In a ferromagnetic solid, the magnetic centers are coupled such that the spins of all unpaired electrons are parallel. As the temperature of the solid increases, the average kinetic energy of the atoms increases until the energy of motion overcomes the force aligning the electron spins. The substance becomes paramagnetic; it still has unpaired electrons, but their spins are no longer aligned.

23.68 (a) Nothing. As noted in Section 20.8, a basic environment (OH^-) inhibits oxidation of Fe^{2+} to Fe^{3+}, even in the presence of $O_2(g)$.

(b) $Cu(NO_3)_2(aq) + 2KOH(aq) \rightarrow Cu(OH)_2(s) + 2KNO_3(aq)$

$Cu(OH)_2(s)$ precipitates. Cu^{2+} forms a soluble complex ion with $NH_3(aq)$, but not $OH^-(aq)$.

(c) The color of the solution changes from orange ($Cr_2O_7^{2-}$) to yellow (CrO_4^{2-}). The equilibrium is

$$Cr_2O_7^{2-}(aq) + H_2O(l) \rightleftharpoons 2CrO_4^{2-}(aq) + 2H^+(aq)$$

As $OH^-(aq)$ is added, it reacts with and removes $H^+(aq)$ from solution, shifting the equilibrium to the right in favor of the yellow CrO_4^{2-}.

23.69 (a) $2NiS(s) + 3O_2(g) \rightarrow 2NiO(s) + 2SO_2(g)$

(b) $2C(s) + O_2(g) \rightarrow 2CO(g)$; $C(s) + H_2O(g) \rightarrow CO(g) + H_2(g)$

$NiO(s) + CO(g) \rightarrow Ni(s) + CO_2(g)$; $NiO(s) + H_2(g) \rightarrow Ni(s) + H_2O(g)$

(c) $Ni(s) + 2HCl(aq) \rightarrow NiCl_2(aq) + H_2(g)$

(d) $NiCl_2(aq) + 2NaOH(aq) \rightarrow Ni(OH)_2(s) + 2NaCl(aq)$

(e) $Ni(OH)_2(s) \xrightarrow{\Delta} NiO(s) + H_2O(g)$

Integrative Exercises

23.72 Recall from the discussion in Chapter 13 that like substances tend to be soluble in one another, whereas unlike substances do not. Molten metal consists of atoms that continue to be bound to one another by metallic bonding, even though the substance is liquid. In a slag, on the other hand, the attractive forces are those between ions. The slag phase is a highly polar, ionic medium, whereas the metallic phase is nonpolar, and the attractive interactions are due to metallic bond formation. There is little driving force for materials with such different characteristics to dissolve in one another.

23.74 (a) $\Delta G° = \Delta H° - T\Delta S°$ (assume $\Delta H°$ and $S°$ are constant with changes in temperature)

$Si(s) + 2MnO(s) \rightarrow SiO_2(s) + 2Mn(s)$

$$\Delta H° = \Delta H_f° \; SiO_2(s) + 2\Delta H_f° \; Mn(s) - 2\Delta H_f° \; MnO(s) - \Delta H_f° \; Si(s)$$

$$\Delta H° = -910.9 + 2(0) - 2(-385.2) + 0 = -140.5 \text{ kJ}$$

$\Delta S° = S°\ SiO_2(s) + 2S°\ Mn(s) - 2S°\ MnO(s) - S°\ Si(s)$

$= 41.84 + 2(32.0) - 2(59.7) - 18.7 = -32.26 = -32.3\ J/K$

$\Delta G° = -140.5\ kJ - 1473\ K(-0.03226\ kJ/K) = -93.0\ kJ$

(b) At 1473 K, the reactants and products are all solids, so they are in their standard states. Since $\Delta G°$ is negative at this temperature, the reaction should be spontaneous and thus feasible.

23.75 (a) According to Section 20.8, the reduction of O_2 during oxidation of Fe(s) to Fe_2O_3 requires H^+. Above pH 9, iron does not corrode. At the high temperature of the converter, it is unlikely to find H_2O or H^+ in contact with the molten Fe. Also, the basic slag (CaO(l)) that is present to remove phosphorus will keep the environment basic rather than acidic. Thus, the H^+ necessary for oxidation of Fe in air is not present in the converter.

(b) $C + O_2(g) \rightarrow CO_2(g)$

$S + O_2(g) \rightarrow SO_2(g)$

$P + O_2(g) \rightarrow P_2O_5(l);\ P_2O_5(l) + 3CaO(l) \rightarrow Ca_3(PO_4)(l)$

$Si + O_2(g) \rightarrow SiO_2$

$M + O_2(g) \rightarrow M_xO_y(l);\ M_xO_y + SiO_2 \rightarrow$ silicates

CO_2 and SO_2 escape as gases. P_2O_5 reacts with CaO(l) to form $Ca_3(PO_4)_2(l)$, which is removed with the basic slag layer. SiO_2 and metal oxides can combine to form other silicates; SiO_2, M_xO_y, and complex silicates are all removed with the basic slag layer.

23.77 $\Delta G° = -RT\ lnK;\ \Delta G° = \Delta H° - T\Delta S°$

Calculate $\Delta H°$ and $\Delta S°$ using data from Appendix C, assuming $\Delta H°$ and $\Delta S°$ remain constant with changing temperature. Then calculate $\Delta G°$ and K at the two temperatures.

$\Delta H° = 2\Delta H_f\ CO(g) - \Delta H_f^\circ\ C(s) - \Delta H_f^\circ\ CO_2(g)$

$\Delta H° = 2(-110.5) - 0 - (-393.5) = +172.5\ kJ$

$\Delta S° = 2S°\ CO(g) - S°\ C(s) - S°\ CO_2(g)$

$= 2(197.9) - 5.69 - 213.6 = +176.5\ J/K = 0.1765\ kJ/K$

$\Delta G^\circ_{298} = 172.5\ kJ - 298\ K(0.1765\ kJ/K) = +119.9\ kJ$

$$\ln K = \frac{\Delta G°}{-RT} = \frac{119.9\ kJ}{-(8.314 \times 10^{-3}\ kJ/K)(298\ K)} = -48.3942 = -48.39;\ K = 9.6 \times 10^{-22}$$

$\Delta G^\circ_{2000} = 172.5\ kJ - 2000\ K(0.1765\ kJ/K) = -180.5\ kJ$

$$\ln K = \frac{-180.5}{-(8.314 \times 10^{-3}\ kJ/K)(2000\ K)} = 10.8552 = 10.86;\ K = 5.18 \times 10^4$$

(log K has 3 decimal places, so K has 3 sig figs.)

23.79 (a) The very low melting and boiling points for VF_5 indicate that it is molecular rather than ionic, and that the intermolecular forces are probably weak London-dispersion forces. In order for the molecule to experience only London-dispersion forces, it must be nonpolar covalent, which requires the symmetrical trigonal bipyramidal structure shown below. PF_5 also has this structure.

F
F
V—F
F
F

(b) $VCl_3(s) + 3HF(g) \xrightarrow{\Delta} VF_3(s) + 3HCl(g)$

(c) V(V) has a relatively small covalent radius. F is the smallest and most electronegative halogen. The steric repulsions associated with placing five larger halogens around the small V(V) central atom would be substantial. Also, the extreme electron attracting nature of F might be required to coax V into the +5 oxidation state.

23.80 Calculate the mass of Zn(s) that will be deposited.

$$2.0 \text{ m} \times 80 \text{ m} \times \frac{(100)^2 \text{ cm}^2}{1 \text{ m}^2} \times 0.49 \text{ mm} \times \frac{1 \text{ cm}}{10 \text{ mm}} \times \frac{7.1 \text{ g}}{\text{cm}^3} \times 2 \text{ sides}$$

$$= 1.113 \times 10^6 = 1.1 \times 10^6 \text{ g Zn}$$

$$1.113 \times 10^6 \text{ g Zn} \times \frac{1 \text{ mol Zn}}{65.39 \text{ g Zn}} \times \frac{2 \text{ F}}{0.90 \text{ mol Zn}} \times \frac{96{,}500 \text{ C}}{\text{F}} = 3.651 \times 10^9 = 3.7 \times 10^9 \text{ C}$$

(2 F/0.90 mol Zn takes the 90% efficiency into account.)

$$3.651 \times 10^9 \text{ C} \times 3.5 \text{ V} \times \frac{1 \text{ J}}{\text{C} \bullet \text{V}} \times \frac{1 \text{ kWh}}{3.6 \times 10^6 \text{ J}} = 3{,}550 = 3.6 \times 10^3 \text{ kWh}$$

$$3.550 \times 10^3 \text{ kWh} \times \frac{\$0.082}{1 \text{ kWh}} = \$291.06 \rightarrow \$291$$

23.81 (a)

$$Ag_2S(s) \rightleftharpoons 2Ag^+(aq) + S^{2-}(aq) \qquad K_{sp}$$

$$2[Ag^+(aq) + 2CN^-(aq) \rightleftharpoons Ag(CN)_2^-] \qquad K_f^2$$

$$Ag_2S(s) + 4CN^-(aq) \rightleftharpoons 2Ag(CN)_2^-(aq) + S^{2-}(aq)$$

$$K = K_{sp} \times K_f^2 = [Ag^+]^2[S^{2-}] \times \frac{[Ag(CN)_2^-]^2}{[Ag^+]^2[CN^-]^4} = (6 \times 10^{-51})(1 \times 10^{21})^2 = 6 \times 10^{-9}$$

(b) The equilibrium constant for the cyanidation of Ag_2S, 6×10^{-9}, is much less than one and favors the presence of reactants rather than products. The process is not practical.

(c)

$$AgCl(s) \rightleftharpoons Ag^+(aq) + Cl^-(aq) \qquad K_{sp}$$

$$Ag^+(aq) + 2CN^-(aq) \rightleftharpoons Ag(CN)_2^-(aq) \qquad K_f$$

$$AgCl(s) + 2CN^-(aq) \rightleftharpoons Ag(CN)_2^-(aq) + Cl^-(aq)$$

$$K = K_{sp} \times K_f = [Ag^+][Cl^-] \times \frac{[Ag(CN)_2^-]}{[Ag^+][CN^-]^2} = (1.8 \times 10^{-10})(1 \times 10^{21}) = 2 \times 10^{11}$$

Since K >> 1 for this process, it is potentially useful for recovering silver from horn silver. However, the magnitude of K says nothing about the rate of reaction. The reaction could be slow and require heat, a catalyst, or both to be practical.

23.82 (a) $M(s) \rightarrow M(g)$. The process of atomization is essentially breaking the "metallic bonds" in the solid metal and separating the particles into isolated gas-phase atoms. This requires relocalizing electrons from the solid lattice onto the individual metal atoms.

(b) ΔH_{atom} is the difference between the energy of a mole of gaseous metal atoms, isolated from one another, and a mole of the metal, with all its metal-metal bonding. The difference will be smaller if: 1) the gaseous atoms have some special stability relative to other metallic elements or 2) the metal-metal bonding in the solid is weaker.

The data indicate that Cr and Mn, in the middle of the first transition series, and Cu at the end, have smaller ΔH_{atom} than their neighbors. The electron configurations for these elements are: Cr, $[Ar]4s^13d^5$ (exception); Mn, $[Ar]4s^23d^5$; Cu, $[Ar]4s^13d^{10}$. The gaseous atoms of each of these elements have special stability due to either full or half-full subshells. Assuming relatively constant metal-metal bond strength, the special stability of the gaseous atoms reduces ΔH_{atom} for these elements, relative to their neighbors.

The lower values of ΔH_{atom} for Fe, Co, and Ni relative to the elements around V (after taking account of the variations in stability of the gaseous atoms) is likely due to decreasing metal-metal bond strength. Moving to the right across the transition series from the middle onward, effective nuclear charge increases, the radial extension of the d-orbitals decreases, and the strength of metallic bonding decreases. This is somewhat of a trend.

24 Chemistry of Coordination Compounds

Visualizing Concepts

24.2 *Analyze.* Given a ball-and-stick figure of a ligand, write the Lewis structure and answer questions about the ligand.

Plan. Assume that each atom in the Lewis structure obeys the octet rule. Complete each octet with unshared electron pairs or multiple bonds, depending on the bond angles in the ball-and-stick model. Black = C, blue = N, red = O, gray = H.

```
                                          -
 [       H   H        H  :O:        ]
 [       |   |        |   ||        ]
 [H—N̈—C—C—N̈—C—C—Ö:]
 [   |   |   |   |    |             ]
 [   H   H   H   H    H             ]
```

There is a second resonance structure with the double bond drawn to the second O atom.

Check. Write the molecular formula, count the valence electron pairs and see if it matches your structure. $[C_4H_9N_2O_2]^-$ (16 + 9 + 10 + 12 + 1) = 48 valence e^-, 24 e^- pair Our Lewis structure also has 24 e^- pairs.

(a) Donor atoms have unshared electron pairs. The potential donors in this structure are the two N and two O atoms.

The ligand is tridentate. (Even though there are four possible donor atoms, the structure would be strained if all four were bound to one metal center. It is likely that only one of the two O atoms binds to the same metal as the two N atoms.)

(b) An octahedral complex has 6 coordination sites. A single ligand has only 4 possible donors, so two ligands are needed. From a steric perspective, the likely donors would be the 2 N atoms and 1 of the carbonyl oxygen atoms. The chelate bite of a carboxyl group is relatively small and would require an O—M—O angle of less than 90°.

24.3 *Analyze.* Given a ball-and-stick structure, name the complex ion, which has a 1- charge.

Plan. Write the chemical formula of the complex ion, determine the oxidation state of the metal, and name the complex.

Solve. $[Pt(NH_3)Cl_3]^-$. Oxidation numbers: [Pt + 0 + 3(–1) = –1, Pt = +2, Pt(II)

Arrange the ligands alphabetically, followed y the metal. Since the complex is an anion, add the suffix -ate, then the oxidation state of the metal: aminotrichloroplatinate(II)

24.5 *Analyze.* Given four structures, decide which are chiral.

Plan. Chiral molecules have nonsuperimposable mirror images. Draw the mirror image of each molecule and visualize whether it can be rotated into the original molecule. If so, the complex is not chiral. If the original orientation cannot be regenerated by rotation, the complex is chiral. *Solve.*

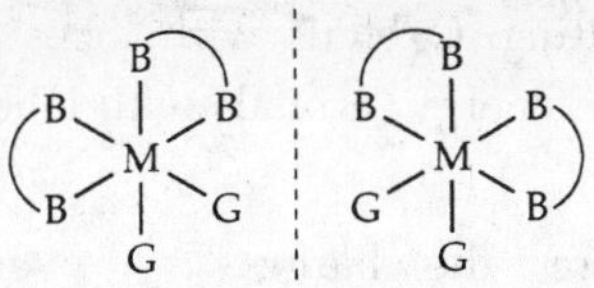

(1) (1) mirror

The two orientations are not superimposable and molecule (1) is chiral.

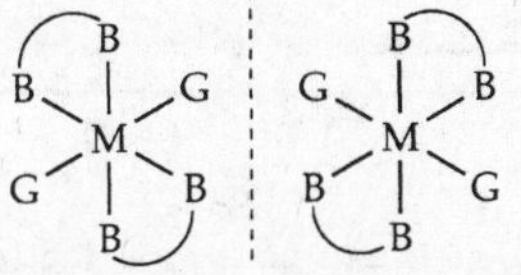

The two orientations are superimposible. Rotate the right-most structure 90° counterclockwise about the B-M-B axis to align the G's; the bidentate ligands then also overlap. Molecule (2) is not chiral.

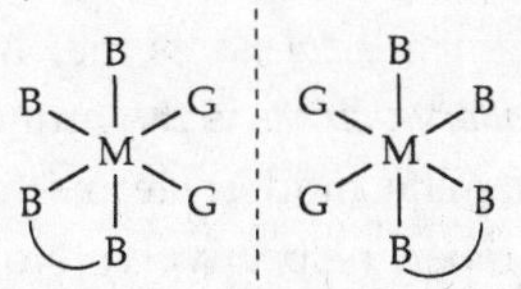

The two orientations are not superimposable and molecule (3) is chiral.

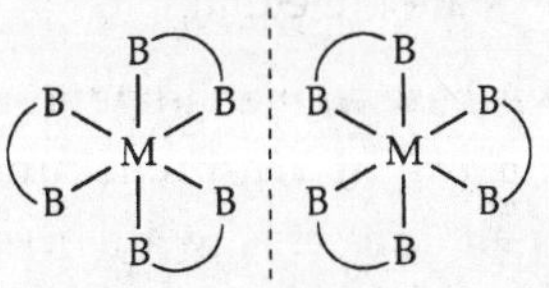

The two orientations are not superimposable and molecule (4) is chiral.

24.7 *Analyze.* Fit the crystal field splitting diagram to the complex description in each part.

Plan. Determine the number of d-electrons in each transition metal. On the splitting diagrams match the d-orbital splitting patterns to complex geometry and electron pairing to the definition of high-spin and low-spin.

Solve. Octahedral complexes have the 3 lower, 2 higher splitting pattern, while tetrahedral complexes have the opposite 2 lower, 3 higher pattern. Low spin complexes favor electron pairing because of large d-orbital splitting. High-spin complexes have maximum occupancy because of small orbital splitting.

(a) Fe^{3+}, 5d-electrons; weak field: spins unpaired; octahedral: 3 lower, 2 higher d-splitting $\therefore$ diagram (4)

(b) Fe^{3+}, 5d-electrons; strong field: spins paired; octahedral: 3 lower, 2 higher d-splitting $\therefore$ diagram (1)

(c) Fe^{3+}, 5d-electrons; tetrahedral: 2 lower, 3 higher d-splitting ∴ diagram (3)

(d) Ni^{2+}, 8 d-electrons; tetrahedral: 2 lower, 3 higher d-splitting ∴ diagram (2)

Check. Diagram (2) was the remaining choice for (d) and it fits the description.

24.8 *Analyze/Plan.* Given the linear diagram and axial labels, answer the questions and predict crystal field splitting. Orbitals with lobes nearest ligand charges (or partial charges) will be highest in energy; orbitals with lobes away from charges are lowest in energy.

Solve. d_{z^2} has lobes nearest the charges. $d_{x^2-y^2}$ and d_{xy} have lobes in the xy-plane farthest from the charges. d_{xz} and d_{yz} point between the respective axes and are intermediate in energy.

———— d_{z^2}

———— ———— d_{xz}, d_{yz}

———— ———— $d_{xz}, d_{x^2-y^2}$

Introduction to Metal Complexes

24.10 (a) In Werner's theory, *primary valence* is the charge of the metal cation at the center of the complex. *Secondary valence* is the number of atoms bound or coordinated to the central metal ion. The modern terms for these concepts are oxidation state and coordination number, respectively. (Note that "'oxidation state" is a broader term than ionic charge, but Werner's complexes contain metal ions where cation charge and oxidation state are equal.)

(b) Ligands are the Lewis base in metal-ligand interactions [see Solution 24.9(b)]. As such, they must possess at least one unshared electron pair. NH_3 has an unshared electron pair but BH_3, with less than 8 electrons about B, has no unshared electron pair and cannot act as a ligand. In fact, BH_3 acts as a Lewis acid, an electron pair acceptor, because it is electron-deficient.

24.12 (a) Yes. There are 6 possible ligands, 3 H_2O molecules and 3 Cl^- ions. Any Cl^- ions that are not coordinated to the metal will form AgCl(s) precipitate when the complex is treated with $AgNO_3$(aq). Absence of AgCl(s) would mean all Cl^- ions were ligands and a coordination number of 6. One mole of AgCl(s) per mole of complex would mean one uncoordinated ligand, and so on. This assumes that all 3 H_2O molecules act as ligands. In fact, they could serve as water of hydration (Section 13.1, A Closer Look: Hydrates). Reaction with $AgNO_3$ gives no information about the nature of H_2O molecules.

(b) Yes. Conductivity is directly related to the number of ions in a solution. The lower the conductivity, the more Cl^- ions that act as ligands. Conductivity measurements on a set of standard solutions with various moles of ions per mole of complex would provide a comparative method for quantitative determination of the number of free and bound Cl^- ions.

24.14 (a) Coordination number = 6, oxidation number = +2

(b) 4, +2 (c) 6, +1 (d) 6, +3

(e) 6, +3 (f) 5, +2

24.16 (a) 6 C (see Solution 24.15(d))

(b) 4 N

(c) 5 C, 1 Br. In CO, both C and O have an unshared electron pair. C is less electronegative and more likely to donate its unshared pair. (This is analogous to the situation in CN^-).

(d) 4 N, 2 O. en is a bidentate ligand bound through N, for a total of 4 N donors. $C_2O_4^{2-}$ is bidentate with 2 O donors.

(e) 6 N. When thiocyanate is written "NCS," it is bound through N. This makes a total of 6 N donors.

(f) 4 N, 1 I. bipy is bidentate bound through N, for a total of 4 N donors.

24.18 (a) 2 coordination sites, 2 N donor atoms

(b) 2 coordination sites, 2 N donor atoms

(c) 2 coordination sites, 2 O donor atoms (Although there are four potential O donor atoms in $C_2O_4^{2-}$, it is geometrically impossible for more than two of these to be bound to a single metal ion.)

(d) 4 coordination sites, 4 N donor atoms

(e) 6 coordination sites, 2 N and 4 O donor atoms

24.20 (a) 4 (b) 4 (c) 6 (d) 6

24.22 (a) Monodentate; py has only one N donor atom.

(b) K for this reaction will be less than one. Two free pyridine molecules are replaced by one free bipy molecule. There are more moles of particles in the reactants than products, so ΔS is predicted to be negative. Processes with a net decrease in entropy are usually nonspontaneous, have positive ΔG, and values of K less than one. This equilibrium is likely to be spontaneous in the reverse direction.

24.24 (a) $[Mn(H_2O)_5I](ClO_4)_2$ (b) $[Ru(bipy)_3](NO_3)_2$

(c) $[Rh(o\text{-phen})_2Cl_2]_2SO_4$ (d) $Na[Cr(NH_3)_2Br_4]$

(e) $[Co(en)_3]_4[Fe(ox)_3]_3$

24.26 (a) dichloroethylenediamminecadmium(II)

(b) potassium hexacyanomanganate(II)

(c) pentaamminecarbonatochromium(III) chloride

(d) tetraamminediaquairidium(III) nitrate

Isomerism

24.28 (a)

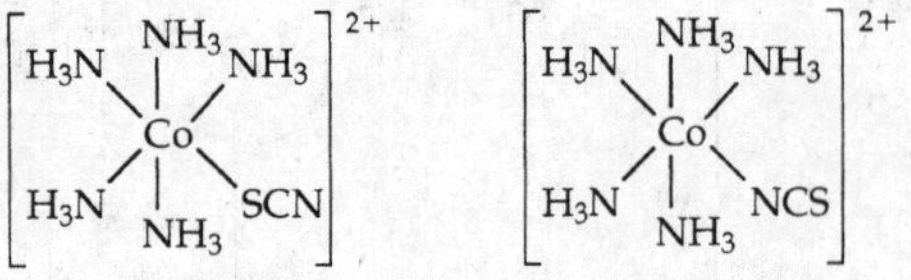

coordination sphere isomerism

(b)

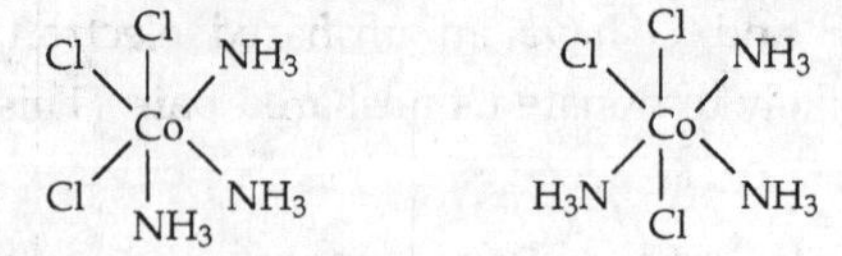

(c) Br^- Cl^-

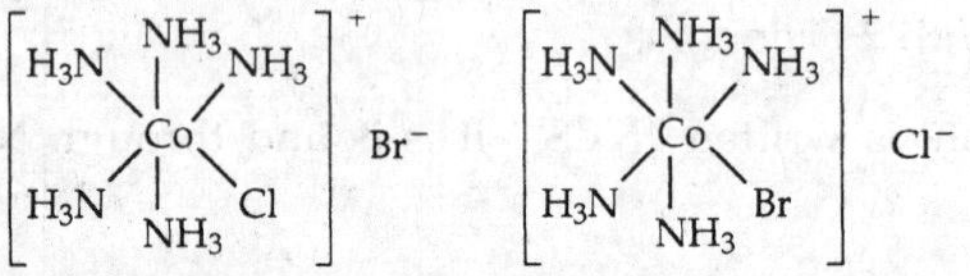

coordination sphere isomerism

24.30 Two geometric isomers are possible for an octahedral MA_3B_3 complex (see below). All other arrangements, including mirror images, can be rotated into these two structures. Neither isomer is optically active.

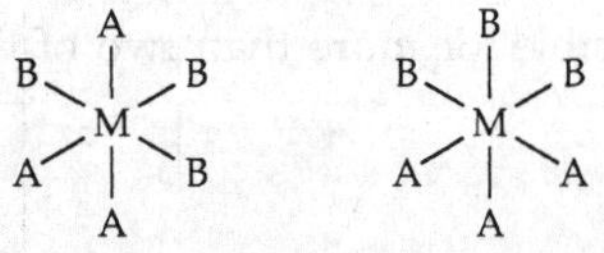

(i) (ii)

24.32

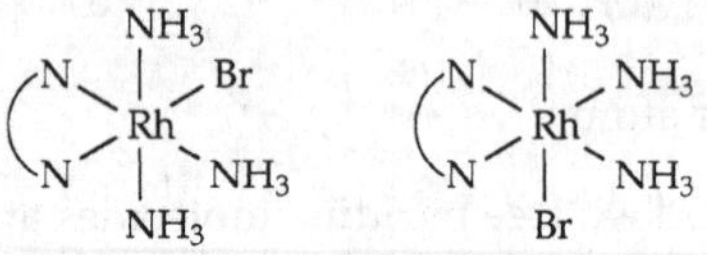

The symbol N⌒N represents the bidentate ligand (bipy).

There are no optical isomers. The mirror image of each structural isomer can be superimposed on the structure above by a 180° rotation.

24.34 (a) (b)

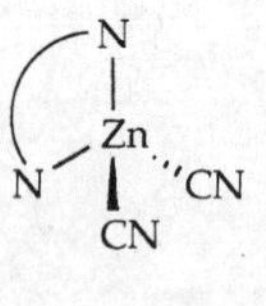

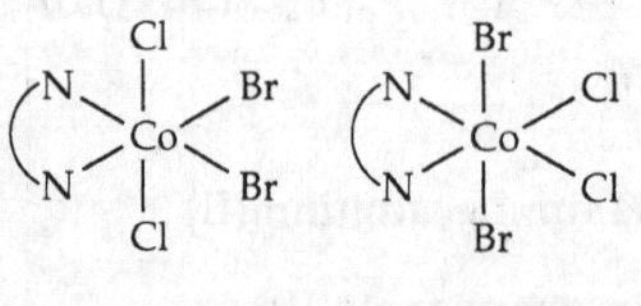

(c)

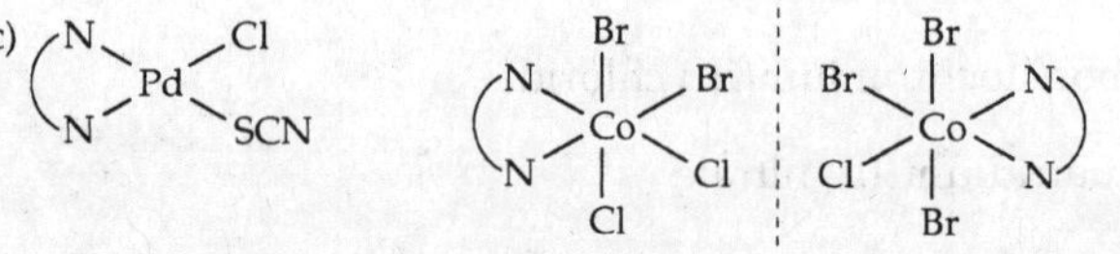

Color, Magnetism; Crystal-Field Theory

24.36 (a) Yes. A complex that absorbs visible light of one wavelength or color will appear as the complementary color. This complex absorbs green light and will appear red.

(b) No. A solution can appear green by transmitting or reflecting only green light (the situation stated in the exercise) **or** by absorbing red light, the complementary color of green.

(c) A visible absorption spectrum shows the amount of light absorbed at a given wavelength. It is a plot of absorbance (dependent variable, y-axis) vs. wavelength (independent variable, x-axis).

(d) E(J/photon) = $h\nu = hc/\lambda$. Change J/photon to kJ/mol.

$$E = \frac{6.626 \times 10^{-34}\ \text{J} \bullet \text{s}}{530\ \text{nm}} \times \frac{3.00 \times 10^{8}\ \text{m}}{\text{s}} \times \frac{1\,\text{nm}}{1 \times 10^{-9}\ \text{m}} = 3.751 \times 10^{-19} = 3.75 \times 10^{-19}\ \text{J}$$

$$3.751 \times 10^{-19} = \frac{1\ \text{kJ}}{1000\ \text{J}} \times \frac{6.022 \times 10^{23}\ \text{photons}}{\text{mol}} = 226\ \text{kJ/mol}$$

24.38 Six ligands in an octahedral arrangement are oriented along the x, y, and z axes of the metal. These negatively charged ligands (or the negative end of ligand dipoles) have greater electrostatic repulsion with valence electrons in metal orbitals that also lie along these axes, the d_{z^2}, and $d_{x^2-y^2}$. The d_{xy}, d_{xz} and d_{yz} metal orbitals point between the x, y, and z axes, and electrons in these orbitals experience less repulsion with ligand electrons. Thus, in the presence of an octahedral ligand field, the d_{xy}, d_{xz} and d_{xy} metal orbitals are lower in energy than the $d_{x^2-y^2}$ and d_{z^2}.

24.40 (a) $$\Delta E = hc/\lambda = \frac{6.626 \times 10^{-34}\ \text{J} \bullet \text{s} \times 2.998 \times 10^{8}\ \text{m/s}}{500 \times 10^{-9}\ \text{m}} = 3.973 \times 10^{-19} = 3.97 \times 10^{-19}\ \text{J/photon}$$

$$\Delta = 3.973 \times 10^{-19}\ \text{J/photon} \times \frac{6.022 \times 10^{23}\ \text{photons}}{1\ \text{mol}} \times \frac{1\ \text{kJ}}{1000\ \text{J}} = 239.25 = 239\ \text{kJ/mol}$$

(b) The *spectrochemical* series is an ordering of ligands according to their ability to increase the energy gap Δ. If H_2O is replaced by NH_3 in the complex, the magnitude of Δ would increase because NH_3 is higher in the spectrochemical series and creates a stronger ligand field.

24.42 The ions absorb the complement of the color they appear. Green $[Ni(H_2O)_6]^{2+}$ absorbs red light, 650-800 nm. Purple $[Ni(NH_3)_6]^{2+}$ absorbs yellow light, 560–580 nm. Thus, $[Ni(NH_3)_6]^{2+}$ absorbs light with the shorter wavelength. This agrees with the spectrochemical series, which indicates that H_2O will produce a smaller d-orbital splitting (Δ) than NH_3. Thus, $[Ni(H_2O)_6]^{2+}$ should absorb light with a smaller energy and longer wavelength.

24.44 (a) Fe^{3+}, d^5 (b) Mn^{2+}, d^5 (c) Ag^{+}, d^{10}

(d) Cr^{3+}, d^3 (e) Sr^{2+}, d^0

24.46 (a) Ru: $[Kr]5s^1 4d^7$ (b) Mo: $[Kr]5s^1 4d^5$ (c) Co: $[Ar]4s^2 3d^7$

Ru^{3+}: $[Kr]4d^5$ Mo^{3+}: $[Kr]3d^3$ Co^{3+}: $[Ar]3d^6$

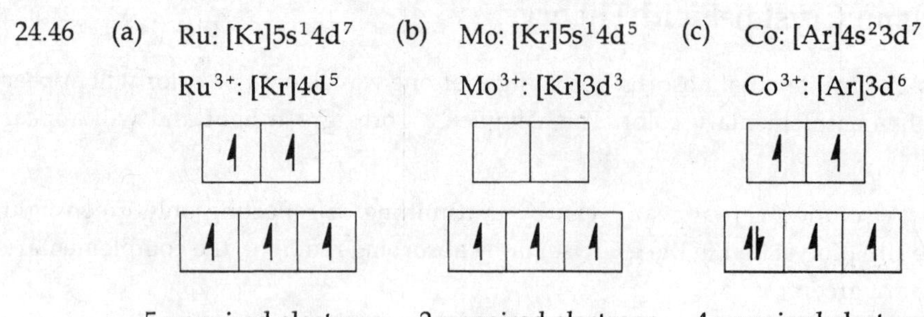

5 unpaired electrons 3 unpaired electrons 4 unpaired electrons

24.48

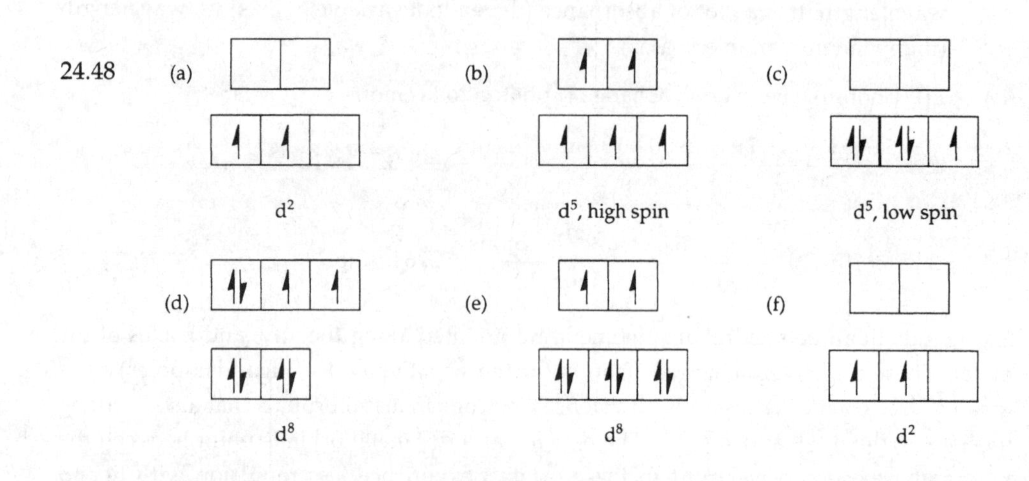

24.50

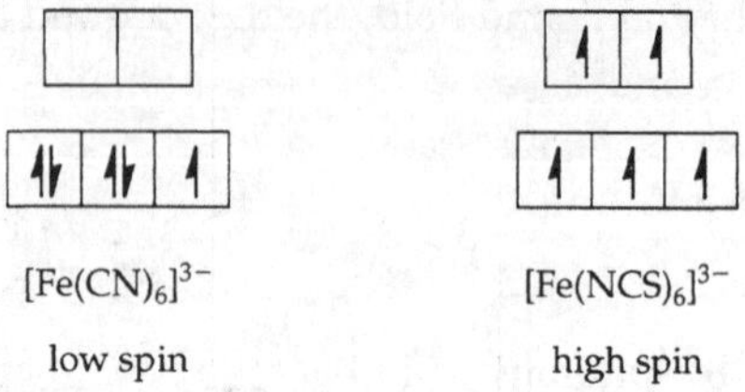

$[Fe(CN)_6]^{3-}$ low spin

$[Fe(NCS)_6]^{3-}$ high spin

Both complexes contain Fe^{3+}, a d^5 ion. CN^-, a strong field ligand, produces such a large Δ that the splitting energy is greater than the pairing energy, and the complex is low spin. NCS^- produces a smaller Δ, so it is energetically favorable for d-electrons to be unpaired in the higher energy d-orbitals. NCS^- is a much weaker-field ligand than CN^-. It is probably weaker than NH_3 and near H_2O in the spectrochemical series.

Additional Exercises

24.52 (a) $[Ru(H_2O)_5Cl]^{2+}$ + $2Cl^-$ (b) $[Ru(H_2O)_6]^{3+}$ + $3Cl^-$

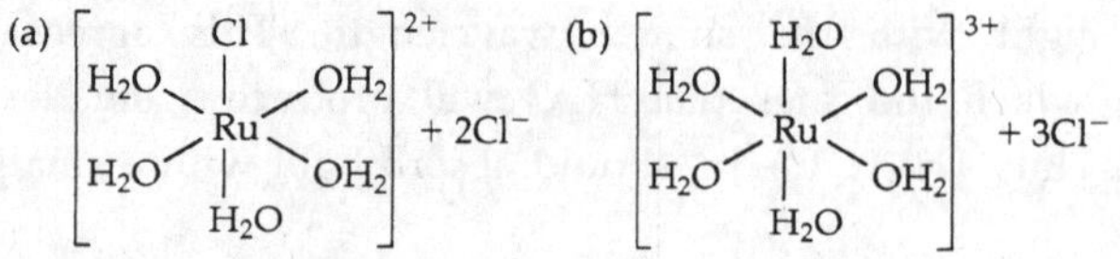

$[Ru(H_2O)_5Cl]Cl_2$ ⟶ $[Ru(H_2O)_6]Cl_3$

24.53 (a)

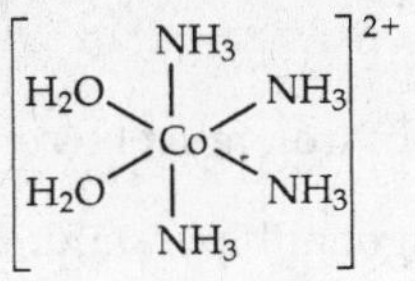

```
        NH3          2+
 H2O\    |    /NH3
      \  Co  /
 H2O/    |    \NH3
        NH3
```

octahedral

(b)

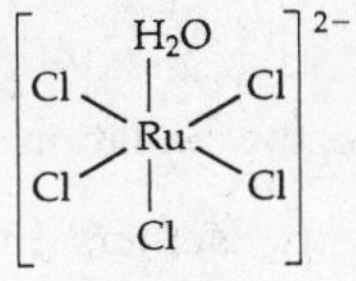

```
        H2O          2-
  Cl\    |    /Cl
      \  Ru  /
  Cl/    |    \Cl
         Cl
```

octahedral

(c)

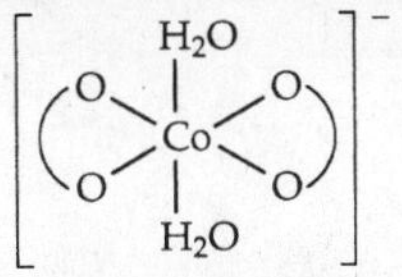

```
        H2O           -
   O\    |    /O
     \   Co  /
   O/    |    \O
        H2O
```

octahedral

(d)

```
         N
  N\     |    /Cl
      \  Ru  /
  N/     |    \Cl
         N
```

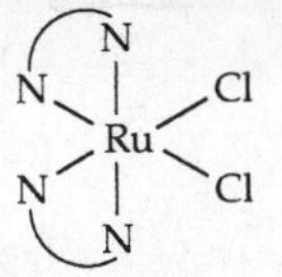

octahedral

24.55 (a) Valence electrons: 2P + 6C + 16H = 10 + 24 + 16 = 50 e⁻, 25 e⁻ pr

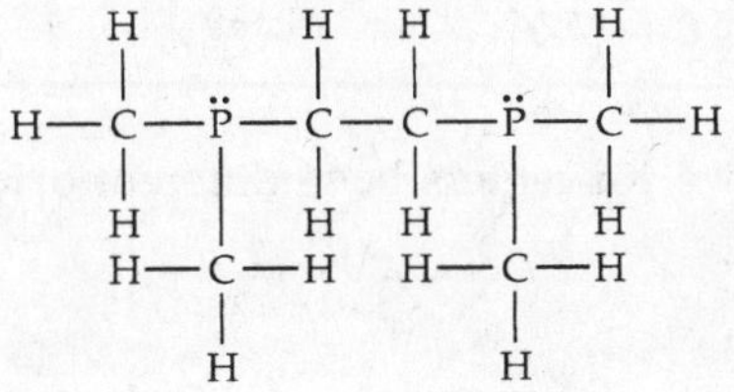

```
     H         H   H         H
     |         |   |         |
H—C—P̈—C—C—P̈—C—H
     |    |    |   |    |    |
     H    |    H   H    |    H
       H—C—H        H—C—H
          |              |
          H              H
```

(b) Both CO and dmpe are neutral molecules, so the oxidation state of Mo must be zero.

(c)

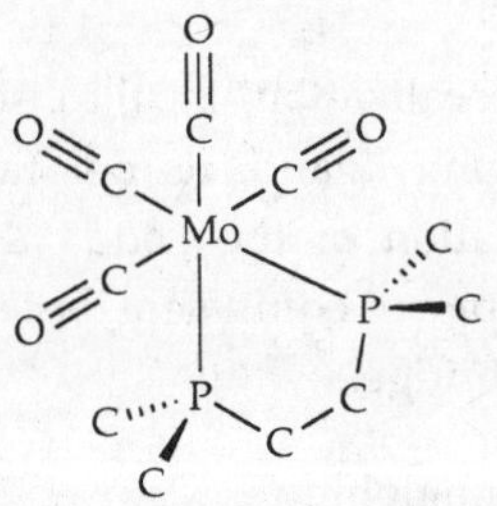

```
            O
            |||
            C
  O≡C       |      C≡O
       \    |    /
          Mo          C
  O≡C /     |     ⋰
            |    P—C
            |    |
   C····P        C
        |  \   /
        C    C
```

C≡O coordinates through C, because it is less electronegative than O, and a better electron pair donor. The molecule has only a single isomer. The dmpe ligand cannot span *trans* positions, so there are no geometric isomers. The mirror image of the structure above is easily superimposable, so there are no optical isomers.

(H atoms omitted for clarity)

24.57 We will represent the end of the bidentate ligand containing the CF_3 group by a shaded oval, the other end by an open oval:

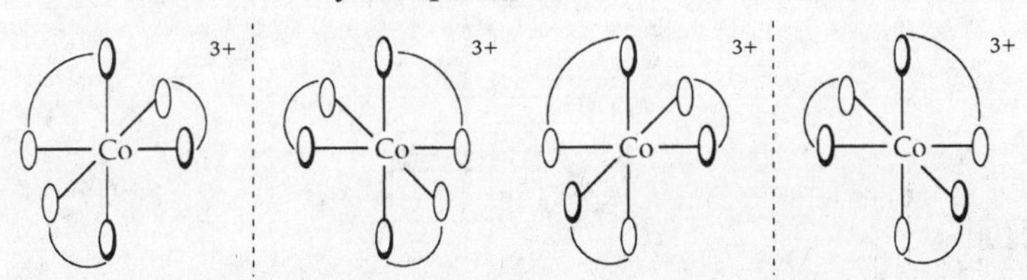

24.58 (a) Hemoglobin is the iron-containing protein that transports O_2 in human blood.

(b) Chlorophylls are magnesium-containing porphyrins in plants. They are the key components in the conversion of solar energy into chemical energy that can be used by living organisms.

(c) Siderophores are iron-binding compounds or ligands produced by a microorganism. They compete on a molecular level for iron in the medium outside the organism and carry needed iron into the cells of the organism.

24.60 (a) pentacarbonyliron(0)

(b) Since CO is a neutral molecule, the oxidation state of iron must be zero.

(c) $[Fe(CO)_4CN]^-$ has two geometric isomers. In a trigonal bipyramid, the axial and equatorial positions are not equivalent and not superimposable. One isomer has CN in an axial position and the other has it in an equatorial position.

CN / OC, OC / Fe—CO / CO CO / OC, OC / Fe—CN / CO

24.61 (a) left shoe

(c) wood screw

(e) a typical golf club

24.63 (a) Formally, the two Ru centers have different oxidation states; one is +2 and the other is +3.

(b) Ru^{2+}, d^6 Ru^{3+}, d^5

(c) There is extensive bonding-electron delocalization in the isolated pyrazine molecule. When pyrazine acts as a bridging ligand, its delocalized molecular orbitals provide a pathway for delocalization of the "odd" d-electron in the Creutz-Taube ion. The two metal ions appear equivalent because the odd d-electron is delocalized across the pyrazine bridge.

24.65

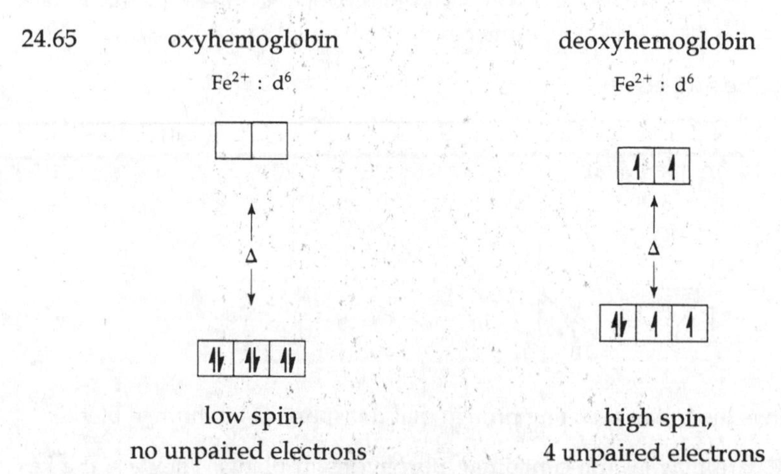

In general, the crystal field splitting, Δ, is greater in low spin than high spin complexes. The energy, Δ, corresponds to the wavelength of light absorbed by the complex and determines its color. Since oxyhemoglobin absorbs higher energy, shorter wavelength light, longer wavelengths remain and the sample appears red. Deoxyhemoglobin absorbs lower energy (orange-red) light, and the sample appears blue.

24.67 (a) The term *isoelectronic* means that the three ions have the same number of valence electrons and the same electron configuration.

(b) In each ion, the metal is in its maximum oxidation state and has a d^0 electron configuration. That is, the metal ions have no *d*-electrons, so there should be no *d*-*d* transitions.

(c) A *ligand-metal charge transfer* transition occurs when an electron in a filled ligand orbital is excited to an empty d-orbital of the metal.

(d) Absorption of 565 nm yellow light by MnO_4^- causes the compound to appear violet, the complementary color. CrO_4^{2-} appears yellow, so it is absorbing violet light of approximately 420 nm. The wavelength of the LMCT transition for chromate, 420 nm, is shorter than the wavelength of LCMT transition in permanganate, 565 nm. This means that there is a larger energy difference between filled ligand and empty metal orbitals in chromate than in permanganate.

(e) Yes. A white compound indicates that no visible light is absorbed. Going left on the periodic chart from Mn to Cr, the absorbed wavelength got shorter and the energy difference between ligand and metal orbitals increased. The 420 nm absorption by CrO_4^- is at the short wavelength edge of the visible spectrum. It is not surprising that the ion containing V, further left on the chart, absorbs at a still shorter wavelength in the ultraviolet region and that VO_4^{3-} appears white.

24.68 Application of pressure would result in shorter metal ionoxide distances. This would have the effect of increasing the ligand-electron repulsions, and would result in a larger splitting in the d-orbital energies. Thus, application of pressure should result in a shift in the absorption to a higher energy and shorter wavelength.

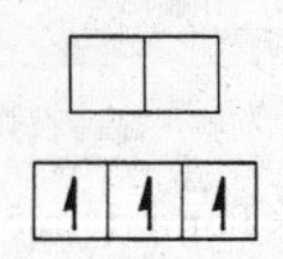

24.70

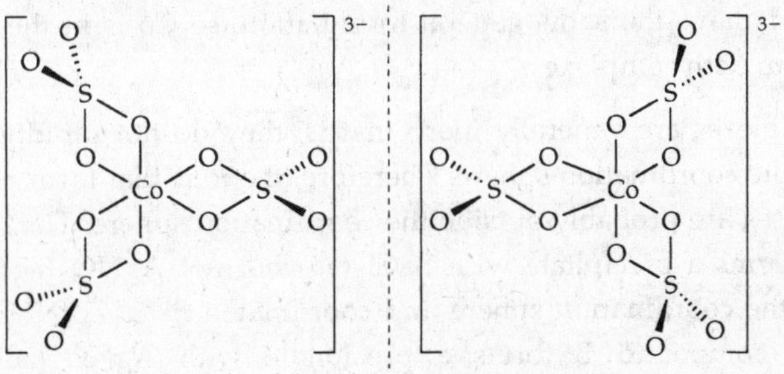

24.71 (a) Only one (b) Two

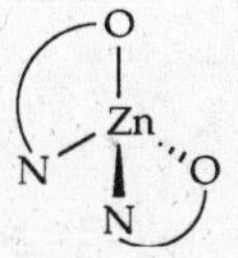

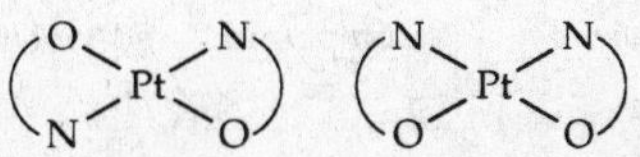

trans cis

(c) Four; two are geometric, the other two are stereoisomers of each of these.

24.72

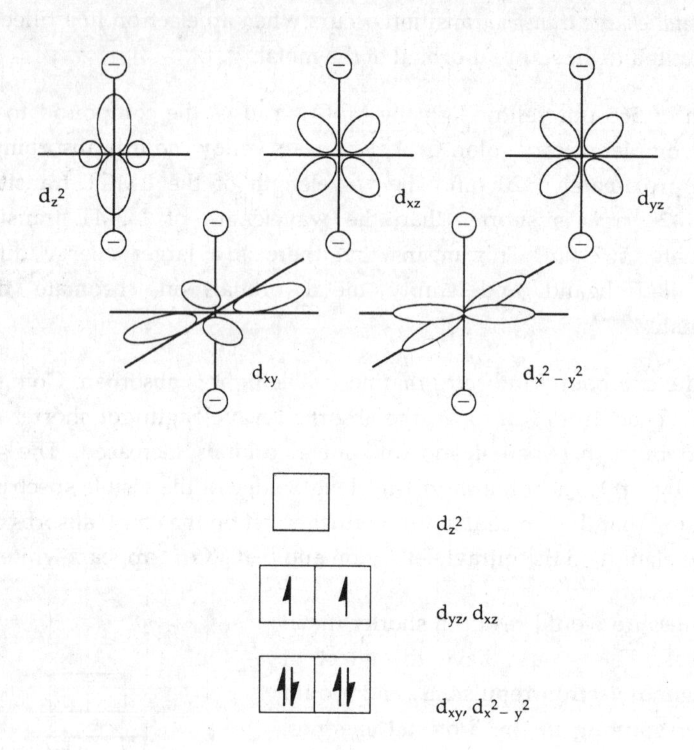

For a d^6 metal ion in a strong ligand field, there would be two unpaired electrons.

Integrative Exercises

24.74 (a) Both compounds have the same general formulation, so Co is in the same (+3) oxidation state in both complexes.

(b) Cobalt(III) complexes are generally inert; that is, they do not rapidly exchange ligands inside the coordination sphere. Therefore, the ions that form precipitates in these two cases are probably outside the coordination sphere. The dark violet compound A forms a precipitate with $BaCl_2(aq)$ but not $AgNO_3(aq)$, so it has SO_4^{2-} outside the coordination sphere and coordinated Br^-, $[Co(NH_3)_5Br]SO_4$. The red-violet compound B forms a precipitate with $AgNO_3(aq)$ but not $BaCl_2(aq)$ so it has Br^- outside the coordination sphere and coordinated SO_4^{2-}, $[Co(NH_3)_5SO_4]Br$.

Compound A, dark violet Compound B, red-violet

(c) Compounds A and B have the same formula but different properties (color, chemical reactivity), so they are isomers. They vary by which ion is inside the coordination sphere, so they are *coordination sphere isomers.*

(d) Compound A is an ionic sulfate and compound B is an ionic bromide, so both are strong electrolytes. According to the solubility rules in Table 4.1, both should be water-soluble.

24.76 First determine the empirical formula, assuming that the remaining mass of complex is Pd.

$$37.6\text{ g Br} \times \frac{1\text{ mol Br}}{79.904\text{ g Br}} = 0.4706\text{ mol Br}; 0.4706 / 0.2361 = 2$$

$$28.3\text{ g C} \times \frac{1\text{ mol C}}{12.01\text{ g C}} = 2.356\text{ mol C}; 2.356 / 0.2361 = 10$$

$$6.60\text{ g N} \times \frac{1\text{ mol N}}{14.01\text{ g N}} = 0.4711\text{ mol N}; 0.4711 / 0.2361 = 2$$

$$2.37\text{ g H} \times \frac{1\text{ mol H}}{1.008\text{ g H}} = 2.351\text{ mol H}; 2.351 / 0.2361 = 10$$

$$25.13\text{ g Pd} \times \frac{1\text{ mol Pd}}{106.42\text{ g Pd}} = 0.2361\text{ mol Pd}; 0.2361 / 0.2361 = 1$$

The chemical formula is $[Pd(NC_5H_5)_2Br_2]$. This should be a neutral square-planar complex of Pd(II), a nonelectrolyte. Because the dipole moment is zero, we can infer that it must be the trans isomer.

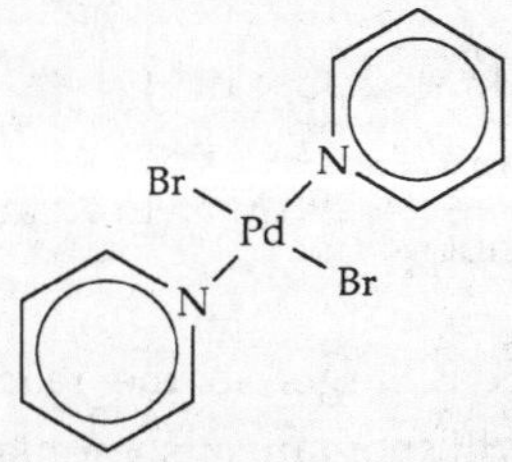

24.77 (a) The reaction that occurs increases the conductivity of the solution by producing a greater number of charged particles, particles with higher charges, or both. It is likely that H_2O from the bulk solvent exchanges with a coordinated Br^- according to the reaction below. This reaction would convert the 1:1 electrolyte, $[Co(NH_3)_4Br_2]Br$, to a 1:2 electrolyte, $[Co(NH_3)_3(H_2O)Br]Br_2$.

(b) $[Co(NH_3)_4Br_2]^+(aq) + H_2O(l) \rightarrow [Co(NH_3)_4(H_2O)Br]^{2+}(aq) + Br^-(aq)$

(c) Before the exchange reaction, there is one mole of free Br^- per mole of complex. mol Br^- = mol Ag^+

M = mol/L; L $AgNO_3$ = mol $AgNO_3$/M $AgNO_3$

$$\frac{3.87\text{ g complex}}{0.500\text{ L soln}} \times \frac{1\text{ mol complex}}{366.77\text{ g complex}} \times 0.02500\text{ L soln used} =$$

$$5.276 \times 10^{-4} = 5.28 \times 10^{-4}\text{ mol complex}$$

$$5.276 \times 10^{-4} \text{ mol complex} \times \frac{1 \text{ mol Br}^-}{1 \text{ mol complex}} \times \frac{1 \text{ mol Ag}^+}{1 \text{ mol Br}^-} \times \frac{1 \text{ L Ag}^+(aq)}{0.0100 \text{ mol Ag}^+(aq)}$$
$$= 0.05276 \text{ L} = 52.8 \text{ mL AgNO}_3(aq)$$

(d) After the exchange reaction, there are 2 mol free Br^- per mol of complex. Since *M* $AgNO_3(aq)$ and volume of complex solution are the same for the second experiment, the titration after conductivity changes will require twice the volume calculated in part (c), 105.52 = 106 mL of 0.0100 *M* $AgNO_3(aq)$.

24.79 Use Hess' law to calculate ΔG° for the desired equilibrium. Then ΔG° = -RTlnK to calculate K.

$$\text{Hb} + \text{CO} \rightarrow \text{HbCO} \qquad \Delta G° = -80 \text{ kJ}$$
$$\text{HbO}_2 \rightarrow \text{Hb} + \text{O}_2 \qquad \Delta G° = 70 \text{ kJ}$$
$$\text{HbO}_2 + \text{Hb} + \text{CO} \rightarrow \text{HbCO} + \text{Hb} + \text{O}_2$$
$$\text{HbO}_2 + \text{CO} \rightarrow \text{HbCO} + \text{O}_2 \qquad \Delta G° = -10 \text{ kJ}$$

$$\Delta G° = -RT\ln K, \ln K = \frac{-\Delta G°}{RT} = \frac{-(-10 \text{ kJ})}{8.314 \text{ J/K} \bullet \text{mol} \times 298 \text{ K}} \times \frac{1000 \text{ J}}{\text{kJ}} = 4.036 = 4.04$$

$K = e^{4.04} = 56.61 = 57$

24.80 (a)

$$[Cd(CH_3NH_2)_4]^{2+} \rightleftharpoons Cd^{2+}(aq) + 4CH_3NH_2(aq) \qquad \Delta G° = 37.2 \text{ kJ}$$
$$Cd^{2+}(aq) + 2en(aq) \rightleftharpoons [Cd(en)_2]^{2+}(aq) \qquad \Delta G° = -60.7 \text{ kJ}$$
$$Cd(CH_3NH_2)_4]^{2+} + 2en(aq) \rightleftharpoons [Cd(en)_2]^{2+}(aq) + 4CH_3NH_2(aq) \qquad \Delta G° = -23.5 \text{ kJ}$$

$\Delta G° = -RT\ln K$; $-23.5 \text{ kJ} = -2.35 \times 10^4 \text{ J}$

$$-2.35 \times 10^4 \text{ J} = \frac{-8.314 \text{ J}}{\text{K} \bullet \text{mol}} \times 298 \text{ K} \times \ln K; \ln K = 9.485, K = 1.32 \times 10^4$$

(b) The magnitude of K is large, so the reaction favors products. The bidentate chelating ligand en will spontaneously replace the monodentate ligand CH_3NH_2. This is an illustration of the chelate effect.

(c) Using the stepwise construction from part (a),

$\Delta H° = 57.3 \text{ kJ} - 56.5 \text{ kJ} = 0.8 \text{ kJ}$

$\Delta S° = 67.3 \text{ J/K} + 14.1 \text{ J/K} = 81.4 \text{ J/K}$

$-T\Delta S = -298 \text{ K} \times 81.4 \text{ J/K} = -2.43 \times 10^4 \text{ J} = -24.3 \text{ kJ}$

The chelate effect is mainly the result of entropy. The reaction is spontaneous due to the increase in the number of free particles and corresponding increase in entropy going from reactants to products. The enthalpic contribution is essentially zero because the bonding interactions of the two ligands are very similar and the reaction is not "downhill" in enthalpy.

(d) $\Delta H°$ will be very small and negative. When NH_3 replaces H_2O in a complex (Closer Look Box), the tighter bonding of the NH_3 ligand causes a substantial negative $\Delta H°$ for the substitution reaction. When a bidentate amine ligand replaces a monodentate amine ligand of similar bond strength, $\Delta H°$ is very small and either positive (part (c)) or negative (Closer Look Box). In the case of NH_3 replacing CH_3NH_2, the bonding characteristics are very similar. The presence of CH_3 groups in CH_3NH_2 produces some steric hindrance in $[Cd(CH_3NH_2)_4]^{2+}$. This complex is at a slightly higher energy than $[Cd(NH_3)_4]^{2+}$, which experiences no steric hindrance, so $\Delta H°$ will have a negative sign but a very small magnitude. Relief of steric hindrance leads to a very small negative $\Delta H°$ for the substitution reaction.

24.82 The process can be written:

$$H_2(g) + 2e \rightarrow 2H^+(aq) \qquad E^\circ_{red} = 0.0\ V$$

$$Cu(s) \rightarrow Cu^{2+} + 2e^- \qquad E^\circ_{red} = 0.337\ V$$

$$Cu^{2+}(aq) + 4NH_3(aq) \rightarrow [Cu(NH_3)_4]^{2+}(aq) \qquad "E^\circ_f" = ?$$

$$H_2(g) + Cu(s) + 4NH_3(aq) \rightarrow 2H^+(aq) + [Cu(NH_3)_4]^{2+}(aq) \qquad E = 0.08\ V$$

$$E = E° - RT\ \ln K;\ K = \frac{[H^+]^2[Cu(NH_3)_4^{2+}]}{P_{H_2}[NH_3]^4}$$

$P_{H_2} = 1$ atm, $[H^+] = 1\ M$, $[NH_3] = 1\ M$, $[Cu(NH_3)_4]^{2+} = 1\ M$, $Q = 1$

$E = E° - RT\ \ln(1);\ E = E° - RT(0);\ E = E° = 0.08\ V$

Since we know E° values for two steps and the overall reaction, we can calculate "E°" for the formation reaction and then K_f, using $E° = \frac{0.0592}{n} \log K_f$ for the step.

$$E_{cell} = 0.08\ V = 0.0\ V - 0.337\ V + "E^\circ_f" \qquad "E^\circ_f" = 0.08\ V + 0.337\ V = 0.417\ V = 0.42\ V$$

$$"E^\circ_f" = \frac{0.0592}{n} \log K_f;\ \log K_f = \frac{n(E^\circ_f)}{0.0592} = \frac{2(0.417)}{0.0592} = 14.0878 = 14$$

$K_f = 10^{14.0878} = 1.2 \times 10^{14} = 10^{14}$

24.83 (a) The units of the rate constant and the rate dependence on the identity of the second ligand show that the reaction is second order. Therefore, the rate-determining step cannot be a dissociation of water, since that would be independent of the concentration and identity of the incoming ligand. The alternative mechanism, a bimolecular association of the incoming ligand with the complex, is indicated.

(b) The relative values of rate constant are a reflection of the kinetic basicities of the three ligands: Pyridine > SCN^- > CH_3CN.

(c) Ru(III) is a d^5 ion. In a low-spin d^5 octahedral complex, there is one unpaired electron.

25 The Chemistry of Life: Organic and Biological Chemistry

Visualizing Concepts

25.1 *Analyze/Plan.* Given structural formulas, specify which molecules are unsaturated. Consider the definition of unsaturated and apply it to the molecules in the exercise. *Solve.*

Unsaturated molecules contain one or more multiple bonds. Saturated molecules contain only single bonds. Molecules (c) and (d) are unsaturated.

25.3 *Analyze/plan.* Given structural formulas, predict which molecule will have the highest boiling point. Boiling point is determined by strength of intermolecular forces; for neutral molecules with similar molar masses, the strongest intermolecular force is hydrogen bonding. The molecule that experiences hydrogen bonding will have the highest boiling point. *Solve.*

Only F—H, O—H, and N—H bonds fit the strict definition of hydrogen bonding. Molecule (b), an alcohol, forms hydrogen bonds with like molecules; it has the highest boiling point.

[Molecules (a) and (d) have dipole-dipole and dispersion forces. The greater molar mass of (d) probably means it has stronger dispersion forces and a slightly higher boiling point than molecule (a). Molecule (c) has only dispersion forces and the lowest boiling point. The probable order of strength of forces and boiling points is: (b) > (d) > (a) > (c).]

25.4 *Analyze.* Given structural formulas, decide which molecules are capable of isomerism, and what type. *Plan.* Analyze each molecule for possible structural, geometric, and optical isomers/enantiomers. *Solve.*

(a) $C_5H_{11}NO_2$. Structural, geometric and optical. There are many ways to arrange the atoms in molecules with this empirical formula, so there are many structural isomers. There is one point of unsaturation in the given molecule, the C=O group; structural isomers with their point of unsaturation at a C=C group could have geometric isomers as well. The C atom to which the $-NH_3^+$ group is bound is a chiral center, so there are enantiomers. All amino acids except glycine have two possible enantiomers.

(b) $C_7H_5O_2Cl$. The most obvious isomers for this aromatic compound are ortho, meta, and para geometric isomers. Because the molecule has several points of unsaturation, the number of structural isomers is limited, but there are a few possibilities with two triple bonds. Switching the —OH and —Cl groups also generates a structural isomer. There are no chiral centers, so no optical isomers.

(c) C_5H_{10}. There are many structural isomers for this empirical formula. The straight-chain alkene shown also has geometric (cis-trans) isomers.

(d) C_3H_8. There are no other structural, geometric, or optical isomers for this molecule.

25.6 *Analyze/Plan.* Follow the logic in Sample Exercise 25.1 to name each compound. Decide which structures are the same compound. *Solve.*

(a) 2,2,4-trimethylpentane (b) 3-ethyl-2-methylpentane

(c) 2,3,4-trimethylpentane (d) 2,3,4-trimethylpentane

Structures (c) and (d) are the same molecule.

Introduction to Organic Compounds; Hydrocarbons

25.8

```
          H   H   H   H   OH  O
          |   |   |   |   |   ||
N≡C — C — C — C = C — C — C — H
  7   6|  5|   4   3  2|   1
       H   H          H
```

(a) C2, C5 and C6 have sp^3 hybridization (4 e^- domains around C)

(b) C7 has sp hybridization (2 e^- domains around C)

(c) C1, C3, and C4 have sp^2 hybridization (3 e^- domains around C)

25.10 From Table 8.4, the bond enthalpies in kJ/mol are: C–H, 413; C–C, 348; C–O, 358; C–Cl, 328. The bond enthalpes indicate that C–H bonds are most difficult to break, and C–Cl bonds least difficult. However, they do not explain the reactivity of C–O bonds, or stability of C–C bonds.

The reactivity of molecules containing C–O and C–Cl bonds is a result of their unequal charge distribution, which attracts reactants that are either electron deficient (electrophilic) or electron rich (nucleophilic).

25.12 All the classifications listed are hydrocarbons; they contain only the elements hydrogen and carbon.

(a) *Alkanes* are hydrocarbons that contain only single bonds.

(b) *Cycloalkanes* contain at least one ring of three or more carbon atoms joined by single bonds. Because it is a type of alkane, all bonds in a cycloalkane are single bonds.

(c) *Alkenes* contain at least one C=C double bond.

(d) *Alkynes* contain at least one C≡C triple bond.

(e) A *saturated hydrocarbon* contains only single bonds. Alkanes and cycloalkanes fit this definition.

(f) An *aromatic hydrocarbon* contains one or more planar, six-membered rings of carbon atoms with delocalized π-bonding throughout the ring.

25 Organic and Biological Chemistry

Solutions to Black Exercises

25.14 cycloalkane, $H_2C—CH_2$, H_2C, CH_2, C, H, CH_3 (ring), C_6H_{12}, saturated

cycloalkene, $HC{=}CH$, H_2C, CH_2, C, H, CH_3 (ring), C_6H_{10}, unsaturated

alkyne, $CH_3—CH_2—C{\equiv}C—CH_2—CH_3$, C_6H_{10}, unsaturated

aromatic hydrocarbon, H—C (benzene ring of six C, each bearing one H) C—H, C_6H_6, unsaturated

25.16 C_nH_{2n-2}

25.18 $CH_3CH_2CH_2CH_2C{\equiv}CH$ $CH_3CH_2CH_2C{\equiv}CCH_3$

$CH_3CH_2C{\equiv}CCH_2CH_3$ $CH_3CH(CH_3)CH_2C{\equiv}CH$

$CH_3CH_2CH(CH_3)C{\equiv}CH$ $CH_3CH(CH_3)C{\equiv}CCH_3$

$CH_3CH_2C(H){=}C(H)—CH{=}CH_2$ $CH_3CH_2C(H){=}C(H)—CH{=}CH_2$

$CH_3C(H){=}C(H)—CH_2CH{=}CH_2$ $CH_3C(H){=}C(H)—CH_2CH{=}CH_2$

$CH_2{=}CHCH_2CH_2CH{=}CH_2$ $CH_3C(H){=}C(H)—C(H){=}C(H)—CH_3$

$CH_3C(H){=}C(H)—C(H){=}C(H)—CH_3$ $CH_3C(H){=}C(H)—C(H){=}C(H)—CH_3$

$CH_2{=}CHCH(CH_3)CH{=}CH_2$ $CH_2{=}C(CH_3)CH_2CH{=}CH_2$

CH_3 CH_3
$CH_2{=}CHC{=}CH$

CH_3 H
$CH_2{=}CHC{=}CCH_3$

CH_3 H
$CH_2{=}CC{=}CCH_3$
H

CH_3
$CH_2{=}CC{=}CCH_3$
H H

CH_3
$CH_2{=}CHCH{=}CCH_3$

CH_2CH_3
$CH_2{=}CHC{=}CH_2$

CH_3 CH_3
$CH_2{=}C{-}C{=}CH_2$

CH_3, H
$C{=}C{=}C$
CH_3, CH_3

CH_3, H
$C{=}C{=}C$
CH_2CH_3, H

H, H
$C{=}C{=}C$
$CH_2CH_2CH_3$, H

H, H
$C{=}C{=}C$
CH_2CH_3, CH_3

CH
HC CH_2
H_2C CH_2
CH_2

CH_3
$CH{=}C$
H_2C CH_2
CH_2

$CH{=}CH$
CH_3
H_2C C H
CH_2

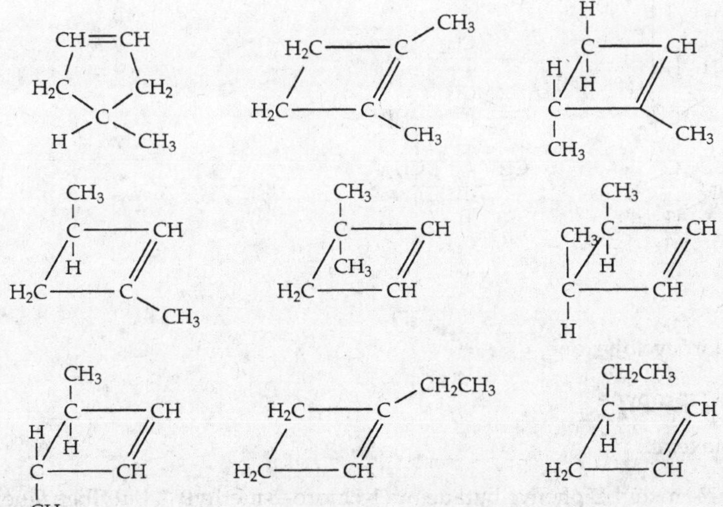

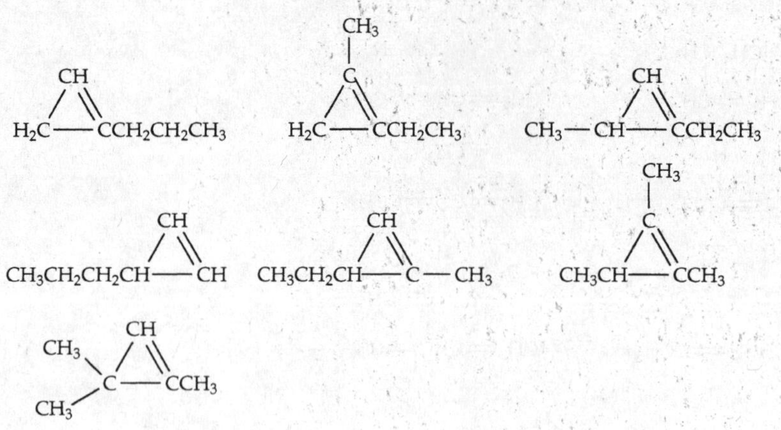

CH_3
C
H_2C—CCH_2CH_3

CH
H_2C—$CCH_2CH_2CH_3$

CH
CH_3—CH—CCH_2CH_3

CH
$CH_3CH_2CH_2CH$—CH

CH
CH_3CH_2CH—C—CH_3

CH_3
C
CH_3CH—CCH_3

CH
CH_3, CH_3 C—CCH_3

25.20 (a) sp^3 (b) sp^2 (c) sp^2 (d) sp

25.22 (a) 3,3,5-trimethylheptane

(b) 3,4,4-trimethylheptane

(c) CH_3—CH_2—CH_2—CH_2—CH(CH_3)—CH_2—CH_2—CH(CH_3)—CH_3

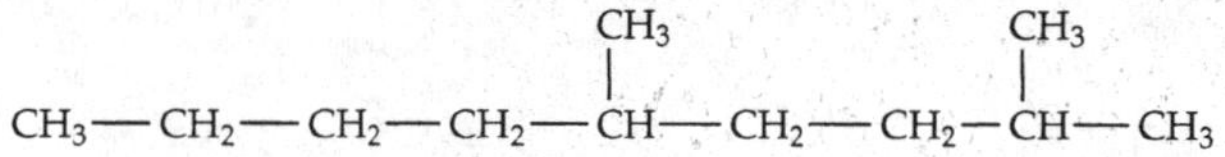

(d) CH_3—CH_2—CH_2—C(CH_3)(CH_3)—CH(CH_2CH_3)—CH_2—CH_3

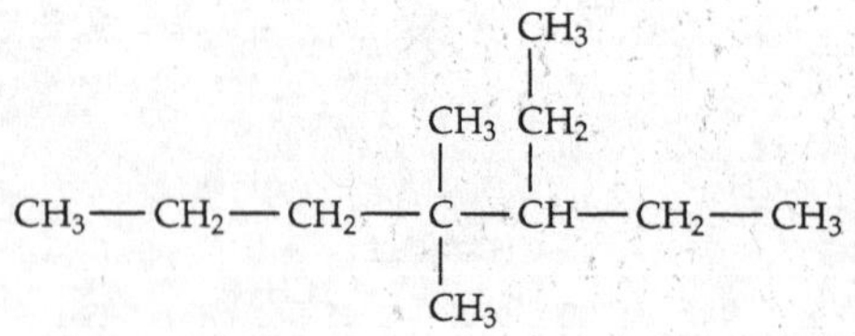

(e)

H
H—C—H
H—C—H H
H C H
H C C H
H C C H
H C H
H
H—C—H
H

CH_3
CH_2
CH
CH_2 CH_2
CH_2 CH_2
CH
CH_3

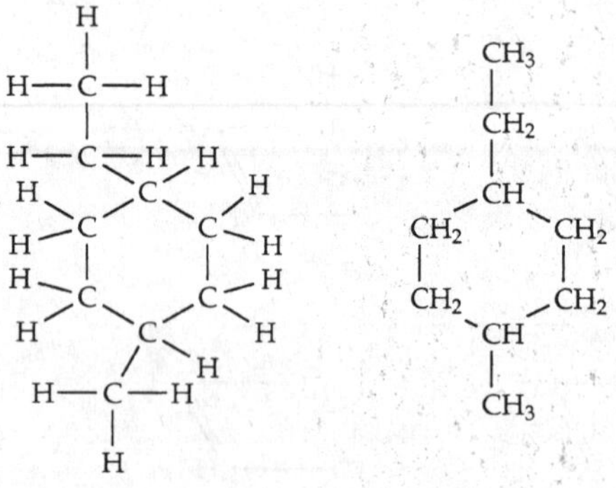

25.24 (a) 1,4-dichlorocyclohexane

(b) 3-chloro-1-propyne

(c) *trans*-2-hexene

(d) 1-chloro-2-methyl-2-phenyl-butane or (1-chloro-2-methyl)-2-butylbenzene

(e) *cis*-5-chloro-1,3-pentadiene

25.26 Butene is an alkene, C_4H_8. There are two possible placements for the double bond:

$$CH_2{=}CHCH_2CH_3 \text{ or } CH_3CH{=}CHCH_3$$

1-butene 2-butene

These two compounds are *structural isomers*. For 2-butene, there are two different, noninterchangeable ways to construct the carbon skeleton (owing to the absence of free rotation around the double bond). These two compounds are *geometric isomers*.

Cl, H / C=C / H_3C, CH_2Cl — *cis*-2-butene

H_3C, H / C=C / Cl, CH_2Cl — *trans*-2-butene

25.28 CH_3, H / C=C / H — C=C — H, CH_3, H

H, H / C=C / CH_3 — C=C — H, CH_3, H

H, H / C=C / CH_3 — C=C — CH_3, H, H

25.30 Octane number can be increased by increasing the fraction of branched-chain alkanes or aromatics, since these have high octane numbers. This can be done by cracking. The octane number also can be increased by adding an anti-knock agent such as tetraethyl lead, $Pb(C_2H_5)_4$ (no longer legal); methyl t-butyl ether (MTBE); or an alcohol, methanol, or ethanol.

Reactions of Hydrocarbons

25.32 (a) cyclohexene + H_2 $\xrightarrow{\text{Ni, 500° C}}$ cyclohexane

(b) H, CH_2CH_3 / C=C / CH_3, H + $\xrightarrow{H_2SO_4}$ $CH_3CH(OH)CH_2CH_2CH_3$ + $CH_3CH_2CH(OH)CH_2CH_3$

(c) benzene + $CH_3CHClCH_3$ $\xrightarrow{AlCl_3}$ $C_6H_5CH(CH_3)CH_3$ + HCl

25.34 (a) The reaction of Br_2 with an alkene to form a colorless halogenated alkane is an addition reaction. Aromatic hydrocarbons do not readily undergo addition reactions, because their π-electrons are stabilized by delocalization.

(b) *Plan.* Use a Friedel-Crafts reaction to substitute a $—CH_2CH_3$ onto benzene. Do a second substitution reaction to get *para*-bromoethylbenzene. *Solve.*

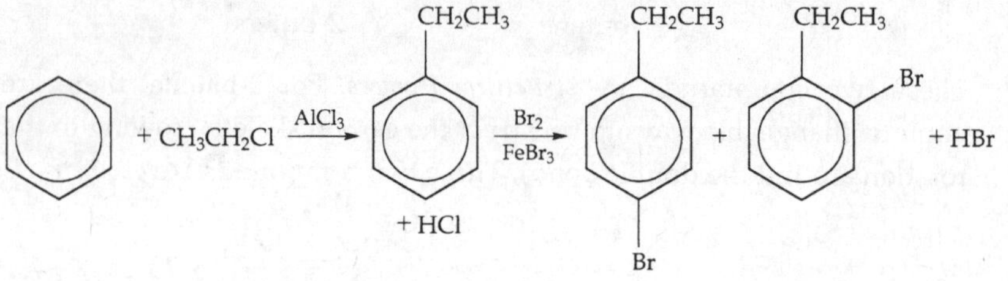

benzene + CH_3CH_2Cl $\xrightarrow{AlCl_3}$ ethylbenzene ($C_6H_5CH_2CH_3$) + HCl $\xrightarrow[FeBr_3]{Br_2}$ para-bromoethylbenzene + ortho-bromoethylbenzene + HBr

It appears that ortho, meta, and para geometric isomers of bromoethylbenzene would be possible. However, because of electronic effects beyond the scope of this chapter, the ethyl group favors formation of ortho and para isomers, but not the meta. The ortho and para products must be separated by distillation or some other technique.

25.36 Not necessarily. That the rate laws are both first order in both reactants and second order overall indicates that the activated complex in the rate-determining step in each mechanism is bimolecular and contains one molecule of each reactant. This is usually an indication that the mechanisms are the same, but it does not rule out the possibility of different fast steps, or a different order of elementary steps.

25.38

	ΔH
$C_{10}H_8(l) + 12O_2(g) \rightarrow 10CO_2(g) + 4H_2O(l)$	-5157 kJ
$-[C_{10}H_{18}(l) + 29/2\ O_2(g) \rightarrow 10CO_2(g) + 9H_2O(l)$	-(-6286) kJ
$C_{10}H_8(l) + 5H_2O(l) \rightarrow C_{10}H_{18}(l) + 5/2\ O_2(g)$	+1129 kJ
$5/2\ O_2(g) + 5H_2(g) \rightarrow 5H_2O(l)$	5(-285.8) kJ
$C_{10}H_8(l) + 5H_2(g) \rightarrow C_{10}H_{18}(l)$	-300 kJ

Compare this with the heat of hydrogenation of ethylene:

$C_2H_4(g) + H_2(g) \rightarrow C_2H_6(g)$; ΔH = -84.7 - (52.3) = -137 kJ. This value applies to just one double bond. For five double bonds, we would expect about -685 kJ. The fact that hydrogenation of napthalene yields only -300 kJ indicates that the overall energy of the napthalene molecule is lower than expected for five isolated double bonds and that there must be some special stability associated with the aromatic system in this molecule.

Functional Groups and Chirality

25.40 (a) $—C{\equiv}C—$, alkyne; $—C(=O)—H$, aldehyde

(b) 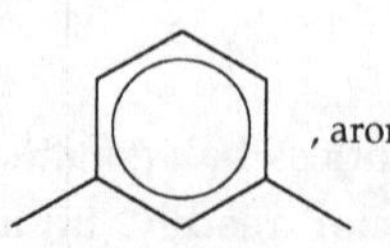 benzene ring, aromatic hydrocarbon (phenol group); —Cl, halogen; $—C{=}C—$, alkene; —COOH, carboxylic acid

(c) [cyclopentane ring], cycloalkane; —Cl, halogen; $\overset{O}{\overset{\|}{C}}$ (with two bonds), ketone

(d) $-C{=}C-$, alkene; $-\overset{O}{\overset{\|}{C}}-O-$, ester

(e) $\overset{|}{N}$ (with two bonds), amine; $-\overset{O}{\overset{\|}{C}}-N\langle$, amide

25.42 (a) C_4H_8O, [ring: H_2C–O–CH_2, H_2C-CH_2]

(b) $CH_3CH_2\overset{O}{\overset{\|}{C}}CH_3$, $CH_3CH_2CH_2\overset{O}{\overset{\|}{C}}-H$

$CH_2{=}CH_2CH_2CH_2OH$, $CH_3CH{=}CHCH_2OH$, (cis and trans)
$CH_2{=}CHCH(OH)CH_3$ (enantiomers)

(Structures with the –OH group attached to an alkene carbon atom are not included. These molecules are called "vinyl alcohols" and are not the major form at equilibrium.)

25.44 (a) $CH_3CH_2\overset{O}{\overset{\|}{C}}-H$ (b) $CH_3CH_2CH_2\overset{O}{\overset{\|}{C}}CH_3$

(c) $CH_3\underset{CH_3}{\underset{|}{C}}H\overset{O}{\overset{\|}{C}}CH_3$ (d) $CH_3CH_2\underset{CH_3}{\underset{|}{C}}H\overset{O}{\overset{\|}{C}}-H$

25.46 (a) $CH_3CH_2CH_2\overset{O}{\overset{\|}{C}}-O-CH_3$

methylbutanoate

(b) $C_6H_5-\overset{O}{\overset{\|}{C}}-O-\underset{CH_3}{\underset{|}{\overset{CH_3}{\overset{|}{C}}}}-H$

2-propylbenzoate

(c) $CH_3CH_2\overset{O}{\overset{\|}{C}}-\underset{CH}{\underset{|}{N}}-CH_3$

N, N-dimethylpropanamide

25.48 (a)

$$CH_3CH_2CH_2CH_2OH + HOCCH_2CH_3 \longrightarrow CH_3CH_2CH_2CH_2OCCH_2CH_3$$

(C=O on the carbonyl carbon of the acid and of the ester)

1-butanol — propionic acid (propanoic acid) — butyl proprionate

(b)

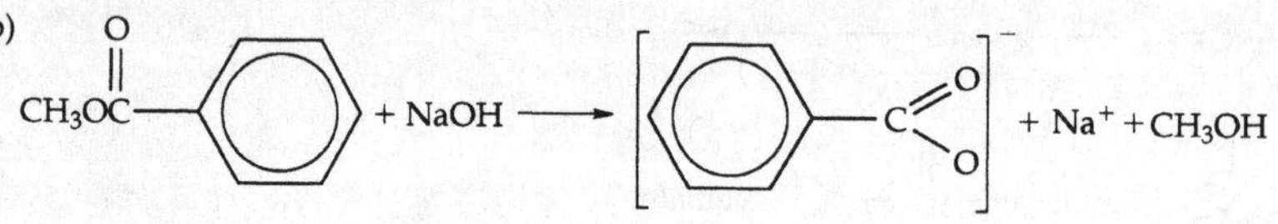

$$CH_3OC(=O){-}C_6H_5 + NaOH \longrightarrow [C_6H_5{-}CO_2]^- + Na^+ + CH_3OH$$

25.50 $2\ CH_3COOH(l) \longrightarrow CH_3COCH_3(l) + H_2O(l)$ (each carbon of the product bearing =O)

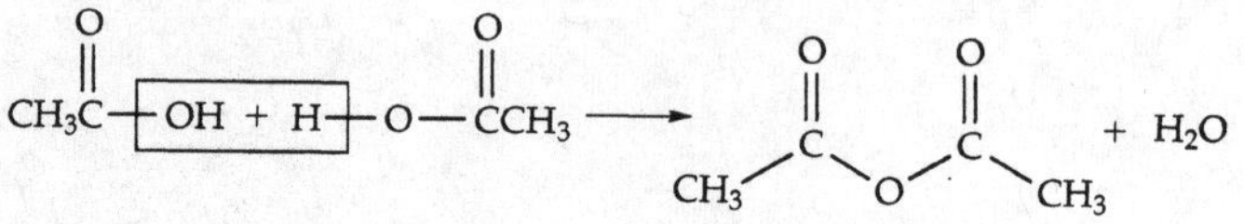

$$CH_3C(=O){-}OH + H{-}O{-}C(=O)CH_3 \longrightarrow CH_3C(=O){-}O{-}C(=O)CH_3 + H_2O$$

25.52 (a) $CH_3{-}CCl_2{-}CH_2{-}C(=O){-}H$ (b) $CH_3{-}C(=O){-}C_6H_5$

(c) $Br{-}C_6H_4{-}C(=O){-}OH$ (d) $CH_3{-}O{-}CH_2{-}CH{=}CH{-}CH_3$

(e) $C_6H_5{-}C(=O){-}N(CH_3)_2$

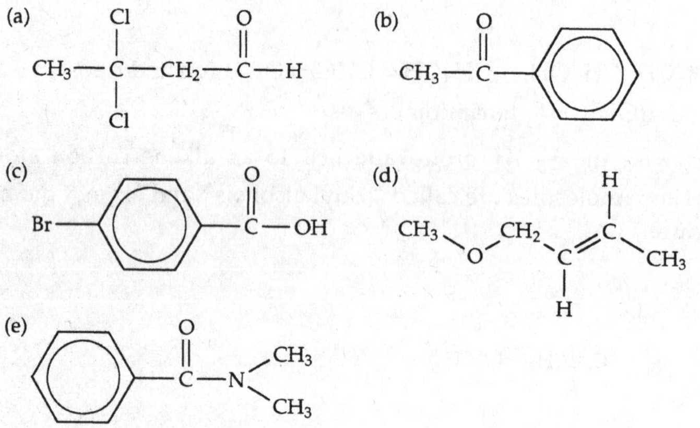

25.54

$$H{-}CH_2{-}CH_2{-}CH_2{-}C^*(Cl)(CH_3){-}CH_2{-}CH_2{-}H$$

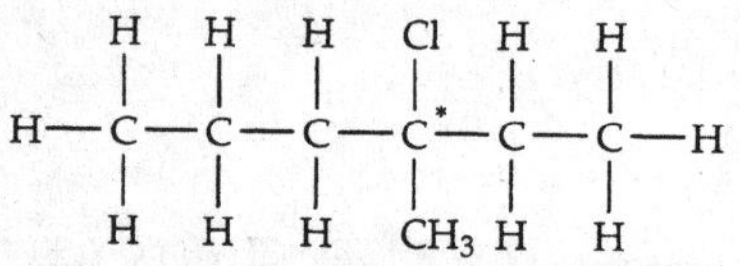

Yes, the molecule has optical isomers. The chiral carbon atom is attached to chloro, methyl, ethyl, and propyl groups. (If the root was a 5-carbon chain, the molecule would not have optical isomers because two of the groups would be ethyl groups.)

Proteins

25.56 The side chains possess three characteristics that may be of importance. They may be bulky (e.g., the phenyl group in phenylalanine) and thus impose restraints on where and how the amino acid can undergo reaction. Secondly, the side chain will be either hydrophobic, containing mostly nonpolar groups such as $(CH_3)_2CH—$ in valine, or hydrophilic, containing a polar group such as —OH in serine. The hydrophobic or hydrophilic nature of the side chain definitely influences solubility and other intermolecular interactions. Finally, the side chain may contain an acidic (e.g., the —COOH group in glutamic acid) or basic (e.g., the $—NH_2$ group in lysine) functional group. These groups will be protonated or deprotonated, depending on the pH of the solution, and determine the variation of properties (including solubility) over a range of pH values. Acidic or basic side chains may also become involved in hydrogen-bonding with other amino acids.

25.58

$$H_3\overset{+}{N}-\underset{(CH_2)_2SCH_3}{\overset{H}{C}}-\overset{O}{\overset{\|}{C}}-O^- + H_3\overset{+}{N}-\underset{H}{\overset{H}{C}}-\overset{O}{\overset{\|}{C}}-O^- \longrightarrow H_3\overset{+}{N}-\underset{(CH_2)_2SCH_3}{\overset{H}{C}}-\overset{O}{\overset{\|}{C}}-\overset{H}{N}-\underset{H}{\overset{H}{C}}-\overset{O}{\overset{\|}{C}}-O^-$$

methionine glycine methionylglycine

25.60 (a) Valine, serine, glutamic acid

(b) Six (assuming the tripeptid contains all three amino acids):

Gly-Ser-Glu; Gly-Glu-Ser; Ser-Gly-Glu; Ser-Glu-Gly; Glu-Ser-Gly; Glu-Gly-Ser

25.62 It is quite evident from Figure 25.26 that the hydrogen bonds between an NH group along the chain and the unshared electron pairs of a carbonyl group further along are responsible for maintaining the helix. Indeed, the pitch and general shape of the helix are determined by what specific interactions produce a good hydrogen-bonding arrangement.

Carbohydrates

25.64 Glucose exists in solution as a cyclic structure in which the aldehyde function on carbon 1 reacts with the OH group of carbon 5 to form what is called a hemiacetal, Figure 25.29. Carbon atom 1 carries an OH group in the hemiacetal form; in α-glucose this OH group is on the opposite side of the ring as the CH_2OH group on carbon atom 5. In the β (beta) form the OH group on carbon 1 is on the same side of the ring as the CH_2OH group on carbon 5.

The condensation product of two glucose units looks like this:

CH_2OH H O H H OH H HO H OH — O — CH_2OH H O H H OH H OH H OH

α-linkage

CH_2OH H O H H OH H HO H H OH — O — CH_2OH H O H H OH H OH H OH

β-linkage

25.66 (a) In the linear form of mannose, the aldehydic carbon is C1. Carbon atoms 2, 3, 4, and 5 are chiral because they each carry four different groups. Carbon 6 is not chiral because it contains two H atoms.

(b) Both the α (left) and β (right) forms are possible.

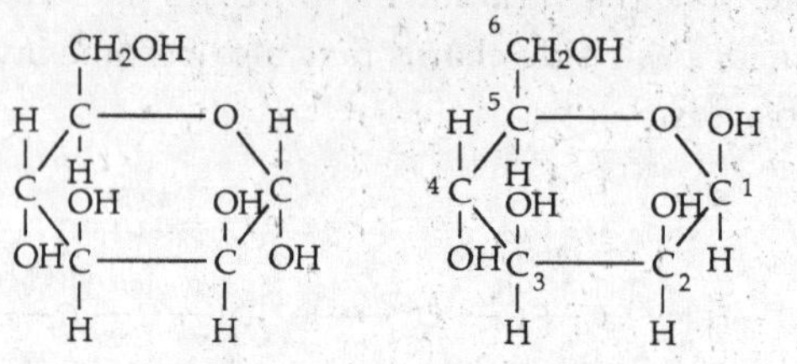

25.68 The empirical formula of cellulose is $C_6H_{10}O_5$. As in glycogen, the six-membered ring form of glucose forms the monomer unit that is the basis of the polymer cellulose. In cellulose, glucose monomer units are joined by β linkages.

Nucleic Acids

25.70

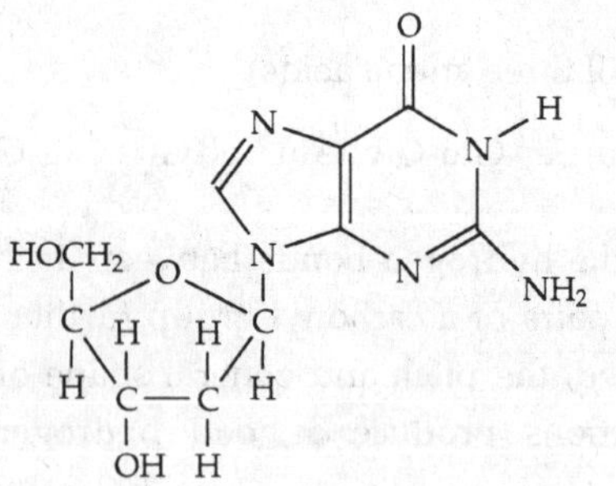

25.72

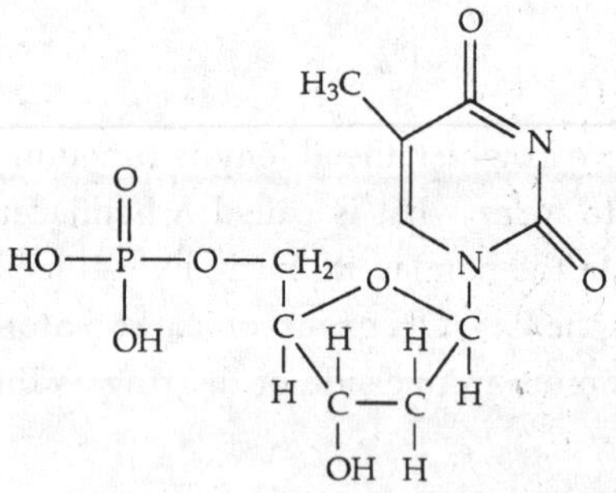

25.74

—A—C—T—C—G—A—

—T—G—A—G—C—T— ← complementary strand

Additional Exercises

25.76 *Analyze/Plan.* We are asked the number of structural isomers for two specified carbon chain lengths and a certain number of double bonds. Structural isomers have different connectivity. Since the chain length is specified, we can ignore structural isomers created by branching. We are not asked about geometrical isomers, so we ignore those as well. The resulting question is: How many ways are there to place the specified number of double bonds along the specified C chain? *Solve.*

5 C chain with one double bond: 2 structural isomers

C=C–C–C–C   C–C=C–C–C

6 C chain with two double bonds: 6 structural isomers

C=C–C=C–C–C   C=C–C–C=C–C   C=C–C–C–C=C

C–C=C–C=C–C   C=C=C–C–C–C   C–C=C=C–C–C

25.77 Because of the strain in bond angles about the ring, cyclic alkynes with less than eight carbons are not stable. Alkyne carbon atoms preferentially have 180° bond angles; this requires a linear four-carbon group in the ring. Three additional carbons in the ring do not provide enough flexibility to make this possible without gross bond length or angle distortions. It is possible that a ring with eight or more carbons could accommodate an alkyne linkage. To test this with models, construct a linear four-carbon group and then add tetrahedral C-atoms until you can complete the ring and it stays together without intervention. This is a good indication of the minimum number of carbon atoms in a stable ring that contains an alkyne linkage.

25.79 In alkanes, carbon forms only single, sigma bonds. Alkenes contain at least one C–C double bond, consisting of one sigma and one pi bond. Alkynes have a least one C–C triple bond, composed of one sigma and two pi bonds. In both alkenes and alkynes,

C atoms are involved in pi overlap. The question is, what feature of Si prevents it from forming double or triple bonds which involve pi overlap.

According to Table 8.5, the average C–C single bond length is 1.54 Å, C=C is 1.34 Å, and C≡C is 1.20 Å. These distances show that pi overlap requires substantially closer approach of the two bonded atoms than sigma overlap alone. The bonding atomic radius of Si is 1.11 Å, while that of C is 0.77 Å (Figure 7.6). The close approach of Si atoms that is required for pi overlap is not possible because of its large bonding atomic radius. Thus, silicon analogs of alkenes and alkynes that involve multiple bonds and pi overlap are virtually unknown. Silicon analogs of alkanes with exclusively sigma overlap are known; the average Si–Si single bond length is 2.22 Å.

25.80 The C–Cl bonds in the trans compound are pointing in exactly opposite directions. Thus, the C–Cl bond dipoles cancel (Section 9.3). This is not the case in the cis compound, as can be seen by drawing the structure:

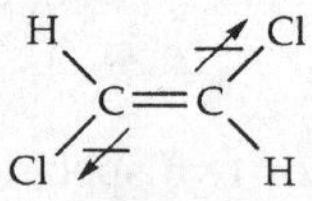

H, Cl / C=C / Cl, H — trans

H, H / C=C / Cl, Cl — net dipole moment — cis

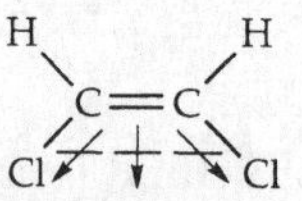

25.82 One. One molecule of HBr would add to the C=C according to the reaction

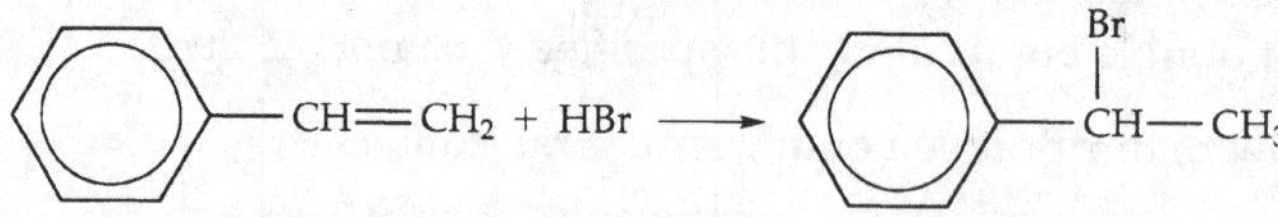

$$C_6H_5-CH{=}CH_2 + HBr \longrightarrow C_6H_5-CH(Br)-CH_3$$

Because the π electrons in the phenyl ring are delocalized, the group is particularly stable and resistant to addition reactions. The phenyl group could undergo substitution with HBr, but only with a catalyst or special conditions.

25.83 Two plausible decomposition reactions are:

(i) $CH_2Cl_2(l) \rightarrow C(s) + H_2(g) + Cl_2(g)$

(ii) $CH_2(NO_2)_2(l) \rightarrow N_2(g) + CO_2(g) + H_2O(g) + 1/2O_2(g)$

Use bond dissociation energies (Table 8.4) to evaluate approximate ΔH values for each reaction.

(i) $\Delta H = 2D(C-H) + 2D(C-Cl) - D(H-H) - D(Cl-Cl)$

$= 2(413) + 2(328) - 436 - 242 = +804$ kJ

(ii) $\Delta H = 2D(C-H) + 2D(C-N) + 2D(N{=}O) + 2D(N-O) - D(N{\equiv}N) - 2D(C{=}O)$

$- 2D(O-H) - 1/2D(O{=}O)$

$= 2(413) + 2(293) + 2(607) + 2(201) - 941 - 2(799) - 2(463) - 1/2(495)$

$\Delta H = -685$ kJ

Clearly, the decomposition of $CH_2(NO_2)_2$ is thermodynamically favorable, while the decomposition of CH_2Cl_2 is not. In particular, this is because of the stability of N_2 and CO_2 relative to $CH_2(NO_2)_2$. For CH_2Cl_2, no oxygen atoms are available to form stable products such as CO_2 and H_2O.

25.85 (a) $CH_3CH_2CH_2C(=O)OH$ or $(CH_3)_2CHC(=O)OH$

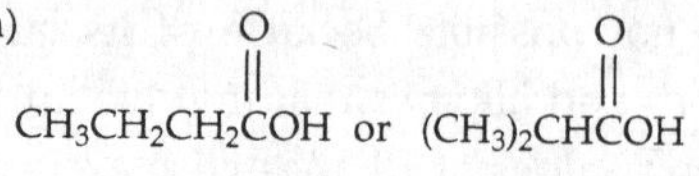

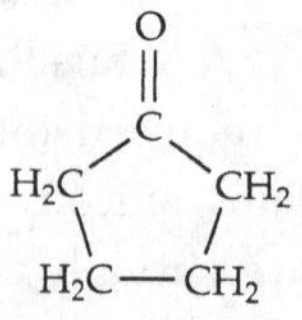

(b) $CH_3-CH(OH)-CH_2(OH)$ or $CH_2(OH)-CH_2-CH_2(OH)$

(c) Cyclobutanone: C=O ring with H_2C, CH_2, H_2C-CH_2

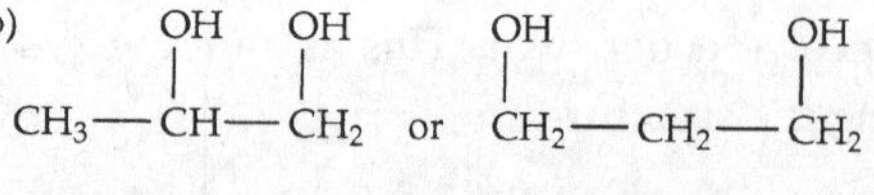

(d) Six-membered lactone: C=O, H_2C, O, H_2C, CH_2, CH_2

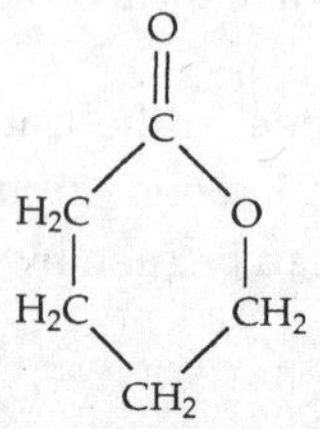

25 Organic and Biological Chemistry

Solutions to Black Exercises

25.87 (a) $CH_3C(=O)—OH$, C_6H_5OH (b) $C_6H_5C(=O)—OH$, CH_3OH

25.88 In order for indole to be planar, the N atom must be sp^2 hybridized. The nonbonded electron pair on N is in a pure p orbital perpendicular to the plane of the molecule. The electrons that form the π bonds in the molecule are also in pure p orbitals perpendicular to the plane of the molecule. Thus, each of these p orbitals is in the correct orientation for π overlap; the delocalized π system extends over the entire molecule and includes the "nonbonded" electron pair on N. The reason that indole is such a weak base (H^+ acceptor) is that the nonbonded electron pair is delocalized and a H^+ ion does not feel the attraction of a full localized electron pair.

25.90 (a) $H_3\overset{+}{N}CH(HC(CH_3)_2)C(=O)—NHCH_2C(=O)—NHCH(CH_2COO^-)C(=O)O^-$

(b) $H_3\overset{+}{N}—CH(CH_2C_6H_5)—C(=O)—NHCH(CH_2OH)C(=O)—NHCH(CH_3)C(=O)O^-$

25.92 Starch, glycogen, and cellulose are all biopolymers built by linking glucose monomers. Starch and glycogen have alpha (α) glucose linkages, where the bridging O atom is on the opposite side of the ring as the CH_2OH group. The smallest repeating unit in starch and glycogen is a single glucose unit. Starch and glycogen can have branched structures, while cellulose is always linear.

Cellulose has beta (β) glucose linkages, where the bridging O atom is on the same side of one of the rings as the CH_2OH group and on the opposite side of the CH_2OH group on the second ring. The geometry of the β linkage requires that the two linked glucose units have different orientations and that the smallest repeating unit in cellulose is two glucose units with a β linkage.

The molecular weight of a polymer is an indication of the number of monomer units present. Starch, glycogen, and cellulose all have a range of molecular weights. Glycogen has the widest range of molecular weights, 5,000–5,000,000 amu, and is potentially the largest polymer. Cellulose is intermediate in size with an average molar mass of 500,000 amu.

Starch and cellulose are produced in plants, while glycogen is produced in animals and serves as an energy storage mechanism.

25.94

```
—G—G—T—A—C—T—
 :  :  :  :  :  :
—C—C—A—T—G—A—  ←— complementary strand
```

Integrative Exercises

25.96 Determine the empirical formula of the unknown compound and its oxidation product. Use chemical properties to propose possible structures.

$$68.1\text{ g C} \times \frac{1\text{ mol C}}{12.01\text{ g C}} = 5.6703;\ 5.6703/1.1375 = 4.98 \approx 5$$

$$13.7\text{ g H} \times \frac{1\text{ mol H}}{1.008\text{ g H}} = 13.5913;\ 13.5913/1.1375 = 11.95 \approx 12$$

$$18.2\text{ g P} \times \frac{1\text{ mol O}}{16.00\text{ g O}} = 1.1375;\ 1.1375/1.1375 = 1$$

The empirical formula of the unknown is $C_5H_{12}O$.

$$69.7\text{ g C} \times \frac{1\text{ mol C}}{12.01\text{ g C}} = 5.8035;\ 5.8035/1.1625 = 4.99 \approx 5$$

$$11.7\text{ g H} \times \frac{1\text{ mol H}}{1.008\text{ g H}} = 11.6071;\ 11.6071/1.1625 = 9.99 \approx 10$$

$$18.6\text{ g O} \times \frac{1\text{ mol O}}{16.00\text{ g O}} = 1.1625;\ 1.1625/1.1625 = 1$$

The empirical formula of the oxidation product is $C_5H_{10}O$.

The compound is clearly an alcohol. Its slight solubility in water is consistent with the properties expected of a secondary alcohol with a five-carbon chain. The fact that oxidation results in a ketone, rather than an aldehyde or a carboxylic acid, tells us that it is a secondary alcohol. Some reasonable structures for the unknown secondary alcohol are:

$CH_3CH(OH)CH_2CH_2CH_3$ $CH_3CH_2CH(OH)CH_2CH_3$ $CH_3CH(OH)CH(CH_3)_2$

25.98 Determine the empirical formula, molar mass, and thus molecular formula of the compound. Confirm with physical data.

$$85.7\text{ g C} \times \frac{1\text{ mol C}}{12.01\text{ g C}} = 7.136\text{ mol C};\ 7.136/7.136 = 1$$

$$14.3\text{ g H} \times \frac{1\text{ mol H}}{1.008\text{ g H}} = 14.19\text{ mol H};\ 14.19/7.136 \approx 2$$

Empirical formula is CH_2. Using Equation 10.11 (MM = molar mass):

$$\text{MM} = \frac{(2.21\text{ g/L})(0.08206\text{ L}\bullet\text{atm/mol}\bullet\text{K})(373\text{K})}{(735/760)\text{ atm}} = 69.9\text{ g/mol}$$

The molecular formula is thus C_5H_{10}. The absence of reaction with aqueous Br_2 indicates that the compound is not an alkene, so the compound is probably the cycloalkane cyclopentane. According to the *Handbook of Chemistry and Physics*, the boiling point of cyclopentane is 49°C at 760 torr. This confirms the identity of the unknown.

25.100 (a) A = adenosine = $C_{10}H_{12}O_3N_5$

$$\left[A{-}O{-}\overset{\overset{O}{\|}}{\underset{\underset{O}{|}}{P}}{-}O{-}\overset{\overset{O}{\|}}{\underset{\underset{O}{|}}{P}}{-}O{-}\overset{\overset{O}{\|}}{\underset{\underset{O}{|}}{P}}{-}O\right]^{4-} + H_2O \longrightarrow \left[A{-}O{-}\overset{\overset{O}{\|}}{\underset{\underset{O}{|}}{P}}{-}O{-}\overset{\overset{O}{\|}}{\underset{\underset{O}{|}}{P}}{-}OH\right]^{2-} + HPO_4^{2-}$$

$$[A{-}P_3O_{10}]^{4-} + H_2O \longrightarrow [A{-}P_2O_6(OH)]^{2-} + HPO_4^{2-}$$

(The placement of the H^+ in these reactions is somewhat arbitrary; H^+ is attracted to the strongest base, but the equilibria are complex.)

(b) If the hydrolysis reaction is spontaneous, the sign of ΔG must be negative.

(c) Adenosine monophosphate (AMP) + inorganic phosphate

$$\left[A{-}O{-}\overset{\overset{O}{\|}}{\underset{\underset{O}{|}}{P}}{-}OH\right]^{-} + HPO_4^{2-}$$

(The placement of the H^+ in these reactions is somewhat arbitrary; H^+ is attracted to the strongest base, but the equilibria are complex.)

25.101 (a) At low pH, the amine and carboxyl groups are protonated. At high pH, the amine and carboxyl groups are deprotonated.

$$H_3\overset{+}{N}{-}\overset{\overset{H}{|}}{\underset{\underset{CH_3}{|}}{C}}{-}\overset{\overset{O}{\|}}{C}{-}OH \qquad NH_2{-}\overset{\overset{H}{|}}{\underset{\underset{CH_3}{|}}{C}}{-}\overset{\overset{O}{\|}}{C}{-}O^-$$

(b)

$$CH_3\overset{\overset{O}{\|}}{C}{-}OH(aq) \longrightarrow CH_3\overset{\overset{O}{\|}}{C}{-}O^-(aq) + H^+(aq)$$

$K_a = 1.8 \times 10^{-5}$, $pK_a = -\log(1.8 \times 10^{-5}) = 4.74$

The conjugate acid of NH_3 is NH_4^+.

$NH_4^+(aq) \rightleftharpoons NH_3(aq) + H^+(aq)$

$K_a = K_w / K_b = 1 \times 10^{-14} / 1.8 \times 10^{-5} = 5.55 \times 10^{-10} = 5.6 \times 10^{-10}$

$pK_a = -\log(5.55 \times 10^{-10}) = 9.26$

In general, a $-COOH$ group is a stronger acid than a $-NH_3^+$ group. The lower pK_a value for amino acids is for the ionization (deprotonation) of the $-COOH$ group and the higher pK_a is for the deprotonation of the $-NH_3^+$ group.

25.102 (a) Because the native form is most stable, it has a lower, more negative free energy than the denatured form. Another way to say this is that ΔG for the process of denaturing the protein is positive.

(b) ΔS is negative in going from the denatured form to the folded (native) form; the native protein is more ordered.

(c) The four S—S linkages are strong covalent links holding the chain in place in the folded structure. A folded structure without these links would be less stable (higher G) and have more motional freedom (more positive entropy).